Professional SAS® Programming Logic

Rick Aster

Also by Rick Aster:
PROFESSIONAL SAS PROGRAMMER'S POCKET REFERENCE

Breakfast Communications Corporation, P.O. Box 176, Paoli, PA 19301-0176 U.S.A.
http://www.pond.com/~finis/bc/
Breakfast is a registered trademark of Breakfast Communications Corporation.

Any updates to this book will be posted at the author's web site at
http://www.rickaster.com/globalstatements/

Publisher and Editor: Nicole Drumheller

Copy editors: Paul Nordquist, Suzanne Garland

Design: Rick Aster

About the cover: The scene was created as a model in Bryce 3D. The pyramid shape represents the engineering design and structure of SAS software. The tiger is a symbol of the dynamic, action-oriented approach that SAS programming calls for. The presence of the Siberian tiger at the Egyptian-style pyramid signifies the global reach of SAS software.

ISBN 1-891957-05-8

1st Edition

Printing 1 2 3 4 5 6 7 8 9

Contents

Introduction

SAS software is a focal point in the search for knowledge. It brings together data with programs that provide answers about that data to create new knowledge in business, science, and other fields. The SAS language in which these programs are written is designed to be easy to work with. You can write SAS programs that supply real answers without really being a programmer — without understanding the actions that the programs take or the objects that they work with. You can create these programs by following the examples of existing programs or by filling in the blanks in program templates. But this imitative approach to SAS programming has limitations akin to those of the paint-by-numbers approach to painting. Following a pattern that someone else has already created makes it possible to produce impressive results quickly, but as you do so, you are limited by what has come before; it is difficult to produce anything new or original. If you want to do something new and different in a SAS program, you need to be able to take on the programmer's point of view and apply the accumulated knowledge of the computer programming field. This book is designed to make that possible. It describes the processes and concepts of computer programming, in the same way that many other books have done, but it does so specifically in the context of the SAS environment.

The ability to do something new in a program, being limited only by your imagination, is not the only reason to be a SAS programmer. Taking on the programming paradigm and mastering the programming process present many advantages. Your programs are more robust and reliable, likely to work even when you are not there to watch them. You have a greater degree of confidence that programs work in the way that you intended. You create more direct, efficient programs that run fast, produce precise results, and are easily adapted to changing requirements or moved to different computers. You can spend much less time writing programs; you make fewer mistakes, and it takes less time to identify and correct whatever mistakes you do make. If you use SAS software and want to gain these benefits of programming mastery, then this book is for you.

This book is also for you if you are already a computer programmer but are new to SAS software. You will understand the possibilities and intricacies of SAS software only after you have seen them from a programmer's point of view. And studying SAS software in this way is the only way for you to become productive quickly in your SAS work. The style of the SAS language is, in some ways, quite different from other programming languages you might have used. After you understand the SAS approach, you will find that the SAS language imposes fewer limitations on the logic of a program than it might appear to have at first glance.

Lexicon

Much of the confusion that seems to surround the computer field comes from its lexicon — from the words it uses and the meanings associated with them. Different people use words differently, and this can create the appearance of distinctions that do not actually exist, and also obscure distinctions that are important to notice.

Notes located throughout this book describe the significant divergences of meaning and usage that you may encounter, along with other interesting points about specific terms. I hope this makes it easier for you to put together information and ideas from different sources.

Of particular concern in a book on SAS programming are the ways in which the common usage of a term in the SAS field differs from its usage in the programming field. In discussing programming concepts that do not depend on any specific programming language, I have tried to employ the terminology of programming.

There are a few points I need to cover right at the start. That is because some accounts of SAS processing, intending to simplify it for computer novices, provide only a fragmentary picture of it. When you see the complete picture, you find out that the processing in a SAS program is not nearly as restricted as the more simplistic accounts of it suggest. As a result, some of the words used to describe SAS processing mean more than what you might have heard.

Input is not merely an input text data file that a data step reads, but includes any input to any step of a SAS program or any other program.

A *file* is not necessarily a text data file or print file. Program files, SAS datasets, and catalogs are other examples of files.

A *program* does not necessarily mean a SAS program, but could be anything that runs on a computer. Within a SAS program, individual steps, views, procs, functions, and other routines are also programs — as are the SAS System, the operating system, and their components.

Data is not merely input data, but also includes output data and the values that a program creates and modifies as it works, along with anything stored in any file.

Output is not merely print output and the output objects that ODS works with. It includes any file that any step creates, anything displayed for the user to see, and any data produced by any program for use by another program.

More than three million people worldwide use SAS software. It is used in 99% of the Fortune 500 companies, 85% of U.S. government agencies, and more than 3,000 colleges and universities. SAS programmers create knowledge that affects everyone. Every day, SAS programs solve problems such as these:

- Find the fastest configuration for a computer processor design
- Measure the safety and effectiveness of new drugs
- Discover patterns in web site traffic
- Predict the profitability of business operations in changing market conditions

The SAS System is an enormous body of software, with a greater range of functionality than any other software in common use. No one book can describe all the capabilities that SAS software has to offer. To give this book some focus, it is limited to the programming capabilities of base SAS software. This is the most effective place for anyone to start to explore SAS software, in any case. Using the diverse capabilities of the other SAS products inevitably also requires you to work with the core functionality of the SAS System that this book describes.

Even with a focus on base SAS software, this book is still much too small to discuss every available feature. Many features that are not described here are covered in other books of mine, particularly *Professional SAS Programmer's Pocket Reference.*

Most programs are more than mere lists of things for the computer to do, arranged into a sequence of actions. A program's logic is its ability to take different actions in different circumstances or for different qualities in the data it works with. If, when you write SAS programs, you work regularly with options, IF-THEN statements, function calls, macro variables, and other elements of program logic, then you are doing real programming work. And if so, the SAS work you do will be easier if you approach it with the perspective of a computer programming specialist — regardless of what your actual specialization might be. That is why this book is called *Professional SAS Programming Logic.*

The gift of imagination has meant more to me
than my talent for absorbing absolute knowledge.

—Albert Einstein

1

SAS Programming

Computer programming is a perfectly natural, obvious process. As soon as you think of work for a computer to do, it only makes sense to decide exactly what it is you want the computer to do and write it down. In other words, you write a computer program. You might as well write it in a language that is meaningful both to the computer and to you — that is, in a programming language.

When your objectives involve working with data, there are good reasons to choose the SAS language as the programming language you will use to write a program. With a SAS program, you can often set forth a computing process in a more clear, concise, and meaningful way than would be possible in any other programming language. This is because the SAS language was designed specifically for working with data and because it is connected to the impressive and diverse data-handling capabilities of SAS software.

It is no great surprise, then, that many of us who had set out merely to answer questions about data have found that we have become SAS programmers — sometimes the result of a thoughtful decision, and sometimes quite by accident. The SAS language is meant to be useful to people who have no special skill or knowledge; it is even more valuable to people who can see it through the eyes of a programmer.

Running a SAS Program

You will need to be able to run SAS programs — both the programs you write and other programs that serve as models and examples. In case you have never run a SAS program before, this section describes how that process works. If I were sitting next to you at your actual computer, I could tell you exactly what steps to follow to run a SAS program. Since I do not know what kind of computer you are using or exactly how SAS software is installed on it, I can only describe the process in general terms. The process I describe here works on any computer that SAS software runs on, but you may have to go to other sources to fill in some of the exact details.

1. SAS software To run a SAS program, you must have SAS software licensed for your use and installed on your computer. The SAS System is a set of related products, but for the kinds of programs we will be considering in this book, only the first SAS product, base SAS, is required. If you need to license SAS software, contact SAS Institute Inc. in the United States or its subsidiaries or distributors in other countries. If you need to install SAS on your computer, follow the process described in the installation instructions that came with the software. If SAS is installed on your computer, make sure you know the operating system command, action, or icon that is required to launch the SAS application.

2. A SAS program The following SAS program has no constructive purpose, but is sufficient to demonstrate the process of running a SAS program.

```
*
 run1.sas
 A demonstration of the execution of a SAS program

 Rick Aster   January 2000
*;

DATA _NULL_;
  PUT 'This line in the log indicates that the program ran.';
  STOP;
RUN;
```

Notice these features of this program.

- The first part of the program, from the first asterisk (*) all the way to the semicolon (;) five lines below, is a comment statement. This comment is not executed as part of the program. Another way to write a comment is to begin it with /* and end it with */. Comments are intended only to identify, describe, or explain a program to any programmer who might read it. The program would run exactly the same way without the comment.
- The program consists of a sequence of statements. Each statement ends in a semicolon (;). These are two of the most fundamental characteristics of SAS syntax.
- Each statement appears on a separate line. This is not a requirement of SAS syntax, but it is something SAS programmers usually do, as a matter of style, in order to make the program easier to read.
- Two of the lines are indented. This is also a matter of style, and it is intended to show that those two statements are secondary statements, related to the DATA statement.
- The DATA statement marks the beginning a data step. A data step is a part of a SAS program that allows general programming statements and the creation of SAS datasets.
- The special name _NULL_ appears in the DATA statement. The name _NULL_ is used to indicate that the data step will not be creating an output SAS dataset.

- Notice that the underscores in the word _NULL_ are part of the word. In SAS, underscores act just like letters. They are often used as the first and last characters of special names in the SAS language.
- The PUT statement writes text to an output file. Because the program did not indicate a particular output file, the PUT statement writes it to the log file, a special print file that is always present in the SAS environment.

  This kind of action happens all the time in SAS: when you do not spell out an action that is required or select an option that is available, SAS fills in the blanks for you, taking a default action.
- The text that the PUT statement writes appears in the PUT statement as a character literal, a character value written in the program and enclosed in quote marks.
- The STOP statement indicates that the data step does not take any further action. In this case, the STOP statement is not really necessary, since the data step would stop at that point anyway.
- The RUN statement marks the end of the data step.

So that you can use a program, you should store it in its own file. Program files are text files, which means they are stored in files that contain nothing but the text characters that make up the program.

To create a program file, type the program in a text editor application and save it as a file, or type the program in the Program Editor window of the SAS windowing environment and select the Save As menu item to save it as a file.

It is common that the names of program files end in a dot (a period) and a word that represents the name of the programming language; SAS program files have names that end in ".sas". The comment at the beginning of this program indicates the program file name run1.sas.

3. Submit After a program file is created, there are various ways to tell SAS to run the program. Among these modes, the most useful are batch mode, in which SAS displays nothing of interest while the program is running, and the SAS windowing environment, which displays messages you can watch as the program runs. I recommend batch mode if you are developing SAS programs on a terminal connected to a multiuser computer, that is, a computer you share with other users. On a single-user computer, it is easier to run programs in the SAS windowing environment, and this is especially true when you are experimenting or developing programs.

In most operating system, to run a SAS program in batch mode, you need to issue a command like this one to the operating system:

```
sas -sysin run1.sas
```

The SYSIN option on this command identifies the program to run — in this example, the program file run1.sas. The exact form of the command may vary from one computer to another; it depends on the way SAS was installed on the computer. The option syntax is different in some operating systems.

If you are working in the OS/390 operating system, you need to do much more than this to run a SAS program in batch mode. You cannot submit a SAS program by itself; you need a specific set of JCL statements to execute the SAS cataloged procedure. These are different at every different installation, but could be the same for every SAS program you run. Find out what JCL is required to run a SAS program on the computer you are using.

To run a SAS program in the SAS windowing environment, you first need to launch the SAS windowing environment. This is sometimes as simple as issuing this command to the operating system:

```
sas
```

However, the command may be a form of this command, or it may be something different — actually, it could be anything at all, depending on the way SAS was installed, so find out what command is required on the computer you are working on. If you are working in a graphical operating system, you may be able to launch the SAS application from a "SAS System" application icon or menu item.

Within the SAS windowing environment, activate the Program Editor window and make sure it is empty. Then open the program file into the Program Editor window using the Open File menu item. Or you can use the command line to enter the INCLUDE command in the Program Editor window with the name of the file, for example:

```
include "run1.sas"
```

If this works, you should see the program in the Program Editor window.

Then you can use the SUBMIT command to submit the program to the SAS supervisor for execution. Select the Submit item from the File menu, or enter the SUBMIT command on the command line:

```
submit
```

The program disappears from the Program Editor window and begins to execute.

It is the SAS supervisor, a small part of the core of the SAS System, that takes charge of executing the SAS program. The SAS supervisor breaks the program up into the words and symbols that make it up, compares them against the rules of SAS syntax to determine what they mean, and calls upon the various routines in SAS to carry out the actions indicated by the program.

4. Results When the SAS supervisor runs a program, it creates a log file that tells the tale of how it ran the program and what happened as a result. This log file is the first place to look when you run a SAS program.

When you run a SAS program in batch mode, the SAS supervisor creates a log file that has the same name as the program file, except that it ends in

".log" instead of ".sas". The log file for run1.sas would be run1.log. You can look at the log file using a text editing program or any other method of displaying a text file.

When you run a SAS program in the SAS windowing environment, the log appears in the Log window. Usually, you can watch the log window as the SAS supervisor is running the program and writing the log. And you can scroll through the log and study it after the program has run.

If you run the demonstration program from before, the log should show something like this:

```
16   *
17     run1.sas
18     A demonstration of the execution of a SAS program
19
20     Rick Aster   January 2000
21   *;
22
23   DATA _NULL_;
24      PUT 'This line in the log indicates that the program ran.';
25      STOP;
26   RUN;

This line in the log indicates that the program ran.
NOTE: DATA statement used:
      real time           0.01 seconds.
```

Notice these features in the log.

- The lines of the program are repeated in the log with line numbers. These line numbers do not correspond to the lines of the program file. Rather, the SAS supervisor assigns sequential numbers to all the program lines it executes. In an interactive SAS session, this includes every program you run in the session. When you run a SAS program in batch mode, a SAS session consists of only one main program, but there are various ways in which the SAS supervisor may execute other lines of SAS code that do not come directly from the program file. When a SAS error or warning message refers to line numbers, look for these line numbers in the log.
- This particular program was intended to write a line in the log, and you can see that line appear just after the statement that marks the end of the step, the RUN statement.
- The log states approximately how long it took to run the step. The log can also contain other information about computer resources used by each step. The information that is available depends on the operating system.
- If you scroll to the top of any SAS log, you will see some additional information that may be important to you as a SAS programmer, including the SAS release number and the number that identifies your SAS site. The log is divided into pages, and the date and time of the SAS session appears at the top of each page.

This demonstration program produced nothing more than a log, but when you run a SAS program that produces something else, messages in the log identify and describe what was created. For example, when a program creates an output text file, the log tells the name of the file, the number of records written to it, and other useful information. When a program creates a SAS data file, the log tells the name of the file and the number of observations and variables. When the program writes pages of print output, the log tells what page numbers were produced by each step. In the SAS windowing environment, the Results window (or the Output Manager window) connects you to the various segments of print output.

5. Problems Sometimes a program runs correctly the first time you try it, but if it does not, that means you have more work to do to find the problems and correct them. When a program fails, the log contains error messages. The most common errors are syntax errors. Slight errors in writing a program, such as leaving out a word or a symbol, arranging statements in the wrong order, or misspelling a word, can result in a program that does not meet the syntax rules of the SAS language. An error of this sort usually means the SAS supervisor does not understand the program and does not know how to execute it. It will attempt to correct only the most minor spelling errors for you. Beyond that, all it can do is write error messages in the log indicating, as best it can, what is wrong with the program.

The SAS supervisor's error messages point you toward the mistake in a program, but the SAS supervisor cannot always find it precisely. The most important error message to look at when a program fails is the first one. A single mistake in a program can result in several error messages, but the first error message is likely to be closest to the location of the mistake. Also, after it encounters one error, the SAS supervisor's subsequent error messages may not be as reliable. When there is a syntax error, the SAS supervisor tries to guess what that part of the program means so it can continue to check the syntax of the program. However, because it is only guessing, it may think it finds errors in subsequent parts of the program that are actually correct. It may also miss errors that are there, and it may find errors but describe them incorrectly.

There are other kinds of errors, besides syntax errors, that can prevent a program from running. Another very common kind of error results when you use the wrong name to refer to a file or other object. This usually results in the SAS supervisor writing error messages saying that the object cannot be found or does not contain the things it expects.

Even when a SAS program runs, it may not run quite the way you intended. Errors in a program that make it run incorrectly are called logical errors. There is often some indication in the log of what went wrong. In particular, you need to be alert to warning messages in the log, messages that begin with the label "`WARNING:`". Warning messages are not error messages, but are conditions the SAS supervisor calls special attention to because they often point to mistakes and problems. Look for possible problems when the log contains error messages or warning messages and when log messages indicate any of the following conditions.

- There was a very long quoted string.
- There appeared to be a reference to a macro language object, but it could not be resolved.
- An input SAS dataset was empty.
- WHERE clause processing resulted in no observations being used.
- No observations were written to (or updated in) an output SAS dataset.
- An INPUT statement reached the end of an input record without finding values for all of its variables.
- A surprisingly small number of records (such as 0 or 1) were read from an input file.
- Missing values were created as a result of an operation on missing values.
- The arguments to a function or CALL routine were invalid.
- The text read by an informat was invalid.
- An informat or format could not be found.
- An attempt to divide by 0.
- In exponentiation, the operands were not mathematically valid, either because the base was negative and the exponent was not a whole number, or because the base was 0 and the exponent was not positive.
- Character values were automatically converted to numeric values, or vice versa.
- An array has the same name as a function.
- A variable that is referred to in a dataset option or BY statement or that is required for some other reason does not exist.
- A variable was never initialized.
- There was an unclosed DO block, or an END statement required for some other reason was missing.
- "Lost card." This note indicates that, at some point in the execution of the data step, an INPUT statement failed to find a record it was expecting to find in the input text file.
- In a match merge operation, more than one input SAS dataset had multiple observations for BY variables.
- The file system or operating system indicated a problem.
- Something was not available, and something else was used in its place.
- Something was "unexpected" or "invalid."
- A message that begins with the word "Assuming," indicating that the SAS supervisor had to guess what the program meant.

When there are problems, the problems are not always located in the SAS program itself. Errors in reading an input file, for example, could indicate problems with the program, but they could also indicate something wrong with the input file. Perhaps the file is formatted incorrectly, has been stored in the wrong place, or contains incorrect data because of errors that occurred when it was created. You may need to fix an input file or something else that the program depends on, then try to run the program again.

When you are doing your first work with SAS programming, understanding the significance of errors and finding the mistakes in the SAS code can be especially difficult. On occasion, you may have to learn something new about the SAS language in order to fix a mistake, and it may not immediately be obvious what you need to learn. Be patient. Tracking down errors is the most difficult part of programming, in any programming language, but as you gain experience it gets easier and easier.

6. Revise and repeat After you have found what may be the problem in the program, you can revise the program and try it again. In essence, this means going through the whole process of editing, saving and submitting the program again.

If you are making substantial changes in a program, or are making changes you are not sure of, consider saving a copy of the previous form of the program, even if it did not work. If it turns out your changes actually made things worse, you will want to be able to go back to the earlier version of the program and work with it again.

When you are working in the SAS windowing environment, you can return the most recently submitted program to the Program Editor window. In the Program Editor window, use the RECALL command or the Recall Text menu item. The program reappears in the Program Editor. You can use this command repeatedly to also recall earlier submitted programs. You can then edit the program, save the revised version of the program, then submit it.

To save a program from the Program Editor window, use the Save As menu item. After the first time you save a program, you can save it again to the same file by selecting the Save menu item. Alternatively, you can use the FILE command with the name of the file, on the command line, for example:

```
file "run1.sas"
```

If you revise and run a program several times in the SAS windowing environment, remember that several versions of the program's results appear in the Log window. Make sure you are looking at the most recent results, at the bottom of the log, to see what happened in the most recent run. If this becomes confusing, you can erase the Log window before you submit the next program. Use the Clear Text menu item in the Log window or the CLEAR LOG command.

The SAS Environment

A program written in a programming language cannot be run directly by a computer. Some other software, such a compiler, interpreter, or assembler, is required to convert the program into instructions that the computer can execute. It is SAS software that is required to interpret and compile a SAS program, and there are other reasons why SAS software is essential to the execution of a SAS program. SAS software provides an environment with many special features that must be present whenever a SAS program is running.

The SAS Supervisor

Programs in most programming languages have to be compiled before they can be run. You run a compiler to convert the entire program into computer instructions, or machine language, which can then be stored in a separate executable file. Then you run the program by running the executable file. Other programming languages are interpreted, rather than compiled. For these programming languages, there is no compiler and no executable file. Instead, an interpreter reads a little bit of the program and carries out the actions it sets forth, then it reads a little bit more and carries out those actions, continuing in that manner until it gets to the end of the file. The SAS language comes closer to fitting the latter model. A SAS program is executed in small pieces, rather than all at once, and there is no way to create a separately executable file from a SAS program. However, it could be misleading to describe the SAS language as an interpreted language. The process of executing a SAS program is somewhat different from the usual process of interpreting a program. Besides, some parts of a SAS program are compiled just before they are run — but here too, the way it works is not exactly like the usual idea of compiling a program. Rather than go into the semantic question of exactly what the words *interpret* and *compile* might mean when applied to the SAS language, I think it is better to simply say that SAS software *executes* SAS programs.

Specifically, it is the SAS supervisor that executes SAS programs. You can think of the SAS supervisor as the central program of SAS software, which makes everything work together as a system and determines the sequence of operations that SAS software takes in executing SAS programs and responding to user actions.

To a SAS programmer, the idea of the SAS supervisor is essential. As long as you try to think of "SAS" as one very big program that runs the SAS programs you write, it can be very hard to understand what "SAS" is doing. And then, when things go wrong, it is hard to say specifically what went wrong and where the problems occurred. The SAS System is not a single piece of software, and it is simply too big to be usefully understood as a single unit when you are writing programs for it. It is easier to describe SAS processes when you can say that the SAS supervisor is doing certain things and specific routines among the thousands of other routines that make up the SAS System are doing the various other things that are happening.

Imagine the SAS supervisor reading your SAS program, determining the correct sequence of actions required to carry out the program, and then handing off as much of the work as possible to other software components. The SAS supervisor calls on other parts of SAS software, and when it can, it also uses routines of the computer's operating system and file system.

This is much like the way a business manager considers the goals of a business unit, sets its work priorities, and delegates objectives to specific workers and suppliers. There are big differences, of course, between people and computer software, and it is possible to be misled when you use this kind of personification to understand software. Still, the metaphor is so apt and so powerful that it is hard to get away from. In fact, in the vernacular of software engineering, composite software components that coordinate the actions of other software, like the SAS supervisor, are often referred to as managers — and it is this same idea that gives the SAS supervisor its name.

Lexicon

Supervisor was once a precise technical term that was part of SAS Institute's own description of the SAS System. They no longer use the term in their user documentation, and with all the changes in SAS software, it would be hard to say exactly what the supervisor includes and does not include. It is still a useful concept.

SAS Software

The SAS System may perhaps be described as the biggest system of integrated software ever developed. It is developed by SAS Institute Inc., a software company in Cary, NC, and is licensed from them in the United States, or from their subsidiaries or distributors in other countries. You will never see the complete SAS System together in one place — except possibly for test or demonstration purposes — so you can refer to the set of SAS software products you happen to be working with simply as SAS, or SAS software.

When compared to other integrated software, there are two things that are particularly noteworthy about SAS software. First, it meant for use by knowledge workers, people who work with data to develop information. This is very different from other integrated software, most of which is built for something related to office automation or business operations. Second, SAS software has a programming language, the SAS language, right at the center of its design. Most integrated software has some ability to be programmed or scripted, but this is often little more than an afterthought. In SAS software, the SAS programming language is essential to the way everything works.

SAS software starts with the SAS supervisor, which is part of what is called the core of the SAS System. The core of the SAS System includes the most essential routines of SAS software, the ones that create the SAS environment and make it possible for SAS programs to execute and for the rest of the pieces of SAS software to function.

The SAS System includes a vast number of products. Each product is licensed separately and installed as a separate installation option. In some cases, a product is divided into several components that you can install separately if necessary. The first and most essential product is base SAS; this product is required before you can use any other SAS product. The core of the SAS System is a part of base SAS. Base SAS encompasses the SAS and SQL languages, the SAS windowing environment's most useful parts, the most generally useful routines in SAS software, and much more. Other SAS products add on such things as specialized analysis routines, development tools, and interactive applications.

Most SAS products are sets of routines that can be used in SAS programs. A SAS program can call on any of these routines, but it cannot run unless the routines belong to a SAS product that has been licensed and installed. When a program cannot run because a SAS routine cannot be found, the SAS supervisor cannot always tell you what product is missing. It is up to you as a SAS programmer to know about such things. When you work with a program or read about a SAS technique, you need to know which SAS products it depends on. You also need to know which SAS products are available to you on any computers for which you are writing SAS programs. The programs and techniques in this book use base SAS only.

Print Output

The SAS environment captures print output from a SAS program so that you can print it on a printer or view it on a computer screen.

The log is a special print file that the SAS environment provides. As the earlier example showed, the SAS supervisor writes messages about how the SAS program runs, and the SAS program can also write text in the log. Both of these features of the log are useful in finding out about problems in SAS programs.

The standard print file is another special print file in the SAS environment. It is the default destination for most print output produced by procs, and data steps can also write output to it.

In version 7, SAS introduced ODS, the Output Delivery System, which can direct output objects from procs to files in various formats. The available destination formats in ODS includes traditional print output; HTML, for display in a web browser; and SAS data files, which can allow you to do further programming with output objects from procs.

Routines

There are five main kinds of routines that can be used directly in a SAS program. Each kind of routine has a different purpose and is used in a different way.

Procs A SAS proc, or procedure, is a specialized application program that is designed to be used in a SAS program. To use a proc, you write a proc step, a separate section of a SAS program. Procs do things like sorting and analyzing data and creating reports — and many, many other things. Most of the specialized SAS products are collections of procs.

Lexicon

Another word for *proc* is *procedure*. Most writers, most of the time, use the word *procedure*. SAS programmers are more likely to use the word *proc*, because PROC is the keyword in the SAS language that identifies a proc.

Both words *procedure* and *proc* are widely used in computer programming for various kinds of routines or parts of programs that have no particular similarity to a *proc* in the SAS environment. In various programming languages and in a widely used model of computer programming called *structured programming*, a *procedure* is one of the structural parts of a program. It is a short, simple routine that a programmer creates and uses within the same program. This is completely different from the way a SAS *procedure* works. The SAS language does not include any statements for creating procs — only for using the procs that already exist.

Informats An informat interprets text to create a data value. Informats are especially used in the process of reading data from an input text file.

Formats You can think of formats as the opposite of informats. Formats are used anywhere you need to show a data value — to display a value on the screen, write it in a report, or create a field in an output text file.

You can create certain kinds of informats and formats yourself in the FORMAT proc.

Functions Most functions are used for mathematical effects or measurements. Functions are used to build expressions, which define values that can be assigned to variables or used in programming logic. A function can take one or several arguments, and it returns a value that results in some way from the arguments.

There are also many utility functions, which take specific actions, such as opening or closing a file, in addition to returning a value.

CALL routines A CALL routine is a simple program you can use as part of the logic of a SAS program. The only real difference between a CALL routine and a utility function is that a function returns a value and a CALL routine does not.

SAS Files

The SAS environment provides support for SAS software's special file types. SAS files — especially SAS datasets — are used all the time in SAS programming because you can use them without having to be concerned with all the details of the way the files are structured. SAS datasets are created in a SAS program and store data that can be used later in the same program or in another SAS program. Usually, you have to put data in a SAS dataset before a proc can do anything with the data. Catalogs, another important kind of SAS file, store various kinds of objects and resources.

The SAS environment also includes several libraries that can contain these SAS files. The most important is the WORK library, which you can use to hold temporary SAS files that will be used within the same program.

System Options

You can fine-tune many of the properties of the SAS environment with system options. There are hundreds of options to let you do such things as reset page numbers, control what kinds of messages appear in the log, and determine the way the SAS supervisor responds to certain error conditions.

You can initialize system options when you start up SAS software in a combination of two ways. You can set system options in the operating system command line. The SYSIN option, which identifies the program for SAS software to run in batch mode, was demonstrated in the earlier example. To set more than a few system options at startup, you can write the system option settings in a text file, which is called a configuration file and which might have a name like `config.sas`.

Most options can also be changed in the OPTIONS statement in a SAS program, or in the Options window of the SAS windowing environment. This lets you use one setting for a system option in one part of a program, then use another setting in another part of the program.

The SAS Session

The objects and properties that make up the SAS environment are initialized when a SAS session begins. Any changes that occur in the SAS environment are maintained for the duration of the SAS session. In batch mode, the SAS session is just the execution of a single SAS program. In the windowing environment, though, the SAS session continues for as many programs as you may submit.

The properties and objects of the SAS session that are significant in a SAS program include system option settings; any temporary files stored in the WORK library; formats and informats that are created in a SAS program; macros and macro variables, which are preprocessor objects that can be used to generate SAS code; and librefs and filerefs, which are short names that are temporarily associated with libraries and files. When you run several programs in the same session in the windowing environment, it is possible for any of these to carry over from one program to the next, resulting in a program not working exactly the same way as it would by itself. If this causes a program to run incorrectly, you need to consider the different things in the SAS environment that may affect the program.

Objects and Names

A computer program indicates a way to reach a certain result by taking a certain sequence of actions. In order to describe the actions in detail, it is necessary to use names to identify the various objects that the program acts on. There are many different kinds of objects that may be available to SAS program. A few kinds of objects — most notably, a program's variables, arrays, and statement labels — exist only within the program. But most kinds of objects a SAS program can address are outside the program in the SAS environment; they exist independently of the program and could also be addressed in another SAS program or in other ways in SAS software.

Words

Most kinds of SAS objects are identified by single-word names. For example, PRINT, SUMMARY, and SORT are procs; SYSVER, SYSSCP, and SYSDATE are macro variables; WORK, SASUSER, and LIBRARY are libraries; and _N_, _ERROR_, and _IORC_ are variables. In the SAS language and throughout the SAS environment, two rules define what qualifies as a word.

1. A word is made up only of letters, digits, and underscores. It includes no other characters. When you write a word, the characters that make up the word have to be consecutive — that is, there can be no punctuation or spaces inside the word.
2. The first character of a word is a letter or underscore. A word does not begin with a digit. If a sequence of letters and digits begins with a digit, it is not a word, but is a numeric constant.

The following table shows examples that do not meet the requirements for a SAS word. It then shows how the same ideas might be written in a way that does meet the requirements to be a SAS word.

Not Words	Words
AVG AMT	AVG_AMT
TAPE-LABEL	TAPELABEL
2x	x2
25	_25

In the SAS environment, words are not case-sensitive. That means you can use uppercase or lowercase letters when you write a word, and it is still the same word. The same is true for other uses of letters in a SAS program.

Name Lengths

Besides meeting the requirements to be a SAS word, the name of a SAS object also has to fit within a length limitation. The maximum length that a name can be is different for different kinds of objects. In addition, the maximum length for names of most kinds of objects changed between versions 6 and 7. Generally, SAS names can be up to 8 characters in length in version 6, or up to 32 characters in length beginning with version 7.

The table on the next page shows the length limitations for the names of specific kinds of objects.

Lengths of Names

Object	Description	Maximum Length of Name	
		Versions 5–6	Versions 7–8
Variable	Variable used inside a SAS program or stored in a SAS dataset	8	32
Array	List of variables in a SAS program	8	32
Statement label	Identifies a data step statement	8	32
SAS file	Any of various special kinds of files	8	32
Generation dataset	SAS dataset stored with generation information		28
Libref	Identifier for directory or collection of SAS files	8	8
Fileref	Identifier for text or binary file or directory	8	8
Catref	Identifier for concatenated catalogs		8
Entry	Object or resource stored in a catalog	8	32
Informat	Converts text to a data value	7[a]	7[a]
Format	Converts a data value to text	8	8
Engine	Routine for accessing SAS data	8	8
Password	Restricts access to SAS file	8	8
Index	Ordered list for a SAS data file	8	32
Integrity constraint	Restricts values in a SAS data file		32
Data step window	Window displayed by SAS program	8	32
Function	Calculation routine	8[a]	16
CALL routine	Simple routine used in data step logic	8[a]	16
Proc	Routine executed in a proc step	8[a]	16[b]
Macro	Stored preprocessor object	8	32
Macro variable	Preprocessor symbol	8	32
Macro label	Identifies a point in a macro	8	32
Macro window	Window displayed by a macro statement	8	32

[a] With exceptions. [b] The first 8 characters must be unique.

Names of Informats and Formats

Two special rules apply to the names of informats and formats. Names of character informats and formats begin with a dollar sign ($). The dollar sign is considered part of the name. The rest of the name must follow the usual rules for a SAS word. Names of informats and formats cannot end in a digit. The last character of the name must be a letter or underscore.

Multi-Level Names

Some SAS objects cannot be fully identified by a one-word name, so they have multi-level names — essentially, two or more names joined together by dots (periods). These are examples of multi-level SAS names:

```
FIRST.STATE
MAIN.DIGITAL.SELECT.SOURCE
```

Name Literals

SAS/ACCESS products allow a SAS program to connect to database management systems, some of which have less restrictive rules about names. They may allow spaces or symbols to appear in names, or their names may be case-sensitive. To let SAS programs use the names of tables and columns as they appear in a database, SAS is beginning to support *name literals*. To write a name literal, write the name in quotes, followed immediately by the letter N. For example, to use "Previous Address" as the name of a variable, write the name literal:

```
'Previous Address'N
```

If necessary, name literals can be used as the levels in multi-level names. Name literal syntax was first introduced in version 7. As of version 8, support for it is still very limited. At this point, the main use for name literals is to create SQL views or SAS data files that show the same data using regular SAS names for the variables. You can then use those views or tables at any other place in a SAS program where you might need to access the data.

Lists of Names

Anywhere SAS calls for a list of names, there is no punctuation between the names, just the spaces that are necessary to keep the names from running into each other. The VAR statement is an example of a place where a list of names is used. This VAR statement lists the variables NAME, CITY, and TITLE:

```
VAR NAME CITY TITLE;
```

Files as SAS Objects

A SAS program usually does not refer to files directly by their physical names, that is, the names that the computer's file system uses for the files.

Instead, the program first creates SAS objects, which have SAS names, as identifiers for the files. There is a good reason for this indirect approach. SAS is designed to work the same way, as much as possible, under many different operating systems. But the different file systems have different rules for naming and identifying files. When you create a SAS name for a file, you can write a program to work with that file in a standard SAS way that will also work on another computer with a different file system.

SAS treats files as belonging to two main categories: text files and SAS files. Binary files are treated the same as text files, as are certain kinds of devices that might be available on a computer. The category of SAS files includes all of SAS software's special types of files.

Filerefs

A SAS identifier for a text file is a *fileref*. You can create, or define, a fileref in a FILENAME statement. This example defines the fileref ORIGINAL:

```
FILENAME ORIGINAL 'original.txt';
```

The new fileref ORIGINAL is associated with a file whose physical name is original.txt. After the FILENAME statement executes, you can use the fileref ORIGINAL to refer to that file in the SAS program, or elsewhere in the SAS environment. If necessary, you can use another FILENAME statement later to clear the ORIGINAL fileref or associate it with a different file. A fileref can also be associated with a directory, or with certain kinds of devices.

When you define a fileref, it does not matter whether the physical file exists. To create a new text file from a SAS program, you can define a fileref for the file, then use that fileref as the output text file in the SAS program. In most operating systems, the SAS supervisor will create the new file for you.

Librefs

The special SAS identifier may be a good idea for a text file, but it is required for a SAS file. SAS software accesses SAS files in groups, or collections, which are called *libraries*, or to be specific, *SAS data libraries*. In most operating systems, a library is a directory. In file systems that do not support directories, notably the native file system of OS/390, a library is a single physical file, but it can still contain any number of SAS files. Before you can use a library, you must define a libref. You can do this in a LIBNAME statement, which works much like the FILENAME statement. This example defines the libref MAIN for the SAS library contained in a particular directory:

```
LIBNAME MAIN '~me/main/';
```

When you create a new library, you should create its directory before you assign the libref. Use the appropriate operating system command or utility program to create the directory. After the libref is assigned, you can use that short name whenever you need to refer to the library in the SAS program or elsewhere in the SAS environment.

Notice that the libref is not the permanent name of the library. You might assign one libref to a library in one program, then assign another libref to the same library in a different program. For example, you might create a library with the libref NEW, then access it later using the libref OLD.

SAS files are referred to as *members* of the library they are stored in, and they are usually referred to with two-level names. The libref is the first level of the name, and the SAS file's member name is the second level. For example, MAIN.ACTION is a SAS file with the member name ACTION, stored in the SAS library associated with the libref MAIN.

Lexicon

The original term for a SAS library, many years ago, was "SAS database."

Member Types

There are several types of SAS files, identified by their different *member types*. The most important member types are DATA, VIEW, and CATALOG.

SAS Data File The member type DATA identifies a SAS data file, a file created by SAS software that stores data in a format that is often referred to as a table. It contains not just data values, but also identifying information that lets you use the data very easily in a SAS program without having to know all the details about the file format.

A SAS data file is a kind of SAS dataset, which means that it is organized in a way that makes it readily accessible in the SAS environment. As a SAS dataset is conceptually organized, it has columns of data, called *variables*, and it can have any number of rows of data, called *observations*. In addition to the values of the variables, it includes identifying information about each variable, such as the name of the variable. Procs are designed to work only with data that is organized as a SAS dataset.

View A view also organizes data, but it is usually data that is physically stored somewhere else, which might be a text file, a database, or one or more SAS data files. A view is also a kind of SAS dataset. It provides a means to organize data from various sources in a way that SAS routines can use, without having to physically rewrite the data as a SAS data file.

Catalog A catalog is a SAS file for storing various kinds of objects and resources that are used in the SAS environment. The objects and resources that are stored in a catalog are called *entries*. Macros and formats are two examples of objects used in a SAS program that are stored as entries in a catalog.

Item store An item store, with the member type ITEMSTOR, is much like a catalog, but with a hierarchical structure. That is, it can contain objects that contain other objects. The item store was introduced in version 7 to make ODS, the Output Delivery System, possible.

It is possible for SAS files of different types in the same library to have the same member name. However, because both views and SAS data files are used as SAS datasets, you cannot have VIEW and DATA members with the same member name. Most SAS statements and options that use SAS files refer only to a particular member type, so it is not necessary to state the member type to identify the correct SAS file. But in utility procs that work with all types of SAS files, it may be necessary to use the MEMTYPE= or MTYPE= option to identify the member type of a SAS file in order to uniquely identify the file.

Catalogs and Entries

When you identify a catalog entry, it is usually necessary to provide the two-level name of the catalog, the entry name, and also the entry type, which indicates what kind of object the entry is. The entry type is especially important because entries of different types often have the same entry name. This results in a four-level name for the entry, which has the form

```
libref.catalog.entry.type
```

where *libref.catalog* is the two-level name of the catalog, consisting of the libref and member name of the catalog; *entry* is the entry name; and *type* is the entry type. For example, the name

```
MAIN.DIGITAL.FRONT.REPT
```

identifies the entry FRONT.REPT, that is, an entry whose name is FRONT and whose entry type is REPT, in the catalog MAIN.DIGITAL.

In some situations it is possible to provide only part of the name of an entry. When a specific catalog is being considered, it is sufficient to provide only the last two levels of the name. When only a certain entry type will do, it may be enough to provide the first three levels of the name. Or, in a situation where a specific catalog and a specific entry type are understood, you can sometimes use only the one-level entry name. There are close to 100 entry types, but most are used in interactive applications and graphics of various specialized SAS products. The following table lists the entry types that are of the greatest interest in SAS programming with base SAS.

Entry Types

Entry type	Description
CATAMS	Text data created by a program
DBCSTAB	DBCS conversion table
FOLDER	Desktop folder
FONT	Font
FORMAT	Numeric format
FORMATC	Character format
INFMT	Numeric informat
INFMTC	Character informat
KEYS	A set of function key definitions
LOG	Text from the Log window
MACRO	Macro
MSYMTAB	Values of local macro variables
OUTPUT	Text from the Output window
PMENU	Menu bar
PROFILE	User settings and preferences
QRYPROF	Profile for the Query window
QUERY	Query stored from the Query window
REPT	Report design for REPORT proc
SOURCE	Text from text editor window
TITLE	Title line or footnote line
TOOLBOX	Tool box
TRANTAB	Translation table
WSAVE	Window size and location

Programming Concepts

Computer programming depends on a small number of key ideas. The simplicity of this set of concepts is deceptive. Individually, it might not seem as if they can do very much — but combined, they create the power of computer programming. An understanding of these ideas makes it possible to see what a program does and how it works.

Variable A computer program does most of its work within the confines of the computer's memory. This is simply because, of all the resources that may be available to a program, memory is the easiest to work with. The idea of variables is to make memory easier for the programmer to work with. A variable is a specific location in memory, organized to hold a certain kind of data, and referred to by name. The actions of a program are reflected in its variables, specifically, in the way the variables have different values at different times. Most of the actions of a program have to do with changing the values of variables.

In the SAS language, there are many different ways to change the value of a variable, but the most simple and direct is the assignment statement, in which the equals sign (=) symbolizes the assignment of a value to a variable, for example:

```
ST = 'PA';
```

In this example, ST is a character variable, and this statement assigns to it the character value "PA". The value of the variable ST stays the same until something else happens to change its value, such as a later assignment statement that assigns a different value:

```
ST = 'NY';
```

Value If the actions of a program are reflected in the changing values of variables, then it is important to know exactly how values work. It might appear that a value could be just any number or text you can think of and write down, but there are limitations on the values you can use. When you assign a value to a variable, the value has to fit into the space that the variable occupies. It also has to fit the way the variable is organized, that is, it has to be the right type of data. And the same limitations apply wherever you might use a value in a program.

The need for a value to fit into a space is easiest to see when you look at a variable that is organized to hold character data. The number of characters the variable can hold depends on the size, or length, of the variable. Assume, for example, that a program uses ST as a character variable with a length of 2. Then the only values you can assign to it are strings of two characters, such as DC and MD. If you would try to assign a three-character value, such as FLA, to this variable, it simply would not fit. The variable, defined with a length of 2, only has room for two characters, and there is no place in it to put that third character. If you really need to be able to put three-character values in the variable ST, then the program would need to define it differently, with a length of 3 rather than 2.

It is even more critical that the value you assign to a variable is the right type for the variable. A programming language usually allows you to organize data as a few different types of values, or *data types*. The SAS language happens to support two data types: the character data type, which primarily holds character data, and the numeric data type, which primarily holds numbers. When you create a variable, you define it as one data type or the other, and then it can hold only that type of data value. There is no way to put a string of characters into a numeric variable, nor is it a good idea to try to squeeze a number into a character variable.

The character data type is based on a character set that mainly includes the characters you see on your computer keyboard. When you are writing with a pen or pencil, you can use any symbol you can write, but to work with character values on a computer, you need to find ways to write things using only the characters in the computer's character set.

Just as the space in a character variable is limited, so is the space of a numeric variable. For most practical purposes, this does not limit the size of a number you can use, but it may limit its accuracy, especially for fractional numbers. When a computer cannot represent a number exactly, it tries to be as close as it can, and usually, that is close enough. For example, 0.3 is a number a computer cannot record exactly — there is no binary value that corresponds exactly to 0.3 — but it is still accurate enough to distinguish 0.3 and 0.300000001. Mathematicians and statisticians need to be aware of the small rounding errors of computer numbers, because they can turn into larger errors when certain kinds of analytical models are applied to certain kinds of data.

Lexicon

Some windows in the SAS interactive environment use the word *number* for the numeric data type and *text* for the character data type.

The simplest kind of value in a program is a specific value that you write directly in the program. This kind of value is called a *constant*, because, unlike a variable, its value never changes. In the earlier examples of assignment statements, a constant value is being assigned to a variable. These are examples of SAS constants:

Numeric constants	Character constants
-.27	'PA'
0	'FLA'
0.3	"run1.sas"
25	'** End of Report **'
1000000	"Rick Aster "
1.0E6	

Notice these features of these SAS constants:

- There are no spaces or commas in numeric constants.
- Character constants are delimited by either single quotes or double quotes.
- Character constants can contain spaces and other special characters.
- If you want to write scientific notation, use the letter E to connect the mantissa to the exponent.

Often, you want to have a program compute a value, perhaps based on the value of several variables and constants. This kind of value is written as an *expression*, which could be a mathematical formula or some other kind of computation. These are examples of SAS expressions:

Numeric expressions	Character expressions
`A + B`	`UPCASE(ST)`
`(A + 1)*(B + 1)`	`ST || ' ' || ZIP`
`ROUND(AMOUNT, .01)`	`STNAME('PA')`
`WIDTH*WIDTH*HEIGHT/3`	

Expressions, and the functions and operators used to build them, are described in detail in chapter 5.

Logic A program sets forth a sequence of actions, but it is not usually a fixed, predetermined sequence. A program's actions usually depend in various ways on conditions of the data the program works with and the program's environment. Some actions are needed only when a certain condition exists in the data values. Groups of actions may need to be repeated a number of times, but not necessarily with the same number of repetitions every time the program is run. The *logic* of a program is the way the program's sequence of actions is determined by the conditions it considers.

To make it easier to see the sequence of action in a program, most programming languages, including the SAS language, divide a program into statements. In the simplest cases, each statement represents one action, and the sequence of statements determines the sequence of actions. But there are frequent exceptions:

- *Declarations* define objects rather than acting on them.
- Some *secondary* statements mean nothing by themselves, but merely modify the actions of other statements.
- *Control flow statements* alter the sequence of execution of action statements.

In a SAS program, issues of logic and sequence are especially important in data steps, which are the sections of a SAS program that allow general programming. The other parts of a SAS program only do things that have to do specifically with the SAS environment and the procs of SAS software.

In the data step, the sequence of actions can be altered by control flow statements, such as the IF-THEN statement, which takes an action only if a certain condition is met, and the DO and END statements, which can be used together to define loops, for repeating actions.

File To be useful, a program must connect to the world outside itself. All the logic and computations of a program mean nothing unless the program gives you some way to see what it did. That is the output of the program, and most programs also need input, usually some kind of data that comes from somewhere. There are various kinds of input and output, or *I/O*, but the simplest and most fundamental kind is reading from and writing to files.

A file is a named location in computer storage. Storage is different from memory. Think of memory as being inside a program — it disappears as soon as the program ends. Storage is outside the program, under the control of the computer's file system. Although connected to the computer, it is really independent of any single computer. Files written by one computer can be read by another computer that can work with the same file system, either by taking the disk or tape that contains the files to the other computer, or by connecting the two computers over a network.

The SAS log and standard print files are files specially designated to communicate the results of SAS programs to you. A SAS program can also create other files for that purpose. And a SAS program can exchange data with other programs by reading files created by other programs and writing files for other programs to read.

In SAS programs, files are used not just to connect the program to the outside world, but also to connect one step to another within the same program. The steps of a SAS program act almost like separate programs, and they connect to each usually only by way of SAS files.

At times, it may also be necessary for a SAS program to communicate directly with a user. A program with a user interface displays information on the computer screen and responds to the user's actions, such as data or commands that the user types on the computer keyboard. SAS programs have some ability to interact with the user in this way. Most interactive applications in the SAS environment, though, are AF applications, developed in SAS/AF and using programs written in another of SAS software's programming languages, SAS Component Language (SCL).

With these four concepts — variable, value, logic, and file — you can understand the cause and effect at work in computer programs. This takes the mystery out of computer programming, and it makes it much easier to learn about a programming language, such as the SAS language, by looking at examples of programs.

2

The SAS Way: Examples

Complex technology often turns out to be not quite so complex when you get to the heart of the matter. The elegance, grace, power, and simplicity of a technological design — if it has any — can be found in its design center, in the way it can be applied to the common, everyday tasks that are central to its design. When you work with it in the way that the designers originally envisioned, you connect to the creative thought and effort behind the design. At the same time, you get the clearest picture of the essential nature of the technology, which can help you understand all the features it offers.

The vision behind SAS software is that you will be able to bring together a set of data and get results from it in a single short, powerful program, such as the ones in this chapter. Many SAS programs are simple enough that you can follow the actions they take even if you have never studied a computer program before. At the same time, they are powerful in the sense that the programs themselves are much shorter than the descriptions of the problems they address. If you want to understand the SAS language, this is the place to start.

From Fixed-Field Text Data to Analysis

If there is one defining problem for SAS software, it is the task of organizing data from a text file in order to analyze it in one way or another. That task is demonstrated in this first example.

SAS Datasets

One of the first things you discover when you work with the SAS language is that it is designed to use SAS datasets — to create them, then work with them. This design concept is the reason behind the two most visible features of the SAS language. The data step is designed to create SAS datasets. The proc step is designed to work with them. In each case, the syntax is designed to make this process as simple as possible. In a data step, for example, you do not have to indicate what variables the new SAS dataset should contain; it can simply take all the variables from the data step. Also, you do not have to indicate how to form observations, or rows, in the SAS dataset; the data step automatically writes one observation for each line of input data. In a proc step, you do not have to indicate what SAS dataset the step works with; by default, it looks for the SAS dataset you created most recently in the SAS program. The result of designed defaults such as these is that you can take data from a text file to analysis in a surprisingly small number of programming statements.

Fixed-Field Text Data Files

The SAS language is admirably suited to working with just about any kind of text data file. One of the easiest kinds of files the SAS language can work with is the traditional lowest common denominator for exchanging data, the fixed-field text data file. Each record in the file is structured the same way — divided into fields, with each field holding a certain kind of data and having a specific length.

Lexicon

In data centers, a fixed-field text data file is usually called a *flat file*, but be careful when using this term — it has several other meanings. Depending on who you talk to, a *flat file* could be any file that is defined with fixed-length records — or it could be any text file at all!

The example program in this section reads the fixed-field text data file shown below. The data represents information about the members of the rock group Yes. You can see that the data elements appear to form columns, as in a table, because of the way each record contains the same fields with the same field lengths.

```
Peter Banks             guitar          United Kingdom          1
Chris Squire            bass guitar     United Kingdom          1
Jon Anderson            vocals          United Kingdom          1
Tony Kaye               keyboard        United Kingdom          1
Bill Bruford            drums           United Kingdom          1
Steve Howe              guitar          United Kingdom          3
Rick Wakeman            keyboard        United Kingdom          4
Alan White              drums           United Kingdom          6
Patrick Moraz           keyboard        Switzerland             7
Trevor Horn             vocals          United Kingdom         10
Geoff Downes            keyboard        United Kingdom         10
Trevor Rabin            guitar          South Africa           11
Billy Sherwood          guitar          United States          16
Igor Khoroshev          keyboard        Russia                 17
```

Record Layout

The way a record in a fixed-field text data file is organized into fields is described in a document called a record layout. To see how the record layout corresponds to the file, it is helpful to look at a record, or a few records, of the file, together with a ruler. The best ruler for this purpose is a text device such as the one shown here. It makes it easy to count the character positions, or columns, in the record.

```
----+----1----+----2----+----3----+----4----+----5----+----6----+-
Peter Banks             guitar          United Kingdom          1
Chris Squire            bass guitar     United Kingdom          1
Jon Anderson            vocals          United Kingdom          1
```

This is the record layout for this file:

Field Description	Format	Start Column	End Column	Length
Name	Character	1	24	24
Instrument	Character	25	40	16
Nationality	Character	41	64	24
Joined at album number	Positive integer	65	66	2

Using the ruler, you can count the characters in the record and see that they correspond to the columns indicated in the record layout.

Preparation for Running Programs

The programs in this book are not complete and ready to run as you see them. They could not be. SAS programs use files, and I cannot tell you where files should be stored on your computer. If you want to run the programs, you will need to add the necessary SAS statements to make the programs refer to the files they use — to define a fileref for each text file and a libref for each SAS library.

This example uses the fileref MEMBERS for the input text data file. To construct a FILENAME statement, you need to add the correct file name, which might look something like this:

```
FILENAME MEMBERS 'members.txt';
```

The SAS files in the examples in this chapter are stored in the library that is identified with libref CENTER. Use the operating system to create a directory for this library — or, if you are working with a file system that does support paths, and a SAS data library is a file, use the operating system to create that file. Then in the SAS program, construct a LIBNAME statement with the directory or file name, such as:

```
LIBNAME CENTER '/test/data/logic/center';
```

Place FILENAME and LIBNAME statements near the beginning of the program, before the first data step or proc step.

If you are running SAS programs in an older version of SAS software, you need to make adjustments to a few of the programs in this book. In particular, if you are running any version of SAS software earlier than version 7, you must make sure variable names are no longer than 8 characters.

Program and Results

This first example of a SAS program reads the fixed-field text data shown above, prints it as a table, and analyzes the data by producing frequency tables.

```
*
   YES.SAS
   Reads data file of Yes members and computes frequencies.
   Rick Aster    January 2000

   Input text file: MEMBERS
   Output SAS data file: CENTER.YES
```

```
*;

*
  Add FILENAME and LIBNAME statements here.
*;


OPTIONS LINESIZE=75 NOCENTER PAGENO=1;
TITLE1 'Yes Members';

*
  Data step creates SAS data file CENTER.YES from
  input text file MEMBERS.
*;
DATA CENTER.YES;
  INFILE MEMBERS TRUNCOVER;
  INPUT @1 NAME $CHAR24.
     @25 INSTR $CHAR16.
     @41 NATIONAL $CHAR24.
     @65 JOINAT 2.
     ;
  LABEL NAME='Name' INSTR='Instrument' NATIONAL='Nationality'
     JOINAT='Joined At';
RUN;

*
 Print SAS dataset.
*;
PROC PRINT DATA=CENTER.YES LABEL;
   ID NAME;
   VAR JOINAT INSTR NATIONAL;
RUN;

*
 Compute frequencies.
*;
PROC FREQ DATA=CENTER.YES;
   TABLES INSTR NATIONAL INSTR*NATIONAL;
RUN;
```

More detailed notes on each part of the program follow. For the moment, notice these major features of the program.

- The many comment statements, which begin with an asterisk (*) and continue until the concluding semicolon (;), are not executed as part of the program. They serve only to describe the program.
- The first step in the program is a data step. It begins with the DATA statement, includes INFILE, INPUT, and LABEL statements, and ends with a RUN statement.
- The final two steps in the program are proc steps. Each begins with a PROC statement.
- To make it easier for you to see this structure in the program, the other statements in the data and proc steps are indented. When a statement

extends for more than one line, the subsequent lines of the statement are indented.

The log describes the execution of the program:

```
19   *
20        YES.SAS
21        Reads data file of Yes members and computes frequencies.
22        Rick Aster    January 2000
23
24        Input text file: MEMBERS
25        Output SAS data file: CENTER.YES
26   *;
27
28   *
29      Add FILENAME and LIBNAME statements here.
30   *;
31
32
33   OPTIONS LINESIZE=75 NOCENTER PAGENO=1;
34   TITLE1 'Yes Members';
35
36   *
37      Data step creates SAS data file CENTER.YES from
38      input text file MEMBERS.
39   *;
40   DATA CENTER.YES;
41      INFILE MEMBERS TRUNCOVER;
42      INPUT @1 NAME $CHAR24.
43          @25 INSTR $CHAR16.
44          @41 NATIONAL $CHAR24.
45          @65 JOINAT 2.
46          ;
47      LABEL NAME='Name' INSTR='Instrument' NATIONAL='Nationality'
48          JOINAT='Joined At';
49   RUN;

NOTE: The infile MEMBERS is:
      File Name=C:\yes.txt,
      RECFM=V,LRECL=256

NOTE: 14 records were read from the infile MEMBERS.
      The minimum record length was 66.
      The maximum record length was 66.
NOTE: The data set CENTER.YES has 14 observations and 4 variables.
NOTE: DATA statement used:
      real time           0.21 seconds


50
51   *
52      Print SAS dataset.
53   *;
54   PROC PRINT DATA=CENTER.YES LABEL;
55       ID NAME;
56       VAR JOINAT INSTR NATIONAL;
57   RUN;

NOTE: There were 14 observations read from the dataset CENTER.YES.
NOTE: PROCEDURE PRINT used:
      real time           0.04 seconds
```

```
58
59  *
60    Compute frequencies.
61  *;
62  PROC FREQ DATA=CENTER.YES;
63      TABLES INSTR NATIONAL INSTR*NATIONAL;
64  RUN;

NOTE: There were 14 observations read from the dataset CENTER.YES.
NOTE: PROCEDURE FREQ used:
      real time           0.06 seconds
```

This is the log of a program that ran successfully, as indicated by:

- An absence of any error or warning messages.
- Notes for the data step that indicate that it read a certain number of records from a specific input file and created a specific SAS dataset with variables and observations — with the number of observations created matching the number of records read.

This is the print output from the program:

```
Yes Members                                      14:04 Monday, July 31, 2000   1

                         Joined
Name                       At      Instrument     Nationality

Peter Banks                 1      guitar         United Kingdom
Chris Squire                1      bass guitar    United Kingdom
Jon Anderson                1      vocals         United Kingdom
Tony Kaye                   1      keyboard       United Kingdom
Bill Bruford                1      drums          United Kingdom
Steve Howe                  3      guitar         United Kingdom
Rick Wakeman                4      keyboard       United Kingdom
Alan White                  6      drums          United Kingdom
Patrick Moraz               7      keyboard       Switzerland
Trevor Horn                10      vocals         United Kingdom
Geoff Downes               10      keyboard       United Kingdom
Trevor Rabin               11      guitar         South Africa
Billy Sherwood             16      guitar         United States
Igor Khoroshev             17      keyboard       Russia
```

```
Yes Members                                      14:04 Monday, July 31, 2000   2

The FREQ Procedure

                                  Instrument

                                                Cumulative    Cumulative
INSTR             Frequency        Percent       Frequency       Percent
-------------------------------------------------------------------------
bass guitar              1          7.14               1          7.14
drums                    2         14.29               3         21.43
guitar                   4         28.57               7         50.00
keyboard                 5         35.71              12         85.71
vocals                   2         14.29              14        100.00
```

```
                        Nationality

                                               Cumulative    Cumulative
NATIONAL           Frequency     Percent        Frequency      Percent
-----------------------------------------------------------------------
Russia                    1        7.14                1          7.14
South Africa              1        7.14                2         14.29
Switzerland               1        7.14                3         21.43
United Kingdom           10       71.43               13         92.86
United States             1        7.14               14        100.00
```

```
Yes Members                                   14:04 Monday, July 31, 2000   3

The FREQ Procedure

Table of INSTR by NATIONAL

INSTR(Instrument)     NATIONAL(Nationality)

Frequency    |
Percent      |
Row Pct      |
Col Pct      |Russia  |South Af|Switzerl|United K|United S|  Total
             |        |rica    |and     |ingdom  |tates   |
-------------+--------+--------+--------+--------+--------+
bass guitar  |      0 |      0 |      0 |      1 |      0 |      1
             |   0.00 |   0.00 |   0.00 |   7.14 |   0.00 |   7.14
             |   0.00 |   0.00 |   0.00 | 100.00 |   0.00 |
             |   0.00 |   0.00 |   0.00 |  10.00 |   0.00 |
-------------+--------+--------+--------+--------+--------+
drums        |      0 |      0 |      0 |      2 |      0 |      2
             |   0.00 |   0.00 |   0.00 |  14.29 |   0.00 |  14.29
             |   0.00 |   0.00 |   0.00 | 100.00 |   0.00 |
             |   0.00 |   0.00 |   0.00 |  20.00 |   0.00 |
-------------+--------+--------+--------+--------+--------+
guitar       |      0 |      1 |      0 |      2 |      1 |      4
             |   0.00 |   7.14 |   0.00 |  14.29 |   7.14 |  28.57
             |   0.00 |  25.00 |   0.00 |  50.00 |  25.00 |
             |   0.00 | 100.00 |   0.00 |  20.00 | 100.00 |
-------------+--------+--------+--------+--------+--------+
keyboard     |      1 |      0 |      1 |      3 |      0 |      5
             |   7.14 |   0.00 |   7.14 |  21.43 |   0.00 |  35.71
             |  20.00 |   0.00 |  20.00 |  60.00 |   0.00 |
             | 100.00 |   0.00 | 100.00 |  30.00 |   0.00 |
-------------+--------+--------+--------+--------+--------+
vocals       |      0 |      0 |      0 |      2 |      0 |      2
             |   0.00 |   0.00 |   0.00 |  14.29 |   0.00 |  14.29
             |   0.00 |   0.00 |   0.00 | 100.00 |   0.00 |
             |   0.00 |   0.00 |   0.00 |  20.00 |   0.00 |
-------------+--------+--------+--------+--------+--------+
Total               1        1        1       10        1      14
                 7.14     7.14     7.14    71.43     7.14  100.00
```

Notice these characteristics of SAS print output:

- Each page of print output begins with a title line that includes the date and page number. There can be additional title lines.
- The proc steps in the program produce the pages of print output.
- SAS print output can be left-aligned or centered. This print output is left-aligned.

Global Statements

Most SAS programs begin with a few global statements to set up the SAS environment. This program starts with an OPTIONS statement, which sets system options, and a TITLE statement, which defines a title line. Consider these statements one term at a time.

`OPTIONS` — The OPTIONS statement sets system options. Each system option setting appears as a separate term in the statement. Any combination of different system options could appear in a single OPTIONS statement, and there is no significance in the order in which they appear within the statement.

System options generally affect the way data and proc steps work. Most options you set stay in effect until you change them later. The options set in this OPTIONS statement affect the appearance of the print output that the proc steps of the program produce.

`LINESIZE=75` — The LINESIZE system option sets the width of the pages of print output. Each line of print output will contain no more than this number of characters.

`NOCENTER` — The CENTER system option centers the print output that a proc step produces. Here, NOCENTER, the negative form of this system option, left-aligns the print output.

`PAGENO=1`
`;` — SAS software automatically numbers the pages in a session starting at page 1. The PAGENO= system option sets the page number to a specific number. This will be the number of the next page produced. Here, the option resets the page number to 1. This is what the page number would be anyway, if you are running this program in batch or as the first program you run in an interactive session. By using this option, this program ensures that its page numbers always start at 1, even when it is not the first program to run in a session.

`TITLE1` — TITLE statements set title lines, which appear at the top of pages in print output. There can be as many as 10 title lines. This statement, with the keyword TITLE1, defines title line 1.

`'Yes Members'`
`;` — This is the text of the title line, written as a character constant. You can see that this text appears in the first title line of each page of print output.

In addition to these global statements, you need to add the necessary FILENAME and LIBNAME statements as described earlier.

Data Step Statements

The first step in the program is a data step to create a SAS data file. Each statement in the step has a different purpose in describing the way the SAS data file is created.

```
DATA CENTER.YES;
```

A DATA statement marks the beginning of a data step and names the output SAS dataset. In this case, the output SAS dataset is CENTER.YES. The two-level name identifies both the library where the SAS dataset is stored, the libref CENTER, and the member of the SAS dataset, YES.

Like most data steps, this one creates only one SAS dataset, and creates it as a SAS data file. A data step that created several SAS datasets would name all of them in a single DATA statement. For a data step to create a view, the DATA statement would also need to have the VIEW= option. The default is for the data step to create a SAS data file. A data step that did not create a SAS dataset would have the special name _NULL_ on the DATA statement.

```
INFILE MEMBERS TRUNCOVER;
```

The INFILE statement identifies the input text file for a data step and sets any necessary options for it. Subsequent INPUT statements can then read the file. In this case, the input file is identified by the fileref MEMBERS. The fileref should have been defined in an earlier FILENAME statement.

The option TRUNCOVER is useful for fixed-field text data. It ensures that the last field in each record is read correctly even if its trailing spaces were omitted when the file was stored. Some text editing programs in some operating systems automatically remove trailing spaces from each line in a text file, but with the TRUNCOVER option in the INFILE statement, the file is read as if those spaces were still present. Without the TRUNCOVER option, a data step might not read all of the data present in a fixed-field text data file.

```
INPUT @1 NAME $CHAR24.
   @25 INSTR $CHAR16.
   @41 NATIONAL $CHAR24.
   @65 JOINAT 2.
   ;
```

The INPUT statement sets forth the details of reading a record from the input file and obtaining values for variables. Writing an INPUT statement for fixed-field data is particularly easy. You can compare the INPUT statement to the record layout and see how they contain the same essential information, which is expressed in different ways.

For each fixed field it read, the INPUT statement contains the following terms, in order:

- A column pointer control, such as @1. This identifies the starting column of the field.
- The variable name of the variable to be created, such as NAME.

- An informat to interpret the field as a data value. The numeric argument that comes between the informat name and the following dot is the width (or length) of the field.

  This INPUT statement uses only two different informats. The $CHAR informat uses a field as a character value without altering the data in contains in any way. The standard numeric informat, which is written without a name, reads numerals in the conventional ways they are written (but without commas) and extracts the intended numeric value. Of the many available SAS informats, these are two of the most useful.

Each term in the INPUT statement represents an action. You can translate the sequence of terms in the INPUT directly to natural language terms you might use to describe the actions:

> At column 1, create a value for the variable NAME by using the informat $CHAR with a width of 24.
>
> At column 25, create a value for the variable INSTR by using the informat $CHAR with a width of 16.
>
> At column 41, create a value for the variable NATIONAL by using the informat $CHAR with a width of 24.
>
> At column 65, create a value for the variable JOINAT by using the standard numeric informat with a width of 2.

None of these variables existed already in the data step, so the INPUT statement defines them. It defines as numeric the variable that it reads with the standard numeric informat. It defines as character those variables that it reads with the character informat $CHAR, and it uses the informat width to determine the length of each variable. This ensures that the character variable is long enough to hold the entire value that is created for it in the execution of the INPUT statement.

```
LABEL NAME='Name' INSTR='Instrument' NATIONAL='Nationality'
   JOINAT='Joined At';
```

The LABEL statement defines labels for the variables. These labels are attributes of the variables that are stored in the SAS dataset that is created for use in later steps or later programs that may print or display the variables.

```
RUN;
```

The RUN statement marks the end of the data step.

Automatic Actions in the Data Step

Just as important as what is stated in the statements of the data step are the actions that are not stated. To begin with, you must have noticed that the actions of the INPUT statement were only sufficient to read a single record from the input file. Then, there is no subsequent statement to write the resulting observation to the SAS dataset that is being created. But we know from the log that the data step did, in fact, read all the records from the input file and write them as observations to the output SAS dataset. How did this happen?

The answer is found in the automatic actions of the data step. When the statements of a data step do not specifically indicate when to write output observations, the data step automatically writes an observation after the last action indicated in the data step statements. When a data step has an INPUT statement, or any other statement to read input data, the actions of the data step automatically repeat. Then the data step stops only when the INPUT statement reaches the end of the file it is reading. This means the data step automatically repeats just long enough to read all the input data and write out all the observations created from it.

Automatic output and the automatic data step loop are two of the many automatic processes that may occur in a data step. They are an important part of the logic of most data steps you will see, so you need to be aware of them every time you read or write a data step.

Proc Steps

After the data step, the program has two proc steps that read and work with the SAS dataset that the data step created. Each proc step runs a proc — a SAS routine specifically designed to be used in a proc step of a SAS program. Coding a proc step is usually a simple, fill-in-the-blanks process, because the proc itself does most of the work. It is different for each different proc you might use, because each proc does different things and has a different set of statements and options.

The PRINT proc is one of the easiest procs to use and one of the most popular, because it prints the data of a SAS dataset in table form, so you can look at it. The `PROC PRINT` statement may include options that control the general appearance of the proc. You can use other statements to control the way variables are used in the report that the proc produces.

A `PROC PRINT` step usually requires only a few statements. This example is typical of the way the PRINT proc is used.

```
PROC PRINT DATA=CENTER.YES LABEL;
```

This `PROC PRINT` statement marks the beginning of the step and includes two options. The DATA= option identifies the SAS dataset to print, in this case, CENTER.YES. The LABEL option tells the proc to use the labels of variables, when they have them, in place of their names. The labels were set in the LABEL statement of the data step when it created the SAS dataset. If they had not been, or if you wanted to use different labels, you could write a LABEL statement in the `PROC PRINT` step.

```
  ID NAME;
```

The ID statement indicates an identifying variable that appears in the first column of the report. If necessary, you can have a few ID variables.

```
  VAR JOINAT INSTR NATIONAL;
```

The VAR statement indicates the other variables that appear as columns in the report and the order in which they appear. If you refer back to the program's print output, you can see that the columns are arranged in the sequence indicated in the ID and VAR statements. However, because of the LABEL option, the column headers are the variable labels, rather than the names. You can refer back to the LABEL statement in the data step to verify that variables in the columns are the variables mentioned in the ID and VAR statements.

```
RUN;
```

The RUN statement marks the end of the proc step.

In this case, the entire SAS dataset fit on a single page. When you have a larger SAS dataset to print, with too many variables to fit across the page or too many observations to fit in the length of the page, the PRINT proc uses multiple pages with headings that allow you to see what data you are reading and put it together easily.

In this program, the purpose of the PRINT proc was to ensure that the data step read the input data correctly. You can compare the print output to the input data file and see that the data values are the same.

SAS software has many procs, most of which analyze a SAS dataset in some way. One of the easiest forms of analysis to understand is an analysis of frequencies, and you can produce frequency tables with the FREQ proc. The essential statements of a `PROC FREQ` step are the `PROC FREQ` statement and the TABLES statement.

```
PROC FREQ DATA=CENTER.YES;
  TABLES INSTR NATIONAL INSTR*NATIONAL;
```

As was the case with the `PROC PRINT` statement in the previous step, the `PROC FREQ` statement includes the DATA= option to identify the input SAS dataset — the SAS dataset for the proc to work with. The TABLES statement indicates combinations of variables to display in frequency tables. The variables INSTR and NATIONAL are displayed in separate frequency tables, then combined in a two-way frequency table. If you refer back to the print output, you can see these three frequency tables in succession.

Again, the RUN statement marks the end of the step.

The output of the FREQ proc includes many more details than are mentioned in the program statements. Usually, that is the case with procs. In a proc step, you should not have to spell out every detail in order to get a standard kind of analysis or report.

The idea of a frequency analysis is that you may find some significance in the different numbers of things in a set of data. Perhaps it is interesting, for example, that while Yes has drawn members from various countries over the years, they have never had two members who originally came from the same country, other than the U.K. A two-way frequency analysis may show how the values of two variables may tend to be related in some way. Is there any meaning, perhaps, in the fact that the Yes members who have come from countries other than the U.K. have all been keyboard and guitar players? Probably there is nothing important you could conclude from this particular analysis, but the program still demonstrates the process that is involved.

Whatever kind of subject matter you draw your data from, you may go through a very similar programming process to analyze it. Imagine modifying this program to read a different file, with different fields, and use a different proc to do a different kind of analysis. You would be changing everything that mattered about what the program would do, but the essential outline of the program would still be the same.

Variation: More Analysis

After you create a SAS dataset, it is easy to analyze it in many different ways using different procs. It scarcely matters whether the analysis proc is in the same program that creates the SAS dataset or in a subsequent program.

With numeric variables, you can do mathematical calculations, such as calculating statistics, that cannot be done with character variables. One of the easiest ways to report simple descriptive statistics is with the SUMMARY proc (also known as the MEANS proc). The following program uses the SUMMARY proc to demonstrate another way to analyze the same SAS dataset.

```
*
   YESSTAT.SAS
   Computes descriptive statistics on Yes members.
   Rick Aster    January 2000

   Input SAS dataset: CENTER.YES
*;

OPTIONS LINESIZE=75 NOCENTER;
TITLE1 'Yes Members';

PROC SUMMARY DATA=CENTER.YES PRINT N MEAN MIN MAX;
  CLASS INSTR;
  VAR JOINAT;
RUN;
```

The OPTIONS, TITLE1, and RUN statements work the same way as in the previous program. The PROC SUMMARY statement includes the DATA= option to identify the input SAS dataset, the PRINT option to indicate that the proc should print a report, and a list of statistics to print. The CLASS statement indicates to classify, or group, observations according to the value of the variable INSTR. The VAR statement indicates to compute the statistics for the variable JOINAT. The proc produces this print output:

```
Yes Members                                    14:04 Monday, July 31, 2000   4

The SUMMARY Procedure

                  Analysis Variable : JOINAT Joined At

                      N
Instrument          Obs    N          Mean          Minimum          Maximum
----------------------------------------------------------------------------
bass guitar           1    1     1.0000000        1.0000000        1.0000000

drums                 2    2     3.5000000        1.0000000        6.0000000

guitar                4    4     7.7500000        1.0000000       16.0000000

keyboard              5    5     7.8000000        1.0000000       17.0000000

vocals                2    2     5.5000000        1.0000000       10.0000000
----------------------------------------------------------------------------
```

The SUMMARY proc shows statistics calculated for each group of observations defined by the CLASS statement. By comparing the statistics for one group to the statistics for another, you may be able to notice differences between groups. This might mean finding a difference between guitarists and drummers, for example.

List Input and Concatenation

The first example showed reading fixed-field text data and analyzing the resulting SAS dataset. This example reads another kind of text data and shows how SAS datasets can be combined for analysis.

Free-Form Text Data

If you have to create a small text data file by typing it yourself, you might choose to type free-form text data — not lining up values in specific columns to create fields, but just leaving spaces between the values. It is the spaces between fields that make it possible for you to see the fields in this kind of data file. This example shows members of another music act, with the same variables that were used in the first example.

```
Keith Emerson  keyboard  Canada  1
Carl Palmer  drums  Canada  1
Greg Lake  bass guitar  Canada  1
```

Spaces separate the fields; even without considering what the variables are, it is easy to see the four fields in each record. Because ordinary character values may contain single spaces, it is necessary to leave two spaces after a character field. If you had several numbers in a row, you could separate them with single spaces.

To read free-form text data in a SAS program, you can use a simplified form of the INPUT statement, known as *list input*. Simply list the variables in order. If a field may contain single spaces between words, and there are at least two spaces to mark the end of the value, write an ampersand (&) after the variable name.

```
INPUT NAME & INSTR & NATIONAL & JOINAT;
```

This form of the INPUT statement does not say enough to define the variables it reads, so a separate statement is usually necessary to declare some of the variables, as seen in the example below.

Combining SAS Datasets

The objective in this example is to extend the analysis that was done in the previous example by adding the members of a second music act, Emerson, Lake & Palmer (ELP). This involves creating a second SAS dataset, then combining the two SAS datasets so they can be analyzed together. The result of the combination is a third SAS dataset that contains all the observations from each of the two SAS datasets. Since it does not matter what order the observations are in in the resulting SAS dataset, we can combine them in a simple process of concatenation, which can be done in a data step with the SET statement listing the SAS datasets to concatenate:

```
SET CENTER.YES CENTER.ELP;
```

As with the INPUT statement, the action of the SET statement is to read a single record, or observation. The effect of that action, with the repetition provided by the automatic data step loop, is to go through all the observations in both SAS datasets.

Program and Results

This program reads the new data file, combines the new data with the previous data, and prints the resulting SAS dataset and a frequency table of the combined data:

```
*
   ELP.SAS
   Reads data file of ELP members, combines with
   data of Yes members and computes frequencies.
   Rick Aster    January 2000

   Input text file: MEMBERS2
   Output SAS data file: CENTER.ELP
   Temporary SAS data file: WORK.BANDS
*;

OPTIONS LINESIZE=75 NOCENTER PAGENO=1;
TITLE1 'ELP Members';

*
  Data step creates SAS data file CENTER.ELP from
  input text file MEMBERS2.
*;
DATA CENTER.ELP;
  INFILE MEMBERS2;
  LENGTH NAME $ 24  INSTR $ 16  NATIONAL $ 24;
  INPUT NAME & INSTR & NATIONAL & JOINAT;
  LABEL NAME='Name' INSTR='Instrument' NATIONAL='Nationality'
```

```
    JOINAT='Joined At';
RUN;

*
 Print SAS dataset.
*;
PROC PRINT DATA=CENTER.ELP LABEL;
   ID NAME;
   VAR JOINAT INSTR NATIONAL;
RUN;

TITLE1 'Yes and ELP Members';

*
  Combine ELP and Yes members in one SAS dataset.
*;
DATA BANDS;
  SET CENTER.YES CENTER.ELP;
RUN;

*
 Print combined SAS dataset.
*;
PROC PRINT DATA=BANDS LABEL;
   ID NAME;
   VAR JOINAT INSTR NATIONAL;
RUN;

*
 Compute frequencies.
*;
PROC FREQ DATA=BANDS;
   TABLES INSTR NATIONAL INSTR*NATIONAL;
RUN;
```

Most of what happens in this program works the same way as in the first program. These notes explain things that are new and different.

- The LENGTH statement is necessary to declare variables that appear in the INPUT statement with the list input style. In the LENGTH statement, the dollar sign ($) indicates the character data type for the preceding variable name. Then, the number indicates the length of the variable. It is not usually necessary to declare numeric variables in a LENGTH statement because a data step gives variables the numeric data type by default, and it is not usually necessary to declare a length for numeric variables. If you did need to declare numeric variables in a length statement, you can declare them with a length of 8, which is the length that allows the greatest accuracy for a numeric variable when it is stored in a SAS dataset.
- The SAS dataset WORK.BANDS is created in the second data step as a concatenation of two existing SAS datasets, as indicated in the SET statement. You can see the effect of the concatenation by comparing the print output of the different PROC PRINT steps.

- The program refers to the SAS dataset WORK.BANDS simply as BANDS. You can use one-level names for SAS datasets in the WORK library; it is not necessary to indicate the WORK libref. This SAS dataset is stored in the WORK library because it would not be useful to store it permanently. The WORK library is automatically deleted at the end of the SAS session.
- The `PROC FREQ` step does the same analysis as in the first program. It produces different results because it is working on a different set of data.

The log describes the execution of the program. As in the first example, the log indicates that the program ran successfully.

```
82   *
83       ELP.SAS
84       Reads data file of ELP members, combines with
85       data of Yes members and computes frequencies.
86       Rick Aster    January 2000
87
88       Input text file: MEMBERS2
89       Output SAS data file: CENTER.ELP
90       Temporary SAS data file: WORK.BANDS
91   *;
92
93   OPTIONS LINESIZE=75 NOCENTER PAGENO=1;
94   TITLE1 'ELP Members';
95
96   *
97      Data step creates SAS data file CENTER.ELP from
98      input text file MEMBERS2.
99   *;
100   DATA CENTER.ELP;
101      INFILE MEMBERS2;
102      LENGTH NAME $ 24  INSTR $ 16  NATIONAL $ 24;
103      INPUT NAME & INSTR & NATIONAL & JOINAT;
104      LABEL NAME='Name' INSTR='Instrument' NATIONAL='Nationality'
105          JOINAT='Joined At';
106   RUN;

NOTE: The infile MEMBERS2 is:
      File Name=C:\elp.txt,
      RECFM=V,LRECL=256

NOTE: 3 records were read from the infile MEMBERS2.
      The minimum record length was 29.
      The maximum record length was 34.
NOTE: The data set CENTER.ELP has 3 observations and 4 variables.
NOTE: DATA statement used:
      real time           0.03 seconds


107
108   *
109     Print SAS dataset.
110   *;
111   PROC PRINT DATA=CENTER.ELP LABEL;
112       ID NAME;
113       VAR JOINAT INSTR NATIONAL;
114   RUN;

NOTE: There were 3 observations read from the dataset CENTER.ELP.
```

```
NOTE: PROCEDURE PRINT used:
      real time           0.00 seconds


115
116   TITLE1 'Yes and ELP Members';
117
118   *
119      Combine ELP and Yes members in one SAS dataset.
120   *;
121   DATA BANDS;
122      SET CENTER.YES CENTER.ELP;
123   RUN;

NOTE: There were 14 observations read from the dataset CENTER.YES.
NOTE: There were 3 observations read from the dataset CENTER.ELP.
NOTE: The data set WORK.BANDS has 17 observations and 4 variables.
NOTE: DATA statement used:
      real time           0.03 seconds


124
125   *
126     Print combined SAS dataset.
127   *;
128   PROC PRINT DATA=BANDS LABEL;
129       ID NAME;
130       VAR JOINAT INSTR NATIONAL;
131   RUN;

NOTE: There were 17 observations read from the dataset WORK.BANDS.
NOTE: PROCEDURE PRINT used:
      real time           0.04 seconds


132
133   *
134     Compute frequencies.
135   *;
136   PROC FREQ DATA=BANDS;
137       TABLES INSTR NATIONAL INSTR*NATIONAL;
138   RUN;

NOTE: There were 17 observations read from the dataset WORK.BANDS.
NOTE: PROCEDURE FREQ used:
      real time           0.09 seconds
```

This is the print output, which you may want to compare to the print output from the first program. Also notice how the two TITLE1 statements in the program result in different first title lines on different pages of print output.

```
ELP Members                                    14:04 Monday, July 31, 2000   1

                     Joined
    Name               At       Instrument     Nationality

Keith Emerson           1       keyboard         Canada
Carl Palmer             1       drums            Canada
Greg Lake               1       bass guitar      Canada
```

```
Yes and ELP Members                           14:04 Monday, July 31, 2000   2

                  Joined
Name                At      Instrument     Nationality

Peter Banks          1      guitar         United Kingdom
Chris Squire         1      bass guitar    United Kingdom
Jon Anderson         1      vocals         United Kingdom
Tony Kaye            1      keyboard       United Kingdom
Bill Bruford         1      drums          United Kingdom
Steve Howe           3      guitar         United Kingdom
Rick Wakeman         4      keyboard       United Kingdom
Alan White           6      drums          United Kingdom
Patrick Moraz        7      keyboard       Switzerland
Trevor Horn         10      vocals         United Kingdom
Geoff Downes        10      keyboard       United Kingdom
Trevor Rabin        11      guitar         South Africa
Billy Sherwood      16      guitar         United States
Igor Khoroshev      17      keyboard       Russia
Keith Emerson        1      keyboard       Canada
Carl Palmer          1      drums          Canada
Greg Lake            1      bass guitar    Canada
```

```
Yes and ELP Members                           14:04 Monday, July 31, 2000   3

The FREQ Procedure

                            Instrument

                                             Cumulative    Cumulative
INSTR           Frequency      Percent        Frequency       Percent
---------------------------------------------------------------------
bass guitar            2        11.76                2         11.76
drums                  3        17.65                5         29.41
guitar                 4        23.53                9         52.94
keyboard               6        35.29               15         88.24
vocals                 2        11.76               17        100.00


                               Nationality

                                                Cumulative    Cumulative
NATIONAL           Frequency      Percent        Frequency       Percent
------------------------------------------------------------------------
Canada                    3        17.65                3         17.65
Russia                    1         5.88                4         23.53
South Africa              1         5.88                5         29.41
Switzerland               1         5.88                6         35.29
United Kingdom           10        58.82               16         94.12
United States             1         5.88               17        100.00
```

```
Yes and ELP Members                              14:04 Monday, July 31, 2000   4

The FREQ Procedure

Table of INSTR by NATIONAL

INSTR(Instrument)     NATIONAL(Nationality)

Frequency    |
Percent      |
Row Pct      |
Col Pct      |Canada  |Russia  |South Af|Switzerl|United K|United S|  Total
             |        |        |rica    |and     |ingdom  |tates   |
-------------+--------+--------+--------+--------+--------+--------+
bass guitar  |      1 |      0 |      0 |      0 |      1 |      0 |      2
             |   5.88 |   0.00 |   0.00 |   0.00 |   5.88 |   0.00 |  11.76
             |  50.00 |   0.00 |   0.00 |   0.00 |  50.00 |   0.00 |
             |  33.33 |   0.00 |   0.00 |   0.00 |  10.00 |   0.00 |
-------------+--------+--------+--------+--------+--------+--------+
drums        |      1 |      0 |      0 |      0 |      2 |      0 |      3
             |   5.88 |   0.00 |   0.00 |   0.00 |  11.76 |   0.00 |  17.65
             |  33.33 |   0.00 |   0.00 |   0.00 |  66.67 |   0.00 |
             |  33.33 |   0.00 |   0.00 |   0.00 |  20.00 |   0.00 |
-------------+--------+--------+--------+--------+--------+--------+
guitar       |      0 |      0 |      1 |      0 |      2 |      1 |      4
             |   0.00 |   0.00 |   5.88 |   0.00 |  11.76 |   5.88 |  23.53
             |   0.00 |   0.00 |  25.00 |   0.00 |  50.00 |  25.00 |
             |   0.00 |   0.00 | 100.00 |   0.00 |  20.00 | 100.00 |
-------------+--------+--------+--------+--------+--------+--------+
keyboard     |      1 |      1 |      0 |      1 |      3 |      0 |      6
             |   5.88 |   5.88 |   0.00 |   5.88 |  17.65 |   0.00 |  35.29
             |  16.67 |  16.67 |   0.00 |  16.67 |  50.00 |   0.00 |
             |  33.33 | 100.00 |   0.00 | 100.00 |  30.00 |   0.00 |
-------------+--------+--------+--------+--------+--------+--------+
vocals       |      0 |      0 |      0 |      0 |      2 |      0 |      2
             |   0.00 |   0.00 |   0.00 |   0.00 |  11.76 |   0.00 |  11.76
             |   0.00 |   0.00 |   0.00 |   0.00 | 100.00 |   0.00 |
             |   0.00 |   0.00 |   0.00 |   0.00 |  20.00 |   0.00 |
-------------+--------+--------+--------+--------+--------+--------+
Total               3        1        1        1       10        1       17
                17.65     5.88     5.88     5.88    58.82     5.88   100.00
```

Variation: Data Lines

Sometimes, it is not so convenient to create a separate data file. In those situations, you can write the data lines directly in the program file. Indicate the special fileref CARDS in the INFILE statement. Write a CARDS statement in place of the RUN statement, and follow that, beginning on the next line, with the data lines, just as they would appear in a separate file.

The SAS supervisor continues to read data lines until it comes to a line that contains a semicolon. If your data lines may contain a semicolon, use the CARDS4 statement instead; see the details in chapter 17, "Execution."

Using the CARDS statement and data lines in the program file, the data step to create the SAS dataset CENTER.ELP is:

```
*
  Create the SAS data file CENTER.ELP from data lines in
  the program file.
*;
```

```
DATA CENTER.ELP;
  INFILE CARDS;
  LENGTH NAME $ 24 INSTR $ 16 NATIONAL $ 24;
  INPUT NAME & INSTR & NATIONAL & JOINAT;
  LABEL NAME='Name' INSTR='Instrument' NATIONAL='Nationality'
    JOINAT='Joined At';
CARDS;
Keith Emerson  keyboard  Canada  1
Carl Palmer  drums  Canada  1
Greg Lake  bass guitar  Canada  1
;
```

When this step is run, the log notes are only slightly different from the notes from reading the same data from a separate file:

```
NOTE: The data set CENTER.ELP has 3 observations and 4 variables.
NOTE: DATA statement used:
      real time           0.02 seconds
```

Computation

Input data does not always correspond so precisely to the form of the SAS datasets you want to create. You can use statements and options in the data step to select a subset of the data, correct values, and create new variables. This example shows how this might be done with some basketball statistics.

Missing Values

This example reads from a fixed-field text data file defined with five fields: team, player, two-point shots, three-point shots, and free throws. A few records are shown here with a ruler for reference.

```
----+----1----+----2----+----3----+----4----+----5
Boston            Antony                   62  18  41
Boston            Banks, A.                11   3  16
Boston            Calvert                  90  31  59
Boston            Case
Boston            Conner                   14  22  21
Boston            Hess                      7       6
Boston            Jones, W.               101   4  72
Boston            Lyte                     50   3  18
Boston            Pierce                   92  11  26
...
```

One of the first things you might notice about the data is that some of the fields are blank. A field is blank when a value is not known or would not apply. In this example, fields are blank when no shots were counted for a particular player. The blank fields in the fourth record indicate that the player Case did not make any shots. The blank field in the sixth record indicates that Hess made no three-point shots. So a blank actually means a zero.

What happens when you read a blank field? If you were reading a character variable, nothing unusual would happen because a space is a normal character, not essentially different from any other character that might form part of a character value. For a numeric value, though, something different has to be done. "Blank" is not a number. What SAS software does is simply note that the value is missing. It gives the variable a missing value. A missing value appears as a period in print output or a visual display of the variable, and you can write it as a period when you need to write it in a program.

Program and Results

The objective in this example is to create a SAS dataset with field goals, three-point shots, free throws, and total points for those players that have points. Field goals and total points are not in the input data, but can be calculated from the variables that are present in the input data.

```
*
  POINTS.SAS
  Create SAS dataset of basketball scoring statistics.
  Rick Aster    January 2000

  Input text file: BASKET
  Output SAS dataset: CENTER.POINTS
*;

OPTIONS LINESIZE=75 NOCENTER PAGENO=1;
TITLE1 'Basketball Scorers';

*
  Create SAS dataset from input data.
*;
DATA CENTER.POINTS (DROP=P2);
  INFILE BASKET TRUNCOVER;
  INPUT
     TEAM $18.
     PLAYER $20.
     FG 4.
     P3 4.
     FT 4.
     ;

  *
    Convert missing values to zeroes.
  *;
  IF FG = . THEN FG = 0;
  IF P3 = . THEN P3 = 0;
  IF FT = . THEN FT = 0;

  *
    Calculate points and field goals.
  *;
  P2 = FG - P3;
  PTS = 2*P2 + 3*P3 + FT;
  FG = P2 + P3;
```

```
  *
    Keep only players that have points.
  *;
  IF PTS > 0;
RUN;

*
  Print first five observations.
*;
PROC PRINT DATA=CENTER.POINTS (OBS=5);
RUN;
```

The log shows these notes when it runs the data step:

```
NOTE: 144 records were read from the infile BASKET.
      The minimum record length was 50.
      The maximum record length was 50.
NOTE: The data set CENTER.POINTS has 128 observations and 6 variables.
NOTE: DATA statement used:
      real time           0.26 seconds
```

and these notes for the proc step:

```
NOTE: There were 5 observations read from the dataset CENTER.POINTS.
NOTE: PROCEDURE PRINT used:
      real time           0.00 seconds
```

This is the output produced by the PRINT proc:

```
Basketball Scorers                                   14:04 Monday, July 31, 2000   1

Obs      TEAM      PLAYER        FG     P3     FT     PTS

  1     Boston     Antony        62     18     41     183
  2     Boston     Banks, A.     11      3     16      41
  3     Boston     Calvert       90     31     59     270
  4     Boston     Conner        14     22     21      71
  5     Boston     Hess           7      0      6      20
```

Notice these new features in the program and its results.

- The option in parentheses in the DATA statement is a dataset option. Dataset options affect the way a SAS dataset is accessed. In this case, the DROP= option is used to keep the SAS dataset from storing the variable P2. By default, when a data step creates a SAS dataset, the SAS dataset stores all the variables available in the data step, so it is often necessary to use dataset options to limit the set of variables that is stored.
- The INPUT statement uses the standard character, or $, informat. This informat is only slightly different from the $CHAR informat. Unlike the $CHAR informat, the $ informat left-aligns values that begin with spaces. The other special feature of the $ informat is that it interprets a field that contains only a period as a blank value.
- In a data step, the IF-THEN statement lets a program take an action only when a certain condition is true. In this case, the statements check the variables to see if their values are missing values. If that condition is true, the statement assigns a 0 value to the variable. Missing values

from these input fields actually do represent zeroes, so it important to correct the values so that the variables actually show zeroes.

- Points, the variable PTS, is calculated at two points per two-point, three points per three-point shot, and one point per free throw. Field goals, the variable FG, is calculated by adding the number of two-point shots and the number of three-point shots.
- The subsetting IF statement checks the condition `PTS > 0` and stops processing for a player who has no points. The effect of this statement is to include in the output SAS dataset only those players who have at least one point.
- The dataset option `OBS=5` appears in the `PROC PRINT` statement. This option limits access to the SAS dataset to the first five observations. The OBS= option is useful when you want to look at the form of a SAS dataset, but you do not need to see all of the observations. Only five observations appear in the output because of this dataset option.
- The PRINT proc appears without any extra options or optional statements. It produces its default report, which includes all the variables it reads from the SAS dataset. The first column it produces shows observation numbers.
- Because of the DROP= dataset option, the SAS dataset does not contain the two-point shot field that was in the original input data. Because of the subsetting IF statement, it does not contain an observation for any record in the input data that did not show any points scored. Other than those changes, if you compare the text data file to the `PROC PRINT` output, you can see the input data accurately reflected in the SAS dataset. In addition, you can see the two computed variables in each observation.

Variation: Creating Multiple SAS Datasets

When it is necessary to create more than one SAS dataset from the data in one input file, it is easiest to do so in a single data step. Suppose, for example, the objective is not to create one SAS dataset of all the scorers, but four separate SAS datasets containing the top players in the categories of field goals, three-point shots, free throws, and points. You can easily create all four SAS datasets in a single data step, as shown here.

```
OPTIONS LINESIZE=75 NOCENTER PAGENO=1;
TITLE1 'Basketball Scorers';

DATA  CENTER.TOPFG (KEEP=TEAM PLAYER FG)
    CENTER.TOPP3 (KEEP=TEAM PLAYER P3)
    CENTER.TOPFT (KEEP=TEAM PLAYER FT)
    CENTER.TOPPTS (KEEP=TEAM PLAYER PTS)
    ;
  INFILE BASKET TRUNCOVER;
  INPUT
    TEAM $18.
    PLAYER $20.
    FG 4.
```

```
    P3 4.
    FT 4.
    ;

  *
   Convert missing values to zeroes.
  *;
  IF FG = . THEN FG = 0;
  IF P3 = . THEN P3 = 0;
  IF FT = . THEN FT = 0;

  *
   Calculate points and field goals.
  *;
  P2 = FG - P3;
  PTS = 2*P2 + 3*P3 + FT;
  FG = P2 + P3;

  *
   Keep only players at or above a certain cutoff
   in each category.
  *;
  IF FG >= 100 THEN OUTPUT CENTER.TOPFG;
  IF P3 >= 30 THEN OUTPUT CENTER.TOPP3;
  IF FT >= 50 THEN OUTPUT CENTER.TOPFT;
  IF PTS >= 250 THEN OUTPUT CENTER.TOPPTS;
RUN;
```

When this data step runs, the log shows these notes:

```
NOTE: 144 records were read from the infile BASKET.
      The minimum record length was 50.
      The maximum record length was 50.
NOTE: The data set CENTER.TOPFG has 16 observations and 3 variables.
NOTE: The data set CENTER.TOPP3 has 16 observations and 3 variables.
NOTE: The data set CENTER.TOPFT has 32 observations and 3 variables.
NOTE: The data set CENTER.TOPPTS has 32 observations and 3 variables.
NOTE: DATA statement used:
      real time           0.16 seconds
```

Notice these different features in this program.

- All the output SAS datasets are listed in a single DATA statement.
- The KEEP= dataset options determine which variables appear in which SAS dataset.
- The IF-THEN statements at the end of the data step determine which observations appear in which SAS datasets. If is possible for the same player to appear in more than one SAS dataset. Most players, though, do not appear in any of the SAS datasets produced in this step.
- The log notes that indicate the number of variables and observations in each output SAS dataset confirm that different observations were stored in different SAS datasets.

The programs in this chapter show the SAS language doing what it does best — accessing, organizing, and analyzing data. Furthermore, these examples show the SAS way of working with data — creating SAS datasets in data steps, then using proc steps for further processing. You can see how steps go together to form a complete SAS program. No matter how complicated SAS programs get, they still have a simple outline that you can see and work with in the way that you see in these examples.

3

Statements and Steps

The details that make up a SAS program are part of a larger picture. They spell out components that go together systematically to carry out the overall actions of the program. Steps are the basic units of execution in a SAS program; statements, the basic units of syntax; and SAS datasets, the form in which data flows through the program. There are also matters of style that unify all SAS code. The common elements of SAS syntax create the distinctively SAS-like qualities that can be seen throughout every SAS program.

Things to Look For — and What They Mean

At first glance, a SAS program might look like a random arrangement of words and symbols. You start to discern the structure of the program as soon as you find these objects and patterns within the program.

Characters Around 90 characters combine to form a SAS program. There are two main sets of characters in SAS syntax — 63 that form words and two dozen that primarily stand alone.

SAS words are written with letters, the underscore (_) character, and digits. Another character, the period, is also used in connection with words. Any consecutive sequence of these characters forms a single token — the smallest unit of meaning in the program.

These special characters form symbols in SAS code:

```
# $ % & ( ) * + , - / : ; < = > ? @ [ ] ^ { | }
```

Usually, each special character is a separate token — that is, other characters immediately before and after it do not affect its meaning. However, there are exceptions. Plus signs (+) and minus signs (the hyphen character, -) can be part of a numeric value. A dollar sign ($) can be the first character of a name, and in some special cases, it is also possible for the at-sign (@) and number sign (#) to be used in a name. And sometimes, a sequence of two or three consecutive special characters has a specific meaning.

Double quotes (") and single quotes (the apostrophe character, ') are also special characters, but they never stand alone. They are used in pairs to form quoted strings. Quoting takes away the special meaning of characters. Inside a quoted string, characters are just text.

Last but not least, the space character is used to separate tokens.

Symbols With the few exceptions that have been noted, special characters are used as symbols. Any special character token, except a period, is considered a symbol. Most symbols are written with just one special character. These characters always form tokens by themselves:

```
( ) , ; [ ] { }
```

Other special characters may combine to form tokens. These characters, whenever they appear in succession, combine to form a token:

```
* : < = > ^
```

They combine to form these symbols, among others:

```
** <: <= <=: <> =* =: > >: >< >= >=:
```

These other symbol tokens are written as double characters:

```
-- ?? @@ ||
```

The caret (^) and brace ([,]) characters are not part of the EBCDIC character set used on IBM mainframe systems. On those computers, you can use the logical not character (¬) anywhere SAS syntax calls for a caret. If you do not know for sure what kind of computer a program might run on, do not use any of these nonportable characters in the program. It is also possible to substitute the exclamation point (!) for the vertical bar (|).

Words A token that begins with a letter or underscore is a word. Words can contain letters, underscores, and digits. Words are used as keywords and as names. If it takes more than one word to form the name of an object, the words are joined by periods to form multi-level names.

Numbers A token that is formed of digits is a numeric constant. Numeric constant values are written without commas. Any token that contains only digits, or digits with a + or - prefix, is a numeric constant:

```
-1 0 25 1000 32768
```

A numeric constant can contain a decimal point, written as a period:

```
-.25 .6 0.64 1. 3.14159 499583.21483407
```

Some special forms of numeric constants can contain letters: the E (or D) that indicates the exponent in scientific notation, or the letters A through F and the final X of hexadecimal constants. However, the numeric constant must begin with a digit. If it begins with a letter, it is not a number at all, because all tokens that begin with letters are words.

Missing values A token that begins with a period, not followed by a digit, is a missing value. A period by itself is a standard missing value. A period followed by a letter or underscore, such as .B or .N, is a special missing value.

Quoted strings Any sequence of characters enclosed in either single or double quotes is a quoted string. A quoted string by itself represents a character constant. When it is immediately followed by a letter, it may represent any of various special kinds of constants or a name.

Tokens and spacing Tokens are the smallest meaningful units of a program. Tokens take on the various forms described in the preceding paragraphs. Spaces are used to separate tokens. Spaces are required between most kinds of tokens, but no space is required between symbol tokens or between a symbol token and any other kind of token.

Invisible control characters, especially tabs and line breaks, are equivalent to spaces in SAS syntax. There is no special significance in the way a program is divided into lines; syntactically, a line break is no different from a space. Most programmers use line breaks and extra spaces in an orderly pattern or style in order to call attention to the parts of the program and make it easy to read.

Statements The most powerful character in a SAS program is the semicolon (;), which marks the end of a statement. To find the statements in a program, you only need to find the semicolons. Every time you write a SAS statement, you must remember to write a semicolon at the end of it.

Most statements begin with keywords that define, in general terms, the action or effect of the statement. This keyword is the name of the statement; for example, the DATA statement begins with the keyword DATA.

Each kind of statement has its own syntactical form, which determines what may appear between the keyword and the semicolon that concludes the statement. Sometimes a statement is only a keyword, as in the case of the STOP statement:

```
STOP;
```

More often, the keyword is followed by other terms. This BY statement, for example, contains a list of variables:

```
BY CLINIC PATIENT VISIT MEASURE;
```

Steps When a SAS program runs, the basic unit of execution is the step. Steps execute in sequence — the first step executes, then the second one, and so on. As soon as you have identified the statements of a SAS program, it is an easy matter to pick out the steps. Each step begins with a DATA statement or a PROC statement. All the related statements that follow are part of the step.

The two statements indicate two different kinds of steps. The data step creates SAS datasets and allows general programming. The statements of a data step spell out the programming logic of a process. Essentially, a data step is like a separate program. A data step can contain various statements, which are put together in an order determined by the programming logic of the data step. Data steps are described in more detail later in this chapter.

A proc step runs an existing program, a proc, and its statements determine specific details of the proc's actions. Every proc has its own rules of syntax. Some procs require the statements of the proc step to follow a specific sequence. For other procs, the order of statements does not matter. For some procs, there are no statements beyond the PROC statement. Proc steps are described in more detail in chapter 19.

The RUN statement marks the end of a step and the point where the SAS supervisor can start to execute the step. For most procs, you can also use the QUIT statement to mark the end of the step. Using a RUN or QUIT statement is not strictly required, but it is a very good idea. It makes it easier to see the steps when you read a SAS program. It also makes it easier for the SAS supervisor to find the steps when it runs the program, which results in better log notes and ensures that the program runs in sequence. When you make the SAS supervisor guess about where a step ends, the rules it follows to determine the order of execution are more complicated, and some statements may run out of sequence. When you explicitly mark the end of each step, you keep things simple, and the program runs in order.

Global statements The statements in a SAS program that can stand on their own are global statements. The SAS supervisor executes these statements as soon as it gets to them in a SAS program. Most global statements modify something about the SAS environment. You can identify global statements by their keywords. The most common global statements are listed in the table below.

Statement	Purpose	Example
FILENAME	Declare a text file.	`FILENAME START 'start.dat';`
LIBNAME	Declare a SAS data library.	`LIBNAME MAIN 'c:\sasdata\main';`
MISSING	List characters for numeric informats to interpret as special missing values.	`MISSING a b c;`
TITLE	Define a title line.	`TITLE1 'Cumulative Results';` `TITLE2 ' ';`
FOOTNOTE	Define a footnote line.	`FOOTNOTE1 'As of July 2000';`
OPTIONS	Set system options.	`OPTIONS NOCENTER PAGENO=1;`
SKIP	Skip lines in the log.	`SKIP 5;`
PAGE	Start a new page in the log.	`PAGE;`
DM	Execute a SAS command.	`DM 'CLEAR LOG; CLEAR OUTPUT';`
X	Execute an operating system command.	`X 'copy file newfile';`
ENDSAS	End the SAS session.	`ENDSAS;`
LOCK	Establish exclusive use of a SAS file or library.	`LOCK SERVER.TRANS;` `LOCK SERVER.TRANS CLEAR;`

Global statements execute between steps. Data steps and most proc steps execute as a single action, so if a global statement appears inside one of those steps, it executes before the step executes. When you write SAS programs, write global statements between steps to emphasize the sequence in which the program executes.

When global statements set or modify properties of the SAS environment, the changes apply to the following step and stay in effect until they are changed again, usually by another instance of the same global statement. For example, if you use the TITLE1 statement to create a title line, that title line appears on any output pages produced by the next step to execute, and on the pages produced by any steps that follow, until another TITLE1 statement executes.

Lists When most kinds of lists appear in SAS syntax, the items are separated only by spaces. You can see this, for example, in the VAR statement of many procs, which lists variables:

```
VAR STYLE SIZE COLOR;
```

In some statements, a list is allowed even though you would usually use only one item. For example, in the OUTPUT statement of the data step, you can write a list of SAS datasets, but most often the statement is used to write an observation to only one SAS dataset.

Usually, spaces separate list items, but there are two important exceptions. In a list of expressions, or a list that may include expressions, commas separate the items. Arguments to functions, for example, are usually expressions, so commas separate arguments:

```
ROUND(PRICE, .01)
```

In lists of constant values, such as a list of initial values for an array, SAS syntax allows the items to be separated by commas or just spaces. Either of these forms is acceptable:

```
ARRAY A{5} / (1, 2, 3, 4, 5);

ARRAY A{5} / (1 2 3 4 5);
```

Abbreviated lists When you have a long list of related names, you can sometimes abbreviate it. This is most useful when a list of names has a common prefix and sequential numeral suffixes, by which I mean a list like this:

```
SENT1999 SENT2000 SENT2001 SENT2002
```

You can abbreviate this list as:

```
SENT1999-SENT2002
```

This kind of abbreviation works when the prefix is the same for all the names and the suffixes are consecutive. The list can go from lowest to highest, as in `PLACE1-PLACE4`, or from highest to lowest, as in `X2-X0`. Write the abbreviation using the first and last variable names with a hyphen between them.

If you write two variable names with two hyphens between them, it represents a different kind of abbreviated variable list, which identifies a group of variables according to their order of position (this corresponds, not always in an obvious way, to the order in which the variables were created). If you created the variables ADDRESS1, ADDRESS2, CITY, ST, and ZIP together, in that order, you might later be able to refer to that list of variables with the abbreviated list `ADDRESS1--ZIP`. You can select variables of a specific data type by position. Write the type keyword CHARACTER (or CHAR) or NUMERIC in the middle of the abbreviation, between the two hyphens, such as:

```
COMPANY-CHARACTER-UNIT
AGE-NUMERIC-WEIGHT
```

In some contexts, you can write an abbreviated variable list of all the variables (or other objects) whose names begin with the same prefix. Write the prefix followed by a colon. For example, this list

```
BALAV:
```

might include the variables BALAVL and BALAVC. This form of abbreviated list can be used in function calls and proc step statements.

There are also a few abbreviated list names. These names can be used in a list or in place of a list:

List Name	Includes
ALL	All available variables, or all available objects of the appropriate kind
CHARACTER _CHAR_	All available character variables
NUMERIC	All available numeric variables

Options An option is a choice about the way a statement works or an extra effect the statement can apply in its actions. Options are an important part of the syntax of many SAS statements. Options may be the only terms in the statement, or they may appear at the end of the statement, after other terms. The syntax of the statement defines a set of options; you can use any of them or a combination of them. When you write a list of options, it usually makes no difference what order you write them in.

When a list of options follows a list of objects in a statement, a slash (/) separates the two lists so that the options are not mistaken for the objects in the statement. A slash is necessary even when there is only one term in each list, as in the DATA statement for a data step view:

```
DATA THIS / VIEW=THIS;
```

There are three main forms of options. In the simplest form, an option is a word that appears in the statement. An example of this is the TRUNCOVER option in the INFILE statement, which allows an INPUT statement to read an input record completely even if the last field is shorter than expected. Alternative options may also be available. For example, instead of

TRUNCOVER, you could use the STOPOVER option to have the INPUT statement create an error condition and stop running if any record is shorter than expected. The TRUNCOVER and STOPOVER options, along with a few other available options, are mutually exclusive; you can use only one of them in the same statement.

Often, you can choose between an option and the negative form of the option, which is the same word with the prefix NO. For example in the `PROC SUMMARY` statement, you can use the PRINT option to print a table of descriptive statistics, or the NOPRINT option not to print that table.

Finally, there are options that take values. The option name is followed by an equals sign and an appropriate value. Sometimes, the value is a whole number, as in the OBS= option of the INFILE statement, which limits the data step to processing a certain number of records from the input file. For example, `OBS=100` would make the data step read the file as if it contained only 100 records. In other cases, the value of an option is taken from a set of words. For example, the TIES= option in the `PROC RANK` statement controls the way the proc computes ranks when values are tied. The values for the option are `TIES=HIGH`, `TIES=MEAN`, and `TIES=LOW`. For other options, the value is the name of an object. The DATA= option in the PROC statement of many procs names a SAS dataset for the proc to work with. The value of the END= option in the INFILE statement is the name of a variable. The option creates that variable, and the INPUT statement sets the value of the variable to 1 when the last record of the input text file is read. Using this option lets you find out when you have reached the end of the file.

Options are, by definition, not required by the syntax of a statement. When you do not indicate an option, there is a default behavior or effect that takes place. For example, if you omit the DATA= option in a PROC statement, the proc works with the SAS dataset created most recently in the session. In the SUMMARY proc, if you do not indicate the PRINT or NOPRINT option, NOPRINT is the default. In the INFILE statement, if you omit the OBS= option, the default is to read as many records as are in the input file.

System options System options give SAS software a way to consider your preferences in the way it does things. For example, the SAS supervisor is capable of creating various kinds of log notes and messages, but it writes or does not write different kinds of messages depending on system options you set. In a SAS program, system options appear as statement options of the OPTIONS statement.

There are other ways you can set system options in the interactive environment, in the operating system command line you use to start the SAS application, and in configuration files, which are special files, often called CONFIG.SAS, that contain system option settings that are applied at SAS startup. In most operating systems, the syntax for system options set at startup is different from the syntax of the OPTIONS statement. It usually follows Unix conventions for writing options. There are some system options that can be set only at startup; they cannot be changed in the middle of a SAS session.

About half of system options have on/off settings. In the OPTIONS statement, write the option name by itself to turn the option on, or with the prefix NO to turn it off. For example, this statement turns the LABEL option on and the CENTER option off:

```
OPTIONS LABEL NOCENTER;
RUN;
```

You can use the OPTIONS proc to find out the settings of system options. This proc does not require any options or additional statements:

```
PROC OPTIONS;
RUN;
```

The default action of the proc is to write a list of system options with their current settings and brief descriptions. This output takes up several pages in the log. The following excerpts show the kind of lines the proc writes.

```
Portable Options:

 NOASYNCHIO        Do not enable asynchronous input/output
...
 WORKTERM          Erase all files from WORK library at SAS termination
 YEARCUTOFF=1950   Cutoff year for DATE and DATETIME informats and
                   functions
 _LAST_=_NULL_     Last SAS data set created

Host Options:

 ALTLOG=           Specifies the destination for a copy of the SAS log
 ALTPRINT=         Specifies the destination for a copy of the SAS
                   procedure output file
...
 XSYNC             Specify to run X-command synchronously. Default is
                   synchronous.
 XWAIT             Specify to run X command waits for 'exit' before
                   returning to SAS (default is wait).
```

The following table lists a few examples of system options, showing their syntax as you would write it in the OPTIONS statement.

System Option	Purpose	Example
CENTER	Centers most print output.	CENTER
COMPRESS=	Applies compression to new SAS data files.	COMPRESS=NO
ERRORS=	Limits the number of complete error messages for data errors.	ERRORS=25 ERRORS=0
FIRSTOBS=	Starts processing an input file at this observation number or record number.	FIRSTOBS=1 FIRSTOBS=101
LABEL	Allows procs to use labels of variables.	LABEL
NUMBER	Writes page numbers in print output.	NUMBER

continued

System Option	Purpose	Example
OBS=	Stops processing an input file at this observation number or record number.	OBS=1000 OBS=MAX
PAGENO=	The page number.	PAGENO=1
WORKTERM	Erases the WORK library at the end of the session.	WORKTERM
YEARCUTOFF=	The interpretation of two-digit year numbers.	YEARCUTOFF=1925

Dataset options Options that appear in parentheses after a SAS dataset name are dataset options. These options modify the way a step reads or writes a SAS dataset. All dataset options are written as the option name followed by an equals sign and a value.

Dataset options can limit the set of variables that are stored when a SAS dataset is created or that are read from an existing SAS dataset. The KEEP= option lists variables to be used. For example, the data step that begins with this statement creates a SAS dataset with just two variables, regardless of how many other variables are available in the step:

```
DATA MAIN.MEMBERS (KEEP=NAME YEAR);
```

Alternatively, you can use the DROP= dataset option to list variables to omit.

Other dataset options can select a subset of observations. The OBS= option gives the number of the last observation to process. You can also use the FIRSTOBS= option to start processing observations at a certain observation number. Alternatively, you can use the WHERE= dataset option to indicate a condition of the data that determines which observations are used. Other dataset options create indexes and labels for the SAS dataset, determine the way the file is compressed, and do various other things to modify the way a SAS dataset is accessed. Chapter 11, "Options for SAS Datasets," describes dataset options in detail.

Comments Comments are text in a program file, but are not part of the program when it runs. Comments are usually used to describe and document a program. They can also be used to prevent a part of a program from running; the affected code is said to be *commented out*. When a program runs, it runs the same way that it would if the comments were not present.

A delimited comment can appear between tokens in a SAS program, anywhere a space can appear. It begins with the characters /* and continues until the characters */ are reached, regardless of how long that takes. The comment can contain any characters at all and those characters, with the single exception of the */ sequence, do not have any special meaning. These are examples of the use of delimited comments:

```
/*
   This is an example of a delimited comment.
*/
```

```
/*    The following statements do not execute because they are commented out.

TITLE1 'Not an Actual Title Line';
OPTIONS PAGENO=1;

*/


  INPUT             /* Comments describing the fields of an input text file. */
     TEAM $18.      /* Team name */
     PLAYER $20.    /* Player name */
     FG 4.          /* Field goals  */
     P3 4.          /* 3-point shots  */
     FT 4.          /* Free throws  */
     ;
```

A comment statement is a comment in the form of a statement. It begins with an asterisk (*) and ends, the same as any SAS statement, with a semicolon. Because a semicolon marks the end of the comment statement, you cannot write a semicolon as part of the comment. However, a comment statement can contain any other characters, and any special characters that appear within the comment do not have their usual special meaning.

The following are examples of comment statements.

```
*
   This is an example of a comment statement.
*;

  *  This is a more compact example of a comment statement. ;
```

Null statements A null statement is a statement with no terms at all, just a semicolon:

```
  ;
```

No action results from a null statement. In most places, it has no effect on the program. However, a null statement can serve as an action in data step logic — to be more specific, it indicates the absence of an action — and that is where null statements are most commonly used.

Statement labels A statement label provides a name for a statement. A statement label is followed by a colon, and provides a name for the statement that follows. This is an example of a statement with a statement label:

```
MORE:
  SET MAIN.LIST;
```

Statement labels are only useful for executable statements in the data step. They are primarily used as the targets of GOTO and LINK statements.

Macro language references Stray occurrences of the characters % and & are a sign of macro language, a separate language that is used to generate SAS statements. Macro language is described in chapter 18.

Steps and SAS Datasets

In a SAS program, each step executes almost like a separate program. Other programming languages have global variables shared by all the routines or units that make up the program. One part of the program can assign a value to a global variable in order to communicate a state or a result to another part of the program. There is nothing like that to hold a SAS program together. When a step ends, all its variables disappear from memory. Steps have no direct way to interact or communicate with each other.

The connections between steps are found in the objects, especially SAS datasets, that are created in one step to be used again later in another step. The SAS dataset is designed to convey tables of data from one step to another. SAS datasets are essential in holding SAS programs together.

The SAS language makes it easy to work with SAS datasets. You can create or use a SAS dataset just by indicating its name. To use SAS datasets in a SAS program is so easy that you might forget that they are files, but they are files, with all the costs and considerations that files entail.

Data Flow

Information systems are often analyzed according to the way data flows between files, programs, and other points in the system. Often, systems analysts draw data flow charts with different shapes of boxes for the programs and files in a system and arrows connecting them to show how the data goes from one point to another.

The data flow of a SAS program should be easy to see even without a chart. The names of SAS datasets appear directly in the steps that access them — or, if the name of an input SAS dataset does not appear, it is usually the SAS dataset that the previous step created. Each step is connected to each SAS dataset named in the step — either by reading or writing it, or sometimes both.

If you want to create a quick data flow chart for a SAS program, follow these steps:

1. Start by drawing a box for each step of the SAS program. Arrange these boxes vertically along the left side of the page.
2. To the right of the box for each step, draw a box for each SAS dataset the step creates. Then draw an arrow to connect the step to the SAS dataset.
3. Add, at the top of the chart, any SAS datasets that are used as input to the program, but are created in an earlier program.
4. Then, draw arrows from each SAS dataset to each step that uses the SAS dataset as input. Draw arrows pointing in both directions when a step modifies a SAS dataset.
5. Finally, if necessary, add any input or output text data files and any reports produced by the program.

If you like to think visually, this kind of chart can show you exactly where the data goes in a SAS program.

Variables

It is not just data values that flow from one step to another in a SAS program. Each data value in a SAS dataset is identified as a variable, almost exactly the same way that data values in the program are identified as variables. When a SAS dataset is created in a data step, the variables of the data step become the variables of the SAS dataset it creates. Usually, when a proc creates a SAS dataset, it works the same way; the variables of the proc step become the variables of the new SAS dataset. The reverse of this process is also true. When a SAS dataset is read in a data step or proc step, the variables of the SAS dataset are used as variables in the step.

In this way, variables flow from one step to another in a SAS program — not directly, but by way of a SAS dataset. In some programs, variables you create in the initial data step may appear in all the subsequent steps of the program, arriving there via one or more SAS datasets. You usually have to define a variable only once, regardless of how many places you use it. This quality of SAS programs makes them easier to maintain. If you subsequently need to add a variable to a program, it is a minor change; you do not have to rewrite the entire program.

As variables transfer between steps and SAS datasets, so do most of their attributes, including the variable name, its data type and length, and its associated informat, format, and label. These attributes are stored in the SAS dataset so that any steps that read the SAS dataset can use them.

Observations and the Observation Loop

A SAS dataset could contain just one value for each variable, but more often, SAS datasets contain many values for their variables, which are organized into observations and stored as records in the file. Each observation in a SAS dataset has a value for each variable. A SAS dataset can have any number of observations. The observations in a SAS dataset have a certain sequence, which is the order in which they appear when you print the SAS dataset with the PRINT proc or access it in some other way. The order of observations, however, does not necessarily correspond to the order in which they are stored in the SAS dataset.

When a step connects with a SAS dataset, variables in the step essentially are the variables in the SAS dataset. The connection between the observations in a SAS dataset and the execution of a step is not quite so easy to describe. The process of a step that reads or writes a SAS dataset contains a sequence of actions that creates or processes an observation. Those actions are repeated in what is called an observation loop. By repeating the actions of the observation loop, once for each observation, the step eventually gets through all of the observations. Each observation in a SAS dataset corresponds to one repetition of the observation loop in the step that created the SAS dataset. There was one specific repetition that created the observation. Later, each observation corresponds to one repetition of the observation loop in a step that reads the SAS dataset.

Most often, a SAS dataset is created in a data step in which an INPUT statement creates the values for the variables. Each time the INPUT statement executes, it reads one record in the input text data file and creates the values of variables for one observation in the SAS dataset. The INPUT statement has to execute one time for each record it reads and each observation it creates. The automatic data step loop, or the observation loop of the data step, is what makes that happen.

The observation loop continues to repeat as long as there is input data. It stops when it reaches the end of the input file for the step, whether it is an input text data file or an input SAS dataset.

Data Steps

Data steps are used to create SAS datasets, but you might not guess that by looking at the many statements available for use in a data step. All in all, it is a comprehensive set of statements for general programming. Only a handful of the statements have to do with output SAS datasets. It is mainly the data step's automatic processes — the hidden parts of the data step — that make it so well suited for creating SAS datasets. When you write data steps, you need to make use of both the statements and the automatic processes.

Automatic Processes

The automatic features of the data step are built around an automatic loop that is present whenever the data step includes any statement that could read input data. This loop repeats all the actions of the data step and continues indefinitely. What finally stops the data step loop? It is one of the statements in the data step, usually a statement that reads from an input file. When the statement gets to the end of the input data, it stops the data step.

The automatic loop is itself a major feature of the data step, and other automatic actions of the data step are connected to the automatic loop. These actions include setting certain variables to missing values at the beginning of each repetition of the loop; maintaining automatic variables, which are reset for each repetition of the loop; and writing observations to the output SAS dataset at the end of each repetition of the loop.

Everything that is automatic about the data step can be overridden in one way or another by statements in the step. For example, you can use the STOP statement to stop the data step, eliminating the automatic loop entirely or cutting it short at any point.

Phases and Types of Statements

The order of statements in a data step is important for two separate reasons: first, it determines the sequence of actions taken by the data step; second, it affects the way the variables of the data step are defined. These two reasons correspond to two stages, or phases, in the execution of a data step.

The first phase has nothing to do with any actual data values. In this data definition stage, the data step creates all of its variables, including all the variables for any new SAS datasets that it creates. Statements in the data step determine the attributes of the variables. When a statement refers to an input SAS dataset, variables from that SAS dataset also become variables in the data step. For the purpose of defining variables, statements are considered in their order of appearance in the data step.

The second phase carries out the logic of the data step. In the logic phase, only executable statements — action statements and control flow statements — have any effect. The sequence of actions comes out of the sequence of action statements, as modified by control flow statements and the automatic control flow features of the data step.

Technically, there is also a third, cleanup phase of data step execution, in which the SAS supervisor cleans up resources, closes output files, and writes final log notes. This phase is not directly related to any statement in a data step, and it almost never has any problems, so you may never need to think about it.

The important point to remember is that when you read a data step, you need to read it two different ways. First read the statements that have to do with variables, to see what variables there are and how they are defined. Then read the executable statements to see the logic of the step, or what the step does with its variables.

Lexicon

The terms *compilation phase* and *execution phase* also describe the distinction between these two phases of data step execution. This terminology is a metaphor drawn from the actions of compiled programming languages; it is not literally accurate when it is applied to the data step. The data definition phase of a data step includes much more than just compilation; among other things, it creates files for any new SAS datasets and writes the header information to those files.

Some data step statements define the variables and other objects of the data step; others set forth the actions and control flow of the data step; some do both; a few do neither. These qualities define the categories of data step statements.

Executable Statements

Executable statements are the statements that set forth the actions and logic of a data step. Knowing which statements are executable is especially important when you consider the automatic data step loop. Only the executable statements are part of the repeating process in that loop.

Statements that affect the flow of execution are control flow statements. I call the other executable statements action statements — but remember that control flow statements also have elements of action about them.

A few executable statements are restricted in the way they can be used, because their entire meaning depends on the preceding statement in the program. Do not use statement labels on any of these restricted statements.

The following table lists and describes selected executable statements.

<table>
<tr><th>Statement</th><th>Description</th><th>Notes</th><th>Example</th></tr>
<tr><td>ABORT</td><td>Ends the SAS session with an error condition.</td><td>C</td><td>IF _ERROR_ THEN ABORT;</td></tr>
<tr><td>Assignment</td><td>Assigns a value to a variable.</td><td>A V</td><td>X = A + C;</td></tr>
<tr><td>BY</td><td>Indicates the sorting or grouping of an input SAS dataset.</td><td>R</td><td>SET MAIN.CONTROL;
BY REGION;</td></tr>
<tr><td>CALL</td><td>Executes a CALL routine.</td><td>C V</td><td>CALL EXECUTE("TITLE1 '"
|| TRIM(TITLE) || "';)";</td></tr>
<tr><td>CONTINUE</td><td>Branches within a DO loop.</td><td>C</td><td>IF FEE = 0 THEN CONTINUE;</td></tr>
<tr><td>DELETE</td><td>Stops processing the observation.</td><td>C</td><td>IF SUBJECT = '' THEN DELETE;</td></tr>
<tr><td>DISPLAY</td><td>Displays a window.</td><td>A</td><td>DISPLAY YOURDATA.MAIN;</td></tr>
<tr><td>DO</td><td>Forms a block or loop.</td><td>C V</td><td>IF _N_ = 1 THEN DO;
  SUM = 0;
  COUNT = 0;
  END;</td></tr>
<tr><td>ELSE</td><td>Alternative action after an IF-THEN statement.</td><td>C R</td><td>IF PAYMENT THEN
    OUTPUT ACTIVE;
ELSE OUTPUT INACTIVE;</td></tr>
<tr><td>END</td><td>Marks the end of a block.</td><td>C R</td><td>END;</td></tr>
<tr><td>FILE</td><td>Identifies an output text file.</td><td>A V</td><td>FILE RESULT;</td></tr>
<tr><td>GOTO</td><td>Branches.</td><td>C</td><td>GOTO NEXT;</td></tr>
<tr><td>IF</td><td>Stops processing observations that do not meet a condition.</td><td>C V</td><td>IF SUBJECT NE '';</td></tr>
<tr><td>IF-THEN</td><td>Conditional action.</td><td>C V</td><td>IF TIME THEN
     SPEED = DISTANCE/TIME;</td></tr>
<tr><td>INFILE</td><td>Identifies an input text file.</td><td>A V</td><td>INFILE NEWDATA;</td></tr>
<tr><td>INPUT</td><td>Reads a record from the input text file.</td><td></td><td>INPUT ID $CHAR5.
    AMOUNT 14.2
    NET 14.2
    ;</td></tr>
</table>

continued

Notes A Action C Control flow R Restricted V May define variables

Statement	Description	Notes	Example
LEAVE	Branches out of a loop.	A	LEAVE;
LINK	Branches and returns.	C	LINK VERIFY;
MERGE	Combines observations from multiple input SAS datasets.	A V	MERGE MAIN.ID (IN=IS) MAIN.NEW (IN=ACTIVE) ; BY IDCODE;
MODIFY	Reads a SAS dataset for editing.	A V	MODIFY MAIN.DIR;
Null	No action.	A	IF FEE > 0 THEN ; ELSE FEE = 0;
OTHERWISE	Alternative action in a SELECT block.	C R	OTHERWISE ;
OUTPUT	Writes an observation to an output SAS dataset.	A	OUTPUT; OUTPUT MAIN.TRANSLOG;
PUT	Writes a record to the output text file.	A V	PUT ID $CHAR5. AMOUNT 14.2 NET 14.2 ;
REMOVE	Removes an observation from a SAS dataset.	A	REMOVE;
REPLACE	Modifies an observation in a SAS dataset.	A	REPLACE;
RETURN	Completes processing of an observation.	C	RETURN;
SELECT	Multiple conditional actions.	C V	
SET	Reads an observation from an input SAS dataset.	A V	SET MAIN.VOTERS;
STOP	Ends the data step.	C	IF COUNT >= 100 THEN STOP;
Sum	Adds a value to a numeric variable.	A V	COUNT + 1;
WHEN	Conditional action in a SELECT block.	C R V	WHEN(1, 2, 3) CLASS = 'L';

Notes A Action C Control flow R Restricted V May define variables

Definitional Statements

Several statements set attributes of variables. The most important of these is the LENGTH statement, which defines the type and length of variables. Usually, you do not specifically declare variables in a data step, because the SAS supervisor can guess the appropriate type of variable, and the length if it is a character variable, from the variable's first appearance in the step. Sometimes it is necessary to declare variables, such as when you compute a character variable or when a character variable should be longer than the input field or variable. Use a LENGTH statement near the beginning of the step to define those variables.

Other statements set other attributes. The FORMAT, INFORMAT, and LABEL statements define those respective attributes, or you can use the ATTRIB statement to set all the attributes of one variable in a single statement.

The RETAIN statement provides initial values for variables and prevents them from being automatically reset to missing. The ARRAY statement defines an array, which often includes defining the variables that are elements of the array. The WINDOW statement defines a window layout. Each of these statements, and many executable statements that can involve variables may define variables if the statement is the first appearance of a variable in the data step.

Other Statements

The few remaining data step statements are not involved in the logic or the objects of a data step. These statements act essentially like proc step statements. In some cases, their sequence in the step is mandated by the form of the step; in other cases, sequence does not matter at all.

A few statements are equivalent to dataset options. The KEEP, DROP, and RENAME statements act the same as dataset options for the output SAS datasets. The WHERE statement acts the same as a dataset option for the input SAS datasets.

The DATA statement appears at the beginning of the step and lists the output SAS datasets. The RUN and CARDS statements mark the end of the step.

Types of Routines

The logic of a data step can employ several of the types of routines in SAS: functions, CALL routines, informats, and formats.

A function returns a value that it derives in some way from the values of the arguments you pass to the function. Usually, the purpose of a function is to do some kind of computation to determine the value that it returns. Functions can be used anywhere an expression appears in an executable statement.

CALL routines are like functions, except that they do not return a value. However, a CALL routine may alter the value of one or more of the variables used as its arguments. CALL routines appear in the CALL statement, which is an executable statement.

Informats and formats are routines that do two kinds of conversions. An informat interprets text to convert it to a data value. Informats are used in the INPUT statement and INPUT function in the data step. Formats do the reverse. A format is a routine that formats a data value to convert it to text that can be seen or written to a file. Formats are used in the PUT statement and PUT function.

Informats and formats are also used to do the same things in proc steps. They can be associated with variables in the INFORMAT and FORMAT statements of the data step to create the informat and format attributes of variables. These variable attributes are stored with the variables in any SAS datasets that are created and are available for use in any subsequent steps that use the SAS datasets, especially proc steps. You can also use the INFORMAT and FORMAT statements in a proc step to use specific informats and formats for variables.

4

Control Flow

The custom logic of a SAS program can be found in its data steps. Only in the executable statements of a data step can you find the conventional control flow devices that make programming logic possible. These are the IF-THEN, ELSE, SELECT, DO, and other statements. The data step also depends on its automatic control flow features, which are based on its observation loop, and these can be shaped and altered to fit the control flow you need for a particular data step.

Data step control flow can be difficult to follow at first because most of it is not directly stated in the statements of the step. Much of it is invisible, happening out of the unstated automatic actions of the data step rather than its statements. Control flow and file I/O are intertwined in data step syntax in a way that may not be readily apparent. These points of potential confusion are not limitations, though. You can program a data step with any kind of logic after you understand the way control flow works.

Logical Constructions

All program logic, no matter how complex, is constructed by combinations of a small set of control flow devices. In data step programming, most of these are implemented as control flow statements.

Conditional Actions

The most basic form of logic in a program is an action that depends on a condition. If the condition is true, the action is taken. In a data step, this kind of conditional action is written with an IF-THEN statement. After the word IF, write the condition that controls whether the action is taken. After the word THEN, write the action. The action in an IF-THEN statement — or in any other control flow statement that indicates an action — can be any executable statement.

```
IF condition THEN action
```

An example of a situation that can require an IF-THEN statement is when you calculate a value by dividing. Dividing by 0 would be an error — both a violation of the rules of arithmetic and a data error that would be reported in an error message in the SAS log. To make sure you do not attempt a division by 0, make a nonzero value the condition of an IF-THEN statement, and make the division part of the action, as in this example.

```
IF VOLUME > 0 THEN DENSITY = MASS/VOLUME;
```

If there is a positive value for VOLUME, the formula that follows calculates a value and assigns it to the variable DENSITY. In this case, the condition checks not only for a value that is not 0, but for a positive value, a value that is greater than 0. For this particular variable, a value less than 0 would violate the laws of physics, which would also invalidate the mathematical formula in question.

Boolean Values

The condition in an IF-THEN statement is a logical expression — notation that describes the computation of a logical value, a value of true or false. The SAS language fits every kind of value into one of its two data types, and it treats logical values as numeric values. The number 1 represents the logical value true, and the number 0 represents the logical value false. This use of 1 and 0 to indicate true and false is a convention that mathematicians have followed for a long time. The numeric representation of true and false make possible a branch of mathematics known as *Boolean algebra*. For our purposes as computer programmers, we need be concerned only with the *Boolean values* of 1 for true and 0 for false.

Boolean values are produced by comparison operators, such as the greater than operator (>) in the previous example. They are also produced by the logical operators AND, OR, and NOT and by a few functions that are designed to check conditions. The example below assigns logical expressions to a variable and prints it to show that logical values really are 1 and 0.

```
DATA _NULL_;
  VOLUME = 6;
  LOGICAL = VOLUME > 0;
  PUT VOLUME= LOGICAL=;
  VOLUME = 0;
  LOGICAL = VOLUME > 0;
  PUT VOLUME= LOGICAL=;
RUN;
```

```
VOLUME=6 LOGICAL=1
VOLUME=0 LOGICAL=0
```

Other Numeric Values As Conditions

The expression you use as a condition in an IF-THEN statement or other control flow statement does not have to have a value of 1 or 0. It could have any numeric value. Control flow statements treat 0 and missing values as false. All negative and positive values are treated as true.

The fact that the value of a condition does not need to be a strict Boolean value often makes it easier to write a condition. A common example of this is division. The value you want to divide by can, itself, be the condition for the division, for example:

```
IF TIME THEN RATE = SIZE/TIME;
```

The divisor TIME is also the condition for the action. The true values of the condition are the same values that make the division possible. If the divisor is 0, the condition is false, and division would be incorrect. If the divisor is a missing value, that also makes the condition false and makes the division impossible. If the value of the divisor is any other numeric value, the condition is true and the division can be carried out.

The rules for writing an expression for use as a condition are the same as the rules for writing any numeric expression. This subject is discussed in detail in chapter 6, "Constants and Expressions."

Actions

You can use any executable statement for the actions in IF-THEN statements and other control flow statements. To make the value of one variable depend on the value of another variable or variables, use an assignment statement as the action. Suppose you find a pattern of errors in a set of data in which the value of CITY is WASHINGTON and the value of STATE is blank, but should be DC. You could make the correction with this IF-THEN statement:

```
IF STATE = '' AND CITY = 'WASHINGTON' THEN STATE = 'DC';
```

Another common use of IF-THEN statements is for counting or summing only selected values, as in the following statements. In these statements, the action is a sum statement, which adds a value to a variable. When these statements are executed repeatedly, in the automatic data step loop, the variable SUM becomes the sum of all positive values of AMOUNT, and the variable COUNT becomes a count of the number of positive values.

```
IF AMOUNT > 0 THEN SUM + AMOUNT;
IF AMOUNT > 0 THEN COUNT + 1;
```

IF-THEN statements are also necessary when you create several output SAS datasets in a data step and each output SAS dataset contains a separately defined subset of the observations that the data step processes. This is illustrated in this sequence of statements, taken from an example in chapter 2.

```
IF FG >= 100 THEN OUTPUT CENTER.TOPFG;
IF P3 >= 30 THEN OUTPUT CENTER.TOPP3;
IF FT >= 50 THEN OUTPUT CENTER.TOPFT;
IF PTS >= 250 THEN OUTPUT CENTER.TOPPTS;
```

Blocks

Several actions can depend on the same condition. Blocks were invented to make this easy to do. A block contains several statements, but it is treated the same as a single statement for control flow purposes. In the SAS language, it is the DO statement that forms an ordinary block. An END statement marks the end of the block.

```
DO;
  several executable statements
```

```
  END;
```

In an earlier example, the same condition appeared in two different IF-THEN statements, so that two separate actions would result:

```
IF AMOUNT > 0 THEN SUM + AMOUNT;
IF AMOUNT > 0 THEN COUNT + 1;
```

That same logic could be written with a DO block, as shown below. The use of the DO block makes it easier to see that the same condition is being used to control the two related action statements.

```
IF AMOUNT > 0 THEN DO;
  SUM + AMOUNT;
  COUNT + 1;
  END;
```

Most SAS programmers indent the statements in a DO block to show that they are secondary to the DO statement. Attention to such matters of style make the control flow structure of a program easier to pick out.

The next example shows another situation that requires DO blocks. It revises a previous example that creates several output SAS datasets. This time, a counter variable is added to count the observations for each SAS dataset. The sum statement that does the counting appears together with the corresponding OUTPUT statement in the same DO block to ensure that the counting process corresponds exactly to the output process that it is counting.

```
IF FG >= 100 THEN DO;
  OUTPUT CENTER.TOPFG;
  COUNTFG + 1;
  END;
IF P3 >= 30 THEN DO;
  OUTPUT CENTER.TOPP3;
  COUNTP3 + 1;
  END;
IF FT >= 50 THEN DO;
  OUTPUT CENTER.TOPFT;
  COUNTFT + 1;
  END;
IF PTS >= 250 THEN DO;
  OUTPUT CENTER.TOPPTS;
  COUNTPTS + 1;
  END;
```

The counter variables COUNTFG, COUNTP3, COUNTFT, and COUNTPTS that are created in these statements might be used to write notes at the end of the data step.

A DO block can contain any combination of executable statements. These often include control flow statements, even other DO blocks. This ability to combine control flow devices makes any kind of programming logic possible, no matter how complex.

A DO block can be modified by adding terms to the DO statement to make it repeat. This forms a loop. You can use DO loops for any kind of repetitive action that occurs within an observation of a data step. DO loops are used most commonly with arrays and are described in chapter 8, "Loops and Arrays."

Lexicon

The word *block* is a generic term that is independent of the specific syntax of a programming language. In the SAS language, I refer to two kinds of blocks: *DO blocks* and *SELECT blocks*. When a DO block repeats to form a loop, I call it a *DO loop*. SAS Institute writers use the terms *DO group* and *SELECT group*; a loop may be referred to as an *iterative DO group*.

Do not confuse these *groups* of program statements with *BY groups*, *class groups*, and other *groups* of data objects.

Nested Blocks

The statements in a DO block may include another DO block. This kind of arrangement is called *nesting* because one block is contained entirely inside another one. Each block has its own END statement. When DO blocks are nested, the first END statement marks the end of the inside block, the one defined by the last DO statement. The last END statement marks the end of the outside block, the one that starts at the first DO statement. The association of END statements to DO statements is commented in this code schematic:

```
DO;  * Beginning of block 1;
  . . .
  DO;  * Beginning of block 2;
    . . .
    END;  * End of block 2;
  . . .
  END;  * End of block 1;
```

To make blocks more visible, most programmers use indenting for the statements in a block, and another level of indenting for a nested block, as shown in this example.

Alternative Actions

The IF-THEN statement lets you use a condition to determine whether or not to execute an action. There are also situations in which you use a condition to select between two actions to execute. You can do this with the IF-THEN statement in combination with the ELSE statement. The ELSE statement comes immediately after the IF-THEN statement and indicates an action to execute when the condition of the IF-THEN statement is false.

```
IF condition THEN action
ELSE action
```

If the condition is true, the THEN action is executed. If the condition is false, the ELSE action is executed. In any case, one and only one of the two actions is executed. An example of the use of an ELSE statement is to assign a value to a variable when the IF-THEN statement cannot determine a value, as when the formula being used depends on a division and the divisor is 0. In this case, the ELSE statement assigns a value of 0 to the variable RATE when the value of RATE cannot be determined because the value of TIME is 0 or a missing value.

```
IF TIME THEN RATE = SIZE/TIME;
ELSE RATE = 0;
```

Another common use of an ELSE statement is to capture observations that are not being stored in the primary output SAS dataset because they are defective in some way. In information systems, such observations are called *exceptions*, and they may be tallied for log notes or be printed in an *exception report* to show what kinds of data errors are occurring. In this example, valid observations are directed to the output SAS dataset BASE.CLINICAL, and any exceptions are stored in another output SAS dataset, UNKNOWN.

```
IF PID AND VISITNO THEN OUTPUT BASE.CLINICAL;
ELSE OUTPUT UNKNOWN;
```

Either or both of the actions in the IF-THEN and ELSE statements can be DO blocks, as shown here:

```
IF PID AND VISITNO THEN DO;
  OUTPUT BASE.CLINICAL;
  VISITS + 1;
  END;
ELSE DO;
  OUTPUT UNKNOWN;
  UNKNOWN + 1;
  END;
```

On occasion there may be no THEN action. You can write a null statement as the action in the IF-THEN statement and still have an alternative action in the ELSE statement. The same logic could be written as a single IF-THEN statement, but the IF-THEN/ELSE form could be easier to understand.

A null statement is just a semicolon. When you use a null statement as an action, it is a good idea to write a space before the semicolon to emphasize that it represents a null action, and is not merely the semicolon to mark the end of a statement.

The following logic is intended to ensure that the e-mail address (the variable EADDR) is stored only for people who authorized the company to send them e-mail (indicated by the Boolean variable EMAILOK) or who are inside the company (indicated by the variable LOC).

```
IF EMAILOK OR LOC = 'INTERNAL' OR EADDR = '' THEN ;
ELSE EADDR = '';
```

The same logic could also have been written this way, but you might consider this harder to read:

```
IF NOT (EMAILOK OR LOC = 'INTERNAL' OR EADDR = '') THEN EADDR = '';
```

It is also possible to have a null statement as the action in an ELSE statement. Such a statement has no effect on the program, but in some cases, you might write it anyway to emphasize that the alternative case was considered and that no action is required for it.

Multiple Conditional Actions

Often, a program needs to select one of a set of actions based on a set of conditions. This kind of logic is written with a SELECT block. There are two kinds of SELECT blocks, based on conditions and comparisons. Any kind of conditions can appear in a conditional SELECT block. The comparison SELECT block is a special case that simplifies the notation for conditions based on equality comparisons.

The conditional SELECT block begins with the SELECT statement. This is followed by several WHEN statements, which include a condition or a list of conditions, in parentheses, followed by an action. An OTHERWISE statement, if there is one, indicates an alternative action. An END statement is required to end the SELECT block.

```
SELECT;
  WHEN (condition, . . . ) action
  . . .
  OTHERWISE action
  END;
```

The following SELECT block assigns a value to the variable XT, which represents the transaction type, based on the value of the variable XAMT, which is the transaction amount.

```
SELECT;
  WHEN (XAMT > 0) XT = 'SALE    ';
  WHEN (.Z < XAMT < 0) XT = 'RETURN  ';
  WHEN (XAMT = 0) XT = 'EXCHANGE';
  OTHERWISE XT = '';
  END;
```

The SELECT block considers each condition in sequence. As soon as it finds a true condition, it executes the corresponding action. Then it does not check the remaining conditions so that only one action is executed. If none of the conditions are true, it executes the alternative action of the OTHERWISE statement. In this example, the condition `XAMT > 0` is considered first. If this condition is true, the action assigning the value `SALE` to the variable XT is executed. If not, the next condition, `.Z < XAMT < 0`, is considered. If this condition is true, the corresponding action assigning `RETURN` to XT is executed. If not, the next condition, `XAMT = 0`, is considered. If this condition is true, the value `EXCHANGE` is assigned to XT. If not, there are no more conditions to consider, so the OTHERWISE action, assigning a blank value to XT, is executed. After you trace through this kind of repetitive logical process, you can see how much easier it is to understand when it is written as a SELECT block.

You can list multiple conditions in each WHEN statement if the same action is associated with several different conditions. Separate the conditions with commas.

It is possible to have a SELECT block with no OTHERWISE statement. The purpose of leaving out the OTHERWISE statement is to create an error condition in the event that none of the conditions in the SELECT block is true. If you do not want the SELECT block to create an error condition, but no alternative action is required, use a null statement as the action in the OTHERWISE statement:

```
OTHERWISE ;
```

The actions in a SELECT block are often DO blocks, because there is more than one statement to execute as the result of a condition. That creates this extended form for a SELECT block:

```
SELECT;
  WHEN (condition, . . . ) DO;
    action
    action
    . . .
    END;
  . . .
  OTHERWISE DO;
    action
    action
    . . .
    END;
  END;
```

With DO blocks inside SELECT blocks, it becomes especially important to use indentation to show how the statements fit into the overall control flow pattern.

When the conditions of a SELECT block all have to do with matching a particular value or variable, you can code the SELECT block more easily as a comparison SELECT block. A comparison SELECT block is based on expressions that are compared to each other to see if they are equal, rather than on conditions that are tested.

A comparison SELECT block works much like a conditional SELECT block, but it has an expression in parentheses in the SELECT statement, and this expression is compared against the expressions that are present in the WHEN statements. As soon as a match is found, the corresponding action is executed. If no match is found, the action of the OTHERWISE statement is executed.

```
SELECT (expression);
  WHEN (expression, . . . ) action
  . . .
  OTHERWISE action
  END;
```

The effect of a comparison SELECT block is easiest to see when there is an obvious connection between the WHEN values and the resulting actions, as in this example:

```
SELECT (STATE);
  WHEN ('PA') OUTPUT NEWPA;
  WHEN ('MD') OUTPUT NEWMD;
  WHEN ('NJ') OUTPUT NEWNJ;
  WHEN ('NY') OUTPUT NEWNY;
  WHEN ('CT', 'RI', 'MA', 'VT', 'NH', 'ME') OUTPUT NEWN_E;
```

```
OTHERWISE OUTPUT NEWOTHER;
END;
```

A comparison SELECT block can be used to assign values to a variable whose values depend directly on the values of another variable. This example translates U.S. men's shoe sizes (the variable USMSIZE) to European sizes (the variable EUROSIZE):

```
SELECT (USMSIZE);
  WHEN (5, 5.5) EUROSIZE = 39;
  WHEN (6, 6.5) EUROSIZE = 40;
  WHEN (7, 7.5) EUROSIZE = 41;
  WHEN (8, 8.5) EUROSIZE = 42;
  WHEN (9, 9.5) EUROSIZE = 43;
  WHEN (10, 10.5) EUROSIZE = 44;
  WHEN (11, 11.5) EUROSIZE = 45;
  WHEN (12, 12.5) EUROSIZE = 46;
  WHEN (13, 13.5) EUROSIZE = 47;
  OTHERWISE ;
  END;
```

It is possible to duplicate the effects of a SELECT block by combinations of IF-THEN and ELSE statements. However, the resulting code is usually not as easy to read, explain, or modify. This example is logically equivalent to the previous example:

```
IF USMSIZE IN (5, 5.5) THEN EUROSIZE = 39;
ELSE IF USMSIZE IN (6, 6.5) THEN EUROSIZE = 40;
ELSE IF USMSIZE IN (7, 7.5) THEN EUROSIZE = 41;
ELSE IF USMSIZE IN (8, 8.5) THEN EUROSIZE = 42;
ELSE IF USMSIZE IN (9, 9.5) THEN EUROSIZE = 43;
ELSE IF USMSIZE IN (10, 10.5) THEN EUROSIZE = 44;
ELSE IF USMSIZE IN (11, 11.5) THEN EUROSIZE = 45;
ELSE IF USMSIZE IN (12, 12.5) THEN EUROSIZE = 46;
ELSE IF USMSIZE IN (13, 13.5) THEN EUROSIZE = 47;
```

Branching

On occasion, you may want the flow of control in a data step to jump from one point to another. This process is called *branching*. The GOTO statement does the simplest form of branching. It jumps to a statement label and execution proceeds from there.

A statement label identifies an executable statement. To write a statement label for a statement, write a name and a colon before the statement. The GOTO statement and other control flow statements and options branch to the points indicated by statement labels. For example, this statement with the label DONE

```
DONE: OUTPUT;
```

is referred to in this GOTO statement:

```
GOTO DONE;
```

The GOTO statement jumps to the statement label DONE. The GOTO statement and the statement label must be in the same data step.

If you used many GOTO statements, your programs might become impossible to read. However, there are times when the GOTO statement is the best way to code a particular kind of logic. When used well, GOTO statements are conditional, executed in unusual or exceptional conditions, such as error conditions and special cases. Use IF . . . THEN GOTO statements to move special cases out of the main flow of the program.

Branching has these restrictions:

- You cannot branch into a block because blocks are treated as a unit. A statement in a DO block can only be the target of a GOTO statement that is contained within the same DO block.
- The statements of a SELECT block, other than the SELECT statement itself, can never be used as the target of a GOTO statement.
- A secondary statement, such as a BY or END statement, cannot be a target because it does not correspond to any point in the execution of the program.

These restrictions apply to GOTO statements and to other control flow devices that branch to statement labels.

Branching and Returning

The LINK statement is similar to the GOTO statement, but with the idea that execution will return to the point of the LINK statement after any necessary processing is done at the branching destination. Like the GOTO statement, the LINK statement jumps to the indicated statement label, and execution proceeds from there. But a LINK statement changes the meaning of the RETURN statement. A RETURN statement branches back to the LINK statement.

The LINK statement is usually executed conditionally as the action in an IF-THEN statement. The statements between the statement label and the RETURN statement act like a subroutine, carrying out an action that is not part of the main flow of the data step. You might use a LINK statement, for example, to branch to a set of statements that look up a value for a variable whose the variable value is missing.

```
IF YEAR = . THEN LINK GETYEAR;

. . .

GETYEAR:
statements to look up a value for YEAR
RETURN;
```

A RETURN statement is implied as the last executable statement in a data step. If no RETURN statement executes after a LINK statement, execution proceeds to the end of the data step, then returns to the point of the LINK statement.

Early Cobol and Basic programmers gave branching a bad name. They used branching for all of their control flow constructions, creating impenetrable code that no one, not even the original programmers, could follow. This experience led many programmers to give up branching and replace it entirely with other control flow devices. However, the best approach is to use each control flow device for its particular purposes, and not to use one control flow device as a substitute for another. When a GOTO or LINK statement makes data step logic easier to see, you should not be afraid to use either statement.

Statement Options for Branching

Two statement options can also produce branching at particular points in the process of reading or writing a file. The EOF= option of the INFILE statement names a statement label, and when the INPUT statement reaches the end-of-file mark at the end of a file, it branches to that statement label for any further processing that may be necessary. The branching action is the same as is produced by the GOTO statement.

The HEADER= option of the FILE statement can be used for branching when creating a print file. When the PUT statement begins a new page, it branches to the indicated statement label. Then, as with the LINK statement, when a RETURN statement is executed, the flow of execution returns to the PUT statement. The statements of a HEADER= branch are usually used to write the first few lines, such as a header, on the new page. While it might seem somewhat strange that branching and returning could take place in the middle of the execution of a PUT statement, the effects are essentially the same as those of the LINK statement.

Multiple Branching and Returning

After branching from a LINK statement, it is possible for another LINK statement to execute before a RETURN statement is reached. When a RETURN statement eventually executes, it returns to the most recent point of branching from a LINK statement. Another RETURN statement is then necessary to return to the previous point of branching from a LINK statement. This works the same way if there is a combination of branching from LINK statements and HEADER= options.

As you might imagine, the use of multiple branching and returning can get confusing very quickly, so it is not seen often in SAS programming.

Shaping the Observation Loop

The major control flow feature of the data step is its automatic loop — the observation loop. The word *observation* usually means a record, or row, of data stored in a SAS dataset. When we discuss the observation loop in the data step, the word *observation* may also, by association, refer to the actions taken in one repetition of the observation loop or to the input data that is processed in those actions. That makes sense because a repetition of the observation loop usually creates one output observation in an output SAS dataset.

The observation loop happens automatically whenever it is needed; no specific statement in the data step is required to make it happen. However, the various features of the observation loop do not always work automatically the way you need them to. The SAS language provides several control flow statements that address specific aspects of the observation loop to let you shape it and alter the way it works.

The Nature of the Observation Loop

Usually, in the code of a data step, there is one statement that reads a record from an input file. The record of input data could be a line from an input text file or an observation from a SAS dataset. In order to read all of the records of data from the input file, the statement that reads a record has to execute repeatedly. A device in a program that executes statements repeatedly is called a loop. In most languages, loops are written with statements that control how many times the loop repeats. You can do this in the SAS language too, with the DO loops that are described in chapter 8. But this particular loop, the observation loop that repeats the process of reading input data, is an automatic feature of the data step. It happens without the need for any statements to specifically indicate it. And it stops automatically at the right time, when the end of the input file is reached.

Subsequent statements in the data step process the data from the input record in whatever way is necessary. These statements too need to repeat along with the statement that actually reads the input record, so that each input record can be read and processed in turn. To make this happen, the observation loop includes all of the executable statements of the data step. A data step executes its executable statements, from beginning to end, then it goes back to the beginning and does it again — and this repeating pattern continues until the processing reaches the end of the input file. The data step itself does not determine how many times its observation loop repeats. That is determined by the extent of the data in the input file.

The same set of statements executes for each observation. It is the nature of the observation loop to process each observation the same way. This approach works well for applications in statistics and business. The essential premise of statistics is that there is a set of observation, and each observation can be considered the same way as all the others. Likewise, successful businesses are built around standards and procedures that allow actions to be repeated efficiently and consistently. The way the observation loop works in the data step is an important part of what makes the SAS language as widely accepted as it is in business and in fields that depend on statistical analysis.

File I/O and Control Flow

In the observation loop, the syntax of control flow is intertwined with that of file I/O — data input from and output to files. Input is what drives the observation loop; output is a default action within the loop. When you consider a data step, you need to see the file I/O and control flow together.

The observation loop happens only when the data step contains a statement that provides input data. The statements that prompt an observation loop are INPUT, which reads from input text files; SET, MERGE, and any other statement that reads from SAS datasets; and DISPLAY, which displays windows that could serve as sources of input data. If any of these statements appears anywhere in a data step, the data step has an observation loop.

These same statements determine when the end of data is reached and the observation loop stops. The data step ends when a statement reaches the end of its input data — not on the last observation itself, but on the next repetition of the observation loop, when the statement attempts to read beyond the end of the input data.

It is possible to write a data step with an observation loop such that it never reaches the end of input data. This could create a data step that would loop indefinitely. The SAS supervisor attempts to keep this from happening by accident. One way it does this is by requiring input for every repetition of the observation loop. If, in any repetition of the observation loop, the statement that supplies the input data for the step does not execute, the data step stops with an error condition. There could be several statements that supply input data for the step; at least one must execute in each repetition of the data step. If it is INPUT statements that supply the input data for a data step, the SAS supervisor attempts to ensure that those statements are moving forward through the input file, and not staying stuck indefinitely on the same input record. However, these are very limited checks. If you use options that make the input process of a data step do something other than reading through a file, record by record, it is your responsibility as the programmer to make sure the logic does bring the data step to an end at some point.

The data step is designed to create SAS datasets, and output is the default action at the bottom of the observation loop. You can code output to SAS datasets explicitly with the OUTPUT statement, but if you do not, output of the current observation occurs automatically at the end of each repetition of the observation loop.

Beginning

The data step begins by executing statements to process the first observation. Often, though, there are statements you need to execute to initialize the data step before processing the first observation. These statements may, for example, contain logic to compute initial values for variables. Execute these statements in a DO block with a condition where you check for the variable _N_ to be 1.

```
IF _N_ = 1 THEN DO;
  actions
  END;
```

_N_ is a variable that counts repetitions of the observation loop. It has a value of 1 for the first observation, so it can be used in a condition to check for the beginning of execution of the data step.

An example of a situation where this kind of logic is required is when an input text file has a header — one or several records at the beginning of the file that do not represent observation, but contain general information about the file. The header might, for example, contain variables that are same for every observation. Or, the header may contain nothing of interest. The code below reads the two-line header of an input text file just to verify that it contains the right file ID codes. If it does not, it stops the data step with a log message.

```
INFILE IN;
IF _N_ = 1 THEN DO;
  INPUT CODE1 $CHAR5.
     / CODE2 $CHAR5.
     ;
  IF NOT (CODE1 = 'ABTLX' AND CODE2 = 'B0000') THEN DO;
    FILE LOG;
    PUT 'File not processed because of header mismatch';
    LIST;
    STOP;
    END;
  END;
```

The statements to execute at the beginning of the data step are often written as the first executable statements of the data step. When these statements read a header, however, the INFILE statement must appear first so that the INPUT statement will read from the correct file. The terms of the INPUT statement indicate the specific details of reading the header records from the input file. Next, the IF-THEN statement checks the fields that were read. If the fields do not contain the expected values, the statements of the subsequent DO block write log messages, including writing the two input records in the log, and stop execution of the data step.

Ending

In a data step with an observation loop, execution ends when a statement that attempts to read input data finds the end of the input data instead. When the source of input data is an input file, the end-of-data condition is the end of the file. If a data step window is the source of input data, it is the execution of the END command in the window that creates the end-of-data condition.

The significant thing to note is that the data step does not end at the bottom of the observation loop, after the last executable statement of the data step. It typically ends at an INPUT statement, a SET statement, or some other statement designed to read input data. These statements are often the first executable statements of the data step, but if they are not, any statements that occur earlier in the step execute again after the last observation has been processed. The following example demonstrates this. The program contains PUT statements to write diagnostic messages in the log to show different points in the execution of the step. PUT statements are often used in this manner when it is necessary to track down logical errors in data steps. This example also demonstrates the automatic variable _N_.

```
DATA _NULL_;
  INFILE CARDS;
  FILE LOG;
  PUT 'Looking for observation number ' _N_;
  INPUT LETTER $CHAR1.;
  PUT 'Data value for observation number ' _N_ 'is ' LETTER $CHAR1. '.';
CARDS;
A
B
C
D
E
;
```

```
Looking for observation number 1
Data value for observation number 1 is A.
Looking for observation number 2
Data value for observation number 2 is B.
Looking for observation number 3
Data value for observation number 3 is C.
Looking for observation number 4
Data value for observation number 4 is D.
Looking for observation number 5
Data value for observation number 5 is E.
Looking for observation number 6
NOTE: DATA statement used:
      real time           0.04 seconds
```

The message "Looking for observation number 6" shows that the data step continues to execute even after it processes the last record of input data. It does not end until the next time the INPUT statement executes, which is after the PUT statement that writes the last diagnostic message in the log. At this point, the INPUT statement looks for more input data to read, finds none, and stops the data step.

When you write a data step, it is sometimes important that some statements at the beginning of the data step may execute after the last observation. Write statements at the beginning of the data step to ensure that they execute again after the last observation, or write them later in the step if it is important that they do not execute again after the last observation.

Not every statement that reads input data reads through a file in a particular sequence. For example, you can use the KEY= option of the SET statement to access observations of a SAS dataset according to key values. The SET statement with the KEY= option never reaches an end-of-data condition that would stop the data step. When you use that kind of statement as the source of input data for a data step, you must have some other logic to bring the data step to a stop at the appropriate point. Use the STOP statement as the action of an IF-THEN statement to end the data step. You can also use the STOP statement in this way when you do not want to read through an entire input file.

Suppose, for example, you want to read data for the largest customers, but only for enough customers to add up to a certain amount of revenue. To do this, you would need a SAS dataset that had variables for size rank and revenue (along with any other necessary variables) with an index on the size rank variable. (See chapters 10 and 11 for the details of creating a SAS dataset and an index.) Then, the example below would create a new SAS dataset that contained only the largest customers, selecting just enough customers to account for at least 1,000,000 in revenue.

```
DATA BIG (DROP=TOTAL);
  SET MAIN.CUSTOMER;
  BY SIZERANK;
  OUTPUT;
  TOTAL + REVENUE;
  IF TOTAL >= 1000000 THEN STOP;
RUN;
```

In this example, the SET and BY statements read the SAS dataset's observations in the order of the variable SIZERANK. The sum statement adds each observation's revenue to the revenue total. Then, the IF-THEN statement checks to see if the total has reached at least 1,000,000 and stops the data step if that has occurred.

When you use a STOP statement, it is often necessary to use an OUTPUT statement to explicitly indicate writing the output observations. Writing the output observation is normally the default action of a data step — but not for an observation for which a STOP statement executes. To get around this, another approach could be to write the conditional STOP statement at the beginning of the data step, like this:

```
DATA BIG (DROP=TOTAL);
  IF TOTAL >= 1000000 THEN STOP;
  SET MAIN.CUSTOMER;
  BY SIZERANK;
  TOTAL + REVENUE;
RUN;
```

The program would work exactly the same written this way. The explicit OUTPUT statement is not necessary because the STOP statement is not executed for the last observation to be used, but for the observation that follows.

There are often additional actions to take at the end of a step. Such actions may include writing log messages that describe the processing of the step, writing a SAS dataset of totals, or writing trailer records, additional records at the end of an output text file.

If the step is being stopped with a STOP statement, the actions to take at the end of the step can be written as part of a DO block that ends with the STOP statement, for example:

```
DATA BIG (DROP=TOTAL AVG);
  SET MAIN.CUSTOMER;
  BY SIZERANK;
  OUTPUT;
  TOTAL + REVENUE;
```

```
  IF TOTAL >= 1000000 THEN DO;
    AVG = TOTAL/_N_;
    FILE LOG;
    PUT 'The ' _N_
       'largest customers were selected, with a total revenue of '
       TOTAL :COMMA9. +(-1) ', and an average revenue of '
       AVG :COMMA9. +(-1) '.';
    STOP;
    END;
RUN;
```

More often, though, a data step ends at a SET, MERGE, or INPUT statement. The END= statement option lets you find out when the statement has reached the last observation in the SAS dataset, so that you can write a condition to execute a block of statements at the end of the data step.

The END= option appears as an option in any statement that reads input SAS datasets. For an INPUT statement, END= is written as an option in the INFILE statement. The option indicates a name for a variable. It creates a numeric variable with a Boolean value that indicates the last record or observation in the file. The value of the variable is initialized to 0. It is set to 1 when the statement reads its last record or observation. Use this variable as the condition for statements to execute after processing the last observation.

As an example, if you simply need to write the values of the last observation in a message in the log, you could do that with these statements:

```
SET MAIN.LATEST END=LAST;
IF LAST THEN PUT 'The most recent event recorded is: ' _ALL_;
```

In some unusual situations, the END= option does not work. In particular, this happens when an input file or device is unbuffered. When a file is unbuffered, the SAS supervisor cannot look ahead when it reads each record in the file, so when it reads the last record, it does not know that it is the last record. In these situations, use the EOF= option of the INFILE statement. The EOF= option indicates a statement label to branch to at the end of the file. When the INPUT statement reaches an end-of-data condition, instead of stopping the data step, as it usually does, it branches to the statement label. The statements there can do any kind of processing you might need to do at the end of the data step. However, the statements must end with a STOP statement to stop the data step, as shown below. When there is an EOF= option, the INPUT statement itself can no longer stop the data step.

```
INFILE AFILE EOF=ENDOFDATA;
INPUT . . . ;
statements to process each observation
RETURN;
ENDOFDATA:
statements to execute after last observation
STOP;
```

The Beginning and End of a Group

Just as you can execute additional statements before the first observation and after the last one, you can execute statements before and after each group of observations you read from a SAS dataset. Name the key variable whose values determine the groups in the BY statement, which comes right after the SET statement or any other statement that reads SAS datasets. The observations have to be read in the order indicated in the BY statement either by being sorted that way or as the result of an index. For example, if the SAS dataset MAIN.LOCATION is sorted by STATE, you could read it with these statements:

```
SET MAIN.LOCATION;
BY STATE;
```

In the data step, the primary effect of the BY statement is to create two automatic variables related to the BY variable. These variables are Boolean variables that indicate the first observation of the group and the last observation of the group. For the BY variable STATE, the two automatic variables are FIRST.STATE and LAST.STATE. FIRST.STATE is 1 for the first observation to have a certain value for STATE, and 0 for all other observations. Similarly, LAST.STATE is 0, except in the last observation to have a certain value for STATE, where its value is 1. If one value of STATE has only a single observation, both variables FIRST.STATE and LAST.STATE have values of 1 in that observation. The following example illustrates how these automatic variables are related to the BY variable.

```
DATA _NULL_;
  SET MAIN.LOCATION;
  BY STATE;
  PUT (_N_ STATE FIRST.STATE LAST.STATE)(=);
RUN;
```

```
_N_=1 STATE=AL FIRST.STATE=1 LAST.STATE=1
_N_=2 STATE=FL FIRST.STATE=1 LAST.STATE=0
_N_=3 STATE=FL FIRST.STATE=0 LAST.STATE=0
_N_=4 STATE=FL FIRST.STATE=0 LAST.STATE=0
_N_=5 STATE=FL FIRST.STATE=0 LAST.STATE=1
_N_=6 STATE=IN FIRST.STATE=1 LAST.STATE=0
_N_=7 STATE=IN FIRST.STATE=0 LAST.STATE=1
```

In the qoutput from this example, you can see how the values for the FIRST and LAST variables are related to the group as a whole.

There can be more than one BY variable. In that case, there are FIRST and LAST variables for each BY variable. For the first BY variable, the FIRST and LAST variables are the same as if there were only one BY variable. For the second BY variable, the FIRST and LAST variables indicate the groups formed by the first two BY variables together. If there is a third BY variable, its FIRST and LAST variables indicate the groups formed by the first three BY variables together, and so on. The way that BY variables are related to each other can be seen in the following example.

```
DATA _NULL_;
  SET MAIN.LOCATION;
  BY STATE COUNTY;
  PUT _N_= STATE= FIRST.STATE LAST.STATE COUNTY= FIRST.COUNTY LAST.COUNTY;
RUN;
```

```
_N_=1 STATE=AL 1 1 COUNTY=FRANKLIN 1 1
_N_=2 STATE=FL 1 0 COUNTY=FRANKLIN 1 0
_N_=3 STATE=FL 0 0 COUNTY=FRANKLIN 0 1
_N_=4 STATE=FL 0 0 COUNTY=NASSAU 1 1
_N_=5 STATE=FL 0 1 COUNTY=LAKE 1 1
_N_=6 STATE=IN 1 0 COUNTY=FRANKLIN 1 1
_N_=7 STATE=IN 0 1 COUNTY=LAKE 1 1
```

The output is arranged here as a table with horizontal rules added to make it easier to read:

_N_	STATE	FIRST.STATE	LAST.STATE	COUNTY	FIRST.COUNTY	LAST.COUNTY
1	AL	1	1	FRANKLIN	1	1
2	FL	1	0	FRANKLIN	1	0
3	FL	0	0	FRANKLIN	0	1
4	FL	0	0	NASSAU	1	1
5	FL	0	1	LAKE	1	1
6	IN	1	0	FRANKLIN	1	1
7	IN	0	1	LAKE	1	1

In this output, you can see that the values of the second-level automatic variables do not merely depend on the values of the second BY variable. The values of FIRST.COUNTY and LAST.COUNTY reflect the groups formed by both BY variables, STATE and COUNTY, together.

Use the FIRST and LAST variables as conditions for actions to execute at the beginning and end of a group:

```
SET MAIN.LOCATION;
BY STATE;
IF FIRST.STATE THEN DO;
  actions at the beginning of the group
  END;
statements to process each observation
IF LAST.STATE THEN DO;
  actions at the end of the group
  END;
```

No Loop

If there is no statement in the data step that could provide input data, there is no observation loop. The executable statements of the data step execute once, then execution stops. Or, if there is a statement in the data step that would prompt an observation loop, you can use a STOP statement without a condition to ensure that the observation loop never loops.

An example of this is when you create a SAS dataset with only one observation. Such a SAS dataset might serve as a parameter file, for example. You might create the variables just by assigning values to them:

```
*
  Create System File.
*;
DATA MAIN.SYSTEM (KEEP=INITDATE EXPDATE SERIAL
             S_ERR U_ERR R_ERR O_ERR);
  * System initialization date is current date at initialization. ;
  INITDATE = DATE();
  * Expiration date is initialized to the same day, one year later;
  M = MONTH(INITDATE);
  D = DAY(INITDATE);
  IF M = 2 AND D = 29 THEN DO; * Special case for leap day. ;
    M = 3;
    D = 1;
    END;
  EXPDATE = MDY(M, D, YEAR(INITDATE) + 1);
  * Serial number initialized to 0;
  SERIAL = 0;
  * All error flags reset. ;
  S_ERR = ' ';
  U_ERR = ' ';
  R_ERR = ' ';
  O_ERR = ' ';
RUN;
```

In the example above, there is no statement that could serve as a source of input data, so the observation loop is not prompted. The step still creates an output SAS dataset and writes one observation to it. You could also write the same step with these statements added before the RUN statement, to emphasize these actions of the step:

```
  OUTPUT;
  STOP;
```

When the purpose of a data step is to display a window with a message, it is usually necessary to have a STOP statement. The DISPLAY statement that displays the window can stop the step if the user enters the END command (or takes an equivalent action, depending on the operating system) in the window. In most cases, though, you do not want to rely on such a specific user response, so you should use the STOP statement to stop the step, as in this example.

```
DATA _NULL_;
  WINDOW VERSION "Running under SAS version &SYSVER";
  DISPLAY VERSION;
  STOP;
RUN;
```

In the WINDOW statement, &SYSVER is a reference to an automatic macro variable. The macro variable has a particular text value, and that text is substituted for the macro variable reference to form the actual SAS statement that is executed.

System Options That Affect the Observation Loop

Three system options have implications for the observation loop. The most important is the `OBS=0` option. The OBS= option limits the number of observations a step will process. The value of the option is the number of the last input observation or record a step will read. When the step attempts to read the next observation or record, it results in an end-of-data condition, just as if the step had reached the end of the file. With the `OBS=0` option, the data step ends the *first* time it reaches a statement that reads from an input file. The effect is the same as if all input files are empty. The step creates its variables and creates any output SAS datasets, but most of its executable statements do not execute. The `OBS=0` option has a similar effect on proc steps; they also create output SAS datasets with variables but no observations. The main use of the `OBS=0` option (or any small value for the OBS= system option) is to check the syntax and test most of the connections in the data flow of a program. A program runs almost instantly with the `OBS=0` option, because it does not have to work with any actual input data values.

The OBS= and FIRSTOBS= options can be used together to limit the amount of input data used by each step. The FIRSTOBS= option skips past records or observations at the beginning of an input data file to start at the indicated observation or record number. The OBS= option, then, indicates the last observation or record number to use. For example, the options

```
FIRSTOBS=22 OBS=27
```

indicate to use 6 input records beginning at record number 22.

These same options can also be used in ways that refer to specific input files. Use the OBS= and FIRSTOBS= options as statement options in the INFILE statement or as dataset options for any input SAS dataset.

Another system option, the ERRORS= option, affects the repetition of data error messages in the log. The option limits the number of observations for which detailed error messages are written for the same data error. An example of a data error is dividing by 0. The detailed error messages let you see the data values involved in the data error. With the `ERRORS=5` option, for example, the detailed error messages are written for only the five occurrences of a specific data error. With the `ERRORS=0` option you can omit the detailed error messages entirely. You still get the summary error messages that describe the error, indicate its location in the program, and count the number of observations it occurs in.

Actions on Observations

The default automatic processes of the data step in processing an observation are designed to keep the observation loop going and create output observations. To alter the way a data step processes an observation, you can use any of several control flow and action statements that are unique to the SAS language.

The unstated final executable statement of any data step is the RETURN statement. A data step executes as if a RETURN statement is written as its last executable statement. The RETURN statement does not always mean the same thing. The effect of a RETURN statement depends on the other statements in the step. After branching from a LINK statement or a HEADER= option, a RETURN statement returns the flow of execution to the point where that branch took place. Otherwise, a RETURN statement completes the processing of an observation. It ends one iteration of the observation loop, and execution proceeds with the next iteration of the observation loop at the first executable statement of the data step.

When the RETURN statement acts to complete processing of an observation, this may or may not include writing an output observation to every output SAS dataset. Output to SAS datasets can be coded explicitly with the OUTPUT statement, or sometimes, with the REPLACE and REMOVE statements. But if there is no explicit output to a SAS dataset in the step, then the RETURN statement implies that action when it completes processing of an observation. It includes the effect of an OUTPUT statement for the output SAS datasets, or that of the REPLACE statement if an output SAS dataset is read in a MODIFY statement. The default action is for the RETURN statement to write the current observation to every output SAS dataset. In all of its actions, the implied RETURN statement at the end of the step works the same as a RETURN statement written in the step.

The RETURN statement is just one of several statements that can conclude the processing of an observation. The table on the next page shows comparisons among several statements that can have similar control flow effects. The subsetting IF and DELETE statements also stop processing of observations. The LOSTCARD and GOTO statements may appear similar, but they actually continue processing of the same observation. The STOP statement is useful in a different way; it stops the processing of not just the current observation, but the entire data step.

The DELETE statement discards the current observation and proceeds to the next one. It is usually executed conditionally, to leave out observations under certain conditions. Often, an input file contains records that would not make sense or do not belong in the SAS dataset you are building. One common example is when an identifying variable is blank in an input record. You can use the DELETE statement to stop processing the observation, which also ensures that it is not written to the output SAS dataset. The DELETE statement might be written:

```
IF NAME = ' ' OR ID = ' ' THEN DELETE;
```

Sometimes it makes more sense to write the condition for observations you want to keep. You can do this in a subsetting IF statement. This special form of the IF statement, with no THEN clause, might look odd at first, but it is simply a short way of writing the action of discarding all observations that do not meet a certain condition. A subsetting IF statement might look like:

```
IF POINTS > 0;
```

This statement is equivalent to:

```
IF POINTS > 0 THEN ;
ELSE DELETE;
```

Comparison of Statements to Stop Processing an Observation

Statement	Continues At	Other Actions/Notes
`RETURN;`	Next Observation	Writes observation to output SAS datasets, unless the step includes a statement such as OUTPUT that explicitly writes output observations. The RETURN statement has a different, unrelated action after branching from a LINK statement or HEADER= option. If the step contains an OUTPUT statement and the RETURN statement does not follow branching from a LINK statement or HEADER= option, the RETURN statement is the same as the DELETE statement.
`DELETE;`	Next Observation	No other actions. Never writes output observations. Can be used even after a LINK statement to force an observation to complete processing.

`IF` *condition*;	Next Observation	Has no effect if the condition is true; processing of the observation continues.
		
		Stops processing of the observation if the condition is false; effect is the same as:
		
		`IF` *condition* `THEN ;`
		`ELSE DELETE;`

`LOSTCARD;`	Repeat Same Observation	Restores the pointer for an input text file to where it was at the beginning of the observation, then advances one record. The objective is to try the same observation again after discarding one input record. Usually, this would be an attempt to recover from an anomaly in the formatting of an input text file in which a record is defective or missing.
`GOTO` *label*;	Repeat Same Observation	Statement label at the first executable statement of the step. Essentially the same as the LOSTCARD statement if the INPUT statement has simply read one record from the input text file in the current observation.
`STOP;`	Stops Processing	Ends the data step. No further processing takes place in the step.

or, it could also be written:

```
IF NOT (POINTS > 0) THEN DELETE;
```

or:

```
IF POINTS <= 0 THEN DELETE;
```

Usually, the subsetting IF or DELETE statement appears immediately after the INPUT statement or the other statement that creates the variables that are used in the condition. When these statements execute immediately after the variables are created, no unnecessary processing is done on observations that are then discarded. The subsetting IF statement and the conditional DELETE statement do the same thing, so use the one that makes the most sense to you. If you naturally describe the action in terms of the observations you want to keep, use the subsetting IF statement. If it is easier to describe it with a focus on the observations that you discard, use the DELETE statement.

The LIST statement is one other statement that is related to the processing of an observation in the observation loop. When you read a record from an input text file, use the LIST statement to have the records written as a message in the log. Usually, the reason to do this is to let the person running the program study the records to look for a possible data error in the input data. This action happens automatically with certain kinds of data errors. The record is written with a ruler to make it easier to find specific columns. It is not written immediately, but only at the end of processing for the observation. This is to make sure that the record appears in the log only once, even if a LIST statement was executed more than once or there are other reasons for it to be written in the log.

Reset to Missing

The data step does something at the beginning of each repetition of the observation loop that is so dramatic and intrusive that you have to be aware of it: it automatically wipes out the values of most variables. The variables start out each observation with missing values. This action, officially known as *reset to missing* (RTM), is generally quite useful. It keeps the values of one observation from accidentally spilling over to the next. But not all variables in a data step really belong to the observations; some are related to the logic of the data step in a way that extends across observations, and you need to make sure that such variables are not reset to missing.

The RTM effect is easy to demonstrate in even the simplest of data steps. In the example below, variables get their values from an INPUT statement and an assignment statement. Diagnostic PUT statements are added to the step to show the values of variables at the beginning and end of each observation.

```
DATA X;
  PUT(_N_ X1-X5 XTOTAL)(=);
  INFILE CARDS;
  INPUT X1-X5;
  XTOTAL = SUM(OF X1-X5);
  PUT(_N_ X1-X5 XTOTAL)(=);
```

```
CARDS;
0 1 0 0 0
1 1 1 0 0
1 0 0 0 1
1 1 1 1 1
;
```

```
_N_=1 X1=. X2=. X3=. X4=. X5=. XTOTAL=.
_N_=1 X1=0 X2=1 X3=0 X4=0 X5=0 XTOTAL=1
_N_=2 X1=. X2=. X3=. X4=. X5=. XTOTAL=.
_N_=2 X1=1 X2=1 X3=1 X4=0 X5=0 XTOTAL=3
_N_=3 X1=. X2=. X3=. X4=. X5=. XTOTAL=.
_N_=3 X1=1 X2=0 X3=0 X4=0 X5=1 XTOTAL=2
_N_=4 X1=. X2=. X3=. X4=. X5=. XTOTAL=.
_N_=4 X1=1 X2=1 X3=1 X4=1 X5=1 XTOTAL=5
_N_=5 X1=. X2=. X3=. X4=. X5=. XTOTAL=.
```

Dots in the output lines represent the missing values. You can see that values of these variables do not carry over in any way from one observation to the next. The variables start out each observation with missing values.

Not all variables in the data step are reset to missing. Generally, the idea is to reset variables that come from input text files and ones created in data step logic, especially in assignment statements and CALL statements. All data step variables are reset to missing except for those that have a specific reason not to be. The variables that are not reset to missing are:

- Any variable read from any input SAS dataset
- Any variable to which a value is added in a sum statement
- Any variable named in a RETAIN statement
- Any variable used in an I/O statement option
- Any variable created by the FGET function
- The variables used as elements of an array, if the array is defined with initial values in the ARRAY statement, if the array is named in a RETAIN statement, or if a reference to the array is the variable to which a value is added in a sum statement
- Temporary variables used as elements of an array
- Automatic variables, including _N_
- All variables, if there is a global RETAIN statement

The RETAIN statement has two purposes. It indicates initial values for variables, and it keeps variables from being reset to missing. There are two forms of the RETAIN statement. When it appears with no terms, the RETAIN statement eliminates the entire RTM process; no variables are reset to missing in the data step. This is called a *global* RETAIN statement because it affects all variables in the data step:

```
RETAIN;
```

Lexicon

Despite the name, a global RETAIN statement is not a global statement. It is a data step statement.

Use the global RETAIN statement when the only variables that belong to observations are variables that come from a single INPUT statement or from SAS datasets, and all calculated variables are used in a way that extends across observations. If you use a global RETAIN statement in any other situation, be very careful that no values accidentally carry over from one observation to the next.

The other form of the RETAIN statement includes variable names and constant values. Use this form of the RETAIN statement to provide initial values for variables and to keep specific variables from being reset to missing. Each constant value in the RETAIN statement is the initial value for the variable or variables that precede it.

The effects of the RETAIN statement are readily seen by adding a RETAIN statement to the previous RTM example and noting the way it changes the results.

```
DATA X;
  RETAIN X1-X5 7 XTOTAL 5000;
  PUT(_N_ X1-X5 XTOTAL)(=);
  INFILE CARDS;
  INPUT X1-X5;
  XTOTAL = SUM(OF X1-X5);
  PUT(_N_ X1-X5 XTOTAL)(=);
CARDS;
0 1 0 0 0
1 1 1 0 0
1 0 0 0 1
1 1 1 1 1
;
```

```
_N_=1 X1=7 X2=7 X3=7 X4=7 X5=7 XTOTAL=5000
_N_=1 X1=0 X2=1 X3=0 X4=0 X5=0 XTOTAL=1
_N_=2 X1=0 X2=1 X3=0 X4=0 X5=0 XTOTAL=1
_N_=2 X1=1 X2=1 X3=1 X4=0 X5=0 XTOTAL=3
_N_=3 X1=1 X2=1 X3=1 X4=0 X5=0 XTOTAL=3
_N_=3 X1=1 X2=0 X3=0 X4=0 X5=1 XTOTAL=2
_N_=4 X1=1 X2=0 X3=0 X4=0 X5=1 XTOTAL=2
_N_=4 X1=1 X2=1 X3=1 X4=1 X5=1 XTOTAL=5
_N_=5 X1=1 X2=1 X3=1 X4=1 X5=1 XTOTAL=5
```

You can see how the variables start out with the initial values from the RETAIN statement, and then, the values of variables carry over from one observation to the next. This example is meant only as a demonstration of the effects of the RETAIN statement; it is probably not a way you would want to use it.

A simple example of a problem that requires a RETAIN statement is when you want a data step to find the maximum value of a variable and write it in a log note at the end of the step. The logic for this process is simple. The program below uses the variable MAX to hold the maximum value. The value of MAX is updated whenever the value of the variable being considered, POINTS, is greater than the value of MAX. For this to work, though, MAX cannot be reset to missing between observations, so a RETAIN statement is required.

```
DATA MAIN.POINTS (KEEP=ID POINTS);
  RETAIN MAX 0;
  INFILE POINT END=LAST;
  INPUT ID $CHAR7. +1 POINTS 6.;
  IF POINTS > MAX THEN MAX = POINTS;
  IF LAST THEN DO;
    FILE LOG;
    PUT 'The highest point value was ' MAX;
    END;
RUN;
```

Note the use of the KEEP= dataset option. It prevents the variable MAX from being stored in the output SAS dataset, where it probably would not be useful or meaningful. This data step is also a good example of the use of the END= option to execute statements at the end of a data step. In this example, the variable LAST serves as a condition to indicate the last input record.

A list of variable names in the RETAIN statement can include an array name, which is the same as listing all the elements of the array, and it can include abbreviated variable lists. However, an abbreviated variable list such as _ALL_ includes only variables that are defined in earlier statements in the data step.

Actions on Automatic Variables

Two automatic variables are closely connected to the functioning of the observation loop. The variable _N_ counts the repetitions of the observation loop. The variable _ERROR_ indicates certain error conditions. Both variables are automatically updated before each repetition of the observation loop.

The variable _N_ starts with a value of 1 at the beginning of the data step and has a value that goes up by 1 for each successive observation. The appropriate value is assigned to _N_ automatically at the beginning of each observation. You can assign values to _N_, but if you do, the value you assign is lost when the next observation begins. The values automatically assigned to _N_ reflect the SAS supervisor's own count of observations and are not affected in any way by any value you might assign to the variable.

The behavior of the variable _N_ does not correspond exactly to any combination of data step statements, but to create a variable that always had the same value as _N_, you could use this sum statement at the beginning of a data step:

```
MY_N_ + 1;
```

The automatic variable _ERROR_ is used as a logical value. It is automatically reset to 0 at the beginning of each observation. Certain data error conditions set the value of this variable to 1. These data errors include such actions as attempting to do arithmetic on missing values, dividing by 0, and applying an informat to text that is not valid for that informat. If _ERROR_ has a value of 1 or any other true value at the end of processing of the observation, detailed notes or error messages may be written in the log. In data step logic, you might check the value of _ERROR_ and take corrective action if an error exists.

You can also assign values to _ERROR_. Most commonly, you would assign a value of 0 so that an action that you determine should not be considered an error does not result in any detailed note in the log. When the value of _ERROR_ is true at the end of a repetition of the data step, the SAS supervisor writes the values of all variables in the log. Avoid this by assigning the value 0 to _ERROR_.

Many functions, for example, generate an error condition when their arguments are not appropriate values. The statements

```
DATA _NULL_;
  ZIP = '00100';
  STATE = ZIPSTATE(ZIP);
RUN;
```

result in an error condition, with the value of the variable _ERROR_ set to 1. Messages in the log read:

```
NOTE: Invalid argument to function ZIPSTATE at line 83 column 11.
ZIP=00100 STATE=  _ERROR_=1 _N_=1
```

To avoid having the list of all the variables in the log when you use arguments that are not certain to be valid for a function, set the value of _ERROR_ to 0 after the function call, such as:

```
  STATE = ZIPSTATE(ZIP);
  _ERROR_ = 0;
```

The note indicating the invalid arguments still appears in the log.

There are several other automatic variables that can appear in a data step, such as the FIRST and LAST variables that are created for each BY variable. These variables are not affected by any automatic processes of the data step. Because they are automatic variables, they are not reset to missing.

Automatic Actions on Text Files

The INFILE and FILE statements are executable statements that identify the input and output text files that INPUT and PUT statements act on. Just as variables are automatically reset to missing between observations to keep values of one observation from getting into the next observation, these file identifiers are also automatically reset between observations. It is as if these statements appeared at the beginning of the data step:

```
  INFILE CARDS;
  FILE LOG;
```

The default input text file, CARDS, is not actually a separate text file, but is the data lines that can appear after the CARDS statement at the end of the step. The default output text file is the log. You do not need to have a FILE statement if the only use you make of PUT statements in a data step is to write messages in the log.

The observation loop can also affect the INPUT statement. When an INPUT statement ends in a single trailing at-sign (@ as the last term of the statement), its pointer remains on that record so that a subsequent INPUT statement can read more fields from the same record. But if the record is still being held at the end of the observation, it is released at that point, so that the next observation can begin reading on a new record.

It is also possible to read multiple observations from each input record. Use a double trailing at-sign in the INPUT statement (@@ as the last term of the statement) to continue reading the next observation on the same record. When you have to do this, though, be careful that the INPUT statements do actually proceed from one input record to the next as you intend them to. If they did not, the data step could loop indefinitely.

Outside Connections

The program units of a SAS program are data steps, proc steps, and global statements, each of which executes independently. If you are familiar with the ideas of structured programming, you will recognize that the SAS language does not have the kind of program units that would make structured programming techniques possible. That is, you cannot use the SAS language to create a set of routines that would work together to implement the logic of a data step.

However, that does not mean that a data step has to be completely self-contained. The logic of a data step can be built on routines that are part of the SAS System, and the computational abilities of these routines can simplify the logic that has to be written in the data step itself. A few specific routines let a data step connect to whole categories of actions that take place outside of its own domain. These include operating system commands and various actions within the SAS environment.

Routines Two important types of routines in SAS software, functions and CALL routines, are designed mainly for use in the data step. A function returns a value, so a function call can appear anywhere an expression can appear in a data step. The function is executed as part of the process of evaluating the expression.

A call to either a function or a CALL routine is written with the routine name followed by its arguments in parentheses, with commas separating the arguments if the routine uses multiple arguments. Several examples of calls to functions and CALL routines can be seen in the discussion that follows.

Unlike a function, a CALL routine does not return a value, so it is executed in a separate statement, a CALL statement. The execution of the CALL statement starts by resolving any expressions used as arguments, and continues by executing the CALL routine itself.

The use of functions in expressions is described in chapter 6, "Constants and Expressions." Individual functions and CALL routines are described in chapter 15.

Informats and formats are mainly used in reading and writing text files. They are used throughout the SAS System, and especially in the data step, where they are part of the INPUT and PUT statements that read and write text files. They are also useful in data step logic anywhere you need to convert text to a data value or vice versa. Chapter 13 describes informats and formats and the functions that allow them to be used in expressions in a data step.

Executing operating system commands The SYSTEM and SYSGET functions let a data step connect to the operating system. The SYSTEM function takes as its argument a character value that contains a command. The available commands are different in each operating system. Operating system commands are typically used to create, delete, rename, and get information about files and directories and to execute programs. There is also a CALL routine, also called SYSTEM, that does the same thing.

The function and CALL routine work exactly the same way. The difference is that the function returns code values from the operating system. These *return codes* are numeric values that characterize the result of the action. Return codes may indicate useful information about the result of an action, such as whether an action was successful, or what kind of problem it encountered.

The specific code values that may be returned and their meanings depend on the operating system and the specific command. Usually, return codes are error codes, in which a value of 0 indicates the successful completion of an action and a nonzero value indicates a possible problem. Each different positive value of an error code usually indicates a different reason for failure. Sometimes, larger error code values indicate more serious problems.

Programmers usually assign the return code to a variable, often a variable called RC, then test this variable in the statements that follow. This is a simplified example:

```
RC = SYSTEM('copy f1 f100');
IF RC > 0 THEN PUT 'Copy of f1 failed for reason ' RC;
```

Another reason you might want to connect to the operating system is to determine the value of an environment variable, which you can do with the SYSGET function. Not every operating system uses environment variables, and different operating systems use them in different ways. Supply the name of the environment variable as a character argument to the function. The function returns the value of the environment variable as a character value.

SAS session information The GETOPTION, SYSPARM, and SYSPROD functions return information about objects in the SAS session. The GETOPTION function returns the value of a system option. Supply the system option as the first argument to the function. The function returns a character value that contains the value of the option. For system options that have numeric values, the function converts the values to character values. For keyword options, it returns '1' and '0'.

For example, GETOPTION('DFLANG') returns the value of the DFLANG system option, which might be 'English'. GETOPTION('PAGENO') returns the page number in the standard print file, converted to a character value, such as '125'. GETOPTION('CENTER') returns '1' if the CENTER option is on, '0' if it is off.

Values from the GETOPTION function are most often used in conditions in data step logic. You can check the value of a system option and take different actions in a data step as a result. The following code fragment determines the column location where it will write a title value based on the values of the CENTER and LINESIZE options.

```
IF GETOPTION('CENTER') = '0' THEN TITLECOLUMN = 1;
ELSE DO;
  LS = INPUT(GETOPTION('LINESIZE'), F12.);
  TITLECOLUMN = CEIL((LS - LENGTH(TITLETEXT))*.5);
  END;
```

The SYSPARM function, which does not use an argument, returns the value of the parameter string in the SYSPARM system option.

The SYSPROD function returns SAS product licensing information. The argument to the function is generally the key word of the product name, not including the prefix "SAS/". The function returns 1 if the product is licensed or 0 if it is not licensed, based on the licensing information entered in the SAS installation process. The function will indicate that a product is not licensed if the license expiration date has passed. If the function does not recognize the argument as the name of a SAS product, it returns –1. A program can call this function before using the features of a SAS product, so that it will not attempt to use a SAS product that is not available. However, a value of 1 returned from the SYSPROD function does not guarantee that a SAS product is installed and available for use. The function checks only licensing information; it does not check installation information.

Macro variables Macro language is a preprocessor language that generates SAS statements for execution. Macro variables are the simplest named objects in macro language. In addition to their use as symbolic objects in generating SAS statements, they can also be used to pass values from one place to another in a SAS session. A data step can assign a value to a macro variable with the SYMPUT routine. Later, a data step can retrieve the value of the macro variable with the SYMGET function.

The SYMPUT routine's two arguments are character values, the first indicating the name of the macro variable, the second indicating its value:

```
CALL SYMPUT(name, text);
```

The PUT, LEFT, and TRIM functions are frequently used in preparing values for the SYMPUT routine — PUT to convert a numeric value to a character value, which the routine requires, and LEFT and TRIM together to remove leading and trailing spaces. This example assigns the value of the automatic variable _N_ to the macro variable NCS:

```
CALL SYMPUT('NCS', TRIM(LEFT(PUT(_N_, 16.))));
```

The SYMGET function retrieves the value of a macro variable. Its argument is the name of the macro variable. You can assign the value returned by the function to a character variable, as in this example:

```
NCS = SYMGET('NCS');
```

Macro processor The RESOLVE function lets you put the macro processor to work to resolve any macro expression and return the value to the data step. The macro expression is the argument to the function. The function passes the text of the argument to the macro processor, which resolves any macro objects in it and returns the resolved value. This lets you make use of macro functions in data step logic. More significantly, you can code function-style macros of your own and use them in data step logic.

Submitting SAS statements to execute after the data step The EXECUTE routine can submit SAS statements to execute after the data step ends. Use this feature when you want parts of global statements or other SAS statements to be determined by data step logic.

An example of this might be a TITLE statement that is based on a data value from a SAS dataset. This step reads the variable SYSTEMNAME from the SAS dataset MAIN.PROFILE and puts together a TITLE1 statement that uses the value of the variable as the text of the title. The resulting TITLE1 statement then executes immediately after the data step, defining a new first title line for use in the steps that follow.

```
DATA _NULL_;
  SET MAIN.PROFILE;
  CALL EXECUTE('TITLE1 "' || TRIM(SYSTEMNAME) || '";');
  STOP;
RUN;
```

In the argument, the || symbol is the concatenation operator, which is used to put strings together.

The argument to EXECUTE can include references to macro objects. The macro processor resolves those references, and the resulting SAS statements are executed. The argument can include more than one SAS statement. Also, a data step can execute the EXECUTE routine more than once to submit a series of SAS statements.

The EXECUTE routine can also be used to execute macro statements. The macro statements execute, but they produce nothing in the log, other than this note:

```
NOTE: CALL EXECUTE routine executed successfully, but no SAS statements
      were generated.
```

Macro statements that might be used in the EXECUTE routine include %GLOBAL, to declare a global macro variable, %KEYDEF, to change the definition of a function key, and %MACRO, to define a macro.

5

Variables and Values

Variables and the values they take on are among the most visible things in a SAS program. They are so fundamental to the actions of a program that to understand what a program does it is often necessary to understand the details of the way variables and values work.

Variables and Attributes

SAS variables are used in programs; the actions of a program are largely reflected in the changes of values of variables. Variables are also stored in SAS datasets. A SAS variable is much more than just a value; there are various other properties, or *attributes*, that a variable has, whether in the execution of a data step or proc step or stored in a SAS dataset.

Looking at Attributes

It is in a SAS dataset that these attributes are easiest to see. You can pick any SAS dataset and see a report of its attributes by running the CONTENTS proc. No options are necessary other than the name of the SAS dataset:

```
PROC CONTENTS DATA=SAS dataset;
RUN;
```

The output of the CONTENTS proc includes a table that shows the attributes of the variables in the SAS dataset. This is an example of this part of the proc output:

```
-----Alphabetic List of Variables and Attributes-----

                  #    Variable    Type     Len     Pos
                  ---------------------------------------
                  2    EXPDATE     Num        8       8
                  1    INITDATE    Num        8       0
                  7    O_ERR       Char       1      27
                  6    R_ERR       Char       1      26
                  3    SERIAL      Num        8      16
                  4    S_ERR       Char       1      24
                  5    U_ERR       Char       1      25
```

In an interactive session, you can see similar information, organized in approximately the same way, in the VAR window. There are also functions you can use to get this kind of information about specific variables in SAS datasets.

Attributes

The important attributes of a SAS variable are its name, position, type, length, informat, format, and label. Variables have these same attributes whether they are being used in a program or stored in a file.

Name A variable's name is the word you use to identify the variable in a program, and a SAS variable name is more than that. It also identifies the variable in log notes and in reports produced by some procs.

The name attribute of a variable may be capitalized differently from the way the name appears in the program. The SAS language is case-insensitive, so the program executes the same way regardless of how you capitalize a name. A variable's name attribute, however, has a specific capitalization, and this affects the way the name appears in reports and in any other place where SAS software displays variable names.

Variable names either can be all capital letters or can take their capitalization from the first appearance of the variable name in the step that creates it. It depends on the system option VALIDVARNAME=. Set this option to ANY or V7 to take the capitalization of variable names from the program. Set it to UPCASE or V6 to convert all variable names to uppercase letters. The VALIDVARNAME= option also limits what is accepted as a valid variable name. Setting it to V6 limits variable name lengths to a maximum of 8 characters. Setting it to V6, V7, or UPCASE requires all variable names to meet the rules for a SAS word, even when name literal notation is used to write the name in the program. With this limitation, names cannot contain spaces or special characters.

Before SAS version 7, the name attribute of variables was always converted to all uppercase letters. The VALIDVARNAME= option was created in version 7 to allow you to continue to use names that way.

Position The DATASETS proc shows the logical position of a variable in the SAS dataset. Technically speaking, this number is an offset from the beginning of the dataset data vector (DDV), the block of data in which the values of the variables in a SAS dataset are accessible to SAS software. The position does not necessarily correspond to the location of the variable in the file — those kinds of details depend on the engine that manages the SAS dataset.

The precise position, as shown by the DATASETS proc, is not important. What matters is the relative position of variables. This affects the sequence in which variables are processed by some procs. For example, if you do not have a VAR statement in the PRINT proc, the proc prints variables in order of position in the reports it produces. The position attribute also affects the selection of variables in the sequential kind of abbreviated variable list. For example, if the variable START is at position 41, and the variable STOP is at position 105, the abbreviated variable list `START--STOP` includes both those variables and all variables that have position attributes between 41 and 105.

Data type SAS's two data types, character and numeric, offer two different ways of organizing data. Throughout SAS software, the dollar sign ($)

identifies the character data type. The character data type can contain character data — values made up of the characters that can be displayed or printed by a computer — or any kind of data. The numeric data type organizes numeric values in a double-precision floating-point format. It is suitable for any kind of numeric data that might be used in arithmetic, including time data.

Length The length of a variable is the number of bytes used to store it in a SAS dataset. The length of a character variable indicates the number of characters the variable can hold. The length attribute of a numeric variable can range from 3 to 8 bytes. In some operating systems, it can be as short as 2 bytes. When the variable is stored in a SAS dataset, using a shorter length reduces the precision of the values and saves storage space.

A program's numeric variables, in memory, always use a length of 8 bytes, regardless of their length attribute. The length attribute of a variable in a program has no affect on that variable itself. It serves only to determine the length of a stored variable that might be created from that program variable.

Informat Informats are routines for converting text to data values. A variable's informat attribute indicates the informat to use when reading new values for the variable in some situations, such as list input, or when editing the SAS dataset in the interactive environment. The informat attribute may optionally contain width and decimal arguments for the informat.

Because the character and numeric data types are different ways of organizing data values, a variable's informat must belong to the right data type. A character variable's informat attribute must indicate a character informat; a numeric variable's informat attribute must indicate a numeric informat. The informat attribute can be left blank, in which case, processes such as list input use a default informat for the variable.

Format Formats are routines for converting data values to text. A variable's format attribute is used when the variable is printed or displayed. The format attribute works much the same way as the informat attribute. It can include width and decimal arguments; the type of the format must match the type of the variable; and the format attribute can be left blank to use the standard formats.

Label The label attribute contains up to 256 characters of text. It can be used in two different ways. Most procs can use the label of a variable instead of the name to identify the variable in display or print output. Variable names are usually short, but labels can be longer and can identify variables in a more meaningful way. This depends on the LABEL system option; when this option is off, procs cannot access or use the labels of variables.

Some programmers do not use labels when displaying or printing variables. Instead, they use the label attribute to contain descriptions of or notes about the variables. Such descriptions or notes may be essential in database applications. When the label attribute is used this way, it is not suitable for identifying variables in reports. Its only significance is in the way it is displayed along with the other attributes in the output of the CONTENTS proc and in the various other ways that variables and attributes can be seen.

Most programmers leave the label attribute blank most of the time. When a variable's label is blank, procs use the variable name to identify the variable.

Setting Attributes

A variable's attributes are determined in the step where the variable is created. Most attributes can be changed at any step along the way.

In a data step, the most critical attributes of a variable are name, type, and for a character variable, length. These essential attributes are set in the first statement in the data step that uses the variable. The SAS supervisor uses any information available in the statement to decide the type and, for a character variable, the length of a new variable name that appears. These are several examples of how this works, in the event that the statement is the first use of a variable in a data step:

Statement `DUE = PREVIOUS;`
Variable PREVIOUS **Attributes** The statement provides no information about PREVIOUS. The SAS supervisor makes it a numeric variable.
Variable DUE **Attributes** DUE is the same type as PREVIOUS. If DUE is a character variable, its length is the same as the length of PREVIOUS.

Statement `IF REASON = 'X' THEN DO;`
Variable TOTAL **Attributes** REASON takes on the type of the value it is compared to. It is a character variable with a length of 1.

Statement `INPUT RT $CHAR4. AMOUNT 11.;`
Variable RT **Attributes** The value for RT is created by a character informat. With a width argument of 4, the informat produces a value with a length of 4. RT is a character variable with a length of 4.
Variable AMOUNT **Attributes** The value for AMOUNT is created by a numeric informat, no AMOUNT is a numeric variable.

Statement `MERGE MAIN.SOURCE MAIN.EVENT;`
Variable All variables in the SAS datasets MAIN.SOURCE and MAIN.EVENT
Attributes Variables take on all the same attributes that they have in the input SAS datasets. If the same variable appears in both SAS datasets, as a character variable in one SAS dataset and a numeric variable in the other, the result is an error condition. If a character variable appears in both SAS datasets, its length attribute is taken from the first SAS dataset, MAIN.SOURCE.

Statement `RETAIN LOC1-LOC4 'OUT' COUNT -1 LENGTH WIDTH;`
Variable LOC1, LOC2, LOC3, LOC4 **Attributes** These variables are character variables with a length of 3, as implied by the three-character initial value.
Variable COUNT **Attributes** COUNT is initialized with a numeric value, so it is a numeric variable.

Variable LENGTH, WIDTH **Attributes** No initial value is indicated for these variables, so the SAS supervisor makes them numeric.

It is not always best to let the SAS supervisor guess about the essential attributes of data step variables. If the first data step statement in which you use a variable does not define it the way you want it to or if there is any possibility of confusion, declare the variable in a LENGTH statement at or near the beginning of the data step. In the LENGTH statement, write the variable names followed by terms to indicate their data type and length. For numeric variables, write a number from 3 to 8 to indicate the length attribute of the variable. In some operating systems, a length of 2 is also allowed. If saving storage space is not critical, use a length of 8. For character variables, write a dollar sign and a number from 1 to 32768 for the length of the variable. Use a length just long enough to hold the longest value the variable might take on.

This is an example of a LENGTH statement:

```
LENGTH NAME $ 24 PHONE $ 10 YEAR 4 DISTANCE LAST 8 NOTES $ 280;
```

The statement declares these variables:

Name	Type	Length
NAME	Character	24
PHONE	Character	10
YEAR	Numeric	4
DISTANCE	Numeric	8
LAST	Numeric	8
NOTES	Character	280

The optional attributes of a data step variable are its informat, format, label, and for a numeric variable, length. Each of these attributes is set in a statement whose keyword is the attribute. The use of the LENGTH statement to set the length attribute has already been mentioned. The other attributes are set in the INFORMAT, FORMAT, and LABEL statements, respectively. These are the syntactical forms of the statements:

```
INFORMAT variable... informat ...;
FORMAT variable... format ...;
LABEL variable='label' ...;
```

In a data step, these statements set attributes of data step variables. The attributes are not necessarily used in the data step itself. Rather, they are stored in an output SAS dataset to be used in subsequent steps.

```
FORMAT COST PRICE MARKUP DOLLAR9.2 MARGIN 9.5 START DATE9.;
LABEL COST='Unit Cost' PRICE='Unit Price' MARGIN='Margin Ratio'
   START='Start Date';
```

You can have as many LENGTH, INFORMAT, FORMAT, and LABEL statements as you want in a data step. Or, for each attribute, you can list all the variables in a single statement.

Variables can also get these optional attributes from statements that read input SAS datasets. If the same variable gets an optional attribute in more than one statement, it is the last such statement to appear in the data step that determines that variable's attribute. This is the opposite of the essential attributes, which are determined according to the first appearance of the variable in the data step.

This is important to know if you read a SAS dataset in a data step and you want an attribute of a variable in the data step to be different from the attribute that variable has in the SAS dataset. To do that, the statement that redefines the attribute must appear later in the data step than the statement that reads the SAS dataset. For example, if the SAS dataset MAIN.EVENT contains the variable DATE with the format DATE9., you could use this step to create a new SAS dataset in which the format attribute of DATE is MMDDYY8.:

```
DATA WORK.EVENT8;
  SET MAIN.EVENT;
  FORMAT DATE MMDDYY8.;
RUN;
```

Because the FORMAT statement appears after the SET statement, it overrides the format attribute that the variable DATE gets from the SET statement. If you wrote the step this way, it would not work:

```
DATA WORK.EVENT9;
  FORMAT DATE MMDDYY8.;
  SET MAIN.EVENT;
RUN;
```

This way, the SET statement appears last, and any format attribute it provides for the variable DATE would override the earlier FORMAT statement.

A program does not directly control the position of variables. However, if the relative position of variables is important, declare all the variables in a LENGTH statement at the beginning of the data step. Then, the order of variables is the order in which you write them in the LENGTH statement.

Instead of writing separate LENGTH, INFORMAT, FORMAT, and LABEL statements, you can define all these attributes together in the ATTRIB statement. The attributes of each of these separate statements become statement options in the ATTRIB statement. Putting them together in the ATTRIB statement may make it easier to see how the attributes are related to each other, as in this example:

```
ATTRIB START LENGTH=4 LABEL='START DATE'
   INFORMAT=DATE9. FORMAT=DATE9.;
```

The same statements that define informat, format, and label attributes in data steps can also be used in proc steps when a proc reads variables from a SAS dataset. In the proc step, these statements give variables attributes that override the attributes stored in the SAS dataset. If often happens, for example, that a variable must be printed with a different format and label in different reports produced by different proc steps. The necessary attributes can be supplied in FORMAT and LABEL statements in the proc steps.

Renaming Variables

Unlike other attributes, a variable's name can change as the variable goes from a step to a SAS dataset or from a SAS dataset to a step. Use the RENAME= dataset option to change the name of the variable you read from or write to a SAS dataset.

With the RENAME= dataset option, you can store a data step variable in two different SAS datasets, with two different names. The options in the DATA statement below indicate that the data step variable CONTACT is stored as the variable CUSTOMER in one SAS dataset and as the variable SUPPLIER in another SAS dataset.

```
DATA MAIN.CUSTOMER (RENAME=(CONTACT=CUSTOMER))
    MAIN.SUPPLIER (RENAME=(CONTACT=SUPPLIER));
```

The RENAME= option can also be used anywhere you read a SAS dataset. You might use it, for example, to avoid conflicts between variables when you merge SAS datasets that have variable names in common.

In the data step the RENAME statement has the same effect as using the RENAME= dataset option on all the output SAS datasets. This is an example of the use of the RENAME statement:

```
RENAME DATE1=DATE TIME1=TIME;
```

This statement is equivalent to the dataset option

```
RENAME=(DATE1=DATE TIME1=TIME)
```

Changing the Attributes of Variables in SAS Datasets

After a SAS dataset is created, the type, length, and position of variables cannot be changed. The only way to change those attributes would be to delete the SAS dataset and create a new one. However, all other attributes of SAS dataset variables can be changed in the DATASETS proc, using the MODIFY statement and its secondary statements.

To identify the SAS dataset to modify, indicate the library in the LIBRARY= option of the `PROC DATASETS` statement, then the member name in the MODIFY statement. The MODIFY statement can be followed by INFORMAT, FORMAT, LABEL, and RENAME statements, which have the same syntax as in the data step. The difference in the DATASETS proc is that the statements directly change variable attributes in an existing SAS dataset.

This example shows the use of the DATASETS proc to change attributes of the variable START in the SAS dataset MAIN.TREND:

```
PROC DATASETS LIBRARY=MAIN NOLIST;
  MODIFY TREND;
    RENAME START=INIT;
    FORMAT INIT MMDDYY10.;
    LABEL INIT='Initialized As Of';
QUIT;
```

...utes of Data Step Variables

The easiest way to find out about the attributes of variables in a data step is to look at the attribute of the variables in an output SAS dataset created by the step. Generally, the attributes of the data step variables correspond to the attributes of the SAS dataset variables. However, if necessary, it is possible to query the attributes of data step variables with either of two sets of functions designed for that purpose.

The VNAME function returns the name attribute of a variable. Its argument is the name of the variable. For example, VNAME(TEXT) returns the name attribute of the variable TEXT. Add an X to the function name, and you can use a character value as the argument, for example, VNAMEX('TEXT'). The most likely use of the name attribute is to write a column header in a report that follows the capitalization indicated in the variable's name attribute.

Other variable information functions return other attributes of a variable name, as indicated in the following table.

Function Call	Type Returned	Attribute Returned
VNAME(*name*) VNAMEX('*name*')	Character	Name
VTYPE(*name*) VTYPEX('*name*')	Character	Data Type: C for Character or N for numeric
VLENGTH(*name*) VLENGTHX('*name*')	Numeric	Length
VINFORMAT(*name*) VINFORMATX('*name*')	Character	Informat
VINFORMATN(*name*) VINFORMATNX('*name*')	Character	Informat name
VINFORMATW(*name*) VINFORMATWX('*name*')	Numeric	Informat width argument
VINFORMATD(*name*) VINFORMATDX('*name*')	Numeric	Informat decimal argument
VFORMAT(*name*) VFORMATX('*name*')	Character	Format
VFORMATN(*name*) VFORMATNX('*name*')	Character	Format name
VFORMATW(*name*) VFORMATWX('*name*')	Numeric	Format width argument
VFORMATD(*name*) VFORMATDX('*name*')	Numeric	Format decimal argument
VLABEL(*name*) VLABELX('*name*')	Character	Label

The following program demonstrates the use of variable information functions.

```
DATA _NULL_;
  A = 1;
  NAME = VNAME(A);
  TYPE = VTYPE(A);
  LENGTH = VLENGTH(A);
  INFORMAT = VINFORMAT(A);
  FORMAT = VFORMAT(A);
  LABEL = VLABEL(A);
  PUT _ALL_;
RUN;
```

```
A=1 NAME=A TYPE=N LENGTH=8 INFORMAT=F. FORMAT=BEST12. LABEL=A _ERROR_=0
_N_=1
```

Data Types

SAS's two data types, character and numeric, are two ways of organizing data values. The numeric data type is based on a particular way of representing numbers in a digital form. The character data type is based on the idea that each byte can represent a symbol you can see.

The Numeric Data Type

Numeric values in SAS are 64-bit floating-point values. That means a numeric value can represent almost any number you might use, with a high degree of accuracy. Floating-point values are not meant to represent numbers exactly, but they are precise enough for most purposes. The precision of numbers can be measured in significant digits; SAS numeric values have about 15 significant digits.

Most programming languages and database management systems have several numeric data types. There may be separate types for integers, dates, times, amounts of money, logical values, and other kinds of numeric values. SAS keeps things simple by using just one data type for all numeric data. This means that you can do arithmetic directly on all kinds of numeric data values without having to be concerned with type conversions, integer truncation, and other such issues.

Floating Point

The mathematical idea behind floating-point numbers is that any positive or negative number can be represented in the form

$$s\, m\, b^e$$

where

s is the sign, 1 or –1.

m is the *mantissa*, which supplies the details of the magnitude of the number. In some systems, the mantissa is a number between 1 and b. In others, it is a number between $1/b$ and 1.

b is a positive integer such as 2 or 16 that is chosen as the *base* of this system.

e is an integer *exponent*, which represents the order of magnitude of the number. The base is raised to this power.

For example, the number 23 can be written as

$1 \cdot 1.4375 \cdot 2^4$

As another example, the number –.15625 can be written as

$-1 \cdot 1.25 \cdot 2^{-3}$

There are tradeoffs to be considered when fitting this floating-point scheme into a fixed number of binary digits, or bits. Any floating-point representation uses one bit for the sign and divides the remaining bits among the two other numbers it has to represent: an exponent and a mantissa. A longer mantissa allows greater precision; a longer exponent allows a greater range of numbers to be represented. A larger base allows a greater range of magnitude, but it also means that some numbers are represented with more precision than others. Using more total bits allows greater range and precision, but requires more computer resources. Recognizing these tradeoffs and other design issues, computer designers over the years have opted to use various ways of representing floating-point numbers. The most common lengths used for floating point values are 32 and 64 bits, or 4 and 8 bytes. Often these are referred to as single-precision and double-precision, respectively. SAS numeric values use the 64-bit, or 8-byte, floating-point format of the computer that SAS is running on. This can vary from one kind of computer to another.

Most computers, unless they are based on designs from the 1970s, follow the IEEE standard for 64-bit floating-point numbers. IEEE, the Institute of Electrical and Electronic Engineers, Inc., is an international membership organization that develops and codifies standards for computer equipment, among its many other activities. IEEE floating-point representations use a base of 2. The IEEE 64-bit floating-point representation uses the first bit for the sign, the next 11 bits for the exponent, and the remaining 52 bits for the mantissa. If you read these binary segments as the binary integer values S, E, and M, S can be 0 or 1; E can range from 0 to 2047; and M can range from 0 to $2^{52} - 1$, or 450,399,627,370,495. The following formulas translate these integer values into the values used for the generic floating-point formula above.

$$s = (-1)^S$$
$$e = E - 1023$$
$$m = 1 + (M/2^{52})$$

If the sign bit is 1, that represents a negative number. The *bias* of 1023 is subtracted from the exponent segment, to make negative exponents possible. Negative exponents are needed for numbers between 0 and 1. The mantissa segment is converted to a fraction by dividing it by the appropriate power of 2, then 1 is added to it to make it a number between 1 and 2.

Not every possible exponent value in a floating-point value actually represents an exponent. Some may be reserved for other purposes. The most important of these is the need to represent 0, a number that does not fit the floating-point scheme described above. In IEEE floating-point representations, the lowest possible exponent value, 0, is reserved to represent the number 0 (along with some other very small numbers). The number 0 is actually represented with all 64 bits set to 0.

The IEEE 64-bit floating-point standard is used on computers running the Windows, OS/2, Unix, Mac, and OpenVMS Alpha operating systems, among others. This includes more than 99 percent of all computers ever made. Still, there are important exceptions — notably, IBM mainframe systems and VAX systems. VAX systems use an 8-bit exponent and a 55-bit mantissa segment. IBM mainframe computers use 16 as a base, with a 7-bit exponent and a 56-bit mantissa, which represents a fraction between 1⁄16 and 1.

Even among computers that do use the IEEE standard, the form of a SAS numeric value may not be exactly the same. Some computers reverse the order of bytes in some or all forms of numeric values. The variability in the form of numeric values is the main reason that a SAS data file created for one kind of computer cannot be used directly on another kind of computer.

Numeric Precision and Length

While floating-point numbers are not meant to represent numbers exactly, they do represent integers exactly — up to a point. On computers that use the IEEE standard, you can count up to 9,007,199,254,740,992 with SAS numeric values; the floating-point form can represent all integers exactly up to that point, but beyond that point, the only values it can represent are multiples of powers of 2.

For situations in which it is important to save storage space, you can use the length attribute to have a variable stored with less than its full 8 bytes. The least significant bytes (of the mantissa) are simply omitted when the variable's values are written to the SAS dataset, a process called *truncation*. The variable is shortened only in the SAS dataset; all numeric variables in SAS programs use the full 8 bytes, regardless of their length attributes. The shortened variable has less precision, but that may not be a problem, depending on the kind of values the variable has.

When a shortened variable is read from a SAS dataset, the least significant bytes are filled in with zeroes. This usually results in a value that is slightly different from the variable's original value, before it was truncated. For some numbers, though, the restored value is exactly the same as the original value. That happens when the original value had all zeroes in its least significant bytes to begin with. This point is not merely a curiosity. There are whole classes of values that you can count on to have all zeroes in their least significant bytes, including the kind of numbers we are all most familiar with, small integers. Integer values that are small enough can be counted on to remain intact through the truncation process. The largest integer value for which this will work depends on the length of the variable, as shown in the following table. If you know a variable will have only integer values, and you know those values will be no greater in magnitude than a certain maximum value, then you can set the length attribute of the variable accordingly.

Maximum Integer Values Represented Exactly

		Type of Computer	
		Most computers	IBM mainframe and VAX
Length of Variable	2	*Not allowed*	256
	3	8,192	65,536
	4	2,097,152	16,777,316
	5	536,870,912	4,294,967,296
	6	137,438,953,472	1,099,511,627,776
	7	35,184,372,088,832	281,474,946,710,656
	8	9,007,199,254,740,992	72,057,594,037,927,936

You might choose to shorten some variables whose values are not necessarily small integers, if you do not need more than a certain amount of precision for that variable. The accuracy of even a 3-byte numeric variable is still within a fraction of a percent, and that might be accurate enough for a variable that represents something like travel distance or estimated television audience size. The table of maximum integer values can give you an idea of the precision afforded by each length a numeric variable can have.

The Character Data Type

The character data type is based on the idea that computer data is something you can see and read. Each byte of a character value represents a character — a letter or other familiar symbol. The length of a character value is the number of bytes it takes up, and it is also the number of characters it can contain.

A byte is, literally, a sequence of 8 binary digits, or bits, and it can be most directly understood as an integer value from 0 to 255. When you see character values, each different character is just the visual representation of a different byte value. For example, a capital K might be what a byte value of 75 looks like, while a byte value of 55 might appear as the digit 7.

The length of a character variable can range from 1 to 32,768 characters. When you create SAS datasets, it is especially important to keep character variables short. Make them just long enough to hold the longest value the variable may take on. The reason this is so important is that each variable appears on each observation of a SAS dataset you create. To determine the total amount of storage space taken up by a variable, you have to multiply the length of the variable by the number of observations. If a character variable has the maximum length of 32,768, it takes only 31 observations for the variable to use up more than 1,000,000 bytes of storage space — and to use up the length of time it takes to read or write 1,000,000 bytes every time you access that file.

Conversely, it is also essential that a character variable is long enough. If the variable is shorter than the value you want it to hold, the variable can take on only part of the value, and that can interfere with the logic of a program and lead to incorrect results. Especially when you create character variables in data step computations, make sure those variables have appropriate lengths. Declare them in a LENGTH statement if necessary.

Character Sets

The way an operating system maps the set of possible byte values to a set of character forms is called a *character set*. Different operating systems use different character sets — and some operating systems support more than one character set.

The foundation of the character set is the ASCII character set standard, which maps numbers 32 to 126 to the character forms you expect to see on a computer keyboard. The ASCII standard was originally developed for digital communications, where its use is almost universal. ASCII characters are also the characters on which most programming languages are built. These are the ASCII characters:

```
space  !  "  #  $  %  &  '  (  )  *  +  ,  -  .  /
0      1  2  3  4  5  6  7  8  9  :  ;  <  =  >  ?
@      A  B  C  D  E  F  G  H  I  J  K  L  M  N  O
P      Q  R  S  T  U  V  W  X  Y  Z  [  \  ]  ^  _
`      a  b  c  d  e  f  g  h  i  j  k  l  m  n  o
p      q  r  s  t  u  v  w  x  y  z  {  |  }  ~
```

The ASCII character set has been used by virtually all new computer operating systems introduced since the 1970s. For some computers, the ASCII character set is the entire set of characters the computer displays. Usually, though, the character set includes other characters, and these added characters may differ from one computer to another.

IBM mainframe computers use a character set that is not based on the ASCII standard. Years ago, IBM designed its mainframe computers to work with *punch cards*, stiff pieces of paper of a certain exact size in which you could store a small amount of data by punching holes at certain exact points. The EBCDIC character set used on IBM mainframe computers is based on the patterns of holes that represented characters in punch cards.

The EBCDIC character set uses most of the same character forms as the ASCII character set but uses different byte values to represent each character. This introduces an extra complication in exchanging data between IBM mainframe computers and the world at large. The IBM mainframe EBCDIC characters need to be translated to the ASCII characters that the rest of the world uses. Usually, the translation is done in communications software. However, if you ever need to translate between ASCII and EBCDIC characters in a SAS program, you can do so with the $ASCII and $EBCDIC informats and formats. Use the $ASCII informat and format on an IBM mainframe computer to read and write ASCII files. Use the $EBCDIC informat and format on other computers to read and write EBCDIC files.

An operating system or printer may let you select the font to use when you display or print text. The font is what determines the specific shapes of the letters and other characters you see. When you choose a font, it is important that the font has a character set that is consistent with the character set of the text you want to display or print. Most fonts have a complete set of ASCII characters, but some may not have any other characters, or the other characters they have may not be the ones you expect.

Binary Data

While the character data type is mostly used for strings of characters, character variables can also be used in other ways, ways that do not involve any direct display of the value. Any of these other uses of the character data type can be described as *binary* data, because they treat a character value just as a sequence of bytes, bytes that do not actually represent characters to be displayed.

As the simplest example, you could use a one-character variable to hold integer values from 0 to 255. When you use a character variable for binary data, however, it is up to you to figure out ways to read, write, display, and compare the binary values. Sometimes, the many SAS informats and formats designed to read and write binary fields can help.

Values

Data step variables take on values in various ways. A variable's value can change as the result of a statement or as the result of an automatic process. Among the various statements that can give a variable a new value, the value may be indicated as a constant or expression or it may be a value that does not appear in the program, but is computed as part of the actions of the statement.

Initial Values

By default, data step variables are initialized to missing values. You can use the RETAIN statement to indicate other initial values for selected variables. In the RETAIN statement, write the name of the variable followed by a constant that indicates the initial value for the variable.

Actions to Change Values

There are several data step statements that can change the values of variables. In addition, the data step includes automatic actions to change the values of some kinds of variables. The following are data step statements whose actions include changing the values of variables.

Assignment The simple, direct way to change the value of a variable is with an assignment statement. The statement indicates the target variable, an expression to assign to it, and between them, an equals sign indicating the assignment action:

```
variable = expression;
```

Ideally, the expression should be of the same data type as the variable.

Sum The purpose of the sum statement is to add a value to a variable. The syntax is different from the assignment statement in that a plus sign replaces the equals sign:

```
numeric variable + numeric expression;
```

The sum statement works only with the numeric data type. Two very common uses of the sum statement are counting and summing:

```
COUNT + 1;
SUMWEIGHT + WEIGHT;
```

CALL The CALL statement executes a CALL routine. Usually, one or two of the arguments to the CALL routine are variables whose values are changed in the execution of the routine or to indicate its results. The RANUNI routine, which generates a random number from a uniform distribution on the interval from 0 to 1, is an example of this. When it executes, the routine changes the values of both variables used as its arguments. The first argument is a seed variable, which maintains the continuity of the random number stream. The second argument is the random number that the routine generates based on the seed.

```
CALL RANUNI(SEED, X);
```

INPUT The INPUT statement reads data from an input text file to create values for variables. It provides values for every variable mentioned in the statement. If a variable appears in the statement, but the statement cannot get a value for it from the file, it gives the variable a missing value.

The INPUT statement also changes the values of variables named in options in the INFILE statement. These options include the END= option, to indicate the end of the file; the FILENAME= option, to indicate the physical file name; and the COLUMN= option, to indicate the column pointer position. Each of these and various other statement options in the INFILE statement indicate a variable name, and the INFILE statement provides values for these variables every time it executes.

Data errors in the INPUT statement may affect the automatic variable _ERROR_. The statement sets the value of this variable to 1 to indicate an error condition. You can use the ?? error control in a variable term in the INPUT statement to keep this from happening.

SAS datasets Data step statements that read from SAS datasets provide values for every variable they read from the SAS dataset every time they execute, with a few specific exceptions. If a variable cannot be read from the

SAS dataset, the statement gives it a missing value. The statements create values for variables even if the variable name never appears in the step.

The important exception to note can occur when using the KEY= option. If the observation being sought is not found, the statement does not change the value of any of the variables. In that situation, you may have to use assignment statements to assign missing values to those variables.

Use the KEEP= dataset option to read only selected variables from the SAS dataset or the DROP= dataset option to avoiding reading specific variables.

If a BY statement follows the statement that reads the SAS dataset, the statement sets values for the FIRST and LAST variable for each BY variable.

Two statement options can create other variables that are affected when the statement executes. The END= option creates a variable whose value is set to 1 to indicate the last observation being read by the statement. The NOBS= option of the SET and MODIFY statements creates a variable that indicates the total number of observations the statement will read. The variables' values are set each time the statement executes.

The IN= dataset option can also be used to create variables; this is useful in statements that read more than one SAS dataset. The variable named in the IN= option for a specific SAS dataset is set to 1 when an observation is read from that SAS dataset. For example, the statement

```
SET DATA1 (IN=IN1) DATA2 (IN=IN2);
```

creates the variables IN1 and IN2. IN1 has the value 1 when an observation is read from the input SAS dataset DATA1. Likewise, IN2 has the value 1 when an observation is read from the input SAS dataset DATA2. You can use these variables for data step logic that depends on knowing where an observation came from.

When you use the KEY= option and the statement fails to read an observation, the automatic variables _IORC_ and _ERROR_ are both set to nonzero values to indicate an I/O error condition. You can test _IORC_ to discover this result and respond accordingly.

DISPLAY The DISPLAY statement displays a window, which can include variables to be entered or altered by the user. The user can directly change the values of those variables.

Two automatic variables are also related to the DISPLAY statement. If the variable _MSG_ is given a value, the DISPLAY statement displays it as a message in the message line of the window, and it clears the value of the variable. If the user enters command text in the window, and the text is not a SAS command, the text is returned to the program in the automatic variable _CMD_.

PUT In the same way that the INPUT statement changes variables that options in the INFILE statement create, the PUT statement changes variables created by options in the FILE statement. The same FILENAME= and COLUMN= options are among the options available in the FILE statement to create variables that reflect the actions of the PUT statement.

DO loop The DO statement can include terms to define a loop, as described in the next chapter. An example of a DO statement that creates a DO loop is:

```
DO I = 0 TO 23;
```

The variable I in this example is an *index variable;* its value is altered for each repetition of the DO loop. Most DO loops use an index variable, although it is not mandatory.

There are also automatic processes, not corresponding to specific statements, that change the values of data step variables. Most kinds of variables are automatically reset to missing at the beginning of each repetition of the observation loop, as described in the previous chapter. At the same time, the automatic variables _N_ and _ERROR_ are set to values. The value of _N_ counts the number of repetitions of the observation loop. The value of _ERROR_ is set to 0 at the beginning of each repetition of the observation loop, and may be set to 1 during the execution of the observation if certain kinds of data errors occur.

Changing Part of a Character Variable

There are times when you want to change the value of just part of a character variable — changing just a single character, or some string of characters within the variable, but leaving other characters unchanged. You can do this in the assignment statement with a special use of the SUBSTR function.

When you use it in an expression, the SUBSTR function extracts a substring from the value of a character variable. But when you use it as the target of an assignment statement, it lets you assign a value to a part of the variable. Use either of these two forms:

```
SUBSTR(character variable, start, length) = character expression;
SUBSTR(character variable, start) = character expression;
```

The second and third arguments are positive integers that indicate what part of the character variable to replace. The second argument is the first character position to replace. Make sure this argument is at least 1 and no greater than the length of the variable. The third argument is also a positive integer value. It indicates how many characters to replace. If you omit the third argument, characters are replaced through the end of the variable.

The following examples demonstrate the use of the SUBSTR function to replace selected characters in character variables.

```
YQ = '1999.1';
PUT YQ=;
SUBSTR(YQ, 1, 4) = '2000';
PUT YQ=;

WORD = 'COVER';
PUT WORD=;
SUBSTR(WORD, 3, 1) = 'D';
PUT WORD=;
```

```
SSN='123456789';
LENGTH SSNH $ 11;
SSNH = '***-**-****';
PUT SSNH=;
SUBSTR(SSNH, 1, 3) = SUBSTR(SSN, 1, 3);
PUT SSNH=;
SUBSTR(SSNH, 5, 2) = SUBSTR(SSN, 4, 2);
PUT SSNH=;
SUBSTR(SSNH, 8) = SUBSTR(SSN, 6);
PUT SSNH=;
```

```
YQ=1999.1
YQ=2000.1
WORD=COVER
WORD=CODER
SSNH=***-**-****
SSNH=123-**-****
SSNH=123-45-****
SSNH=123-45-6789
```

Automatic Variables

Variables that the data step creates on its own, separate from any specific action stated in the program, are *automatic* variables. Several automatic variables have already been mentioned. Each automatic variable has its value changed automatically in very specific circumstances. You can test the values of these variables to learn things about the processing of the data step. There are reasons you might assign values to some of these automatic variables. The essential facts about the automatic variables are summarized here.

_N_ The purpose of this automatic variable is to count the repetitions of the observation loop.

Type: Numeric (positive integer)

Values set automatically: Set at the beginning of each repetition of the observation loop. Set to 1 on the first repetition, and a value 1 higher on each successive repetition.

Uses for value: Find out what repetition of the observation is executing.

Reasons to assign a value to the variable: None

ERROR This variable serves as an error flag.

Type: Numeric (Boolean)

Values set automatically: Reset to 0 at the beginning of each repetition of the observation loop. Set to 1 on certain data errors, especially when an informat, function, CALL routine, or operator attempts to work with values that are not valid for it.

Uses for value: Find out whether an error occurred.

Reason to assign a value to the variable: Set to 0 to indicate that a situation should not be treated as an error condition.

INFILE This variable, actually a pseudo-variable, contains the current input record of the current input text file.

Type: Character

Values set automatically: The value changes to reflect the current input record when an INPUT statement advances to a new record or when an INFILE statement selects a different input text file.

Uses for value: Gives you direct access to the input buffer that the INPUT statement uses. You can also use the variable to copy the input record to an output file.

Reason to assign a value to the variable: Modify the input buffer before the INPUT statement reads fields from the record.

IORC This variable contains the file system return code from an action on an observation in a SAS data file.

Type: Numeric (return code)

Values set automatically: Reset to 0 at the beginning of each repetition of the observation loop. Set to a nonzero value when an observation is not found in the execution of a SET or MODIFY statement with the KEY= option.

Uses for value: After a SET or MODIFY statement with the KEY= option, always test this value to find out whether the observation was found or not.

Reason to assign a value to the variable: Set to 0 after handing the condition in which the observation was not found.

FIRST.*variable* and LAST.*variable* These variables indicate groups in input data when there is a BY statement. They are created for each BY variable.

Type: Numeric (Boolean)

Values set automatically: The FIRST variable is set to 1 when reading the first observation in a BY group. It is set to 0 when reading any other observation. The LAST variable is set to 1 when reading the last observation in a BY group. It is set to 0 when reading any other observation.

Uses for value: Use as a condition for actions to take at the beginning and end of each BY group.

Reason to assign a value to the variable: None.

MSG To display text in the message line of a data step window, assign the text to this variable.

Type: Character

Values set automatically: Set to blank when the window is displayed.

Uses for value: None.

Reasons to assign a value to the variable: Assign the text of a message to the variable before displaying the window. The message appears in the message line of the window.

CMD This variable returns command text entered by the user in a data step window.

Type: Character

Values set automatically: Contains text of a command entered by the user when a window is displayed, if the command was not a SAS command that the SAS supervisor could recognize.

Uses for value: With this variable, you can have a data step program respond to a set of user commands or function keys. After displaying the window, parse the value of this variable to find any commands entered by the user.

Reason to assign a value to the variable: You might clear the variable after processing the command.

Lengths of Character Values

It is important to declare character variables with the right length — the length of the values that you intend them to have. However, when values are assigned to variables, the values are not always the same length as the variables. The length of the value is adjusted so that it fits the variable.

When the value is longer than the variable, the beginning part of the value is used, and any extra bytes at the end of the value are discarded. This process is called truncation, and it is essentially the same process as the truncation of numeric values when they are stored with lengths shorter than 8.

When a character variable is shorter than you mean it to be, truncation can result in the loss of data that you need to make a program's logic work. The following simple example shows how easily truncation can occur by accident.

```
PAGECT = 0;
IF PAGECT THEN FORMAT = 'PRINT';
ELSE FORMAT = 'AUDIO CASSETTE';
PUT FORMAT=;
```

```
FORMAT=AUDIO
```

In this example, the length of the variable FORMAT is 5. This could be either because it was previously declared with a length of 5 or because its first appearance is in the assignment statement that is the action of the IF-THEN statement, which gives it a length of 5. When the ELSE action executes, assigning a longer value to the variable, that value is truncated to 5 characters, the length of the variable. An important part of the value is lost in the process.

To avoid this problem, use a LENGTH statement early in the data step to declare the variable with an appropriate length, such as:

```
LENGTH FORMAT $ 18;
```

After this declaration, any value up to 18 characters long can be assigned to the variable FORMAT intact. Only a value longer than 18 characters would be truncated.

In other situations, truncation may be just the effect you want. This example uses truncation to extract the first initial of a name:

```
NAME = 'ADAMS';
LENGTH INITIAL $ 1;
INITIAL = NAME;
PUT NAME= INITIAL=;
```

```
NAME=ADAMS INITIAL=A
```

If it is not clear from the code itself that the truncation is deliberate, write a comment at that point in the program to describe the purpose or effect of the truncation to any programmer who might read the program. In the following example, notice how the comment statement calls attention to the effect of truncation, which could easily escape your attention if you read just the assignment statement by itself.

```
* Two-digit year is the first two characters of date field. ;
YEAR2 = DATEFLD;
```

When the value assigned to a variable is shorter than the variable, the value is extended to the length of the variable by adding spaces to the end. This process is called *blank padding*. Blank padding is usually not a problem. Trailing spaces are not visible when you display or print the variable, and they rarely have any affect on the logic of a program.

However, there are a few situations when you need to be aware of the trailing spaces in a character value. Trailing spaces show up when you use the concatenation operator (||) to combine character values. The following example of concatenation is intended to convert a text value to an HTML object, specifically, a second-level heading. This requires adding an HTML tag before and after the text.

```
LENGTH TEXT $ 50 OBJECT $ 60;
TEXT = '2001';
OBJECT = '<h2>' || TEXT || '</h2>';
PUT OBJECT;
```

```
<h2>2001                                              </h2>
```

While the resulting object is not actually incorrect in this case, you might prefer to put it together without all the extra spaces between the text "2001" and the end tag "</h2>". These spaces are the 46 trailing spaces that are added to the variable TEXT when a 4-character value is assigned to it, to complete the variable's 50-character length. To take away trailing spaces, you can use the TRIM function:

```
LENGTH TEXT $ 50 OBJECT $ 60;
TEXT = '2001';
OBJECT = '<h2>' || TRIM(TEXT) || '</h2>';
PUT OBJECT;
```

<h2>2001</h2>

The TRIM function removes all trailing spaces so that you can concatenate or do other processes with a character value where you do not want to include trailing spaces.

One of the classic puzzles in SAS programming involves adding a character to the end of a character variable. When you attempt to do this with simple concatenation, as in the example below, it can never work.

```
LENGTH WORD $ 18;
WORD = 'lion';
PUT 'Before: ' WORD=;
WORD = WORD || '*';
PUT 'After:  ' WORD=;
```

```
Before: WORD=lion
After:  WORD=lion
```

The asterisk that was concatenated to the end of the word disappeared! How did that happen? The answer is obvious as soon as you see the lengths of all the values involved. The variable WORD has a length of 18, as declared in the LENGTH statement. Then, the value created by the concatenation has a length of 19 — the combined lengths of the two values being concatenated. The length of the variable is 18, and the length of the constant is 1, and the sum of these two lengths, 19, is the length of the value that results from the concatenation. The asterisk is actually the 19th character in this value. But the value is then assigned to the variable WORD, which has a length of 18. The 19-character value has to be truncated. Only the first 18 characters are assigned to the variable, and the 19th character — which is the character that was intended to be added — is dropped.

The intended effect can be produced with the use of the TRIM function. Then, the value that results from the truncation is only 5 characters, and it can be assigned to the variable intact, with the necessary blank padding to complete the value of the variable.

```
LENGTH WORD $ 18;
WORD = 'lion';
PUT 'Before: ' WORD=;
WORD = TRIM(WORD) || '*';
PUT 'After:  ' WORD=;
```

```
Before: WORD=lion
After:  WORD=lion*
```

Type Conversion

Although I do not recommend it, it is possible to assign a value of one data type to a variable of the other data type — to assign a numeric value to a character variable or to assign a character value to a numeric variable. The SAS supervisor attempts to carry out the action by converting the value to the type of the variable.

Assignment statements are not the only situations in which an automatic type conversion can occur. The SAS supervisor automatically converts numeric values to character values when a numeric expression:

- is assigned to a character variable.
- is an argument to a function or CALL routine for which a character value is expected.
- is an operand for the concatenation operator.
- is an index value in a DO statement that uses a character index variable.

The SAS supervisor uses the BEST format to do the conversion. Usually, this produces the best possible representation of the numeric value in the length available, which might be in standard or scientific notation. If the context of the conversion does not indicate a specific length to the SAS supervisor, it uses the BEST format with a width of 12. If the length is too short to hold any representation of the number, the BEST format fills the available space with asterisks. For example, if you assign the value –10 to a two-character variable, the resulting value of the variable is "**".

The SAS supervisor automatically converts character values to numeric values when a character expression:

- is assigned to a numeric variable.
- is an argument to a function or CALL routine for which a numeric value is expected.
- is an operand for an arithmetic or logical operator.
- is an operand for a comparison operator or the MIN or MAX operator and the other operand is a numeric value.
- is used in a sum statement.
- is a condition in a control flow statement.
- is a value being compared in a SELECT or WHEN statement of a SELECT block, if any of the values being compared in the SELECT block is a numeric value.
- is an index value or a starting or stopping value in a DO statement that uses a numeric index variable.
- is an array subscript.

The SAS supervisor uses the standard numeric informat to do the conversion. This informat can read ordinary numerals and scientific notation with the letter E, provided there are no embedded spaces anywhere inside the numeral. Of course, most character values cannot be interpreted as numbers; in that case, the SAS supervisor produces a missing value as the result of the conversion.

Every time you have the SAS supervisor do an automatic type conversion, it produces a log message indicating the point in the program where the type conversion occurred.

Avoid type conversions by using variables and values of the appropriate data type. When you need to do type conversions, it is best to code them explicitly. To convert a numeric value to a character value, use the PUT function and choose the appropriate format and width. This is an example of using the PUT function to convert a numeric value to a character value and assigning the result to a character variable:

```
LENGTH Y4 $ 4;
YEAR = 2004;
IF YEAR <= 9999 THEN Y4 = PUT(YEAR, 4.);
PUT YEAR= Y4=;
```

```
YEAR=2004 Y4=2004
```

Similarly, to convert a character value to a numeric value, use the INPUT function and choose the appropriate informat and width. If it is possible that the character value may not translate successfully into a numeric value, use the ?? error control in the INPUT function to avoid creating an error condition.

```
DO TEXT = '1E4', '1EX', '2.3';
  NUMBER = INPUT(TEXT, ?? 3.);
  PUT TEXT= NUMBER=;
  END;
```

```
TEXT=1E4 NUMBER=10000
TEXT=1EX NUMBER=.
TEXT=2.3 NUMBER=2.3
```

In this example, the DO statement indicates a DO loop, which executes the statements between the DO and END statements once for each value indicated for the index variable TEXT. The INPUT function successfully interprets the first and third values. It fails to interpret the second value, which is not valid text for the standard numeric informat, but because of the ?? error control, there is no error condition and no resulting note in the log.

6

Constants and Expressions

Computation can take place before the point at which a value is assigned to a variable. A segment of code that puts together a value is called an expression. Constants are values that are the starting points of expressions, and they have other uses at other points in a program.

Constants

A data value that is written directly in a program is a *constant*. Different kinds of constants are used at different points in a SAS program.

Whole Number Constants

Many places in SAS statements require a whole number — a number that is written as a sequence of digits, with no other symbols allowed. In these places, fractions are not allowed, and even writing a decimal point would be a syntax error. The allowable whole numbers are 0, 1, 2, 3, and so on.

Whole number constants are used in various places throughout the SAS environment, especially outside the data step. These are examples of places where whole numbers are required in SAS statements:

- the width or decimal argument of an informat or format
- the location or distance of a pointer control in the INPUT, PUT, or WINDOW statement
- values of some statement options, system options, and dataset options

Some options take integer values, allowing a value to be positive or negative.

Numeric Constants

Numeric constants are constant values of the numeric data type. In addition to whole number values, this kind of constant allows any real number you can write, and there are several notational forms you can use to write them.

To write a standard numeric constant, follow the conventional way of writing a numeral, but without any commas between digits. These are examples of standard numeric constants:

```
37      -25      0.025      -.64      10287.455834
```

Very large and very small numbers can be easier to write in scientific notation. Scientific notation begins with a number in standard numeric notation and appends the letter E and an integer. This integer represents a power of 10. The number is multiplied by 10 raised to the indicated power. So, for example, `2.5E5` is 2.5 times 100000, or 250000. The process works the same with negative exponents. For example, `2.06E-8` is 2.06 times .00000001, or .0000000206.

The scientific notation form allows for some variability of notation. You can write the letter D instead of E. When the exponent is positive, you can write a positive sign between the letter and the exponent.

In addition to its use as a numeric constant, scientific notation can be used in input and output fields. The standard numeric informat can read this kind of scientific notation, and the default BEST format for numeric output writes scientific notation for values that are not as easy to write in standard notation.

Hexadecimal notation is a base 16 system of writing integers. It is often used for numeric codes and sizes in computer systems because of the way it translates so easily to binary. Each hexadecimal digit corresponds to four binary digits. Two hexadecimal digits can represent one byte.

The hexadecimal system uses the digits 0–9 and A–F (or a–f). After counting to 9, the letter A is the digit that represents 10, B is 11, C is 12, D is 13, E is 14, and F is fifteen. Values of 16 or more are written with more than one digit. The last digit in a hexadecimal numeral represents the actual value of the digit. Preceding digits represent successively higher powers of 16. So, for example, the hexadecimal numeral 2BC can be interpreted as

$$(2 \times 16 + 11) \times 16 + 12$$

or 700.

A hexadecimal constant is a hexadecimal numeral followed by the letter X (or x). So that the SAS supervisor will recognize it as a number and not as word, a hexadecimal constant must begin with one of the digits 0–9. If a hexadecimal numeral begins with one of the digits A–F, write a leading 0 before it. These are examples of hexadecimal constants, along with the same value in standard notation:

```
2BCX        700
0FX         15
400X        1024
0A0X        160
```

A minus sign is not allowed as part of the hexadecimal constant notation. There is no way to write a negative or fractional number as a hexadecimal constant.

A numeric constant can be used as an expression or in an expression. RETAIN and ARRAY statements also use numeric constants to define subscript ranges of arrays and to indicate initial values of numeric variables.

There are other ways of writing numeric constants that have to do with special uses of numeric values. Time, datetime, date, and missing constants are described in the next chapter. Any of the different forms of numeric constants can be used in any of the places where a numeric constant is expected.

Character Constants

A standard character constant is nothing more than a quoted string. The characters of the quoted string are the characters of the character value. These are examples of the use of character constants in SAS statements:

```
IF NAME = '' THEN NAME = 'Anonymous';
TITLE1 "Professional SAS Programming Logic";
FILENAME IN "/data/in/latest.txt";
PUT 'Beginning processing.';
```

Text values that use characters outside of the standard character set and values that are binary data rather than text cannot always be written as literal values. For these kinds of constant values, there are character hexadecimal constants. In a character hexadecimal constant, each byte, or character, of the value is written as two hexadecimal digits. The number of hexadecimal digits you write in a character hexadecimal constant is twice the length of the value. To write a character hexadecimal constant, write an even number of hexadecimal digits, quoted, followed by the letter X, for example, `'48656C6C6F'X`. For clarity, you can write commas between bytes in a character hexadecimal constant, such as `'4040,60606060'X`.

Character constants are used in expressions in data step programming and in various other ways throughout the SAS environment. These are examples of uses of character constants in SAS statements:

- the value of various system and statement options
- an initial value of a character variable
- a title or footnote line in a TITLE or FOOTNOTE statement
- the label of a variable, SAS file, or entry
- text to be written in a PUT statement
- text to be displayed in a WINDOW statement
- the physical name of a file
- a command to execute in the X statement

Expressions

In a data step, when it is necessary to compute a value, the computation process is expressed as a combination of terms known as an *expression*. An expression is a combination of values and actions for the computation of a value. Expressions are expected at specific places in the syntax of statements, and they are executed in a predictable way, following a simple set of rules. Expressions are used most often in data step statements. Expressions formed with slightly different rules are used as the conditions in WHERE clauses, as in the WHERE= dataset option and WHERE statement.

SAS expressions are much like the expressions of most programming languages. The syntax of computer expressions is based on the notation of mathematical expressions, but it is adapted to the different needs of programming and the limitations of plain text and the standard character set.

Elements of Expressions

An expression is built out of values and actions. A value in an expression is a variable, a constant, or an expression. An expression used as a value in a larger expression is a *subexpression*. In some cases, it is necessary to enclose a subexpression in parentheses to ensure that it is treated as a unit. Constants and variables are written the same way in expressions as they are in other SAS statements.

Actions in expressions result from operators and functions. An operator is a word or a symbol that represents a simple action. A function is a routine that might take any kind of action; it is written as a name, followed by a list of arguments in parentheses. Each argument to a function is itself an expression. If there is more than one argument, the arguments are separated by commas.

Operators

An operator is a symbol or word that represents a simple action in an expression. The values that an operator acts on are called *operands*. A *unary* operator has only one operand, which is written after the operator:

operator operand

A binary operator has two operands, which are written before and after the operator:

left operand operator right operand

Most binary operators are *commutative*, which means that the order of the operands does not matter. Switching the positions of the two operands would result in the same value.

Lexicon

SAS Institute writing uses the term *prefix operator* for the kind of unary operator described here and *infix operator* for a binary operator.

Concatenation

The concatenation operator uses character operands to produce a character value as a result. The operator, written as ||, produces a value that combines the two operands. The resulting value contains all the characters of the first operand, then all the characters of the second operand. Its length is the sum of the lengths of the operands.

Expression	Resulting Value
'T' || 'O'	'TO'
'100' || 'GA' || '4'	'100GA4'
'Alan ' || 'Weeks '	'Alan Weeks '
'return ' || 's'	'return s'

The value that results from concatenation includes the trailing spaces of the two values that are concatenated. Any trailing spaces in the left operand end up in the middle of the concatenated value. To concatenate without those spaces, use the TRIM or TRIMN function to remove them:

Expression	Resulting Value
`TRIM('Alan      ') || ' ' || 'Weeks      '`	`'Alan Weeks      '`
`TRIM('return    ') || 's'`	`'returns'`

Selection Operators

The value that results from the binary operators MIN and MAX is always one of their two operands. These operators can be used with either character or numeric operands. When both operands are character values, the operator results in a character value. With numeric operands, the operator results in a numeric value.

The MIN operator selects the lesser of its two operands. The MAX operator selects the greater of its two operands. In executable statements in the data step, these operators can also be written as symbols, MIN as >< and MAX as <>. The following examples demonstrate the use of these operators. In practice, the values used with operators would often be variables or expressions based on variables. However, the examples shown here use constant values as operands so that you can easily see the effects of the operands.

Expression (keyword form)	Expression (symbol form)	Resulting Value
'AA' MAX 'A'	'AA' <> 'A'	'AA'
'A' MIN 'Z'	'A' >< 'Z'	'A'
.1 MIN .01 MIN .001	.1 >< .01 >< .001	.001
1 MAX -100	1 <> -100	1
-5 MIN -37	-5 >< -37	-37
(-10000) MAX 0	(-10000) <> 0	0

Arithmetic Operators

Arithmetic operators calculate numbers by doing the familiar operations of arithmetic. The SAS language has seven arithmetic operators:

Operator	Name	Operation
a + b	plus	addition
a - b	minus	subtraction
*a***b*	times	multiplication
a/b	divided by	division
*a****b*	raised to the power	exponentiation
-a	negative	negation
+b	positive	identity

These are examples of uses of the arithmetic operators:

Expression	Resulting Value	Expression	Resulting Value
1 + 1	2	10**5	100000
4 - 9	-5	4**0.5	2
1000*-.1	-100	--10	10
24/3	8	+23	23

The positive operator — a plus sign written before a value — has no effect on a numeric operand, but it is included in the language for the sake of completeness. The negative operator is the same symbol as the negative sign that may be part of a constant value, but it does not have exactly the same effect. In an expression, the SAS supervisor recognizes the symbol as an operator rather than as part of a constant value.

This makes a difference if you use a negative constant as the left operand of the exponentiation, MAX, or MIN operator. You must enclose the constant value in parentheses. For example, in the expression

```
-2**EXP
```

it might look as if the constant value –2 is the base of the exponentiation operator, but it does not parse that way. The SAS supervisor understands the symbols -2 as two separate terms, the negation operator and the constant value 2. The expression, then, is evaluated as

```
-(2**EXP)
```

Enclose the constant value –2 in parentheses to make that the base:

```
(-2)**EXP
```

Similarly, parentheses may be necessary around the operands of the MAX and MIN operators. The expression

```
-2 MAX -1
```

results in the value –2, not because the operator does not work correctly, but because the initial minus sign is not part of its left operand. The expression determines the maximum of 2 and –1, then applies the negation operator to the result. Whenever an operand of the MAX or MIN operator is anything more than a single term, enclose it in parentheses. With parentheses, you can use values like –2 and –1 as operands:

```
(-2) MAX (-1)
```

Comparison Operators

Whereas the selection operators compared two values in order to select the lowest or highest one, the comparison operators compare two values just to see how they compare. The value that results from a comparison operator is a Boolean value — a value of 1 for true or a value of 0 for false.

A comparison operator can compare two numeric values to each other or two character values to each other. Either way, the result is a Boolean value. When character values are compared, the comparison is based on the binary value of each byte in the character value. The first characters are compared to each other, and if they are the same, the second characters are compared, and so on, until a difference is found.

The way two character values compare to each depends on the character set. In this context, a character set may be called a *collating sequence* — "collate" is a verb that refers to the action of putting things in order. A comparison of the same character values may produce different results in a computer that uses the EBCDIC character set than in a computer that uses the ASCII character set. When ASCII characters are compared, digits are lower than letters, and capital letters are lower than lowercase letters. With EBCDIC characters, it is the reverse; letters are lower than digits, and lowercase letters are lower than capital letters.

When two character values are not the same length, the shorter value is treated as if it is extended by adding spaces to the end until it is the length of the longer value. This means that trailing spaces do not affect the way two values compare.

The classic set of comparison operators test whether two values are equal or one is greater than the other:

Operator	Name
<	is less than
<=	is less than or equal to
=	equals
NE	is not equal to
>=	is greater than or equal to
>	is greater than

There are alternate forms of each of these operators:

Operator	Alternate forms
<	LT NOT>= NOT GE
<=	LE NOT> NOT GT
=	EQ
NE	NOT= NOT EQ
>=	GE NOT< NOT LT
>	GT NOT<= NOT LE

In addition to the forms listed in the preceding table (and for several other operators mentioned in the discussion that follows), operators can be written differently by replacing the word NOT with the caret symbol (^). The caret is not part of the EBCDIC character set, so the logical not symbol (¬) is permitted as the equivalent of NOT on EBCDIC computers. And because some computers, at least at one time, did not support either the caret or logical not, you may also be able to use the tilde (~) as a substitute. However, I do not recommend any of this. The caret, logical not, and tilde symbols are not as portable as the word NOT. If you use them in a program, you may someday have to change them to run the program on a different kind of computer. By using only the portable forms of the operators, you can ensure that you will not have that extra work to do.

Comparison operators can be strung together. That is, you can use the same value as an operand for two comparison operators at once. Most often, this is done to make sure a value falls between two other values. This kind of comparison is made more compact by stringing the comparison operators together. The resulting comparison expression is true only if all the comparisons in it are true. For example, the comparison

```
0 < PROB < 1
```

indicates the two separate comparisons

```
0 < PROB
PROB < 1
```

Using the logical AND operator, the comparison expression above would be equivalent to

```
0 < PROB AND PROB < 1
```

The IN operator lets you compare a value against a list of constants. The operator results in a true value if the value compares equal to any of the constants in the list.

value IN (*constant*, *constant*, . . .)

Using the IN operator is easier than making a series of equality comparisons, which would be equivalent. For example, this expression

```
STATE IN ('NY', 'NJ', 'PA', 'MD')
```

would be equivalent to the expression below, written using the logical OR operator.

```
STATE = 'NY' OR STATE = 'NJ' OR STATE = 'PA' OR STATE = 'MD'
```

There is also a negative form of the IN operator. The NOTIN operator returns a true value if the value does not compare equal to any of the constants in the list.

The following examples show the effects of these comparison operators.

Expression	Resulting Value
1 = 1	1
2 + 2 = 4	1
-1 = 1	0
'X' = 'X '	1
1 NE 1	0
2 NE 3	1
'A' NE 'a'	1
8 > 0	1
8 >= 1.6	1
8 >= 8	1

Expression	Resulting Value
8 < -8	0
'OO' < 'OO'	0
'b' < 'bee'	1
0 <= 0	1
'p' <= 'pea'	1
5 IN (1, 2, 3, 4, 5)	1
2.5 IN (1, 2, 3, 4, 5)	0
'M' IN ('MA', 'ME', 'MI', 'MS')	0
-1 NOTIN (1, 2, 3, 4, 5)	1

These eight comparison operators can be modified with a colon (:) to change the way character comparisons are done when the two character values have different lengths. Ordinarily, the shorter operand is treated as if it is extended with spaces to the length of the longer operand. With a colon written after the operator, however, the longer operand is truncated to the length of the shorter operand. This can allow you to make comparisons that do not consider any extra characters that one operand may have. Trailing spaces may be significant in these comparisons, because the trailing spaces in the shorter operand affect the way the longer operand is truncated. The following examples demonstrate the use of comparison operators with the colon modifier.

Expression	Resulting Value
'X' =: 'X '	1
'A' NE: 'Aaa'	0
'A ' NE: 'Aaa'	1
'OO' <: 'OO'	0
'b' <: 'bee'	0

Expression	Resulting Value
'p' <=: 'pea'	1
'Box' IN: ('A', 'B', 'C')	1
'M' IN: ('MA', 'ME', 'MI', 'MS')	1
'AX' NOTIN: ('A', 'AT', 'ASK')	0

Additional Comparison Operators for WHERE Clauses

Five additional comparison operators, which cannot be used in data step expressions, are available for use in WHERE expressions. Most of these operators are also binary operators, but one is a unary operator in which the operand is on the left, and another is the SAS language's only *ternary* operator, which uses two keywords because it has three operands.

These operators are more complicated than the classic comparison operators and require specific explanations.

Range test The BETWEEN-AND operator tests to see if a value falls into a range. The second operand is written between the two keywords BETWEEN and AND. If the first operand falls into the range between the second and third operands, the result is a true value.

Test for missing The IS MISSING or IS NULL operator tests whether its operand, which is written to its left, is a missing value. If the operand is a character value, the operand is the same as comparing to a blank value.

Substring The CONTAINS operator tests to see if one character operand is a substring of the other. If the right operand is contained within the left operand, the result is a true value. This operator can also be written as a question mark.

Wild-card The LIKE operator is similar to the = operator for comparing two character values, except that it allows wild-card characters. In the right operand, the character _ matches any character that might appear at that position in the left operand, and ? matches any sequence of characters.

Sounds-like The =* (or EQ*) operator uses a variation of the Soundex algorithm to categorize words according to the approximate way they sound. Generally, values are considered to sound alike if they have the same first letter and similar-sounding consonants. Usually, this operator is used to find a name in a SAS dataset when the exact spelling of the name is not known.

Negative forms of these operators are formed by adding the keyword NOT. The following table summarizes the additional comparison operators for WHERE clauses, showing the operators and their negative forms.

Description	Operator	Negative Form
range test	*a* BETWEEN *b* AND *c*	*a* NOT BETWEEN *b* AND *c*
test for missing	*a* IS MISSING *a* IS NULL	*a* IS NOT MISSING *a* IS NOT NULL *a* NOT IS MISSING *a* NOT IS NULL
substring	*a* CONTAINS *b* *a* ? *b*	*a* NOT CONTAINS *b* *a* NOT ? *b*
wild-card	*a* LIKE *b*	*a* NOT LIKE *b*
sounds-like	*a* =* *b* *a* EQ* *b*	*a* NOT =* *b* *a* NOT EQ* *b*

Logical Operators

There are three logical operators: AND, OR, and NOT. These operators are taken from symbolic logic. They take logical values as operands and produce Boolean values as results. The AND operator results in a true value only if both of its operands are true. The OR operator results in a true value if either of its operands is true. It results in a false value only if both operands are false. The NOT operator is a unary operator that results in the logical opposite: a true value if its operand is false or a false value if its operand is true.

It is permissible to write logical operators as symbols, but I do not recommend it, mainly because the keywords are easier to recognize. There are also particular problems with the symbols used for the AND and NOT operators. The ampersand is so important in macro language that it can cause confusion when used as the AND operator. The symbol used for the NOT operator is not portable between computers with different character sets. The following table summarizes the logical operators.

Description	Operator	Symbol (not recommended)
logical and	*a* AND *b*	*a* & *b*
logical or	*a* OR *b*	*a* | *b*
logical not	NOT *a*	^ *a* ¬ *a* ~ *a*

The following table lists all the different logical operands for logical operators and shows the resulting values. The operands may not necessarily be 0 and 1. Logical operators treat any true value, that is, any positive or negative number, the same as a 1.

Expression	Resulting Value	Expression	Resulting Value
0 AND 0	0	0 OR 1	1
0 AND 1	0	1 OR 0	1
1 AND 0	0	1 OR 1	1
1 AND 1	1	NOT 0	1
0 OR 0	0	NOT 1	0

Most of the time, the operands for logical operators are subexpressions formed with comparison operations. Comparison operators produce the logical values that the logical operators can work with. The following examples show more typical uses of logical operators.

```
UPCASE(COUNTRY) =: 'CAN' OR STATE = 'AK'
AGE >= 21 AND LICENSED
YEAR = 2001 AND MONTH IN (4, 5, 6, 7, 8, 9)
NOT (MONTH = 2 AND DAY = 29)
TYPE = 'CAR' AND (MODELYR >= 1997 OR (MODELYR >=1995 AND PRICE <= 3350))
```

Parentheses are almost always required for the operand of the NOT operator, and they are often needed when an expression contains both AND and OR operators.

Adding Conditions to a WHERE Expression

The SAME AND operator is designed to add conditions to a WHERE expression. This operator is useful only in situations where there is an existing WHERE expression, which may occur, for example, in WHERE statements used in the run-group processing of some procs. Start the WHERE expression with SAME AND, followed by an additional condition:

```
WHERE SAME AND condition;
```

The SAME AND operator is always evaluated last, after any other operators that appear in the same expression.

Parentheses and Priority of Operators

When an expression contains more than one operator, you can use parentheses to indicate the order of evaluation of operators. The subexpression enclosed in parentheses serves as the operand of the adjacent operator. It is completely evaluated before any operator adjacent to the parentheses is evaluated.

When there is an expression with more than operator and no parentheses to indicate priority, the operators are evaluated according to a set of rules of priority. Operators are categorized into priority classes. When several operators are in the same class, they are evaluated either from left to right or from right to left, according to the rule that applies for that class. When operators are in different classes, the operator with the highest priority (indicated by the lowest class number) is evaluated first. The priority rules are expressed in the following table.

Operators	Priority	Order
Prefix (NOT, +, -), MIN, MAX, **	1 (first)	right to left
*, /	2	left to right
+, - (addition, subtraction)	3	left to right
||	4	left to right
Comparison	5	left to right
AND	6	left to right
OR	7 (last)	left to right

The following table shows examples of the ways the rules of priority are applied.

Expression	Evaluated As	Resulting Value
3*3 + 4*4	(3*3) + (4*4)	25
-2*10**6 + 5	(-2*(10**6)) + 5	-1999995
NOT 4 > 5	(NOT 4) > 5	0
2 >= 2 OR 'C' >= 'E'	(2 >= 2) OR ('C' >= 'E')	1
3 MIN 5 + 2 = 7	((3 MIN 5) + 2) = 7	0
17 MIN 5 MAX 25	17 MIN (5 MAX 25)	17
0 < 1 <= 8 OR 8 > 8	(0 < 1 <= 8) OR (8 > 8)	1

Function Calls

A function call is an expression, and it can be used as part of a larger expression. Each argument to a function — each of the values that the function works with — is also an expression (with a few exceptions). Like operators, functions take actions to create values. But where operators are part of the syntax of the programming language, functions are separate programs, routines with a looser connection to the programming language. This gives functions more flexibility in the way they work. They use various number of arguments and do a wider range of things.

In the program, the notation for a function call is very different from that of an operator. The function name comes first, and then, in parentheses, the various arguments of the function:

```
function(argument, argument, . . . )
```

The order of arguments is almost always important with functions. Commas separate the various expressions used as arguments. Despite the different appearance, the actions of a function could be much like those of an operator. This is most apparent when you consider the SUM function. The expression `SUM(1, 2)`, based on the SUM function, is essentially the same as the expression 1 + 2, based on the addition operator.

Because a function is a separate routine, it is said to *return* the value that results from the actions it takes. The data type of a function is the data type of its return value. Each of a function's arguments is also expected to be of a certain type. Each function has its own rules about the type and number of arguments and the allowable values of the arguments.

The following examples demonstrate some of the most useful functions for calculating numeric values.

Function Call	Return Value	Description
INT(5.1)	5	integer part of value
FLOOR(5.1)	5	round down to integer
CEIL(5.1)	6	round up to integer
ROUND(14.431, .1)	14.4	round to nearest multiple
MOD(51, 7)	2	modulo
ABS(-14)	14	absolute value
SQRT(729)	27	square root
MIN(6, 0, 14, 18)	0	minimum
MAX(6, 0, 14, 18)	18	maximum
MEAN(6, 0, 14, 18)	9.5	mean
SUM(6, 0, 14, 18)	38	sum
LOG10(1000)	3	common logarithm

These are examples of functions that process character strings.

Function Call	Return Value	Description
UPCASE('Cover')	'COVER'	uppercase
LOWCASE('Cover')	'cover'	lowercase
REVERSE('abcde')	'edcba'	reverses string
LEFT(' S ')	'S '	left-aligns
RIGHT(' S ')	' S'	right-aligns
COMPRESS(' t h e r e ')	'there'	removes spaces
COMPRESS('-T*I*M*E-', '-*')	'TIME'	removes characters
TRANSLATE('short', 'akp', 'oth')	'spark'	replaces characters
QUOTE('abcde')	'"abcde"'	produces quoted string

These examples show constant values as arguments to functions in order to demonstrate the effects of the functions. In practice, most function arguments are variables or expressions based on variables.

Variable Lists As Function Arguments

Ordinarily, the arguments to functions are expressions and are separated by commas. Sometimes, however, it might look better to write the arguments as a list of variables. To do that, write the keyword OF at the beginning of the arguments. Then write a variable list. As in any variable list, the variable names are separated by spaces.

```
function(OF variable list )
```

For example, the function call

```
MEAN(OF LOW HIGH)
```

is equivalent to

```
MEAN(LOW, HIGH)
```

Variable lists are especially useful when you are writing the arguments to statistic functions, such as MEAN, SUM, and MAX. These functions work with any number of arguments, often a large number, and the order of arguments does not matter. If the variables belong to a set of variables that you named with a sequence of numeric suffixes, you may be able to abbreviate the variable list, for example:

```
SUM(OF PRECIP1-PRECIP31)
```

In this case, the variable-list notation is better than writing

```
SUM(PRECIP1, PRECIP2, PRECIP3, PRECIP4, PRECIP5, PRECIP6, PRECIP7,
PRECIP8, PRECIP9, PRECIP10, PRECIP11, PRECIP12, PRECIP13, PRECIP14,
PRECIP15, PRECIP16, PRECIP17, PRECIP18, PRECIP19, PRECIP20, PRECIP21,
PRECIP22, PRECIP23, PRECIP24, PRECIP25, PRECIP26, PRECIP27, PRECIP28,
PRECIP29, PRECIP30, PRECIP31)
```

Other forms of variable lists you can use with the OF keyword include the prefix form, which selects all variables whose names begin with the same characters, and the array form, in which the special subscript * indicates the variable list that was used to define the array:

```
function(OF prefix:)
function(OF array{*})
```

7

Special Kinds of Values

SAS can get by with just two data types because the two data types are as versatile as they are. The numeric data type includes the ability to represent various indicators and measurements that are not simply numbers. Values of either data type can be treated as binary values for bit testing.

Missing Values

There are many different ways a value can be unknown, unavailable, or meaningless, but the process of creating computer data forces you to supply a value for each variable, in each observation that you create. In some other software, that requires making up values that do not mean anything just to fill in the necessary value for a variable. But SAS software has the ability to mark a numeric variable as not having a value. The value you assign to a variable to make this happen is a *missing value*. A missing value is a special value that says, in effect, "There is no value."

Support for missing values is found throughout SAS software. Most numeric informats can read a blank field or one that contains only a period as a missing value. Formats that write visual numerals write missing values. They write a missing value as a period or as another character that you select with the MISSING= system option. For example, you can change the option to `MISSING=' '` to leave fields blank when a value is missing.

Statistical procs process missing values in a statistically appropriate way. Some procs allow you to include or exclude missing values in processing as determined by an option on the PROC statement.

A missing constant is written simply as a period. For example, this statement assigns a missing value to a variable:

```
RESPONSE = .;
```

Missing Values in Expressions

You can compare missing values to the values of numeric variables using the comparison operators. In comparisons, missing values compare less than numbers. Because of this, the MIN operator results in a missing value if a missing value is one of its operands.

To logical operators, and in any situation in data step programming where a missing value is used as a condition, missing values are treated as false values, the same as a 0 value.

Arithmetic operators cannot do anything with a missing value as an operand. When a missing value is used with an arithmetic operator, the result is a missing value, and a log note states that missing values were generated as a result of operations on missing values at a specific point in the program.

Most functions also cannot use missing values as arguments. If an argument is a missing value, the function returns a missing value, and it may generate an error condition. However, every function has different rules, and there are some functions that work well with missing values. In particular, the statistic functions, such as the MIN, MAX, SUM, and MEAN functions, simply ignore any missing arguments. They return the same value as if the missing arguments were not present. Some other functions expect missing arguments; a missing argument may have a specific meaning for a function.

Generating Missing Values

There are numerous situations in which SAS software generates missing values as part of its regular processing. At the same time, it generates blank values for character variables. In this context, a blank value may be referred to as a missing value.

SAS software generates missing values when:

- Missing values are used as operands for an arithmetic operator.
- A zero value is the right operand of the division operator.
- The exponentiation operator has a negative number as the left operand, and the right operand is less than 1, but not 0.
- Both operands for the exponentiation operator are 0.
- Operations on very large or very small numbers result in a number that is outside the range that floating-point values can represent.
- The arguments to a function are not valid.
- A numeric informat reads a blank field or a field that contains only a period.
- An informat reads a field that is not valid text for that informat.
- An INPUT statement does not find values for all of its variables.
- An automatic type conversion from character to numeric fails because the character value does not represent a numeral.
- A variable is created in an option of an I/O statement, but the value that the variable is supposed to represent is not available because of the type of file or the engine being used. For example, the variable created by the NOBS= option of the SET or MODIFY statement ordinarily indicates the number of input observations, but it is a missing value if the engine that reads one of the input SAS datasets cannot determine the number of observations in the SAS dataset.
- At the beginning of data step execution, all regular variables are initialized to missing values unless other initial values are indicated in a RETAIN statement, except that variables that are used as the target of a sum statement are initialized to 0.

- At the beginning of each repetition of the data step loop, certain variables are reset to missing. Generally, these are regular variables that are created in the data step, are not named in a RETAIN statement, and are not the target of a sum statement.
- In a statement that reads observations from more than one SAS dataset, in which different SAS datasets have some different variables, variables that are not available for a particular observation are set to missing. This happens when the variables come from a SAS dataset that does not supply any part of the observation being read or assembled.
- Many procs that create output SAS datasets create missing values in those SAS datasets.

Special Missing Values

There might be various different reasons why a value is unavailable. In case it is necessary to distinguish among different reasons, SAS allows you to use several *special missing values* for numeric variables in addition to the standard missing value that has been discussed so far. The 27 different special missing values correspond to the 26 letters of the alphabet and the underscore character.

Special missing values are usually assigned to variables in data step statements, perhaps in an assignment statement, assigning the special missing value as a constant. Write a special missing value constant as a period followed by a letter or underscore.

Numeric informats can find special missing values in input data. In an input field, a special missing value must be written as just a letter or underscore. In order for informats to be able to interpret a text character as a special missing value, the specific character must be declared in a MISSING statement.

The MISSING statement is a global statement that lists characters for informats to interpret as special missing values. Both capital and lowercase letters can be read as missing values, but the exact character must be listed in the MISSING statement; this is one SAS statement in which uppercase and lowercase letters are not interchangeable. For example, if you want to be able to read both capital N and lowercase n as special missing values, list them both in the MISSING statement:

```
MISSING N n;
```

This allows you to read input data such as the following, interpreting any field that contains a capital N or lowercase n as the special missing value .N:

```
4.0    N     11.2   5.8
n      1.0    9.9   N
```

Without the MISSING statement, informats would not be able to be read the letters in the fields, and they would create an error condition because of the invalid data.

Numeric formats write special missing values as single characters, either a capital letter or an underscore. This ensures that you can see the different special missing values in any print output that shows the data values. You might include footnotes in a report you create to indicate the meaning of any special missing value characters that appear in the report.

In expressions, special missing values act the same as standard missing values. However, when different missing values are compared to each other, they are not equal. Missing values compare in the following order, from lowest to highest:

```
._ . .A .B .C .D .E .F .G .H .I .J .K .L .M .N .O .P .Q .R .S .T .U .V .W .X .Y .Z
```

All missing values compare less than all numbers. This may be important to remember when you want to use a comparison expression to identify negative numbers. If there are no missing values, a comparison such as the following would work.

```
VALUE < 0
```

However, that expression is also true when the variable has a missing value. If you do not know what missing values might be present, use a comparison like the following to select negative values.

```
.Z < VALUE < 0
```

If you know that no special missing values are being used, you could also write:

```
. < VALUE < 0
```

Values That Measure Time

Measurements of time are not usually reported as single numbers. They may be counted in years and months, or minutes and seconds, or some other combination of units. However, time can be measured in a single number. Time, as we measure it, is a linear phenomenon, and that means it can be counted in numbers just by picking a point of reference and a unit of measure. SAS includes support for time measured with two units of measure in particular: seconds and days.

Definitions and Constants

SAS defines SAS time values, SAS date values, and SAS datetime values as its standard way of measuring times of day and dates.

A SAS time value is the time of day measured in seconds, starting with 0 at midnight. A SAS datetime value is the date and time measured in seconds, starting with 0 at the beginning of 1960. A SAS date value is the date counted in days, also starting with 0 at the beginning of 1960. SAS informats, functions, and formats support the use of these values.

For example, the DATE informat can read a date written as 27AUG2002 and interpret that text as a SAS date value. The DATE format can write the resulting SAS date value as the same text, or you can use other formats, such as the YYMMDD format, can write it in other conventional ways. Various functions allow you to extract the year, month, and day from the SAS date value and to work with it in other ways.

SAS time, datetime, and date values can be written as constants. These constants are quoted strings immediately followed by the letters T, DT, and D, respectively. These are examples of these special kinds of constants:

```
'07:00'T
'31DEC1999 12:00'DT
'27AUG2002'D
```

SAS time values use the 24-hour clock, with hours starting at 0 and going through 23. Write a SAS time constant as hours, hours and minutes, or hours, minutes, and seconds. Use colons to separate the numbers, as shown in the example above. You can have fractional seconds as part of the value. Enclose the entire value in quotes, and write the letter T immediately after the closing quote.

```
'hh'T
'hh:mm'T
'hh:mm:ss'T
'hh:mm:ss.ssss'T
```

A SAS time constant can also be written with a 12-hour clock, using the letters AM or PM at the end of the string to indicate the day half.

To write a SAS date constant, write the day of the month, the three-letter abbreviation for the month, and the year, quoted, followed by the letter D.

```
'ddMONyyyy'D
```

The SAS datetime constant combines the notation of the SAS date constant and the SAS time constant. Write a space or colon between the date and the time of day.

```
'ddMONyyyy hh:mm:ss'DT
```

Dates in SAS date and datetime values belong to the Gregorian calendar. You can have dates as far back as 1582, when Pope Gregory XIII established the calendar. Earlier years, if you tried to use them, would create a data error. SAS date and datetime values in the years from 1582 to 1959 are negative numbers. Dates can go very far into the future. However, formats can write four-digit years only through 9999, and functions handle date calculations accurately only through 19900.

Two-Digit Years

If you want to, you can use two-digit years with SAS. Informats and functions and SAS date and datetime constants interpret two-digit year numbers according to the setting of the YEARCUTOFF= system option. The option indicates the first year of the 100-year span that can be represented with two-digit year numbers. For example, with `YEARCUTOFF=1925`, two-digit years belong to the century from 1925 to 2024; when you write the year 25, it means 1925, and when you write the year 24, it means 2024. Date formats write any year as a two-digit year, simply writing the last two digits of the year.

Selected Informats and Formats for Date Fields

Form[†]	Example	Informat	Format
mm/dd/yyyy	04/15/2000	MMDDYY10.	MMDDYY10.
mm/dd/yy	04/15/00	MMDDYY8.	MMDDYY8.
mm/dd	04/15		MMDDYY5.
ddMONyyyy	27AUG2002	DATE9.	DATE9.
ddMONyy	27AUG02	DATE7.	DATE7.
ddMON	27AUG		DATE5.
dd/mm/yyyy	15/04/2000	DDMMYY10.	DDMMYY10.
yyyy-mm-dd	1998-10-14	YYMMDD10.	YYMMDD10.
yy.mm.dd	98.10.14	YYMMDD8.	YYMMDDP8.
yy.mm	98.10		YYMMP5.
MONyyyy	JUN1999	MONYY7.	MONYY7.
MONyy	JUN99	MONYY5.	MONYY5.
yyQq	99Q4	YYQ4.	YYQ4.
Month dd, yyyy	September 7, 2004		WORDDATE18.
Day	Thursday		DOWNAME9.
Month	July		MONNAME9.
yyyy	1999		YEAR4.
yy	99		YEAR2.

[†] *yy*=two-digit year; *yyyy*=four-digit year; *mm*=month number; *MON*=three-letter abbreviation for month; *dd*=day of month

Selected Informats and Formats for Time Fields

Form[†]	Example	Informat	Format
PM	PM		TIMEAMPM2.
hh	16		HHMM2. TIME2. HOUR2.
hh PM	11 PM		TIMEAMPM5.
hhmm	1645		HHMM4.
hh:mm	16:45	TIME5.	TIME5. HHMM5.
hh:mm PM	11:00 PM		TIMEAMPM8.
hh:mm:ss	16:45:00	TIME8.	TIME8.
hh:mm:ss PM	11:00:00 PM		TIMEAMPM11.
hh:mm:ss.sss	16:45:00.000	TIME12.	TIME12.3

[†] *hh*=hour; *mm*=minute; *ss*, *ss.s...*=second; *PM*=AM or PM

Routines

SAS software has a large set of routines for working with time measurements. In addition to the SAS date, time, and datetime values, these routines work with the more familiar measures of time: hour, minute, second, day of the week, day of the month, month, year, and quarter. Routines that work with SAS time values also work with measurements of elapsed time. SAS routines count days of the week starting with Sunday as day 1 and going through Saturday as day 7.

The two tables on the previous page show some of the ways that dates and times of day may be written. The first table shows informats and formats for reading and writing SAS date values. The second table shows an informat and formats for reading and writing SAS time values.

SAS datetime values are usually read and written with the DATETIME informat and format. The DATETIME informat reads the same kind of values that you can write in a SAS datetime constant. The DATETIME format writes values in the same way, and you can use it with a width as short as 7 to write just the date of the value. You can also use the TOD format to write just the time of day of a SAS datetime value. The TOD format works the same way as the TIME format, but it works with SAS datetime values instead of SAS time values.

A set of functions is available to translate between different ways of measuring time. There are also functions that return the current date and time. The following table shows the arguments and values returned by these functions.

Function Call	Return Value
YEAR(*SAS date value*)	Year
QTR(*SAS date value*)	Quarter
MONTH(*SAS date value*)	Month
DAY(*SAS date value*)	Day of month
WEEKDAY(*SAS date value*)	Day of week (number)
DATEPART(*SAS datetime value*)	SAS date value
HOUR(*SAS datetime value or SAS time value*)	Hour
MINUTE(*SAS datetime value or SAS time value*)	Minute
SECOND(*SAS datetime value or SAS time value*)	Second
HMS(*hour, minute, second*)	SAS time value
DHMS(*SAS date value, hour, minute, second*)	SAS datetime value
DATE() TODAY()	The current date as a SAS date value
TIME()	The current time as a SAS time value
DATETIME()	The current date and time as a SAS datetime value

Duration

By using seconds as their unit of measurement, SAS time and datetime values let you measure the time of day or the date and time of day as a single number. For the same reason, it makes sense to use seconds as the unit of measurement when you are measuring the duration of time, or how long something lasts.

Duration is measured, for example, when you show the results of a race. Times are often written in hours, minutes, and seconds, or for an event that is under an hour, in minutes and seconds.

SAS time constants are not limited to values that represent the time of day. They can be used for any time measurement in hours, minutes, and seconds — or, for that matter, for anything that might be measured in 3,600s, 60s, and units.

The TIME informat and format work with duration values the same way that they work with values that represent the time of day. If an event might take more than 99 hours, use the TIME informat or format with a larger width to allow for more digits in the hour part of the value. For shorter lengths of time, use the MMSS format to write time in minutes and seconds.

For longer durations, you might want to write the length of time in days, hours, minutes, and seconds. As long as the unit of measurement is seconds, you can use the DATEPART function to extract the numbers of days and the TIMEPART function to extract the number of seconds that make up the fractional part of a day. Then, write the number of days using the standard numeric format and the fractional part of a day using the TIME format.

Time Comparisons and Arithmetic

You do not have to do anything special to use SAS software's special time measurements in computations. You can compare them using any comparison operator. They use days and seconds as their units of measure, so you can use them in arithmetic that is based on those units.

You can compare a SAS date value to another SAS date value. In such a comparison, the lesser value is earlier; the greater value is later. This kind of comparison works the same way when you compare a SAS datetime value to another SAS datetime value.

Be careful, however, that you do not accidentally compare a SAS date value to a SAS datetime value, because that comparison would not be valid. If you need to make that comparison, you can multiply the SAS date value by 86400, the number of seconds in a day, to convert it to a SAS datetime value. Then you have two SAS datetime values you can compare.

When you compare two SAS time values, the lesser value is earlier in the day than the greater value. Be careful, though: earlier in the day does not necessarily mean earlier. For example, sometimes `'00:15'T` is actually one half hour later than `'23:45'T`. Do not use a comparison of SAS time values to determine the sequence of a set of events unless you know that the events all fall into the same calendar day.

SAS date values use days as their unit of measure, so you can add or subtract values that represent numbers of days. This is demonstrated in the following program.

```
*
  THISWEEK.SAS
  A demonstration of SAS date arithmetic.
*;
DATA _NULL_;
 TODAY = DATE();
 YESTERDAY = TODAY - 1;
 TOMORROW = TODAY + 1;
 SUNDAY = TODAY - WEEKDAY(TODAY) + 1;
 MONDAY = SUNDAY + 1;
 TUESDAY = SUNDAY + 2;
 WEDNESDAY = SUNDAY + 3;
 THURSDAY = SUNDAY + 4;
 FRIDAY = SUNDAY + 5;
 SATURDAY = SUNDAY + 6;
 PUT 'Today is ' TODAY : WORDDATE.
   / 'Yesterday was ' YESTERDAY : WORDDATE.
   / 'Tomorrow is ' TOMORROW : WORDDATE.
   / 'This week, Sunday is ' SUNDAY : WORDDATE.
   / 'This week, Monday is ' MONDAY : WORDDATE.
   / 'This week, Tuesday is ' TUESDAY : WORDDATE.
   / 'This week, Wednesday is ' WEDNESDAY : WORDDATE.
   / 'This week, Thursday is ' THURSDAY : WORDDATE.
   / 'This week, Friday is ' FRIDAY : WORDDATE.
   / 'This week, Saturday is ' SATURDAY : WORDDATE.
   ;
RUN;
```

You can also do subtraction with SAS date values. When you subtract one SAS date value from another, the result is the length of time in days between the two values.

SAS datetime values work the same way but with seconds as their unit of measure. Add a length of time in seconds to a SAS datetime value to compute a later SAS datetime value. Subtract a length of time in seconds from a SAS datetime value to compute an earlier SAS datetime value.

Subtract one SAS datetime value from another to determine the length of time, in seconds, between the two values. This is the calculation you would do to determine how long a program took to run.

You can also add seconds to or subtract seconds from a SAS time value, but with the caution that the value that results may fall outside the range of SAS time values, which is from 0 to but not including 86400. If the resulting value is 86400 or greater, you have to subtract 86400 from it to make it a SAS time value. Similarly, if the resulting value is negative, you have to add 86400 to it. The easiest way to do this is with the TIMEPART function, which forces a value to be a proper SAS time value.

The INTNX and INTCK functions do arithmetic with time measurements based on all the common units of the calendar and clock, such as hours and months. These and other functions are described in chapter 15.

Bitfields

The whole idea of a data type is to organize binary data into meaningful units that can represent such things as numbers and character strings. However, there may also be times when you want to consider binary data simply as a group of binary digits, or bits, with each bit having an independent meaning. Binary data considered this way is called a *bitfield*. SAS provides a way to test the individual bits of a bitfield, a process called *bit testing*.

Bit testing compares a value to a *bit mask*, a special kind of constant value that you write as a quoted string followed by the letter B. The characters of a bit mask are 0s, 1s, and periods, which represent the different bits of the value. A 0 or 1 in a bit mask represents a bit that must have that value for the comparison to be true. A period is used a placeholder in a bit mask for a bit that is not being tested. The bit mask can contain spaces to make it easier to read. A bit mask can only be used as an operand of the = operator.

This is an example of an expression in which a value is compared against a bit mask.

```
VALUE = '0.......'B
```

If the specific bit in VALUE that is being tested is a 0, the comparison is true. If it is a 1, the comparison is false.

Bit Testing in Character Values

When the SAS supervisor compares a character value to a bit mask, it proceeds from left to right. The first 8 bits of the bit mask are compared against the 8 bits of the first character of the character value, the next 8 bits are compared against the second character, and so on. The comparison stops, with a false result, as soon as a 0 or 1 in the bit mask does not match the correspond bit in the value. If there are no mismatches, the result of the comparison is true.

The process of bit testing can be easier to see if you write out the value in binary form, which is something you can do with the $BINARY format. The following example shows bit testing with log notes that let you see how the value lines up with the bit mask. The results in the example are based on the ASCII character set.

```
STRING = '*+';
RESULT = STRING = '0...1...0...1...'B;
PUT STRING $BINARY. +4 '(value)'
   / '0...1...0...1...' +4 '(bit mask)'
   // RESULT=;
```

```
0010101000101011    (value)
0...1...0...1...    (bit mask)

RESULT=1
```

The result of the comparison is true because all the 0 and 1 bits in the bit mask match the corresponding digits in the value. The result is also true if you run this program with the EBCDIC character set, although a few of the bits in the binary value of the variable STRING are different.

Bit Testing in Numeric Values

Bit testing can also be done with numeric values. It can be especially useful when you need to interpret operating system codes that, although presented as integers, are actually bitfields. SAS software itself creates bitfields; when the SUMMARY proc creates output SAS datasets, it creates a numeric variable _TYPE_, which is a bitfield.

Bit testing for numeric values is a different process. The SAS supervisor starts by taking the integer part of the numeric value and converting it to a 32-bit signed integer format, as if by writing the value with the S370FIB4. format. The first bit is a sign bit, which is 1 for negative numbers, 0 for positive numbers. The remaining 31 bits represent an integer value.

If the numeric value is 2^{31} or greater, it requires more than 32 bits to write it as a signed integer. In that case, the 31 least significant bits of the number are used, along with the sign bit.

The SAS supervisor works from right to left in comparing the bit mask to the 32-bit signed integer form of the value. It compares the rightmost digit in the bitfield to the rightmost digit in the value, and works its way left from there. If all the 0 and 1 digits in the bitfield match the digits in the value, the result of the comparison is a true value. Otherwise, the result is a false value.

The bitfield used to test bits in a numeric value should not be longer than 32 bits; if it is, only its last 32 bits are used. If the bitfield is shorter than 32 bits, it is compared against the least significant bits of the value.

As mentioned, the sign bit is the first bit of the 32-bit signed integer form of the value. The following example tests the sign bit to determine whether the value is positive or negative.

```
IF VALUE = '0....... ........ ........ ........'B THEN
     PUT 'The value is positive or 0.';
ELSE PUT 'The value is negative.';
```

Double Byte Character Set Data

Many languages of east Asia, including Japanese, Korean, and Chinese, are written with thousands of characters. One byte can contain only 256 distinct values, so it takes two bytes to represent a character in these languages. A character set that may use two bytes for a character is called a *double byte character set*, or *DBCS*.

There are several languages that use DBCS text, and a language can have more than one writing system. In addition, there are several character sets, or encoding systems, for each language, there may be various fonts available, and there can be other variations in DBCS data. The options that are available for DBCS data in the SAS System are different in each different environment. System options control the specific way DBCS data can be used. The DBCS option enables the SAS System to recognize and display DBCS data. The DBCSLANG= option selects the language, and the DBCSTYPE= option specifies the encoding system for that language. The values available for the DBCSLANG= and DBCSTYPE= system options vary by operating system and are likely to change from one SAS release to the next.

A variable that holds DBCS data is a regular character variable. As with any variable, its length is measured in bytes. However, with DBCS data, the number of bytes is not the same as the number of characters. The length required for a variable may be twice the number of characters or slightly more than that, depending on the encoding system.

DBCS values are most important as values of variables. They are also used as labels for variables and for SAS datasets. In a SAS program, DBCS values can appear as character constants. DBCS constants can be used in data step expressions and WHERE expressions. They can be used for title and footnote lines, the definitions of formats and informats in the FORMAT proc, and certain system and statement options.

DBCS values can be assigned to a variable in an assignment statement. In some encoding systems, they can be compared successfully using the comparison operators or sorted with the SORT proc or other procs. They generally work well with the more transparent informats and formats, such as the $CHAR informat and format. However, they cannot necessarily be used with other SAS language elements that work with character values. To make it possible to do routine processing of DBCS values correctly, there are informats, formats, and functions specifically designed to work with DBCS data. The $KANJI and $KANJIX informats and formats add and remove the shift codes that some DBCS encoding systems use and others do not use. Various functions (whose names begin with K) allow you to process, measure, and update DBCS values.

It is sometimes necessary to convert values from one character set to another within the same DBCS language, especially when moving data from one computer to another. The SAS System does that with conversion tables stored in DBCSTAB entries, which you can create with the DBCSTAB proc. The INTYPE= and OUTTYPE= options of the `PROC CPORT` statement apply a conversion while creating a transport file.

The SAS System writes text values from left to right, and that is also true in its display of DBCS values, even though many DBCS languages are conventionally written from right to left. Use the KREVERSE function if necessary to create a reversed version of a character value which reads from right to left. But do not use the $REVERS or $REVERJ format to reverse DBCS values, because those formats reverse bytes rather than characters.

8

Loops and Arrays

In a data step, when you want to do the same or similar processing on a set of variables, you can define a list of the variables as an array. Repetitive processing, such as the processing of the elements of an array, can be programmed in a loop that is controlled by optional terms in the DO statement.

Arrays

The idea of an array is that you can write a list of variables, give the list a name, then use that name to refer to any of the variables in the list. An array, in the SAS language, is a named list of variables. The variables are called *elements* of the array. You can use an array reference in place of the name of a variable that is an element of the array. Array references can be written in the most important places in data step syntax where variables are used.

Because arrays are used like variables, the name of an array cannot be the same as the name of any variable in the data step. Also, because the syntax of an array reference is similar to the syntax of a function call, array names should not be the same as the names of functions.

Arrays exist only in the data step in which they are defined. When you create a SAS dataset that includes the variables of an array, it does not include the array definition. To use the same variables in an array in a later data step, repeat the same array definition in that later step.

Defining an Array

Use the ARRAY statement to define an array. The ARRAY statement indicates the name of the array, its dimensions, and the variables that are the elements of the array. The ARRAY statement can also declare the type and length of the elements and can indicate initial values for them. The ARRAY statement should usually appear near the beginning of the data step.

The simple form of the ARRAY statement is

```
ARRAY name{dimension} variable list;
```

where the number in braces is the number of elements in the array. This is an example of a simple array definition:

```
ARRAY PLACE{3} START MIDDLE END;
```

The array PLACE is defined with three elements, which are the variables START, MIDDLE, and END.

The ARRAY statement can be abbreviated in either of two ways. Rather than indicating the number of elements, you can write an asterisk between the braces in the statement. The asterisk means that the dimension of the array is the number of variables listed as elements of the array.

```
ARRAY name{*} variable list;
```

Alternatively, you can omit the variables that are elements of the array.

```
ARRAY name{dimension};
```

If you do not list elements of the array, the statement generates variables using the name of the array with numeric counting number suffixes. For example, this array definition

```
ARRAY NODE{5};
```

is equivalent to

```
ARRAY NODE{5} NODE1-NODE5;
```

defining the variables NODE1, NODE2, and so on, as elements of the array.

It is possible to use the same variable as more than one element of an array. Simply list the variable more than once in the ARRAY statement.

Array References

The purpose of defining an array is to allow you to use the array to refer to the elements in the array. An array reference consists of the array name followed by a subscript value enclosed in braces. The subscript value is a numeric expression whose value is within the subscript range defined for the array. The subscript value indicates a particular element of the array.

Consider again the example of the definition of the array PLACE.

```
ARRAY PLACE{3} START MIDDLE END;
```

After this definition, the array reference `PLACE{1}` refers to the variable START, which was declared as the first element of the array. The array reference `PLACE{2}` refers to the variable MIDDLE, and the array reference `PLACE{3}` refers to the variable END. This is demonstrated in the following example.

```
ARRAY PLACE{3} START MIDDLE END;
PLACE{1} = 0;
PLACE{2} = 5;
PLACE{3} = 10;
PUT _ALL_;
```

```
START=0 MIDDLE=5 END=10 _ERROR_=0 _N_=1
```

The subscript value can be calculated by any kind of expression. If the resulting value is not an integer, the integer part of the value is used. If the value is not in the subscript range of the array, the array reference results in an error condition. If you use an expression to calculate a value in a way that is not certain to be within the subscript range of the array, you can use the MIN and MAX operators to restrict the range of the value. Or you can use an IF-THEN statement with a condition to check the subscript value and execute the action that contains the array reference only if the value is within the subscript range.

Most often, the subscript is either a constant or the index variable of a DO loop. Use a DO loop, as described later in this chapter, to give a variable a series of values that cover the subscript range of the array.

Usually, braces enclose subscripts in array references and the ARRAY statement. If you are working on an old-fashioned keyboard or terminal that does not support the brace characters, use brackets or parentheses instead.

Array references can take the place of variable names in most of the places where you can refer to variables in data step programming. An array reference can be used as a value in an expression, as the target of an assignment or sum statement, or in a variable term in an INPUT, PUT, or WINDOW statement. It can be an argument of a function or CALL routine. An array reference can even be the subscript of an array.

Subscript Ranges

The default for a subscript range is for it to go from 1 through the number of elements in the array. That is the subscript range you get when you use an asterisk or a single number to indicate the subscript range in the ARRAY statement that defines an array. However, subscripts do not have to start at 1. They can start at any integer value and count up from there.

Indicate the subscript range by writing the lowest subscript value, a colon, and the highest subscript value:

```
ARRAY name{lowest:highest} variable list;
```

If you omit the variable list, the variable names that are generated still have their suffixes starting at 1 regardless of the subscript values.

An example of a case where you would want to use subscripts starting at a value other than 1 is when the array elements represent values for a range of years, as shown in the declaration below. When you can use meaningful numbers for subscript values, it makes the array references easier to understand than when you just use the numbers 1, 2, 3, and so on.

```
ARRAY PRECIP{1991:2000} PRECIP1991-PRECIP2000;
```

It is as easy to declare an array with a million elements as it is to declare an array with two elements. However, it is important to limit the declaration of an array to include only the array elements you are actually using or not too many more than that. Like any variables or other data objects used by a program, array elements occupy space in memory, and if you write a data step so that it requires more memory than is available, it will not be able to run. And even with the very fast processors of current computers, a data step that uses an inordinate amount of memory could take a long time to run.

In a data step that creates SAS datasets, consider whether the variables of an array should be stored in the output SAS datasets, especially if the array contains dozens of variables. If they should not be stored in the output SAS datasets, be sure to exclude them with KEEP= or DROP= dataset option (or perhaps a DROP statement). A data step that writes many extra, unneeded variables to an output SAS dataset can take several times as long to run as it needs to.

Optional Declarations for Arrays

When it defines an array, an ARRAY statement can also declare the type and length of the variables that are elements of the array, and it can provide initial values for them.

The declaration of type and length comes between the subscript range and the list of elements. Declare the type and length using the same terms as in the LENGTH statement. For a numeric array, write a number indicating the length attribute of the variables. For a character array, write a dollar sign and a number indicating the length of the variables. All the elements in an array must be the same type. For character variables, the length declaration applies only to any elements of the array that have not already been defined as variables in earlier statements in the data step. This ARRAY statement simultaneously defines an array and declares its elements as character variables with a length of 14:

```
ARRAY ROOM{*} $ 14 ROOM1-ROOM15;
```

To declare initial values for each element of the array, list constant values in parentheses at the end of the array statement. List one value for each element of the array. The initial values may be separated by commas or just spaces. The following ARRAY statement is an example of providing initial values for the elements of an array.

```
ARRAY BASE{20} BASE1-BASE20
   (0 0 0 0 0  1 1 2 3 4  5 10 15 20 25  30 33 35 36 36);
```

This statement declares, for example, an initial value of 0 for BASE1, an initial value of 4 for BASE10, and an initial value of 25 for BASE15.

When every element has the same initial values, it is easier to indicate in a RETAIN statement. Write the RETAIN statement after the ARRAY statement. In the RETAIN statement, write the array name and the initial value. All elements of the array are initialized to that value. This is an example:

```
ARRAY LIMIT{12};
RETAIN LIMIT 0;
```

The RETAIN statement gives the elements of the array LIMIT, which are the variables LIMIT1, LIMIT2, and so on, an initial value of 0.

When you declare initial values for elements of an array, none of the elements of the array are automatically reset to missing at the beginning of each repetition of the observation loop. This is also true if any reference to the array is a target of a sum statement. This allows the array elements to retain their values from one observation to the next as the data step executes.

The elements of an array do not have to be variables with names. If there is no reason to store the variables in an output SAS dataset or to refer to them by variable names, declare the array with the keyword _TEMPORARY_ in place of the list of elements. This creates an array of unnamed elements, which are called *temporary variables* because they are the same as the unnamed variables that the data step uses to hold the values of expressions and subexpressions. This is an example of the definition of an array with temporary variables:

```
ARRAY EL_NAME{102:109} $ 18 _TEMPORARY_
   ('Nobelium' 'Lawrencium' 'Rutherfordium' 'Dubnium'
   'Seaborgium' 'Bohrium' 'Hassium' 'Meitnerium');
```

You must include the type and length terms to create a character array of temporary variables.

Temporary variables are not reset to missing in the automatic processing of the data step, even if they do not have initial values. They are not part of the abbreviated variable list _ALL_, so they do not appear if you use the statement `PUT _ALL_;` for debugging purposes. They cannot be written to an output SAS dataset.

Multidimensional Arrays

You can define arrays that require more than one subscript to identify an element. These arrays are called multidimensional arrays; each subscript is a different dimension of the array. When you define the subscript ranges in the ARRAY statement, separate the dimensions with commas. Similarly, in an array reference, separate the subscripts with commas.

A chessboard, for example, would be most easily represented as a two-dimensional array:

```
ARRAY BOARD{8, 8} $ 1 SQUARE1-SQUARE64;
```

The number of elements in a multidimensional array is the product of all the dimensions. In the example above, there are two dimensions of 8, so the number of elements is 8×8, or 64.

The order of elements is determined according to the first dimension first. So, in the example above, the array reference `BOARD{1, 1}` is the variable SQUARE1, then `BOARD{1, 2}` is SQUARE2, and so on through `BOARD{1, 8}`, which is SQUARE8. Then `BOARD{2, 1}` is SQUARE9, `BOARD{2, 2}` is SQUARE10, and so on. Follow this order of elements if you supply initial values for the elements in the ARRAY statement.

Subscript Range Functions

It is often necessary to write the boundary values of the subscript range of an array in the program. You can write them as constant values, making sure that the values match the boundary values you defined in the ARRAY statement. Or you can use function calls to the LBOUND and HBOUND functions that return the subscript boundary values for the array.

Writing subscript boundaries as function calls instead of constant values can make it easier to code and maintain a program. With the function calls, you do not have to compare the values to the array statement to make sure they match. If you revise the program to change the dimensions of the array, you would have to change any constants you used as subscript boundaries, but you do not have to change the function calls. The function calls help to document the program; they explicitly indicate that you mean the values to be subscript boundaries, something that might not be so obvious if you write constant values. Writing the function calls in place of constants also produces a more efficient program that uses slightly less memory and time when it executes.

The functions LBOUND and HBOUND return low and high subscript boundary values, respectively. For a one-dimensional array, the functions take the array as their argument.

```
LBOUND(array)
HBOUND(array)
```

Other functions use expressions as arguments, but the array functions require an array as an argument. Write the array name as the argument. For example, the low subscript value of the array PLACE is:

```
LBOUND(PLACE)
```

The LBOUND and HBOUND functions are used most often in the DO statement. As described in the next section, the DO statement can define an index variable that takes on a sequence of values. Then, the statements in the DO block repeat for each value of the index variable, in what is called a DO loop. When you use a DO loop to process an array, the index variable should usually take on each of the subscript values of the array. The DO statement, then, can be written

```
DO index variable = LBOUND(array) TO HBOUND(array);
```

Another use of the LBOUND and HBOUND functions is to force a subscript value to fall within the subscript range. If you use an expression to calculate a value that could possibly be too low or too high, you can use the following form of expression to substitute the low subscript value for values that are too low and the high subscript value for values that are too high.

```
LBOUND(array) MAX (subscript expression) MIN HBOUND(array)
```

When an array has more than one dimension, use a second argument to the LBOUND and HBOUND functions to select the dimension for which you want the subscript boundaries. The dimension argument is 1 for the first dimension, 2 for the second dimension, and so on.

```
LBOUND(array, dimension)
HBOUND(array, dimension)
```

Alternatively, you can write the dimension as a numeric suffix to the function name.

```
LBOUNDdimension(array)
HBOUNDdimension(array)
```

The DIM function also takes an array as an argument. It returns the dimensions of the array, that is, the number of subscript values in each dimension. The arguments are the same as for the LBOUND and HBOUND functions.

```
DIM(array)
DIM(array, dimension)
DIMdimension(array)
```

For arrays whose subscript values start at 1, the value returned by the DIM function is the same as the value returned by the HBOUND function. Therefore, you will often see the constant value 1 and the DIM function used in place of the LBOUND and HBOUND functions, such as:

```
DO index variable = 1 TO DIM(array);
```

Loops

The big loop in data step processing is the automatic observation loop. Within each repetition of the observation loop, you can have other loops. These loops are DO loops — DO blocks that repeat according to optional terms in the DO statement. There are three different kinds of terms that can control the repetition of a DO loop. The most common involves the use of an index variable.

Index Variable

An index variable takes on a different value for each repetition of a loop. The loop executes once for each value listed for the index variable. Write the index variable, an equals sign, and the list of values in the DO statement:

```
DO index variable = value, value, . . . ;
```

The index values are written as expressions, and they can be any expression that is valid in the data step. Most often, they are simply constant values.

An END statement marks the end of the DO loop. Between the DO and END statements you can write any number of executable statements. This is a demonstration of the execution of a DO loop:

```
DO I = 5, 10, 90, 95;
  PUT I=;
  END;
```

```
I=5
I=10
I=90
I=95
```

The DO loop repeats four times because there are four values listed for the index variable. As a result, the PUT statement executes four times, with a different value for the index variable in each iteration.

An index variable can be either a numeric variable or a character variable. For index values, use expressions of the same type as the index variable.

If the index variable is a character variable that has not been previously declared, its length is determined from the length of the first index value. Make sure the first index value is at least as long as all the remaining index values so that the index variable is long enough to hold all the index values.

Extend the first index value with trailing spaces if necessary.

If the index variable is a numeric variable and the index values, or some of them, form a regularly spaced sequence of values, you can indicate the sequence by writing the starting value, the stopping value, and the increment value. Write the starting value, the keyword TO, the stopping value, the keyword BY, and the increment value. For example, this statement:

```
DO I = 0 TO 100 BY 10;
```

results in the same index values as:

```
DO I = 0, 10, 20, 30, 40, 50, 60, 70, 80, 90, 100;
```

You can omit the word BY and the increment value if the increment is 1, as it most often is. The increment value can be negative if the stopping value is less than the starting value. No index values are generated, and the loop never executes, if the starting, stopping, and increment values are not consistent — if the stopping value is less than the starting value with a positive increment value, or if the stopping value is greater than the starting value with a negative increment. An error condition results if an increment value is 0 or if a starting, increment, or stopping value is a missing value.

You can change the way the loop executes by changing the value of the index variable during a repetition of the loop. The DO statement works the same way with the changed index value as it does with an index value it generated. At the beginning of the next repetition of the loop, it adds the increment value to the index variable. If the resulting index value is beyond the stopping value, the loop stops at that point.

You can use a combination of single values and sequence terms for an index variable, as in this example:

```
DO I = 0, 1 TO 10 BY .5, 11 TO 20, 25 TO 50 BY 5;
```

All the expressions that determine the values of the index variable are evaluated at the beginning of the loop. Each expression is resolved to a value before the execution of the loop begins. If the expression is a variable or contains variables, its value is not affected by any changes in the values of those variables that might take place during the execution of the loop. You cannot, for example, use a variable as a stopping value, then assign a value to that variable to change the stopping point of the loop. If you want to have an expression reevaluated for each repetition of a loop to determine when to stop the loop, use a WHILE or UNTIL condition.

Lexicon

There is also a kind of *index* that keeps track of the location of observations in a SAS data file. There is no connection between an *index variable* and the *index* of a SAS data file.

Stopping Conditions

Instead of using an index variable to determine how many times a loop repeats, you can use a condition. The loop repeats until the condition determines that it should stop. This is especially useful when you do not know in advance how many times it will be necessary to repeat a loop.

There are two different kinds of conditions: WHILE conditions and UNTIL conditions. To use either kind of condition, write the keyword followed by the condition in parentheses.

```
DO WHILE (condition);
DO UNTIL (condition);
```

With a WHILE condition, a DO loop executes as long as the condition is true. The condition is checked at the beginning of each repetition of the loop. If the condition is false, execution of the loop stops. If the condition is false to begin with, the loop never executes.

With an UNTIL condition, a DO loop stops executing when the condition becomes true. The condition is checked at the bottom of the loop — at the end of each repetition of the loop. If the condition is true to begin with, the loop still executes at least once.

The following example finds the lowest power of 2 that is at least 1000. The loop generates successive powers of 2 until one of them is 1000 or more. At that point, the UNTIL condition stops the loop.

```
POWER = 1;
DO UNTIL (POWER >= 1000);
  POWER = POWER*2;
  END;
PUT POWER;
```

```
1024
```

Whenever you write a loop, it is your responsibility to make sure that the loop does stop one way or another. If a loop does not end, the program that contains it never ends either, and you have to interrupt it to regain control of the computer. If it is a WHILE condition that you expect to stop the loop, make sure the condition eventually becomes false for any kind of data values that might be involved. Similarly, if you rely on an UNTIL condition to stop the loop, make sure that the condition eventually becomes true in every case.

When you use an index variable to control a loop, there is usually no problem about the loop stopping after a certain number of repetitions. But if, while the index variable is in the middle of a sequence of values, you assign values to the index variable inside the loop, that could possibly keep the index variable from passing the stopping value. If you write a program that uses that kind of logic, analyze it carefully to make sure it works.

A Logical Variable as a Stopping Condition

If the condition for stopping a loop is too complicated to write as a single expression, use a logical variable to stop the loop. Use the variable as the WHILE or UNTIL condition. Use as many statements as necessary, inside the loop, to assign a true or false value to the variable. When you use a logical variable as a WHILE condition, use an assignment statement to initialize the variable to 1 before the beginning of the loop.

Index Variable With Stopping Condition

A loop can use an index variable along with a WHILE or UNTIL condition. Either the index variable or the condition can stop the loop. It might be that the loop is expected to take a certain number of repetitions, but the condition is there to stop the loop early in certain special cases. Or it could be that it is the condition that is supposed to stop the loop, and the index variable is used to limit the number of repetitions of the loop, just to make sure that it does not repeat indefinitely.

Another possibility is that the condition stops the loop, and the index variable is present not to stop the loop, but because it is part of the calculations that take place in the loop. That is the way the following example works. This example, from mathematics, uses a series to approximate the value of the number *e*. The index variable I is part of the calculation, but it is the UNTIL condition that stops the loop as soon as the approximation reaches a certain degree of accuracy.

```
*
  E.SAS
  Demonstration of a series approximation of e.
*;
DATA _NULL_;
  E = 1;
  TERM = 1;
  DO I = 1 TO 1000 UNTIL (TERM < 1E-6);
    TERM = TERM/I;
    E + TERM;
    PUT (I TERM E)(=);
    END;
RUN;
```

```
I=1 TERM=1 E=2
I=2 TERM=0.5 E=2.5
I=3 TERM=0.1666666667 E=2.6666666667
I=4 TERM=0.0416666667 E=2.7083333333
I=5 TERM=0.0083333333 E=2.7166666667
I=6 TERM=0.0013888889 E=2.7180555556
I=7 TERM=0.0001984127 E=2.7182539683
I=8 TERM=0.0000248016 E=2.7182787698
I=9 TERM=2.7557319E-6 E=2.7182815256
I=10 TERM=2.7557319E-7 E=2.7182818011
```

The result of the program is not the exact value of *e*, but it is close. You could calculate a more accurate approximation by changing the constant in the UNTIL condition to a smaller value, perhaps `1E-15`. Then the loop would repeat a few more times and would add more terms to create a more precise approximation of *e*.

Branching in a Loop

Branching in a loop is slightly different from branching elsewhere in data step programming. You can use the familiar GOTO statements for branching, but depending on how you use it, it could affect the execution of the loop. There are also two statements, CONTINUE and LEAVE, that are designed specifically for branching in a loop.

If the GOTO statement and the label it branches to are both in the same loop, the GOTO statement works the same way it usually does. The GOTO statement, when it executes, transfers control to the location of the label. But a GOTO statement can also branch out of a loop. If the GOTO statement is inside a loop and the label is outside the loop, it stops the loop and transfers control to somewhere else. Use a GOTO statement if it is necessary, under certain conditions, to stop a loop and skip ahead to a later point in the observation.

The LINK statement, by contrast, does not stop a loop, even if you use it to transfer control to a point outside the loop. The RETURN statement, when it executes, returns control to the inside of the loop, where execution resumes with the next statement.

Several other control flow statements stop execution of a loop, along with their other effects. In some situations, you might use a RETURN, DELETE, or subsetting IF statement in a loop. These statements stop the processing the loop and, at the same time, stop processing of the current observation. Or, if you complete the processing of a data step in the middle of a loop, you can use the STOP statement at that point.

The CONTINUE statement is a control flow device just for branching inside a loop. It branches to the end of the current repetition of the loop, in order to proceed to the next repetition. Use the CONTINUE statement to skip over statements at the bottom of the loop.

The following program demonstrates the use of the CONTINUE statement. The second PUT statement in the loop executes only for the index values "Oak" and "Pine" because the CONTINUE statement executes for the other index values.

```
DATA _NULL_;
  LENGTH WORD $ 11;
  DO WORD = 'Oak', 'Hickory', 'Maple', 'Pine', 'Spruce', 'Orange';
    PUT '* ' @;
    IF LENGTH(WORD) > 4 THEN CONTINUE;
    PUT WORD @;
    END;
RUN;
```

```
* Oak * * * Pine * *
```

You can use the LEAVE statement to stop execution of a loop at any point in the loop. When the LEAVE statement executes, it stops the loop, and execution continues at the first statement after the end of the loop. Use the LEAVE statement when you want to stop processing at a point other than at the end of a repetition of the loop.

The program below demonstrates the effect of the LEAVE statement. Although the DO statement indicates index values going from 1 to 100, the loop stops executing during the repetition in which the index value is 11.

```
DATA _NULL_;
  DO I = 1 TO 100;
```

```
      PUT I @;
      IF I > 10 THEN LEAVE;
      N = -I;
      PUT N @;
      END;
   PUT '*';
RUN;
```

```
1 -1 2 -2 3 -3 4 -4 5 -5 6 -6 7 -7 8 -8 9 -9 10 -10 11 *
```

The following example shows a practical use of the CONTINUE and LEAVE statements. The purpose of the loop is to change the second occurrence of a hyphen in the character variable ID to a period.

```
HYPHENS = 0;
DO C = 1 TO 16;
   IF SUBSTR(ID, C, 1) NE '-' THEN CONTINUE;   * Not a hyphen;
   HYPHENS + 1;  * Count hyphens;
   IF HYPHENS NE 2 THEN CONTINUE;  * Not the second hyphen;
   SUBSTR(ID, C, 1) = '.';
   LEAVE;
   END;
```

Note these properties of the logic in this example:

- In the DO statement, the ending index value should be the same as the length of the variable ID. The index variable C is used as a character position in the value of that variable.
- The function call `SUBSTR(ID, C, 1)` isolates one character of the variable.
- When the character is not a hyphen, the first CONTINUE statement executes, to move on to the next character. The subsequent statements execute only when the character is a hyphen.
- The variable HYPHENS is a count of the hyphens in the value. The count is initialized to 0 before the beginning of the loop.
- After the first hyphen is counted, the second CONTINUE statement executes to move on to the next character. After the second hyphen is counted, the assignment and LEAVE statements execute.
- The assignment statement demonstrates a special use of the SUBSTR function as a target of an assignment statement. The statement assigns a new value to the single character of the character variable.
- The LEAVE statement stops processing of the loop so that no further characters are considered.

If a LEAVE statement is the only way you stop a particular loop, you do not need a stopping condition in the DO statement. You might write the DO statement this way:

```
DO WHILE (1);
```

The constant value 1 is a true condition, allowing the loop to continue to execute for any number of repetitions that might be required. This kind of logical device is sometimes called an *infinite loop* because the loop controls do not, themselves, stop the loop at any point, and there is no set limit on the number of repetitions the loop can have. When you rely on statements inside the loop to stop the loop, and the loop does not use an index variable, it is a good idea to code the DO statement this way to clarify that the controls in the DO statement do not stop the loop. As always, inspect the loop logic carefully to ensure that the loop does end no matter what data values are used.

Nested Loops

You can write one loop inside another, nesting them the same way you can nest DO blocks generally. You might nest loops because the processing you do in one loop includes an action that requires another loop. Or, more often, you would nest loops in order to repeat a process for every combination of index values for two or more index variables.

The following example uses nested loops to print a table of loan payment amounts. The MORT function computes monthly payments for fixed nominal annual interest rates from 7.4 to 8.2 percent and for loan terms from 5 to 30 years.

```
DATA _NULL_;
  RETAIN A 100000;  * A: principal amount;
  PUT 'RATE' @10 '  5 YR' @20 ' 10 YR' @30 ' 15 YR'
     @40 ' 20 YR' @50 ' 25 YR' @60 ' 30 YR' /;
  DO INT = 7.4 TO 8.2 BY .1; * INT: annual interest rate percent;
    PUT INT 5.2 @;
    R = INT/1200;  * R: monthly interest rate fraction;
    DO TERM = 5 TO 30 BY 5;  * TERM: term in years;
      N = TERM*12;  * N: number of monthly payments;
      P = MORT(A, ., R, N);  * P: monthly payment amount;
      PUT @(TERM*2) P 7. @;
      END;
    PUT ;
    END;
RUN;
```

RATE	5 YR	10 YR	15 YR	20 YR	25 YR	30 YR
7.40	1999	1182	921	799	732	692
7.50	2004	1187	927	806	739	699
7.60	2009	1192	933	812	746	706
7.70	2013	1197	938	818	752	713
7.80	2018	1203	944	824	759	720
7.90	2023	1208	950	830	765	727
8.00	2028	1213	956	836	772	734
8.10	2032	1219	961	843	778	741
8.20	2037	1224	967	849	785	748

Branching in Nested Loops

When you use CONTINUE and LEAVE statements for branching inside nested loops, you need to know that SAS syntax treats those statements as being inside the innermost loop that contains them. The CONTINUE statement jumps to the next repetition of the innermost loop that contains it. The LEAVE statement stops execution only of the innermost loop — execution continues with the next statement in the loop that contains that loop.

The following code schematic shows two nested loops with various points marked to describe specifically how the branching statements work.

```
DO I = 1 TO 10;  * Beginning of outer loop.;
  . . .
  point A
  . . .
  DO J = 1 TO 10;  * Beginning of inner loop;
    . . .
    point B
    . . .
    point C
    END;  * End of inner loop;
  point D
  . . .
  point E
  END;  * End of outer loop;
point F
```

In this context, the CONTINUE and LEAVE statements work as follows.

- If a CONTINUE statement executes at point B, inside the inner loop, it branches to point C, the point just before the END statement of the inner loop.
- If a LEAVE statement executes at point B or C, inside the inner loop, it ends the inner loop and branches to point D, the point just after the inner loop.
- If a CONTINUE statement executes at point A or D, inside the outer loop, it branches to point E, the point just before the END statement of the outer loop.
- If a LEAVE statement executes at point A, D, or E, it branches to point F, the point just after the outer loop.
- A CONTINUE statement at point C or E would have no effect.

For any other kind of branching in nested loops, use GOTO statements and, if necessary, logical variables. If you need to end more than one nested loop at the same time, use a GOTO statement that branches to a point after the loops.

Programming With Loops and Arrays

More often than not, loops and arrays go together. The purpose of a loop is to do the same action for each element of an array — or the purpose of an array is to hold the values used in each repetition of a loop. There might be several arrays that are used together in the same loop.

Index as Subscript

Typically, the index values of the loop match the subscript values of the array. The index variable is used as the subscript expression for the array. You can see this in the simplest loops, such as this one, which sets the values of the elements of an array to 0:

```
DO I = LBOUND(COMP) TO HBOUND(COMP);
  COMP{I} = 0;
  END;
```

When a DO loop processes all the elements of an array, using the index variable as the subscript, you can use the LBOUND and HBOUND functions, as in this example, to set the starting and stopping values of the index variable.

There can be several arrays, declared with the same dimensions, in a loop. You can use the arrays much like you would use individual variables elsewhere, as in this example:

```
DO QUARTER = 1 TO 4;
  PROFIT{QUARTER} = REV{QUARTER} - COST{QUARTER} - EX{QUARTER};
  END;
```

This example depends on all four of the arrays having the same subscript range.

Multidimensional Arrays in Nested Loops

In simple cases, arrays of two or more dimensions can be processed almost as easily as arrays of one dimension. This processing requires nested loops. Write the loop for the first array subscript first, as the outermost loop. Inside that loop, write the loop for the second array subscript. Continue with more nested loops if there are more dimensions in the array.

This example revises the previous example to work with an array with two dimensions:

```
DO YEAR = LBOUND(PROFIT, 1) TO HBOUND(PROFIT, 1);
  DO QUARTER = 1 TO 4;
    PROFIT{YEAR, QUARTER} = REV{YEAR, QUARTER}
       - COST{YEAR, QUARTER} - EX{YEAR, QUARTER};
    END;
  END;
```

Often, two-dimensional arrays are used together with one-dimensional arrays. That is the case in the example below, which creates an output SAS dataset of a network flow model. The one-dimensional array, LOCATION, indicates the meanings for the index values of the two-dimensional array, FLOW. The assignment statements in the example create the output variables SRC, DST, and VOL (source, destination, and volume) from the elements of the arrays.

```
DO S = LBOUND(LOCATION) TO HBOUND(LOCATION);
  SRC = LOCATION{S};
  DO D = LBOUND(LOCATION) TO HBOUND(LOCATION);
    IF D = S THEN CONTINUE;
    DST = LOCATION{D};
    VOL = FLOW{S, D};
    OUTPUT MAIN.FLOW;
    END;
  END;
```

In this example, the CONTINUE statement is used to skip array elements for which the two index values are equal. Those particular elements are not used in the network flow model because they would represent a flow from a place to the same place, and that would not be a meaningful part of the model.

Sorting an Array: an Example of Nested Loop Logic

The following example of sorting the values in an array shows some of the complex logic that can be required when programming with nested loops. The bubble sort in this example is simple, but slow compared to other sort algorithms. It works by swapping consecutive values when they are out of order. Its work is complete when it gets through the entire array without having to swap any values.

```
*
  Bubble sort of the array A.
*;
DO UNTIL (COUNT = 0);
  COUNT = 0;
  * Consider each pair of consecutive elements in the array.;
  DO I = LBOUND(A) + 1 TO HBOUND(A);
    * No action required if the two values are already in order.;
    IF A{I} > A{I - 1} THEN CONTINUE;
    * For two values that are out of order: count, and swap the two values.;
    COUNT + 1;
    SWAP_VALUE = A{I - 1};
    A{I - 1} = A{I};
    A{I} = SWAP_VALUE;
    END;
  END;
```

Note the following points about this example.

- Because the LBOUND and HBOUND functions are used to determine the beginning and ending values of the index variable, this example works for any subscript range the array A might have.
- As a result of the way the subscripts are calculated, the starting index value has to be 1 more than the low subscript value.
- The variable COUNT counts the number of times pairs of elements are swapped. The UNTIL condition of the outer loop, `COUNT = 0`, is not true until the inside loop goes through the array without finding any elements out of order. That happens after the array is entirely sorted. The outer loop repeats the whole process until that point is reached.
- The three assignment statements at the end of the inner loop swap elements `I` and `I - 1` of the array A.
- The IF-THEN statement tests two consecutive elements to see if they are in order. If they are, it executes the CONTINUE statement to skip ahead to the next pair of elements.

9

I/O

A computer program does most of its work in the computer's memory, working with such things as variables and values. However, for a program to do much of anything useful, it also has to work with data that comes from somewhere, usually a file — and for its results to have any value, they have to be delivered somewhere, usually to another file. It is a program's input and output, or *I/O*, that makes it relevant to the world.

In summary, these are the I/O capabilities of a SAS program:

- Data steps can read and write text files. Other kinds of files and other devices can be treated as text files.
- Data steps can create SAS datasets.
- Data steps can read and update SAS datasets.
- Proc steps can read, create, modify, update, and delete SAS datasets. They also work with other kinds of SAS files.
- Data steps and proc steps can create print output.
- Data steps and macros can display information in and get user input from windows.
- Using I/O functions and CALL routines, a data step can work with SAS datasets, text files, and directories.

Storage and Other Devices

Most I/O is file I/O — reading and writing files. There are also other kinds of devices that can be involved in I/O.

Memory and Storage

If you see computers the way computer manufacturers present them — as boxes — the use of the terms *input* and *output* might seem confusing. In the average computer, storage and memory are both inside the box, so why should a transfer from storage to memory be considered input and a transfer from memory to storage be considered output? The answer is easier to see if you have ever taken the cover off a computer and worked with the components inside. A storage device is connected to the computer only by a cable. It could just as well be outside the box, or even in another room. In fact, computers are designed to work with external storage devices and network storage in almost exactly the same way they work with their internal storage devices. Memory chips, by contrast, are right on or at the computer's main digital board. They have to be physically close to the computer's central processor. If they were more than a certain distance away, the time it takes for electrical signals to pass back and forth between the memory and the central processor would become large enough that they would get out of sync, and then the computer would no longer work.

The essential difference between memory and storage is a difference in speed. Memory is fast enough that its operation can be compared to the speed of light. Storage is also fast, but it is slow enough that its operation can be compared to an automobile engine. A computer might access a location in storage in an average of 10 milliseconds. By contrast, it might be guaranteed to access a location in memory within 10 nanoseconds, or .00001 milliseconds. Hardware speeds vary from one computer to another, and each year's computer hardware is faster than the year before, but memory is orders of magnitude faster than storage.

The reason memory can be so much faster is that it is a semiconductor, a purely electrical device. By contrast, a storage device, such as a hard disk drive, is an electromechanical device; it has moving parts. The time a storage device takes to retrieve data is, for the most part, the time it takes for those moving parts to move from one place to another.

Because storage is relatively slow already, it is practical for storage devices to be connected to a network and shared among many computers. Depending on the speed of the network connections, network storage might be nearly as fast as local storage.

Speed is not the only difference between memory and storage. Most memory used in computers is *volatile,* meaning it loses its data when the power is turned off. Storage it non-volatile — it can keep its data for long periods of time, whether power is supplied or not. Because storage is non-volatile, it is possible for storage volumes to be removable, able to be stored separately from the computer or shipped off to another computer.

Another practical difference between memory and storage has to do with size. Because of the relative cost of memory and storage, most computers have at least 100 times as much storage space as memory space. With memory space limited, memory is generally reserved for the current activities of the computer — for objects in use by programs that are running at the moment.

Files

The largest logical unit of storage is a *volume*. Often, a volume is a physical unit of storage — a disk or a tape. Sometimes, though, a hard disk might be partitioned into several volumes, or several hard disks might be used together, as a disk array, to make a single volume.

A volume encompasses more storage space than you would ordinarily use at one time, so volumes are divided into files. Each file contains a certain kind of data or data on a particular subject, organized in a specific way. To make it easier to organize and manage files, most file systems let you arrange files as groups, which are called directories or, sometimes, folders. To identify a file, you have to name both the directory and the file.

A *file type*, or *file format*, is a specific way of organizing data in a file so programs can use it. SAS software works with a few file types in particular. Text files are files that contain records of text characters. The INPUT and PUT statements are designed specifically to read and write text files. SAS programs are also stored as text files. Print files are a special kind of text file that also contains print control characters that prepare a file for printing. Most procs can write output as print files. The log and standard print files are the two automatic output print files of SAS software.

There are also several types of SAS files. These are files that are organized specifically for use by SAS software. The various kinds of SAS files are identified by different member types.

File types are often indicated by a part of a file name, the *extension*. An extension is a short abbreviation or alphabetic code that follows a dot (that is, a period) at the end of the file name. TXT and DAT are extensions that often identify a text data file. So, for example, a file named PLACE.TXT is likely to be a text data file. SAS is an extension customarily used for a text file that contains a SAS program. PRN, LOG, and LST extensions usually indicate a print file. Each type of SAS file has its own extension, but these extensions vary from one operating system to another.

Devices

A file system is primarily concerned with storage, but it may also provide other kinds of access. In a file system, various devices that can be connected to the computer may be treated as files. The operating system may also provide ways to exchange data with other programs or other computers following certain protocols for exchanging data. Within certain limitations, these other connections can be treated like files. You use device names in statements such as the FILENAME and LIBNAME statements to tell SAS what kind of device you are connecting to.

Some examples of device names are PRINTER, PLOTTER, TAPE, PIPE, FTP, TERMINAL, and DUMMY. Many of these devices require some sort of file name to use with them. Others are used without a file name. The details vary considerably among operating systems.

Access Methods

The easiest way to read or write a file is to start at the beginning and proceed through the file to the end. This method of file access is called *sequential* access, because it proceeds through the data in the file in sequence. Most of the time, when SAS accesses a file, it is using sequential access.

Another important access method is *direct* access. In direct access, the program determines the sequence it follows when it reads the file. It might read the entire file or only a small part of it. Not every kind of file and not every device supports direct access.

Some *linear* storage devices, generally those that work with tape, support only sequential access. Various other devices, such as printers and plotters, also have to be treated as linear devices.

There is another method for accessing a SAS data file, which involves the use of an index. Indexed access is direct access, but it appears sequential to the program. The index presents the data to the program in a sequence that is different from the sequence the data is stored in.

Access Modes

The access method a program uses in working with a file depends not just on the access method, but also on the actions the program takes with respect to the file. The actions of a SAS program on a file can generally be described by one of the following categories.

Input An access mode is described as input if the program is reading and using data from the file without changing the file in any way. This is the access mode of SAS statements such as SET, INPUT, and MERGE.

Output An access mode is described as output if the program is creating a new file or writing data to an existing file without considering any data that the file might already contain. Output tends to be sequential by nature, because without reading the file, the program has no way to select a location in the file to write to. For an existing file, output can either start at the beginning of the file, replacing the previous contents of the file, or at the end, adding on to what the file already contains.

Copy Copying involves input from one file simply for the purpose of output to another file. The program simply copies data from one place to another without doing anything else with it along the way. The source file is unaffected; the destination file is created, replaced, or appended to.

Update Often it is necessary to read and write data in the same file — for example, to read data, then change it if certain conditions are true. This kind of process can be described as updating or editing.

Modify A program may be said to be modifying a SAS file when it changes the attributes in the file, rather than changing data values.

Sort Sorting in place is a particular kind of action on a file that involves changing the order of records or observations, without changing any of the data values. In sorting, the records of the file are put into a particular order, but the data values in the records are not changed.

Delete Deleting, or erasing, is removing a file from a storage volume. Deleting a file prevents any further access to the file. It frees up the space that was used by the file. It also frees up the file name, making it possible for a new file to use that name. In routine computing, it is often necessary to delete an old file that is no longer needed so that you can create a new file.

Some files cannot be modified in any way. These *read-only* files are on physical media that cannot be rewritten, such as a CD-ROM; or they are on volumes that are *locked*, protected from changes by the file system or storage hardware; or the individual file or the directory that contains it is locked; or the file was declared with read-only access in the SAS program. For a read-only file, the only access methods that are possible are input and copying.

Names for Files

In SAS statements, it is possible for files to be identified by their physical file names, the names that the file system uses for the files. Usually, though, the physical file name is used only once, to declare the file and assign a SAS name to it. In subsequent statements in the program, the SAS name is used to identify the file.

Two Classes of Files

SAS software creates and uses SAS names in two different ways for two classes of files. For SAS I/O purposes, files are classified as SAS files and text files.

SAS files are the special file types SAS software uses. There are several types of SAS files, including SAS datasets. SAS software includes high-level support for access to SAS files. SAS files are stored in collections called *libraries*. In most operating systems, a library is a directory. In operating systems that do not support directories, a library is a single physical file. The SAS name for a library is a *libref*. It is defined in a LIBNAME statement. The SAS name for a SAS file is a two-level name, which combines the libref and the member name of the SAS file.

All other files are treated as text files in SAS I/O. A SAS name for a text file is a *fileref*. It is defined in a FILENAME statement. Binary files can also be treated as text files for I/O purposes. You can also define filerefs for directories. You can use such a fileref to identify the directory, or to identify text files in the directory as members of the directory.

Physical File Names

Physical file names are the names by which the computer's file system identifies files. Usually, the physical file name must include the directory that contains the file. Although SAS names are usually used to identify files in a SAS program, you can use physical file names instead. Whenever you write a physical file name in a SAS program, write it as a quoted string.

The specific form of physical file names varies considerably from one file system to another. The form of the physical file names you use in SAS programs is the same as the form you use for file names in operating system commands and in other applications on the same computer.

There are several advantages in using SAS names for files. SAS names are shorter and easier to read and write. The SAS names do not have to be changed if the physical file names change. Using SAS names for files makes it much easier to move a SAS program from one computer to another or from one operating system to another.

Filerefs

To define a fileref, all you need is the physical file name of a file or directory. In the FILENAME statement, write the fileref and the physical file name.

```
FILENAME fileref 'physical file name';
```

This is an example of a FILENAME statement:

```
FILENAME CODES 'codes.txt';
```

This statement associates the fileref CODES with the physical file name codes.txt. After the FILENAME statement executes, you can use the fileref CODES to refer to that physical file. The physical file does not have to exist before you execute the FILENAME statement; often, you might define a fileref for a new output file, which will not be created until a subsequent step writes to the file.

A fileref is a SAS name. It can be no more than 8 characters long. You can subsequently change the physical file that a fileref refers to. Execute another FILENAME statement with the fileref and the new physical file name. Conversely, you can use more than one fileref for the same physical file.

There are a few optional terms you can use in the FILENAME statement. If necessary, you can indicate a device name, such as TAPE or PRINTER, between the fileref and the physical file name. For some devices, a physical file name is not necessary. For example, the device name DUMMY is used as a dummy file; it acts the same as an empty file. You do not need a physical file name with the DUMMY device.

At the end of the FILENAME statement, there are a few options you can use, including the LINESIZE=, PAGESIZE=, and MOD options. Other options are available in specific operating systems. These options have the same meaning as in the INFILE and FILE statements; see the descriptions of those statements in chapter 12, "Text File I/O," for details.

After they are defined in the FILENAME statement, filerefs are most often used in the INFILE and FILE statements, the data step statements that identify input and output text files. These statements sometimes simply indicate a fileref, for example:

```
INFILE CODES;
```

After you define a fileref for a directory, you can use that fileref to specify a file in that directory. Write the file name in parentheses after the fileref. For example, if the fileref NEW is associated with a directory that contains the file WIRE.TXT, you could refer to that file in either of these two ways:

```
INFILE NEW(WIRE.TXT);
```

```
INFILE NEW('WIRE.TXT');
```

In the OS/390 operating system, you can use this notation to refer to members of partitioned datasets.

You can concatenate several physical files in a FILENAME statement. List the physical file names in parentheses:

```
FILENAME fileref ('physical file name' 'physical file name' . . .);
```

When you use the fileref for input, the INPUT statement reads the files as if they were concatenated.

Librefs

Librefs are defined in much the same way that filerefs are, but in the LIBNAME statement instead of the FILENAME statement.

```
LIBNAME libref 'physical file name';
```

Like filerefs, librefs can be no more than 8 characters long.

Access to each SAS data library is controlled by a routine called an *engine*. Usually, you should use the default library engine for the release of SAS software you are running. However, if you need to use a different library engine for a particular library, write the engine name after the libref.

There are a few options that you can write at the end of the LIBNAME statement to control or limit the way SAS files in the library are accessed. Most of these options depend on the operating system and engine you are using. These two options are of more general interest:

`ACCESS=READONLY` This option prevents you from changing any of the files in the library.

`ACCESS=TEMP` With this option, SAS treats the library as *scratch* storage, intended for use only during the execution of the program. SAS is less thorough about protecting the files in the library from possible damage or corruption.

After you define a libref, you can use it in options in some procs to refer to the library. You also use it to form names for the SAS files that are members of the library. A SAS file has a two-level name. The first level of the name is the libref. The second level of the name is the member name of the SAS file. For example, MAIN.REVIEW refers to a SAS file named REVIEW in the library MAIN.

You can make a libref refer to a different library just by executing another LIBNAME statement. You can also use more than one libref to refer to the same library. However, errors may result if you use more than one libref that refers to the same library in the same step in a SAS program.

Operating System Commands to Define Filerefs and Librefs

In some operating systems, you can use operating system commands or statements to define filerefs and librefs. For example, in batch mode in OS/390, you can use DD statements to define DDnames that can be used as either filerefs or librefs in the SAS program. An advantage of using the operating system to define these names is that it lets you run the same program on different files without having to change the program itself at all.

Name Conflicts for SAS Files

Any time a new SAS file is created, it replaces an existing SAS file of the same name and type. When a step replaces a SAS data file, it deletes the existing file of that name upon the successful completion of the step; the existing file is not replaced if the step stops executing because of an error. The old SAS data file can be used as input to the same step. Even though it has the same name, it is not the same file.

Similarly, a view can replace an existing view of the same name, if it is the same kind of view. However, it is an error to attempt to create a SAS dataset when a different kind of SAS dataset already exists with that name. A SAS data file cannot replace a view of the same name, and a view cannot replace a SAS data file or a view of a different kind. To avoid this error, delete the existing SAS dataset before you create the new one.

Concatenating Libraries

There may be times when you want to treat several libraries as a single library. You can do this by concatenating several libraries in the LIBNAME statement. List the libraries in parentheses in the LIBNAME statement. You can use the physical file name or a previously defined libref for each library. You can optionally write commas between the libraries in the list.

```
LIBNAME libref (library, library, . . .);
```

If you use a concatenated library only for the purpose of reading data from its members, and if all the members have different names, then working with a concatenated library is just like working with a regular library. However, if you create new members in a concatenated library, or if there are members that have conflicting names among the libraries that are concatenated, the details of the actions are more complicated than with a single SAS library.

If two or more of the concatenated libraries contain members of the same name and type, you can access only the first such member by name. An action to open a member for input or update or to delete or rename a member affects only the first member of that name among the concatenated libraries, according to the order the libraries are listed in.

When you create a new SAS file in a concatenated library, it is stored in the first library in the list. This is true even if a member of the same name already exists in another library in the list. The new SAS file replaces an existing SAS file of the same name only if the existing SAS file is in the first library. There are no conflicts with an existing member of the same name in any other library in the concatenation.

Any options in the LIBNAME statement for concatenated libraries apply only to libraries listed by their physical file names in the statement. They do not apply to libraries listed as librefs. Similarly, if you name an engine in the LIBNAME statement, it applies only to the libraries listed by their physical file names.

Concatenating Catalogs

Just as you can concatenate libraries, you can concatenate catalogs. Catalogs are SAS files that contain various data objects, called entries. When you concatenate catalogs, two or more separate catalogs are treated as a single catalog. You can find an entry without having to know specifically which catalog it is in. Catalogs are automatically concatenated when you concatenate libraries that contain catalogs of the same name. You can also define a *catref* to specifically concatenate a set of catalogs. A catref is defined in the CATNAME statement, which works almost the same way that the LIBNAME statement works for concatenating libraries.

```
CATNAME libref.catref (catalog, catalog, . . . );
```

You can omit the libref from the CATNAME statement. Then, the catref is created in the WORK library. You can use the ACCESS=READONLY option at the end of the CATNAME statement or in parentheses after any catalog in the list. This option prevents any changes to the entries in the catalog.

Catalogs are also concatenated if they have the same member name and are in libraries that are concatenated.

When you concatenate two or more catalogs, access to entries of the concatenated catalog is similar to access to members of a concatenated library. All new entries are created in the first catalog of the concatenation. Any other action on an entry affects only the first such entry name found among the concatenated catalogs.

Entry as Text File

Several entry types can contain text. You can use these entry types in a SAS program as if they were text files. To do this, define a fileref with the CATALOG device name and an entry name.

```
FILENAME fileref CATALOG 'catalog';
```

You can use the resulting fileref to read any entry type — or you can define the fileref for a catalog and read the entire catalog. However, if you use the fileref for output, the entry must be of a type that contains text. You can only write entries of the types CATAMS, SOURCE, LOG, and OUTPUT.

Using a catalog entry instead of a separate text file can be useful when you use a data step to write a text file just for use later in the same program. It might also be useful as a way to collect certain kinds of notes written by data steps.

Clearing Filerefs, Librefs, and Catrefs

If you need to clear an existing fileref, libref, or catref, you can do it in the corresponding statement with the CLEAR option:

```
FILENAME fileref CLEAR;
LIBNAME libref CLEAR;
CATNAME libref.catref CLEAR;
```

If you are not sure of the definition of a fileref, libref, or catref, you can use the LIST option instead to write a log note describing the definition.

To clear all currently defined filerefs, librefs, or catrefs, you can use the abbreviated list _ALL_:

```
FILENAME _ALL_ CLEAR;
LIBNAME _ALL_ CLEAR;
CATNAME _ALL_ CLEAR;
```

Similarly, to list all currently defined filerefs, librefs, or catrefs in the log, use the abbreviated list _ALL_ with the LIST option.

SAS Files in Memory

There are times when you can speed up the execution of a program significantly by temporarily locating a SAS dataset in memory. Memory is much faster than storage, but it is limited in size, so you should be selective about which SAS datasets you place in memory. Consider situations in which at least two of these conditions are true:

- The computer has a large amount of physical memory.
- The SAS dataset is not especially large.
- The program accesses the SAS dataset repeatedly. Three or more steps use it, or a step reads it with a direct access technique.
- Speed is critical because a user is editing the SAS dataset or is waiting for an interactive application to process it.
- The SAS dataset is temporary and can be discarded before the end of the program.

Use the SASFILE statement to locate a SAS dataset in memory. The SASFILE statement works by opening the SAS dataset and allocating enough buffers for the entire file.

Automatic Librefs

Several librefs are defined automatically when the SAS session starts up. The most important of these is the WORK library, which is designed to hold SAS files that are only used within a single SAS session.

You can omit the WORK libref when writing the names of SAS datasets and some other files in the WORK library. One-level names for SAS datasets refer to the WORK library.

By default, the WORK library is erased at the beginning and end of the SAS session. However, this is controlled by system options, and you can override the default behavior. Use the NOWORKINIT option at startup to keep the WORK library from being erased then. Use the NOWORKTERM option to keep the WORK library from being erased at the conclusion of the SAS session.

Another automatic library, the SASUSER library, is designed to contain files that are related to a specific SAS user. It contains your user profile, which includes your preferences about the way certain things work in an interactive SAS session. The SASHELP library is also related to interactive SAS sessions. It contains online help, interactive applications, and some default settings. It may also contain sample data and programs.

Several other libraries are defined automatically for SAS datasets used in SAS/GRAPH and various other SAS products.

In addition to these libraries, there are a few librefs that have special meanings in SAS programs. The LIBRARY library is the usual location for the SAS supervisor to look for informats and formats you define. The USER library, if you define it, is the location of SAS files with one-level names, instead of the WORK library. DICTIONARY is not, strictly speaking, a libref, but it can be used like a libref in SQL statements to get information about objects in the SAS environment.

System Options for File Access

There are a few system options that have to do with accessing files. The ERRORCHECK= option determines the SAS supervisor's response when, in batch mode, an error occurs with a file named in a LIBNAME or FILENAME statement or a few other global statements. With `ERRORCHECK=NORMAL`, the SAS supervisor continues processing as well as it can. With `ERRORCHECK=STRICT`, the SAS supervisor stops processing as soon as this kind of error occurs.

The FIRSTOBS= and OBS= options can limit the number of records used when a step reads an input file, as described in chapter 4, "Control Flow."

Engines and SAS Dataset Access

A SAS dataset is not really a single type of file; rather, it is a uniform way of organizing data. The data of a SAS dataset can be stored in any way at all, so long it can appear in the form of a SAS dataset. For every different way a SAS dataset can be stored, there has to be a routine, an engine, to convert between the stored form of the data and the logical form of a SAS dataset. Engines are provided as part of base SAS, and it is possible to develop custom engines using SAS/TOOLKIT.

Every library has an engine associated with it. These library engines keep track of the members of a library, and they also include all the details needed to access SAS data files and catalogs. Separate view engines are needed to access the data of a view. A library can have only one library engine, but it can use any number of view engines. These are some library engines of note:

- BASE, a full-featured engine used the default engine for new libraries.
- TAPE, used for libraries on sequential storage volumes.
- XML, to create XML files of SAS data, or to import data from XML files.
- XPORT, to create transport files. Transport files can be moved to other computers that may have different file structures for SAS files.

Views

The idea of a view is to organize data that may be stored in any of various ways so that it can be used as a SAS dataset. Unlike a SAS data file, a view does not contain its own data; instead, it stores a program that executes to access the data.

Views can be created in several ways. A data step view is a special use of a data step program, and it can read any kind of data that an ordinary data step can read. Views can be created in SQL statements to extract or combine data from SAS datasets, which are used as SQL tables. With SAS/ACCESS, SQL views can also access data in various kinds of databases. SAS/ACCESS also creates its own kind of views, which access data in databases or other file formats. In each case, the program that creates the view is only slightly different from a program to create a SAS data file. One advantage of a view, compared to a SAS data file, is that it can refer to data that may change frequently. A view is always up to date, because its data is not put together until it is requested.

Because a view is a program that executes to access data, there are times when it is not the most efficient way to access input data. When you are using the same set of data several times in the same program, it is usually faster to create a SAS data file of the data. Also, writing the data as a SAS data file helps to ensure that the data values will not change while you are working on them, which is an important consideration for some kinds of processing.

Some views can be used to update the data they refer to. Data step views are read-only, but many SAS/ACCESS and SQL views provide read-write access to data. SQL views can be updated within these limitations:

- The view's query expression can refer to only one table, and that table must be able to be updated. The query cannot contain a subexpression.
- The view cannot be defined with an `ORDER BY` clause.
- Only view columns that are taken directly from table columns can be updated. Any derived columns in a view, those that are based on expressions other than simple table columns, cannot be updated.

SAS/ACCESS views are similar to the SQL views that can be updated. They can be updated to the extent that database security settings permit it.

Access Capabilities for SAS Data Files

SAS data files can be read and written in many different ways, but not every library engine can do all the things that a SAS program can do with a SAS data file. All library engines have these capabilities of accessing SAS data files:

- Reading
- Sequential access
- Member name
- Attributes of variables
- BY variables

The BASE engine, which is the default library engine for new libraries if you do nothing to select a library engine, has all the access capabilities listed below, but some dataset options take away some of these capabilities. Other library engines may not have all of these capabilities.

- Writing
- Updating
- Modifying
- Creating new SAS data files
- Direct access of observations
- Observation numbers
- Determining the number of observations
- Identifying the last observation
- Deleting observations
- Sorting in place
- Compression, encryption, indexes, integrity constraints, audit trails, and other optional properties and objects that a SAS data file might have

If you need to use one of these access capabilities for a SAS data file, and you plan to use a library engine other than the BASE engine, test the program to make sure that the engine can do what you expect from it. If it cannot, a log message will explain the limitations that apply.

Observation Numbers

Most library engines keep track of observation numbers of a SAS data file. These observation numbers can appear, for example, in reports of data from the SAS data file.

When observations are deleted from the middle of a SAS data file, it complicates the meaning of observation numbers. The *physical observation number*, which is a number that corresponds to the observation's location in the file, may no longer match the *logical observation number*, which indicates the order in which the observation is accessed when the SAS data file is accessed sequentially. Most engines keep track of both physical and logical observation numbers, but some can maintain only the physical observation numbers and are unable to provide the logical observation numbers.

Remote and Shared Libraries

Creating a transport library with the XPORT engine is the simplest way to take SAS data from one kind of computer to another, but it is not always the best or the fastest way. If a network connects the two computers, you might want to create a live connection to the data. You can do that with SAS/CONNECT or SAS/SHARE.

SAS/CONNECT creates a connection to a SAS session on another computer, which is called a remote session. The various capabilities of SAS/CONNECT include the ability to access SAS libraries located on the remote computer, which are called remote libraries. The REMOTE option and other options in the LIBNAME statement make it possible to declare a remote library. The SAS files in that library can then be used in much the same way as SAS files of any other library in the SAS session.

SAS/SHARE adds the ability to serve the same library to multiple users at the same time. For example, you can edit a SAS dataset at the same time as several other SAS users. SAS/SHARE operates a SAS server, and the libraries you access through the SAS server are called shared libraries. The SAS server provides access to the data while making sure that conflicts do not occur. For the most part, this means that while one user is editing an observation, no other user can access that same observation at the same time. There are several options in the LIBNAME statement that are necessary to connect to a shared library.

10

SAS Dataset I/O

With SAS datasets, SAS makes it as easy as possible to access data in a program. To access the data of a SAS dataset, you only need to indicate the name of the SAS dataset; the SAS supervisor keeps track of all other details. Both data steps and proc steps can read and write SAS datasets.

Input SAS Datasets in the Data Step

The data step provides four statements for reading SAS datasets: SET, MERGE, MODIFY, and UPDATE. The actions of these statements can be modified by statement options, the BY statement, and the use of multiple SAS datasets to give you many different ways to read SAS datasets in a data step.

In each case, there is a single executable statement that reads an observation from one or more SAS datasets. The statement reads an observation from a single SAS dataset, or, in the case of merging and updating, it creates an observation by combining observations from two or more SAS datasets. Ordinarily, the statement is repeated in the data step's automatic loop to read all the input observations. The statement reads all the variables from the input SAS datasets, assigning the values of those variables to data step variables of the same names.

You can use dataset options on the input SAS datasets in any of these statements. The dataset options follow the SAS dataset name and are enclosed in parentheses. The various dataset options are described in detail in the next chapter.

Sequential Input

To read a SAS dataset sequentially in a data step, name it in a SET statement:

```
SET SAS dataset;
```

The SET statement reads through the SAS dataset in order.

If you omit the SAS dataset name in the SET statement, the special name _LAST_ is implied. _LAST_ is the most recently created SAS dataset, or the value of the _LAST_ system option.

You can read more than one SAS dataset in a SET statement. The statement reads all the observations from the first SAS dataset before beginning to read from the second SAS dataset. This kind of combination of two or more SAS datasets is called *concatenation*.

```
SET SAS dataset SAS dataset . . . ;
```

The SAS datasets being concatenated do not have to have the same set of variables. If any SAS dataset does not have a variable that is in another SAS dataset, that variable is given missing values in the observations that come from that SAS dataset.

The diagram below demonstrates the way concatenation works with two SAS datasets, EARLY and LATE.

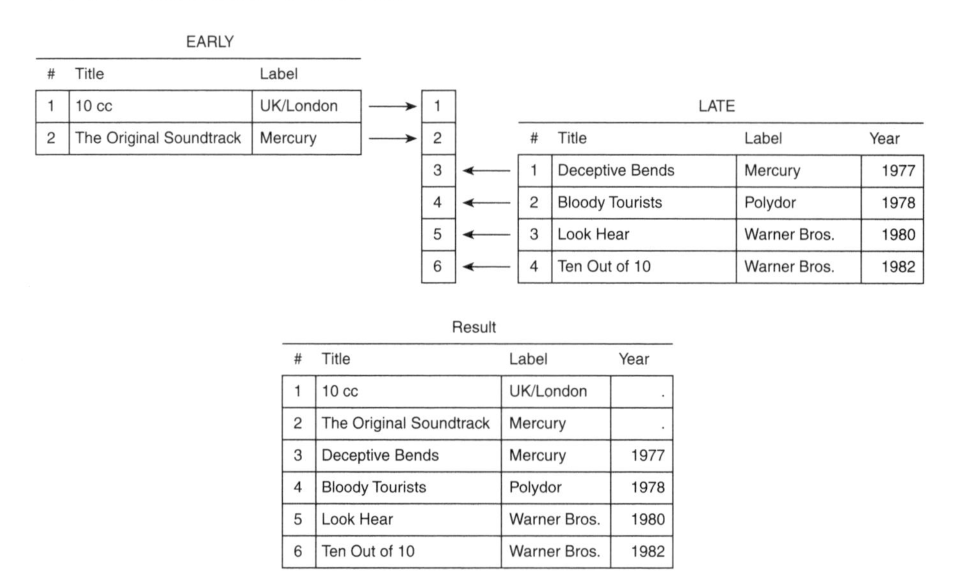

EARLY

#	Title	Label
1	10 cc	UK/London
2	The Original Soundtrack	Mercury

LATE

#	Title	Label	Year
1	Deceptive Bends	Mercury	1977
2	Bloody Tourists	Polydor	1978
3	Look Hear	Warner Bros.	1980
4	Ten Out of 10	Warner Bros.	1982

Result

#	Title	Label	Year
1	10 cc	UK/London	.
2	The Original Soundtrack	Mercury	.
3	Deceptive Bends	Mercury	1977
4	Bloody Tourists	Polydor	1978
5	Look Hear	Warner Bros.	1980
6	Ten Out of 10	Warner Bros.	1982

Options for Observations

There are two statement options and a dataset option you can use in the SET statement to get more information about the observations that the statement reads. Each of these options creates a numeric variable in the data step. The variable gets its value from the SET statement. The value of the option is a name for the variable. Supply a name that is not already being used as a variable in the data step.

The NOBS= and END= options are statement options, which you can write at the end of the SET statement. The NOBS= option creates a variable that contains the number of observations in the SAS dataset. If there are two or more SAS datasets in the SET statement, the NOBS= variable contains the total number of observations in all the SAS datasets combined.

Not every engine is able to determine the number of observations in a SAS dataset. If you use the NOBS= option, and the number of observations is not available, the variable has a missing value.

The END= option creates a Boolean variable that indicates the last observation that the statement reads. The value of the variable is usually 0. It is set to 1 when the SET statement reads its last observation. If there are multiple SAS datasets in the SET statement, the END= variable is set to 1 for the last observation from the last SAS dataset.

Not every engine is able to determine when it is reading the last observation from a SAS dataset. If the last observation comes from a SAS dataset that uses an engine that cannot identify the last observation, the END= variable is always 0.

You can use the IN= dataset option in a statement with multiple SAS datasets to find out which SAS dataset an observation came from. The option creates a Boolean variable whose value is 1 when the current observation came from that SAS dataset, and 0 otherwise. Like any dataset option, the IN= option is written in parentheses after the SAS dataset name. Use the IN= option on each SAS dataset in the statement to create a separate IN= variable for each SAS dataset.

The statement below shows an example of a SET statement with two SAS datasets, using the IN= dataset option. In this statement, the IN= options create the variables PAST and PRESENT.

```
SET MAIN.PRIOR (IN=PAST) MAIN.CURRENT (IN=PRESENT);
```

For observations that come from the SAS dataset MAIN.PRIOR, the variable PAST has a value of 1. For observations that come from the SAS dataset MAIN.CURRENT, the variable PRESENT has a value of 1. You can use these variables as conditions in IF-THEN statements to take actions on the observations that come from a particular SAS dataset. Often it is useful to create a new variable that indicates which SAS dataset an observation came from, with statements such as:

```
IF PAST THEN PERIOD = -1;
IF PRESENT THEN PERIOD = 0;
```

Indexed Input

One of the purposes of an index on a SAS data file is to let you read its observations in a different order. To read a SAS data file in a data step using the sort order indicated by an index, name the SAS dataset in a SET statement and the sort order in a BY statement.

```
SET SAS dataset;
BY sort order;
```

For this purpose, the sort order in the BY statement is just a list of key variables. Often, it is a single identifying code variable that indicates the sort order, as in this example:

```
SET CORP.NEWHIRE;
BY EEID;
```

The observations of the SAS dataset CORP.NEWHIRE are read in ascending order of EEID. That is, the observation with the lowest value of EEID is read first, then the observation with the second lowest value of EEID, and so on.

The BY statement modifies the operation of the SET statement, so it must be written immediately after the SET statement.

This use of the BY statement works only if an appropriate index exists on the input SAS dataset — or if the SAS dataset happens to be sorted in that same order. Otherwise, an error condition occurs as soon as the SET statement reaches an observation that is out of order — probably on the second or third observation that the statement reads.

The sort order may be indicated by more than one variable, as in this example:

```
SET CORP.NEWHIRE;
BY START EEID;
```

When there is more than one sort order variable, the observations are put in order according to the first variable, in this case, START. Then, wherever there is a group of observation with the same value of the first variable, the second variable is used to put the observations in order within that group.

Grouping Input Observations

The BY statement can be used with the SET statement when there is no index on the SAS dataset. In this case, the BY statement merely indicates the order of the observations in the SAS dataset, and there are more options you can use in the statement.

Before any variable in the sort order clause, you can use the option DESCENDING to indicate that the variable is sorted with the highest values first. Or you can use the GROUPFORMAT option to indicate that the variable forms groups according to the formatted values of the variable. That is, the variable's format attribute is applied to the variable's values, and the resulting text is used to define the groups of observations. The formatted values are not necessarily in any particular order.

If the observations are in some order other than simply ascending or descending, use the NOTSORTED option. You can write the option anywhere in the BY statement, but it is usually best to write it at the end of the statement. The NOTSORTED option indicates that the observations are in some kind of sorted order, but not necessarily in a strictly ascending or descending order of the key variables. With the NOTSORTED option, the SET statement does not check to see what order the observations are in.

When used in this way, with a SET statement that reads a single SAS dataset without a corresponding index, the BY statement has only these two effects:

- It creates the automatic FIRST and LAST variables for each of the BY variables. The FIRST and LAST variables are Boolean variables that mark the beginning and end of a group of observations, as described in chapter 4, "Control Flow."

- It creates an error condition if the SAS dataset is not sorted in the order indicated. Use the NOTSORTED option at the end of the BY statement to avoid creating this error condition.

The BY statement can be used whenever a data step statement reads SAS datasets. It always has the effects listed above. Most of the time, it also helps to determine the order of observation or the way observations are assembled from multiple SAS datasets. When there are multiple SAS datasets, the NOTSORTED option usually cannot be used.

Interleaving

You can use the SET statement with a BY statement to *interleave* two or more SAS datasets. The interleaving process reads all the observations from all the SAS datasets, but in sorted order.

```
SET SAS dataset SAS dataset . . . ;
BY sort order;
```

The SAS datasets must have an appropriate index or be sorted in the order that the BY statement indicates.

The diagram below demonstrates the way the interleaving process combines two SAS datasets in sorted order. The statements to interleave the two SAS datasets shown are:

```
SET EAST CENTRAL;
BY RATING;
```

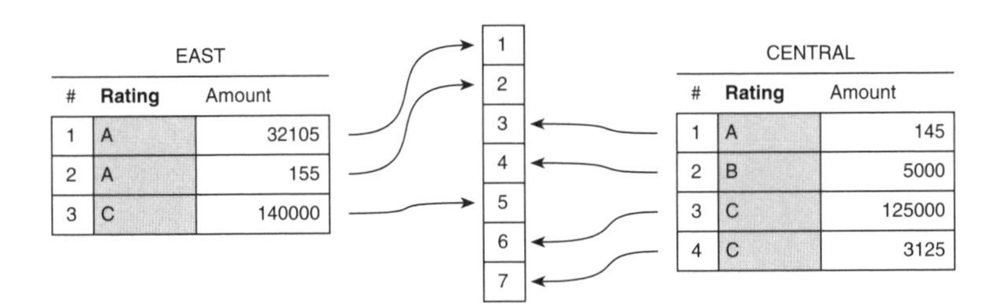

EAST

#	Rating	Amount
1	A	32105
2	A	155
3	C	140000

CENTRAL

#	Rating	Amount
1	A	145
2	B	5000
3	C	125000
4	C	3125

Result

#	Rating	Amount
1	A	32105
2	A	155
3	A	145
4	B	5000
5	C	140000
6	C	125000
7	C	3125

Access by Observation Number

The POINT= option of the SET statement allows you to access observations of a SAS dataset by observation number. The value of the POINT= option is the name of a numeric variable. To read a specific observation, assign the observation number to that variable in an assignment statement, then read the observation with a SET statement with the POINT= option. If the name of the observation number variable is N, the statements are:

```
N = observation number;
SET SAS dataset POINT=N;
```

You can use the POINT= option with ordinary SAS data files, but most other kinds of SAS datasets do not support direct access. Generally, direct access does not work with compressed SAS data files, views, transport engines, or sequential engines. It does not make use of indexes, and it is incompatible with the WHERE= dataset option and the BY statement.

The observation number variable is treated as an integer value. If the value is not an integer, it is truncated to the next lower integer value. A value between 0 and 1 is treated as a 1. If the value is 0, negative, missing, or greater than the number of observations, an error condition results.

If you name more than one SAS dataset in the statement, the SAS datasets are concatenated for the purpose of determining observation numbers. An observation number larger than the number of observations in the first SAS dataset would belong to an observation in the second SAS dataset, an observation number larger than the number of observations in the first two SAS datasets would belong to an observation in the third SAS dataset, and so on. Use the IN= dataset option if you need to determine which SAS dataset an observation comes from.

Lookup

Another form of direct access for a SAS data file lets you look up a specific observation based on the values of key variables. This requires a SAS data file that has an index on the key variable or the set of key variables you want to use. First use assignment statements to assign the values you want to look up to the key variables. Then, in the SET statement, name the index in the KEY= option, followed by the /UNIQUE option to indicate that you are looking for just one observation.

For example, if CHRON is an index of the variables START and EEID on the SAS data file MAIN.NEWHIRE, then the statements below look up an observation with these specific key values in that SAS dataset.

```
DATE = '21JUN1999'D;
EEID = '10383';
SET MAIN.NEWHIRE KEY=CHRON/UNIQUE;
```

If there is an observation that matches the values of the key variables, the statement reads that observation. If more than one observation matches, the statement reads the first matching observation. If there is no observation that matches, the statement does not read any values, and it assigns a nonzero error code to the automatic variable _IORC_.

When you use the KEY= option, you should check the value of _IORC_ and take appropriate actions if the value is nonzero, indicating that the SET statement failed to read an observation. When the SET statement fails, you need to assign missing values to the variables you had been hoping to read.

Use a KEEP= dataset option on the SAS dataset in the SET statement to list all the variables you read from the SAS dataset. After the SET statement, if there is a nonzero value for _IORC_, assign missing values to the variables from the SAS dataset, other than the key variables. Also assign 0 to the automatic variables _IORC_ and _ERROR_ so that there is no error condition. The complete code to correctly look up an observation might look like this:

```
DATE = '21JUN1999'D;
EEID = '10383';
SET MAIN.NEWHIRE (KEEP=DATE EEID DOB NAME) KEY=CHRON/UNIQUE;
IF _IORC_ THEN DO;
  DOB = .;
  NAME = '';
  _IORC_ = 0;
  _ERROR_ = 0;
  END;
```

If you need to distinguish among the various possible error conditions that the error codes of the variable _IORC_ might represent, the function IORCMSG might help. The function, which does not have any arguments, returns the text of the error message.

Subset Lookup

Used in a slightly different way, the KEY= option can let you read a subset of observations defined by key variables. Omit the /UNIQUE option to indicate that you want to retrieve all the matching observations, one at a time. Repeat the SET statement in a loop until a nonzero value of _IORC_ is reached.

In the example below, the SAS dataset LIVE.LOG has an index named DATEUNIT on the variables DATE and UNIT. The DO loop reads the observations in LIVE.LOG that match on those variables and adds up the total for the variable VOLUME across those observations. If there are no matching observations, the loop results in a total of 0.

```
DATE = DATE();
UNIT = '4040';
N = 0;
TOTAL = 0;
```

```
DO UNTIL (_IORC_);
  SET LIVE.LOG (KEEP=DATE UNIT VOLUME) KEY=DATEUNIT;
  IF _IORC_ THEN LEAVE;
  N + 1;
  TOTAL + VOLUME;
  END;
_IORC_ = 0;
_ERROR_ = 0;
```

Important Caution About Stopping

The direct access techniques of the KEY= and POINT= options never reach an end-of-data condition. Ordinarily, the automatic data step loop depends on an end-of-data condition to stop the loop. If you use a SET or MODIFY statement with the KEY= or POINT= option as the primary source of input data in a data step, you must use a STOP statement at the appropriate point to stop the data step.

Merging

A merge combines two or more SAS datasets by combining observations. The idea is to have each SAS dataset contribute some of the variables to the combined observation. Name the SAS datasets in the MERGE statement. Then, name the key variable or variables that are used to match observations in the BY statement.

```
MERGE SAS dataset SAS dataset . . . ;
BY sort order;
```

The BY statement works the same way as with a SET statement. It indicates the sequence of observations. Each SAS dataset in the merge has to follow that sequence, either by having an appropriate index or by being sorted that way. If not, an error condition results. Because key variables are used to match observations in the merge, this form of merge is called a *match merge*.

Every SAS dataset must contain the key variables. In addition, each SAS dataset usually contains other variables that it adds to the combined observation that the MERGE statement produces. In the simplest case, these are different variables in each different SAS dataset.

The MERGE statement considers each key value in sequence. It reads observations from every SAS dataset listed that has observations for that key value, combining them to form a single observation. The combined observation has values for every variable that is in any of those SAS datasets. If a SAS dataset does not have values for a particular key value, the MERGE statement gives its variables missing values.

The diagram below shows how the matching process works in merging two SAS datasets. The statements to merge these SAS datasets are:

```
MERGE RANKING ASSIGN;
BY ID;
```

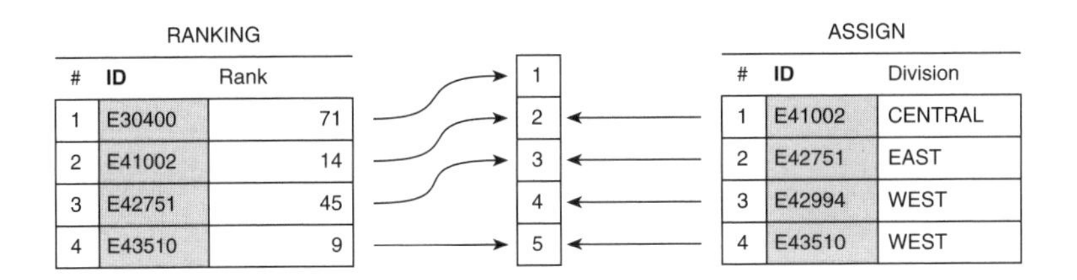

RANKING

#	ID	Rank
1	E30400	71
2	E41002	14
3	E42751	45
4	E43510	9

ASSIGN

#	ID	Division
1	E41002	CENTRAL
2	E42751	EAST
3	E42994	WEST
4	E43510	WEST

Result

#	ID	Rank	Division
1	E30400	71	
2	E41002	14	CENTRAL
3	E42751	45	EAST
4	E42994	.	WEST
5	E43510	9	WEST

In contrast to the match merge that has been described so far, it is also possible to have a merge without using key variables to match the observations. There is no BY statement. Instead, the observations are matched on observation number — just as if the observation number were the key variable. The practical uses for this unmatched form of merging are rare. If you ever do use it, you should comment it so that a programmer reading the program will know that it is not just a case of a mistakenly omitted BY statement.

You can use the END= option in the MERGE statement, just as in the SET statement, to identify the last observation that the statement produces.

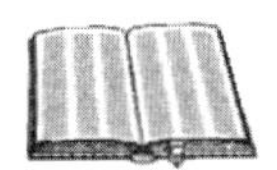

Lexicon

SAS Institute's official term for an unmatched merge is "one-to-one merge," but this is a term I try not to use, because it tends to confuse people. *One to one* has a specific meaning in database theory that denotes a kind of matching — the exact opposite of the meaning intended here. Another term I have used at times, "sequential merge," is not much better. Both kinds of merging, after all, are sequential by nature. My suggestion, at this point, is to use the term *unmatched merge*.

Issues in Merging

Match merges are not usually so tidy as the scenario described previously. There can be duplicate variables across the SAS datasets being merged, or multiple observations in BY groups in one or more of the SAS datasets. In either of these conditions, the merge process is slightly more complicated. Still, the process works with one BY group at a time. One BY group has no effect on the way observations are formed in the next BY group.

A SAS dataset may not have any observations in a particular BY group. In that case, it has no effect on the combined observations that are formed for that BY group. If you use the IN= dataset option, the value of the IN= variable for that SAS dataset is 0 for that BY group. Any variables that are supplied only by that SAS dataset are missing for that BY group.

When there are common variables in a match merge, the value of the variable in a BY group comes from the last SAS dataset in that BY group that has that variable.

Consider, for example, a match merge of the SAS datasets A, B, and C, on the key variable ID. All three SAS datasets contain the common variable NAME, so the last SAS dataset, C, ordinarily provides the value of NAME. In BY groups where C does not have an observation, B provides the value of NAME. For a BY group in which only A has an observations, the value comes from A.

The number of observations formed in a BY group is the largest number of observations in that BY group in any of the SAS datasets. When a SAS dataset has a smaller number of observations in a BY group, its last observation in the BY group is repeated for the remaining observations that the MERGE statement forms for that BY group.

In any case, the first observation in the BY group is formed by combining the first observations in the BY group in each SAS dataset. Similarly, the last observation in the BY group is formed by combining the last observations in the BY group.

Suppose, in the example, for the BY group defined by a value of 1211 in the variable ID, there are five observations in A, one in B, and two in C. The MERGE statement forms five observations for the ID value of 1211, as follows.

- The first observation combines the first observations: the first observation from A, the observation from B, and the first observation from C.
- The second observation combines the second observation from A, the observation from B, and the second observation from C.
- The third observation combines the third observation from A, the observation from B, and the second observation from C.
- The fourth observation combines the fourth observation from A, the observation from B, and the second observation from C.
- The fifth observation combines the last observations: the fifth observation from A, the observation from B, and the second observation from C.

As you can see in this example, observations from the input SAS datasets may be repeated in a rather arbitrary way when there are multiple observations in a BY group. To produce the most meaningful results, you should usually arrange the input data in one of these two ways:

1. No more than one observation in a BY group in any of the input SAS datasets. This allows a kind of matching that in database theory is called *one-to-one* matching.
2. Multiple observations in BY groups in one input SAS dataset, and no more than one observation per BY group in any of the other input SAS datasets. This allows a kind of matching that in database theory is called *one-to-many* matching.

Match Merge Compared to Interleave

Match merge and interleave are similar processes and are sometimes confused with each other. In their syntax, the difference is the use of the keyword MERGE for a match merge and the keyword SET for an interleave. In their actions, the difference is the way observations are formed.

Both match merge and interleave start with sorted input data and produce a sorted set of observations. Both process each BY group separately. An interleave, however, tends to produce more observations than a match merge. Because a match merge matches observations when it can, its number of observations per BY group is the largest number of observations in that BY group in any of the input SAS datasets. An interleave does not match observations, so its number of observations per BY group is the sum of the number of observations in the BY group across the input SAS datasets.

In one special case, match merging is the same as interleaving. This happens when there are no matching key values. When there is no key value in any SAS dataset that matches a key value in any other SAS dataset, then there is no way to match observations in a BY group, and match merging produces the same results as interleaving.

Editing

You can use data step logic to edit a SAS dataset in place — to read observations from the SAS dataset and conditionally modify them or remove them. The first part of the process is reading the observation. Do this with the MODIFY statement.

The MODIFY statement works the same way as the SET statement with a single SAS dataset. You can use all the same options, including the POINT= or KEY= option for direct access. You must name the same SAS dataset in the DATA statement.

After you read an observation with the MODIFY statement, you can modify it with the UPDATE or REMOVE statement, as described later in this chapter.

Updating

In data step logic, *updating* is a process that lets you apply specific changes to a SAS dataset. The SAS dataset that contains the data you want to change is called the *master dataset*. The SAS dataset that contains observations that list the specific changes to apply is called the *transaction dataset*.

Both SAS datasets must be indexed or sorted according to the same key variable or variables. There should be only one observation for each key value in the master dataset. In the transaction dataset, a key value can have multiple observations. Usually, the transaction dataset contains many or all of the same variables that are in the master dataset.

There are two different ways to do updating. Usually, it makes sense to update the master dataset in place. You can do that with the MODIFY statement.

```
MODIFY master dataset  transaction dataset  options;
BY sort order;
```

Also name the master dataset in the DATA statement.

You cannot add variables or observations when you update a SAS dataset in place. To add data in the updating process, use the UPDATE statement instead. A new SAS dataset is created with the updated data.

```
UPDATE master dataset  transaction dataset  options;
BY sort order;
```

In the DATA statement, name the new SAS dataset you want to create.

In the update process, values from the transaction dataset replace values from the master dataset. If there are no transaction observations for a particular master observation, then that observation is unchanged. If there is a transaction observation, then its values replace the values in the master observation. If there are multiple transaction observations, then each transaction is applied in turn, and the updated output observation is not created until the last transaction has been processed.

If there is no master observation for a transaction observation, an observation is added, using the transaction values. When updating in place, the new observations are added at the end of the SAS dataset.

The way transaction values are used depends on an option in the MODIFY or UPDATE statement. If you use the MISSINGCHECK option, the statement does not read standard missing values or blank character values in the transaction dataset. When a variable has a standard missing or blank value in the transaction observation, the value of the variable in the master observation is kept. This lets you use a transaction dataset that contains only the key variables and the values you want to change; other values can be set to missing.

In addition, with the MISSINGCHECK option, transaction values of ._ and '_' are special values that you can use to update a value to missing. When these transaction values appear, the statement changes the value of the variable to a standard missing or blank value, regardless of the value that was present in the master observation.

The MISSINGCHECK option is the default, which means this is the behavior that results if you do not indicate a statement option.

With the NOMISSINGCHECK option, all transaction values are used, including missing values. Variables from the transaction dataset completely replace the values from the master dataset for any key value for which there is a transaction observation. With this option, there is no point in having multiple transactions for a key value, because only the last one is used.

The UPDATE or MODIFY statement can also use the END= option. This option works the same way as in the SET or MERGE statement to identify the last observation that the statement puts together.

It is possible to update in place with the NOTSORTED option in the BY statement. For this to work correctly, the key values must be in the same order in the transaction dataset as in the master dataset.

In a data step that contains the update process, some of the automatic behavior of the data step is changed. Ordinarily, if a data step does not contain a statement to write an output observation, the data step writes one at the end of each repetition of the observation loop. The action is as if an OUTPUT statement were present at the end of the data step, or if there is a MODIFY statement, it is as if there were a REPLACE statement for the SAS dataset being modified. With the update process, the output observation is written only on the last observation of a BY group, in order to create output that still has only observation per BY group. The following two examples show how this works.

This is a simple example of a data step with an UPDATE statement.

```
DATA UPTODATE;
  UPDATE MASTER TRANS;
  BY KEY1 KEY2;
RUN;
```

This step executes as if a conditional OUTPUT statement were present:

```
DATA UPTODATE;
  UPDATE MASTER TRANS;
  BY KEY1 KEY2;
  IF LAST.KEY2 THEN OUTPUT;
RUN;
```

Similarly, a data step that updates in place with a MODIFY statement executes as if a conditional REPLACE statement were present. Follow this pattern if you ever need to explicitly code the statements that write output observations in a data step with an update process.

Conflicts

Statements that read input SAS datasets in the data step are supposed to create program variables that have the same names and values as the SAS dataset variables. Various kinds of conflicts are possible if the variable already exists in the data step, or if the same variable is in multiple input SAS datasets.

The most serious conflict is a conflict of data type. A variable can have only one data type. If a data step defines a variable with one data type, and an input SAS dataset has it with the other data type, an error conditions occurs and the step cannot run. The same problem occurs if a variable has different types in two input SAS datasets. To avoid this conflict, use the

DROP= dataset option to skip the input variable, or use the RENAME= dataset option to change the name of the input variable to a name that is unique in the step.

A conflict in the length of a character variable can be just as serious, but harder to detect. A character variable gets its length from its first use in a data step. If a character variable comes from input SAS datasets, its length is determined by the first SAS dataset that contains it. However, in the merge process, the *value* of the variable comes from the *last* SAS dataset that contains the variable for that BY group. Potentially perplexing problems can occur whenever the length of the variable is shorter than the values assigned to it. This can be an especially serious problem when a key variable is affected. The data step treats the variable as having the length it has in the data step, even though it might be longer in an input SAS dataset. Among other things, this can result in an error condition with a message that says the input observations are not in the correct order, even though they are in order. With the BY statement or the KEY= option, it can result in key values matching even though they have different values. To solve any and all problems of conflicts in the length of a character variable, use a LENGTH statement near the beginning of the data step to declare the variable with the appropriate length.

Another way an error condition can occur is if the name of a variable read from an input SAS dataset has the same name as an array in the data step. Use a different name for the array, or use the RENAME= dataset option to rename the variable.

A more routine kind of problem can occur when two unrelated variables have the same name. The values of one variable replace the values of the other. Use the RENAME= option on the input SAS datasets to give each different variable a name that is unique in the step.

Summary of Statements and Options

The two tables that follow summarize the statements available for reading SAS datasets in a data step and the options you can use in those statements.

Statement Options for Data Step Statements that Read SAS Datasets

Option	Value	Description
END=	*numeric variable*	Boolean variable that indicates the last observation read by the statement
NOBS=	*numeric variable*	Variable that indicates the number of observations available
MISSINGCHECK NOMISSINGCHECK		Whether standard missing values and blank values in the transaction dataset are disregarded

Data Step Statements for Reading SAS Datasets

Statement	Description	Options
SET *SAS dataset* . . . ;	Sequential input	END=*variable* NOBS=*variable*
SET *SAS dataset*; BY *sort order*;	Indexed input, or sequential input in sorted order	END=*variable* NOBS=*variable*
SET *SAS dataset* *SAS dataset* . . .; BY *sort order*;	Interleave	END=*variable* NOBS=*variable*
variable = *number*; SET *SAS dataset* . . . POINT=*variable*;	Direct access by observation number	NOBS=*variable*
key variable = *value*; . . . SET *SAS dataset* KEY=*index*/UNIQUE;	Lookup	NOBS=*variable*
key variable = *value*; . . . SET *SAS dataset* KEY=*index*;	Subset lookup	NOBS=*variable*
MERGE *SAS dataset* *SAS dataset* . . . ; BY *sort order*;	Match merge	END=*variable*
MERGE*SAS dataset* *SAS dataset* . . .;	Unmatched merge	END=*variable*
MODIFY *SAS dataset*;	Sequential access for editing	END=*variable* NOBS=*variable*
MODIFY *SAS dataset*; BY *sort order*;	Indexed access for editing, or sequential access for editing in sorted order	END=*variable* NOBS=*variable*
variable = *number*; MODIFY *SAS dataset* . . . POINT=*variable*;	Direct access for editing by observation number	NOBS=*variable*
key variable = *value*; . . . MODIFY *SAS dataset* KEY=*index*/UNIQUE;	Lookup for editing	NOBS=*variable*
key variable = *value*; . . . MODIFY *SAS dataset* KEY=*index*;	Subset lookup for editing	NOBS=*variable*
MODIFY *master dataset* *transaction dataset*; BY *sort order*;	Update in place	MISSINGCHECK END=*variable*
UPDATE *master dataset* *transaction dataset*; BY *sort order*;	Update	MISSINGCHECK END=*variable*

Creating SAS Datasets in the Data Step

The data step makes it easy to create SAS datasets. Only a few statements are involved, and usually, only the DATA statement is necessary.

Naming the Output SAS Datasets

A data step starts with a DATA statement, which is a list of the output SAS datasets of the step. Most often, there is just one output SAS dataset in a data step, but there can be any number of them. This could be the DATA statement for a step that creates the SAS datasets WORK.STOCK and WORK.ONORDER:

```
DATA WORK.STOCK WORK.ONORDER;
```

If you omit the SAS dataset name in the DATA statement, the special name _DATA_ is the default. When _DATA_ is indicated as an output SAS dataset name, the SAS supervisor generates a name by appending a sequential numeric suffix to DATA in the WORK library. That is, it creates the SAS dataset WORK.DATA1 the first time you have it generate a name, WORK.DATA2 the second time, and so on. To run a data step that does not create a SAS dataset, use the special name _NULL_ in the DATA statement:

```
DATA _NULL_;
```

An output SAS dataset named in the DATA statement is created as a new SAS dataset. You can use dataset options on any of these SAS datasets. Write the dataset options in parentheses after the SAS dataset name.

A SAS dataset that is being updated is also listed in the DATA statement, but it is an existing SAS dataset, not a new one, and it should not have dataset options in the DATA statement. Details of updating a SAS dataset can be found later in this chapter.

Output Variables

By default, the output SAS datasets have all the variables that are available in the data step. However, not every variable in a data step is available to be used as an output variable. These variables are never written to output SAS datasets:

- Automatic variables
- Variables used in I/O statement options or the IN= dataset option
- Temporary variables used as elements of an array
- Variables named in a DROP statement or, if there is a KEEP statement, variables not named in the KEEP statement

If you need to write one of these variables to an output SAS dataset, use an assignment statement to assign the value of the variable to another variable.

You can select the variables that are stored in a new SAS dataset. Use the KEEP= dataset option in the DATA statement, for example:

```
DATA WORK.STOCK (KEEP=SKU QUANTITY DATE)
   WORK.ONORDER (KEEP=SKU QUANTITY DATE SUPPLIER)
   ;
```

Output variables are stored in the output SAS dataset with the attributes they have in the data step, including the informat, format, and label attributes. The length attribute or each variable determines the number of bytes used to store the variable in the SAS dataset. If you use the RENAME statement in the data step, variables are stored with the new names indicated in the RENAME statement, rather than the names they have in the data step.

Writing Observations

The OUTPUT statement writes observations to output SAS datasets — or if you prefer, the data step can write output observations automatically. Output observations are added to the end of the output SAS dataset.

Use the OUTPUT statement with no additional terms to add an observation to every output SAS dataset:

```
OUTPUT;
```

To write an observation to one or more selected SAS datasets, name the SAS datasets in the OUTPUT statement, for example:

```
OUTPUT WORK.FLOOR WORK.FIELD;
```

You cannot use dataset options in the OUTPUT statement. Write all dataset options for an output SAS dataset in the DATA statement.

If there is no OUTPUT statement to explicitly write observations, the data step automatically writes the current observation at the end of each repetition of the observation loop. In the data step code, that is at the bottom of the data step or where a RETURN statement indicates the end of processing for an observation.

Data Step Views

You can use a data step to define an input data step view. A view created in a data step can be used as an input SAS dataset in a later step. When you read from the view, its program executes to create the observations of the view. But the program does not execute until you read from the view.

A data step that defines a view is very much like an ordinary data step. The only difference in the way it works is that instead of storing the output observation in a file, it passes the observation along to the step or program that is accessing the data in the view.

To write a data step that creates a view, write the name of the view as the output SAS dataset and again as the value of the VIEW= option in the DATA statement. There must be a slash between the SAS dataset name and the option. Then, write the rest of the statements of the data step in the usual way, usually to create an output observation based in some way on input data. Data step views can be difficult to debug, so test the data step to make sure it works before you add the VIEW= option to the DATA statement.

The step below is an example of a data step that defines an input data step view.

```
DATA WORK.PLACES / VIEW=WORK.PLACES;
  INFILE NEW;
  INPUT PLACE $18.;
RUN;
```

Subsequently, you can read from the view in the same way you can read from any kind of SAS dataset, for example:

```
PROC SORT DATA=WORK.PLACES OUT=WORK.LIST NODUPKEY;
  BY PLACE;
RUN;
```

When you define an input data step view, it is possible for the DATA statement to name another output SAS dataset. The other output SAS dataset is stored as a regular SAS data file when the view executes — that is, every time you read from the view. Usually, that would be a bad idea, but there are situations for which it makes sense. For example, you might need to store an exception list or transaction log that comes out of the processing of the data for the view.

Updating a SAS Dataset in the Data Step

Usually, the data step is used to create SAS datasets, but you can also use a data step to update, or edit, an existing SAS dataset, using data step logic to change values and add and delete observations.

Name the existing SAS dataset in the DATA statement, and read observations from it using any of the forms of the MODIFY statement, as described earlier in the chapter. After you read an observation, you can work with the values of the variables, testing them or assigning new values to the variables.

To store an updated observation in the SAS dataset after you change the values of variables, use the REPLACE statement. To delete an observation, use the REMOVE statement. You can also add observations to the end of the SAS dataset with the OUTPUT statement, which works the same way as it always works (but do not execute an OUTPUT statement between a MODIFY statement and a REPLACE or REMOVE statement). If you do not write any of these statements in the data step, the data step automatically writes the updated observation at the end of each repetition of the observation loop, equivalent to the action of the REPLACE statement.

This is an example of editing a SAS dataset, using IF-THEN statements to make several conditional changes in the values of the variables:

```
DATA MAIN.EVENT;
  MODIFY MAIN.EVENT;
  IF TIME = . THEN TIME = '20:00:00'T;
  IF STATUS = 'POSTPONED' THEN DATE = .;
  IF STATUS = 'CANCELED' THEN REMOVE;
  ELSE REPLACE;
RUN;
```

Not every kind of SAS dataset that a SAS program can read can be opened for update, as the MODIFY statement requires. If a SAS dataset is in a read-only library, is an input data step view or another kind of read-only view, or is in a library that uses a sequential or transport engine, you cannot use a data step to edit it.

Generally, because the SAS data set already exists, there is no need to use dataset options in the DATA statement. However, you can use most dataset options in the MODIFY statement. It is the MODIFY statement that accesses the file, so when you read the SAS dataset with dataset options in the MODIFY statement, the options also affect the way observations are written. For example, if you use the KEEP= option to read only certain variables from the SAS dataset, then those are the only variables you can write to the SAS dataset. If you use the RENAME= dataset option, it works in reverse when you write the variable back to the SAS dataset.

You can use the WHERE= dataset option in the MODIFY statement to limit the scope of the editing to observations that meet a certain condition. The other observations are unaffected. This can create a very concisely coded and efficient data step to make small changes or correct errors in a SAS dataset, as in this example:

```
DATA GROUP.PERSONNEL;
  MODIFY GROUP.PERSONNEL
    (WHERE=(LAST='Aster' AND FIRST IN ('Rich', 'Richard'))
     WHEREUP=NO);
  FIRST = 'Rick';
RUN;
```

When you use the WHERE= dataset option in the MODIFY statement, it can apply either to the input observations or to both the input and output observations, depending on the WHEREUP= dataset option. Use the option WHEREUP=YES to apply a WHERE clause on both input and output, preventing you from changing an observation in a way that would make it no longer meet the WHERE condition. Use WHEREUP=NO to use the WHERE clause only to select observations for input, as in the previous example. This allows observations that meet the condition to be changed so that they no longer meet the condition.

Exercise caution when you use the MODIFY and OUTPUT statements in the same step. If you use the sequential form of the MODIFY statement, and you use the OUTPUT statement to add observations to the SAS dataset you are updating, the new observations are added to the end of the SAS dataset. The MODIFY statement reads the new observations when it gets to that point. With an ordinary IF . . . THEN OUTPUT statement, using a condition based on values of variables in the SAS dataset, the data step logic would create an infinite number of observations. Such a data step would run indefinitely, ending in an error condition only when the available storage space is exhausted. If you use an IF . . . THEN OUTPUT statement, be sure that its condition will not also be true for the new observation the OUTPUT statement creates.

SAS Datasets in the Proc Step

Most procs require a SAS dataset as their source of input data. Many procs also produce output SAS datasets. Each proc has its own rules of syntax, but much of the syntax to refer to SAS datasets is consistent from one proc to another. The DATA= option in the PROC statement identifies the input SAS dataset. For example, the input SAS dataset for this PROC PRINT step is WORK.TALLY:

```
PROC PRINT DATA=WORK.TALLY (KEEP=TOPIC COUNT);
RUN;
```

You can use dataset options almost anywhere a proc step uses a SAS dataset name. The example above shows the KEEP= option to select variables from the SAS dataset. The WHERE= option is used frequently in proc steps, sometimes written as the WHERE statement. The KEEP= and RENAME= options can be useful for the output SAS dataset of a proc step when it creates more variables than you need or you want to give variables different names than the proc gives them.

In the proc step, you can use the BY statement to divide the input SAS dataset into groups that are processed separately. There are two differences in the syntax of the BY statement for the proc step, when compared to the data step. First, the GROUPFORMAT option is not available, because procs handle the values of categorical variables, such as BY variables, in a different way. Second, because the BY statement applies to the proc step as a whole, it can be written anywhere in the step.

The common elements of proc step syntax and execution, including its I/O syntax, are described in more detail in chapter 19.

If many procs produce output SAS datasets, even more produce print output. To create print output, a proc produces a set of objects that ODS, the Output Delivery System, converts to a print file. ODS options let you use the output objects in other ways. You can create SAS datasets from output objects by directing them to the Output destination of ODS. ODS is described in chapter 14, "Print Output."

The SQL proc uses SAS datasets in a way that is different from anything else you could find in a SAS program. It fits SAS data into the SQL data model, in which SAS data files are tables and variables are columns. SAS datasets can be read as tables in SQL. Conversely, the tables that SQL statements create are ordinary SAS data files, and SQL views are also SAS datasets that can be used in data and proc steps. SQL, the SQL proc, and SQL views are described in chapter 23.

11

Options for SAS Datasets

Options, indexes, integrity constraints, and password protection can affect the way you access a SAS dataset. Dataset options help determine the form of a SAS dataset and can change the way the dataset appears when you read it. Indexes speed up access to a SAS data file by keeping track of the location of the observations, and they can present the observations in a different order. Integrity constraints limit the values that can be stored in a SAS data file, so that invalid values cannot be added. You can use passwords to limit access to SAS files. There are various system options that can fine-tune the way SAS files work.

Dataset Options

Dataset options are options for the way a SAS dataset is accessed. To use dataset options, list them in parentheses after the SAS dataset name. Write each option as the option name, an equals sign, and the option value. The values of a few dataset options are expressed with equals signs. Values for these options — the INDEX=, WHERE=, and RENAME= options — have to be enclosed in parentheses. If there are multiple options, separate them with spaces.

```
SAS dataset (option=value  option=value ...)
```

You can use dataset options in most places where you write a SAS dataset name in a SAS program, including data step statements that mention SAS datasets, other than the OUTPUT statement, and in options in the proc step that identify input and output SAS datasets.

Many of the things that can be done with dataset options, such as selecting observations and renaming variables, could also be done in a separate data step or proc step. However, the dataset options use fewer computer resources than are used in a separate step. Dataset options should be your first choice for doing any of the things they can do, because they are more direct and usually more efficient than any other way of doing the same thing.

Dataset options can be used with most kinds of SAS datasets, but not with SQL views. If you indicate dataset options for an SQL view, the SAS supervisor writes a warning message and ignores the options.

Options for Observations

These dataset options give you various ways to control the way you use observations in a SAS dataset.

FIRSTOBS= and OBS= The FIRSTOBS= and OBS= options let you limit the number of observations you can access in an input SAS dataset. Both options take observation numbers as their values. The OBS= option indicates the observation number of the last observation to process. The FIRSTOBS= option indicates the observation number of the first observation to process.

IN= The IN= option lets you mark observations that come from an input SAS dataset in a data step statement. This option can only be used in data step statements and is only useful in data step statements that read two or more SAS datasets. The value of the IN= option is a numeric data step variable. The value of the variable is set to 1 for observations that come from the SAS dataset. It is set to 0 for other observations.

WHERE= The WHERE= option lets you limit the observations you read from or write to a SAS dataset to those that meet a certain condition. That condition is the value of the option. The condition must be enclosed in parentheses. A WHERE condition is written much like an expression in the data step, but it can only refer to variables of the SAS dataset; it cannot use any other variables. The application of a WHERE condition to a SAS dataset can be referred to as a *WHERE clause.*

On an input SAS dataset, a WHERE clause skips over observations that do not meet its condition rather than reading them. When appropriate indexes exist, they are used to speed up the processing of the WHERE clause. In a data or proc step, you can use the WHERE statement as the equivalent of the WHERE= dataset option for all the input SAS datasets in the step. The statement can appear anywhere in the step. Write the word WHERE followed by the condition:

```
WHERE condition;
```

For an output SAS dataset, a WHERE clause discards observations that do not meet its condition. Only observations that meet the condition are stored in the output SAS dataset.

WHEREUP= The WHEREUP= option determines how a WHERE clause applies when you are updating a SAS dataset. This option is especially useful when you access a SAS dataset with a MODIFY statement in the data step. The value of the option is either YES or NO. When the value is YES, the WHERE clause applies on both input and output. Observations that do not meet the condition are not read, and changed or new observations that do not meet the condition cannot be updated in or added to the SAS dataset. When the value is NO, the WHERE clause applies only on input. You can write output observations to the SAS dataset regardless of whether they meet the WHERE condition. This is the default behavior.

SORTEDBY= The SORTEDBY= option lets you describe how the SAS dataset is sorted. The value of the option is a sort order clause, the same as you would use in the BY statement of a PROC SORT step. You can use this option when you create a new SAS dataset that is already sorted. SAS routines such as the SORT and REPORT procs check the SORTEDBY= option, and they do not sort the SAS dataset if the option indicates that it is already sorted in the correct order.

Options for Variables

The KEEP=, DROP=, and RENAME= options change the way you access variables in a SAS dataset.

KEEP= and DROP= These two options let you select the set of variables you use in a SAS dataset. The value of either option is a list of variables. The default action is to read all variables from an input SAS dataset and to write all available variables in a data step or all variables that are part of a result in a proc step to the output SAS dataset. When you use the KEEP= option, only the variables listed in the option are read from or written to the SAS dataset. When you use the DROP= option, all variables other than the variables listed are read from or written to the SAS dataset. You cannot use both the KEEP= and DROP= options at the same time.

RENAME= Use the RENAME= option to change the names of variables. With the RENAME= option, you can store a variable with a name that is different from the name it has in the step, or you can read a variable and use it with a name that is different from the name it has in the SAS dataset.

The value of the option is a list of name changes, enclosed in parentheses. Each term in the list is an old name, an equals sign, and a new name, or a list of old names, an equals sign, and a list of a new names.

```
RENAME=(old=new ...)
```

An example of a situation where the RENAME option is necessary is when you want to merge two or more parts of the same SAS dataset. For example, suppose the SAS dataset MAIN.EMP contains employment counts and payroll totals for both 1973 and 1974, and you want to rearrange the data so that those two years appear on the same observation. A data step that did that could be written like this:

```
DATA MAIN.EMP2 (KEEP=REGION SECTOR EMP1973 EMP1974 PAY1973 PAY1974);
  MERGE
     MAIN.EMP (WHERE=(YEAR=1973) RENAME=(EMP=EMP1973 PAY=PAY1973))
     MAIN.EMP (WHERE=(YEAR=1974) RENAME=(EMP=EMP1974 PAY=PAY1974))
     ;
  BY REGION SECTOR;
RUN;
```

It might look odd to write a merge that uses the same SAS dataset twice, but it is a useful way to rearrange the data in a case like this. As a result of the RENAME= options, the variables EMP and PAY turn into EMP1973 and

PAY1973 for the 1973 data, and EMP1974 and PAY1974 for the 1974 data. Using distinct names for the two years lets you include values for both years in the observation that the MERGE statement puts together.

The RENAME= option is the last dataset option to be applied to the variables of an input SAS dataset. If you use the RENAME= option at the same time as any other dataset option that refers to the same variables, use the old names of the variables in the other dataset option. For an output SAS dataset, the RENAME= option is applied after the KEEP= and DROP= options, but before the WHERE= and INDEX= options. For the WHERE= and INDEX= options, use the variable name that is actually stored in the SAS dataset.

Options for File Management

Several dataset options are designed to let you use different kinds of SAS datasets in different ways.

COMPRESS= Any way of storing data in less space than the data normally occupies can be called compression. SAS includes two different ways to compress the observations in SAS data files. One compression algorithm is designed to compress character data. The other is better suited for binary data. Either can reduce the stored size of some SAS data files, but the amount of compression depends on the nature of the data in the SAS data file. For many SAS data files, neither algorithm can reduce their size significantly.

The way a SAS data file is compressed is determined when the SAS data file is created. Use the COMPRESS= dataset option when you create the SAS data file to set the kind of compression the SAS data file will use. With the value CHARACTER, or YES, character compression is applied. With the value BINARY, binary compression is applied. Use the value NO to store the SAS data file without compression.

REUSE= If a SAS data file is stored with compressed observations and some of those observations are subsequently deleted, the library engine can either leave that space empty or reuse it for any observations that might be added. Use the REUSE= option when you create a SAS dataset to control the way this works.

With a YES value for the option, the engine reuses the space of deleted observations. This saves storage space, and it is usually more efficient for SAS data files in which observations are frequently deleted or are usually added one at a time. With a NO value, the engine does not reuse the space of deleted observations. This tends to be more efficient for SAS data files in which not many observations are deleted or to which many observations are added at once.

This option affects only SAS data files that are compressed and in which observations may be deleted. Observations can be deleted only when a SAS data file is edited or otherwise updated. Most SAS data files are never updated, so this option does not usually matter.

REPLACE= When you create a new SAS dataset, it automatically replaces an existing SAS dataset that has the same name. There may be times when you want to prevent an existing SAS dataset from being replaced by a new SAS dataset. Use the REPLACE= dataset option on the new SAS dataset for that purpose. With the value YES, the new SAS dataset replaces an existing SAS dataset of the same name. With the value NO, if a SAS dataset of the same name already exists, the new SAS dataset is not created and an error condition results.

POINTOBS= For a compressed SAS data file, the POINTOBS= option determines whether you can access observations directly by observation number. With the value YES, you can use the POINT= option to access the SAS data file. With the value NO, the SAS data file is slightly smaller.

LABEL= The LABEL= option provides a descriptive label for a SAS dataset. If you use this option when you create a SAS dataset, the label is stored in the SAS dataset. Write the label as a character constant. The label can be up to 256 characters long.

CNTLLEV= Control level is sometimes important when you access an existing SAS dataset. When possible, the SAS supervisor lets more than one program access a SAS dataset at the same time while still avoiding conflicts between the programs. The programs that can access a SAS dataset include data steps, proc steps, AF applications, and windows in the interactive environment. In a SAS session, it is possible for many such programs to be running at the same time. When one program accesses a SAS dataset, it is the control level that the program uses that determines whether another program is able to access that same data at the same time.

When it uses the record control level, the program accesses the SAS dataset one record, or observation, at a time. Another program can access the SAS dataset at the same time, but it cannot access the same observation at exactly the same time.

Some processes, however, require the member control level. With this control level, the program uses the entire SAS dataset, and no other program can access the same SAS dataset until the program is finished with it. Sorting is an example of a process that requires the member control level. The sorting process would not work if another program could be changing the data values while the sort was going on.

Ordinarily, procs and data step statements use the record control level when they can, and they use the member control level when that is required. However, you can override that and set a different control level with the CNTLLEV= dataset option.

For example, a direct access SET statement ordinarily uses a record control level. That would allow another program to change observations in the same dataset. But the logic of the data step might depend on the values throughout the SAS dataset staying the same for the duration of the data step. If so, use the CNTLLEV= dataset option in the SET statement to use a member control level.

The values allowed in the CNTLLEV= dataset option are RECORD, MEMBER, and LIBRARY. The library control level prevents any other program from accessing any member of the same library while you are accessing a SAS dataset.

TYPE= The TYPE= option indicates the dataset type. The type is a short code that identifies a specialized use of a SAS dataset, with specific variables and observations. Several procs use the type code to indicate a particular kind of data, especially when a SAS data file is created by one proc, then used by another proc. If you use a data step to create such a SAS dataset, use this dataset option to set the dataset type.

Other Options

There are many other dataset options with more specialized uses. The other dataset options are listed here according to the general subject areas that they fall into.

Subject	Dataset Options	
File Access	BUFNO=	OUTREP=
	BUFSIZE=	TOBSNO=
	DLDMGACTION=	TRANTAB=
	FILECLOSE=	
Indexes	INDEX=	IDXWHERE=
	IDXNAME=	
Generation Datasets	GENNUM=	GENMAX=
Passwords	READ=	PW=
	WRITE=	PWREQ=
	ALTER=	ENCRYPT=

Modifying SAS Datasets

The DATASETS proc is a utility proc that lets you do various things with SAS files. Most of the things you can do in the DATASETS proc are described as *modifying* a SAS dataset; *modify,* in this context, includes all changes to a SAS dataset other than changes in its data values.

The DATASETS proc is designed to work with all the members of a particular library. Name the library in the LIBRARY= option in the PROC statement. Name the members you want to work with in the individual action statements of the proc. By default, the DATASETS proc produces a list in the log of all the members of the library. In a production program or a program that runs in a batch environment, the list of members is not usually helpful, so you can use the NOLIST option to suppress it. If you want to restrict processing to SAS data files, use the MEMTYPE=DATA option in the PROC DATASETS statement.

This is an example of a PROC DATASETS statement:

```
PROC DATASETS LIBRARY=WORK NOLIST;
```

Renaming SAS Files

There are three statements in the DATASETS proc that are designed for renaming SAS files: the CHANGE, EXCHANGE, and AGE statements. The AGE statement also deletes SAS files.

In the CHANGE statement, write the current member name of the SAS file, an equals sign, and the new member name you want to give it. For example, this statement changes the name of the member PERSONNEL to ER:

```
CHANGE PERSONNEL=ER;
```

In the EXCHANGE statement, write two member names you want to exchange between two files, with an equals sign between them. This statement, for example, changes the name of TEMP to MAIN while at the same time changing the name of MAIN to TEMP:

```
EXCHANGE TEMP=MAIN;
```

The AGE statement contains a list of names. It changes the name of each member to the next name on the list. It deletes the last member listed. This statement, for example, changes the name of the member NEW to OLD, changes the name of OLD to ARCHAIC, and deletes ARCHAIC:

```
AGE NEW OLD ARCHAIC;
```

Modifying SAS Datasets

In the DATASETS proc, the MODIFY statement and its many secondary statements modify specific SAS datasets in various ways. Write the member name of the SAS dataset to modify in the MODIFY statement. You can write some dataset options after the SAS dataset name in the MODIFY statement. For example, use the LABEL= dataset option in the MODIFY statement to change the label of the SAS dataset. After the MODIFY statement, you can write any combination of the secondary statements.

These secondary statements include the INFORMAT, FORMAT, and LABEL statements, which you can write the same way as in the data step, to change the attributes of variables in the SAS dataset. You can also write variable names without attributes in these statements to clear the attributes of variables. You can use the RENAME statement, which also works the same way as in the data step, to rename variables in the SAS dataset. There are additional secondary statements for working with indexes, integrity constraints, and audit trails.

Deleting SAS Files

To delete SAS files, you can use the DELETE statement of the DATASETS proc. List the members to delete in the DELETE statement.

```
DELETE member ...;
```

To delete all the members of a library, you can use the KILL option in the PROC DATASETS statement. This step, for example, deletes all the members of the WORK library:

```
PROC DATASETS LIBRARY=WORK NOLIST KILL;
RUN;
```

Repairing SAS Files

It is possible for a SAS dataset or catalog to be damaged if a software or hardware failure occurs while the file is open or if the storage device on which the file is located is damaged. SAS software can sometimes repair the damaged file. List the library members to repair in the REPAIR statement of the DATASETS proc:

```
REPAIR member ...;
```

The proc attempts to repair the files. It writes log messages that describe its actions and indicate the success or failure of the repair.

Indexes

An index is a list of the observations in a SAS data file. It contains the values of selected key variables of the SAS data file, along with the locations of the observations that have each different key value. A SAS data file's indexes are stored in a separate file in the same location as the SAS data file, but there is no way to access an index by addressing it directly. You use an index only by using the SAS data file that the index belongs to.

The Purpose of Indexes

Indexes have several purposes, each of which has something to do with accessing the observations of SAS data files.

- Indexes allow faster processing of WHERE clauses.
- An index is necessary for the lookup style of access to a SAS data file, using the KEY= option in the SET and MODIFY statements.
- When you read a SAS data file with a BY statement, an index can present the observations in an order different from the order they are stored in.
- With the UNIQUE option, an index can prevent duplicate observations from being added to a SAS data file.
- Integrity constraints often require indexes. An integrity constraint limits the kinds of values that a variable in a SAS data file can take on, and to do this, the integrity constraint may require one or more indexes to

keep track of the observations.

Indexes use computer resources in two ways. They take up a small amount of storage space, and a small amount of processing is required to maintain an index every time an observation is added, deleted, or changed in the SAS data file. Because of these costs of an index, you should create an index only if there is a reason to. However, as long as there is any reason to have an index, the costs of creating and maintaining the index tend to be small, often too small to notice.

Simple and Composite Indexes

Two different kinds of indexes, simple and composite indexes, are created and named in different ways. A simple index is an index on one key variable of the SAS data file. Its name is the same as the name of the variable. A composite index is an index on two or more key variables in the SAS data file. Its name cannot be the same as the name of any variable in the file.

Creating Indexes for a New SAS Data File

You can create indexes at the same time that you create a new SAS data file. Use the INDEX= dataset option. The indexes are built as each observation is added, and they are complete as soon as the SAS data file is complete.

The value of the INDEX= option is a list of index definitions. The list must be enclosed in parentheses. An index definition can be either of the following:

```
variable
index=(variable variable . . .)
```

For a simple index, the index definition is just the variable name. For a composite index, the index definition is the index name, an equals sign, and a variable list in parentheses.

The example below is a data step that creates a SAS data file with two indexes: a simple index on the variable SIZE and a composite index ITEM on the variables SIZE and COLOR.

```
DATA MAIN.STOCK (INDEX=(SIZE ITEM=(SIZE COLOR)));
  INFILE X;
  INPUT SIZE COLOR QUANTITY;
RUN;
```

Creating Indexes for an Existing SAS Data File

If a SAS data file already exists, you can add indexes to it using statements in the DATASETS and SQL procs.

In the DATASETS proc, use the INDEX CREATE statement to create an index. The INDEX CREATE statement must follow the MODIFY statement that indicates the SAS data file for which the index is being created. The necessary statements to create an index in the DATASETS proc are:

```
PROC DATASETS LIBRARY=libref options;
MODIFY member;
```

```
  INDEX CREATE index definition . . . / index options;
RUN;
```

Write the index definition in the INDEX CREATE statement the same way you would write it in the INDEX= option. To create multiple indexes, you can write more than one index definition in a single INDEX CREATE statement, or you can write a separate INDEX CREATE statement for each index.

There are index options you can use in the INDEX CREATE statement. These options are described below.

The SQL statement that creates an index is the CREATE INDEX statement. This statement indicates the index name, the SAS dataset name, and the list of variables (or columns) used in the index. Following the SQL style, use commas between the variables in the list. The statements to create an index are:

```
PROC SQL;
CREATE INDEX index ON SAS dataset (variable, variable, . . . );
```

These are examples of SQL statements to create indexes:

```
CREATE INDEX ID ON CORP.PERSONNEL (ID);
CREATE INDEX TRANSFER ON MAIN.NETWORK (SRC, DST);
```

Index Options

There are three options you can use when you create an index to change the way the index works.

UNIQUE The UNIQUE option creates a *unique index*, an index that does not allow duplicate observations. The index allows only one observation for each key value. If the SAS dataset has duplicate observations, you cannot create the unique index, and an error condition results. After the index is created, it prevents you from adding observations or changing values in the SAS data file if that action would result in a duplicate observation.

NOMISS The NOMISS option excludes missing key values from the index. This is a useful thing to do if all the following conditions are true: many or most of the key values in the SAS data file are missing values; the purpose of the index is for processing of WHERE clauses or for the KEY= option in the SET and MODIFY statements; and you will not be looking for missing key values when you use the index. In this special case, the /NOMISS option can speed up index processing significantly.

UNIQUE and NOMISS combined Use the UNIQUE and NOMISS options together to create an index that forces nonmissing key values to be unique but also allows missing values.

UPDATECENTILES= As part of keeping track of the observations in a SAS data file, an index divides the key values into 100 ranks, called *centiles*. It updates these centiles when values in the SAS data file change and observations are added. However, it does not usually update centiles after every single change in an observation. For many SAS data files, that would take an excessive amount of time. The UPDATECENTILES= option indicates how often to update the centiles. The value of the option is a percent, and the centiles are updated after the number of changed observations adds up to that percent of the total number of observations in the SAS data file.

The default value for this option is `UPDATECENTILES=5`. That means that the centiles in the index are updated after 5 percent of the observations are changed.

Values for the option can range from 0 to 101. A value of 0 means that the centiles are updated every time an update to the SAS data file results in a change in the index. You can also write this value of the option as the word ALWAYS. A value of 101 means that the centiles are never automatically updated, not even if every observation changes. You can also write this value of the option as the word NEVER. For most applications, values between 1 and 20 work well.

You can shorten the name of this option by using its alias, UPDCEN.

All three of these options can be used at the end of the `INDEX CREATE` statement in the DATASETS proc. Write a slash between the index definitions and the options.

The UPDATECENTILES option can be changed in the `INDEX CENTILES` statement. In the `INDEX CENTILES` statement, write the name of one or more indexes, then a slash, then either or both of the REFRESH and UPDATECENTILES= options. The REFRESH option updates the centiles of the index immediately.

In the INDEX= dataset option, you can use the UNIQUE and NOMISS options after an index definition. An option must be written with a slash before it so that it cannot be mistaken for a simple index definition. If you use the two options together, write a slash before each option, for example:

```
INDEX=(ITEM=(SIZE COLOR)/UNIQUE/NOMISS)
```

When you create an index in an SQL statement, only the UNIQUE option is available. Write the word UNIQUE before the word INDEX. That is, the statement begins `CREATE UNIQUE INDEX` . . .

Using, Maintaining, and Deleting Indexes

Using and maintaining indexes tends to be easier than creating them. Most of the time, the SAS supervisor automatically selects the appropriate index to use when you access the SAS data file. The SAS supervisor automatically maintains and updates indexes as needed when it adds and changes observations.

These are statements and options that can prompt SAS to use an index when reading a SAS data file:

BY statement When you use a BY statement to indicate the ascending sorted order of a SAS data file — that is, a BY statement with no options — the SAS supervisor can use an index, if it exists, to read the observations of the SAS data file in that order.

To be used in this way, the index has to include the BY variables. If there is only one BY variable, it can be the only key variable of the index or the first key variable of the index. If there are multiple BY variables, they have to appear, in the same order, as the key variables of the index. However, an index can still be used for a BY statement if it has additional key variables after the BY variables. For example, if the list of variables in the BY statement is `A B`, an index on `A B`, `A B C`, or `A B C D` could be used for that BY statement.

Indexes with the NOMISS option are not used for processing BY statements.

WHERE clause When a SAS dataset is used with a WHERE clause, an index may be used to speed up processing of the WHERE clause. The SAS supervisor decides whether to use an index or not based on its estimate of the amount of time it will take to process the BY clause with any of the available indexes, compared to not using an index. Generally, an index is helpful if the index contains the more important variables in the WHERE clause and the result of the WHERE clause is a selection of relatively few of the observations in the SAS data file.

There are two dataset options you can use to control the selection of an index when you read a SAS dataset with WHERE clause processing. Use the IDXWHERE= option to indicate whether an index should be used when reading a SAS data file with a WHERE clause. The value of the option can be YES or NO. The YES value tells the SAS supervisor to use the best available index, even if it might be faster to access the data without an index. The NO value tells the SAS supervisor to ignore indexes and process the WHERE clause without the use of an index.

Use the IDXNAME= option to tell the SAS supervisor to use a specific index for accessing the SAS data file with a WHERE clause. That index is used provided that it meets two conditions. The main condition is that the first variable of the index must appear in the WHERE condition. Also, if you use a WHERE clause and a BY statement at the same time, the index selected must also be suitable for the BY statement. This requires that the first BY variable is part of the WHERE condition.

An index created with the NOMISS option can be used for WHERE clause processing only if the WHERE clause selects only nonmissing values of the key variables. If it is possible for the WHERE clause to select an observation with missing key values, the index is not used.

KEY= option When you use the KEY= option in the SET or MODIFY statement, as described in the previous chapter, the statement uses the index you indicate to find input observations.

If, in the KEY= option, you indicate an index created with the NOMISS option, make sure you only look up observations with nonmissing key values. If you look for an observation with a missing key value, the SET or MODIFY statement will fail to find that observation.

The SAS supervisor automatically updates indexes when you update a SAS data file. This is true whether you use a data step, proc step, AF application, or window in the interactive environment to change, add, or delete observations or to change the names of variables.

However, there are some processes in which the SAS supervisor is unable to maintain indexes. If you sort a SAS data file in place, all its indexes are automatically deleted. If you remove a variable that is a key variable of an index, the index is deleted. Whenever you delete a SAS data file, all its indexes are also deleted. This is true even if the deletion results from creating another SAS dataset of the same name.

When you copy a SAS data file, you can elect to copy its indexes or not.

You can use either the DATASETS or SQL proc to delete an index. Either way, you only need to identify the SAS dataset and the index to delete. In the DATASETS proc, use the INDEX DELETE statement.

```
PROC DATASETS LIBRARY=libref options;
MODIFY member;
  INDEX DELETE index . . . ;
RUN;
```

In the SQL proc, use the DROP INDEX statement.

```
PROC SQL;
DROP INDEX index FROM SAS dataset;
```

It takes some time for SAS to maintain indexes whenever you update a SAS data file. A program that makes extensive changes to a SAS data file can often run faster if you delete the indexes first, then create new indexes after the update is complete.

Integrity Constraints

Like an index, an integrity constraint is something added to a SAS data file. The purpose of an integrity constraint is to maintain the integrity of the data in the SAS data file by constraining, or limiting, the values that can be stored in the SAS data file. Each integrity constraint applies a specific rule of data integrity to the values in the SAS data file. If you attempt to change or add an observation in a way that would violate that rule, the integrity constraint prevents you from doing so and generates and error condition.

The most common kind of data integrity rule is a rule that restricts the values of a specific variable. Physical measurements are often subject to these limitations. A variable that measures distance, for example, cannot have a negative value. A distance can be 0 or positive, or perhaps a missing value, but a negative value would be wrong. You could express this restriction as a condition:

```
  DISTANCE >= 0 OR DISTANCE <= .Z
```

The first comparison, `DISTANCE >= 0`, tests for numbers that could be valid measures of distance. The second comparison, `DISTANCE <= .Z`, tests for missing values, including special missing values.

If you create an integrity constraint based on this condition, it prevents you from accidentally adding an observation that has a negative value for

DISTANCE or changing an existing observation to make DISTANCE negative.

Rules

When you create an integrity constraint, there are five types of rules you can choose from.

CHECK A CHECK rule lets you use a condition to validate the value of an individual variable or any combination of the variables. When you write a CHECK rule, it takes this form:

```
CHECK(WHERE condition)
```

where *condition* is an expression that you can write according to the same rules you use to write any WHERE condition.

NOT NULL The NOT NULL rule for a variable prevents the variable from having a missing value. Write the rule with the name of the variable in parentheses:

```
NOT NULL(variable)
```

UNIQUE or DISTINCT The UNIQUE rule for a variable requires unique values for the variable in every observation. No two observations can have the same value for the variable. The UNIQUE rule for a set of variables requires a unique combination of values for those variables in every observation. Write this rule with the variable or list of variables in parentheses. You can use the word UNIQUE or DISTINCT.

```
UNIQUE(variable)
DISTINCT(variable)
UNIQUE(variable variable . . . )
DISTINCT(variable variable . . . )
```

PRIMARY KEY The PRIMARY KEY rule defines a primary key based on one or several key variables. Each variable in the primary key is required to have a nonmissing value. The combination of key values is required to be unique in each observation, the same as in a UNIQUE rule. There can be only one primary key for a SAS data file.

Defining a primary key allows it to be referenced in a FOREIGN KEY rule of another SAS data file. Options in the FOREIGN KEY rule can prevent you from deleting certain observations from the SAS dataset that contains the primary key.

List the key variables when you write a PRIMARY KEY rule:

```
PRIMARY KEY(variable . . . )
```

FOREIGN KEY A FOREIGN KEY rule defines a foreign key based on one or several variables. These variables are the same as the variables of a primary key of a different SAS dataset. The foreign key values are required to match the values in an observation of the primary key that the foreign key refers to.

A FOREIGN KEY rule must indicate the key variables and the name of the SAS dataset of the primary key that the foreign key refers to. Often, other terms are needed.

```
FOREIGN KEY(variable ...)
  REFERENCES SAS dataset optional variable  action options
```

A foreign key of one variable can refer to a primary key that has a different variable name. If so, write the primary key variable name after the name of the SAS dataset that contains the primary key.

There are several action options you can use at the end of the FOREIGN KEY rule. These options determine the reactions of the foreign key and primary key when you try to update or delete a primary key value that the foreign key refers to. For each of the two actions, UPDATE and DELETE, there are two alternatives, RESTRICT and SET NULL, resulting in four possible options:

Action Attempted in Primary Key	Action Option in Foreign Key	Response of Integrity Constraints
Change a value in an observation that the foreign key refers to	ON UPDATE RESTRICT	Prevent the change and generate an error condition
	ON UPDATE SET NULL	Set the foreign key values to missing values
Delete an observation that the foreign key refers to	ON DELETE RESTRICT	Prevent the deletion and generate an error condition
	ON DELETE SET NULL	Set the foreign key values to missing values

Creating Integrity Constraints

Integrity constraints can be created and managed by three statements in the DATASETS proc. These statements are similar to the statements that work with indexes. As with indexes, use the PROC and MODIFY statements to identify the SAS data file that the integrity constraints belong to.

The `IC CREATE` statement creates an integrity constraint. The form of the statement is:

```
IC CREATE name = rule MESSAGE='message text';
```

In this statement, the name of the integrity constraint is a SAS name of your choice. The rule is written as described in the previous few pages. The MESSAGE= option provides the text of an error message that the integrity constraint displays when it generates an error condition. You can omit the MESSAGE= option to have the integrity constraint display a generic error message.

The only names you cannot use for an integrity constraint are the words CHECK, DISTINCT, FOREIGN, MESSAGE, NOT, PRIMARY, and UNIQUE. You also cannot use any name that is already being used for another integrity constraint for the SAS data file.

This is an example of a statement to create an integrity constraint:

```
IC CREATE VALIDDISTANCE = CHECK(WHERE DISTANCE >= 0 OR DISTANCE <= .Z)
  MESSAGE = 'A value for distance must be positive or 0.';
```

Managing Integrity Constraints

There are various error conditions that can result in a foreign key being dwactivated, such as when the SAS data file that contains the primary key cannot be found. If this occurs, you can reactivate the foreign key with the IC REACTIVATE statement. The statement indicates the name of the integrity constraint and the libref of the SAS data file that contains the primary key that the foreign key refers to.

```
IC REACTIVATE integrity constraint REFERENCES libref;
```

To delete integrity constraints, name them in the IC DELETE statement.

```
IC DELETE integrity constraint . . . ;
```

Integrity Constraints in SQL

There are also SQL statements and elements to create and manage integrity constraints. You can use the DESCRIBE TABLE CONSTRAINTS statement to list the integrity constraints of a SAS data file, for example:

```
PROC SQL;
DESCRIBE TABLE CONSTRAINTS MAIN.CUSTOMER;
```

Use the ALTER TABLE statement to create a new integrity constraint for an existing SAS data file, for example:

```
PROC SQL;
ALTER TABLE MAIN.CUSTOMER ADD CONSTRAINT U_CUSTOMER UNIQUE(CUSTOMER);
```

When you write a CHECK constraint in the ALTER TABLE statement, leave out the word WHERE.

You can also create integrity constraints when you create SAS data files in the CREATE TABLE statement of SQL. List each constraint definition, starting with the keyword CONSTRAINT, at the end of the list of table columns.

Generation Datasets

The idea of generation datasets is that you can keep multiple versions of the same SAS dataset. When you define a SAS dataset as a generation dataset, the SAS supervisor does not delete the old versions of the SAS dataset when you create (or generate) a new SAS dataset of the same name. You can use the GENNUM= option to access previous versions of the SAS dataset.

It is the GENMAX= dataset option that defines a SAS dataset as a generation dataset. Use this option when you create a generation dataset. The value of the option is a whole number that indicates the maximum number of generations that are kept. For example, up to 8 generations are kept for the SAS dataset created in this DATA statement:

```
DATA CORP.EVENTS (GENMAX=8);
```

The default value for the GENMAX= option is 0. This value indicates a standard SAS dataset, for which generations are not stored.

To change the number of generations for a generation dataset, use the GENMAX= dataset option with a new value in the MODIFY statement of the DATASETS proc.

By default, when you access a generation dataset, you get the newest generation. You can use the GENNUM= dataset option to access a specific generation of a generation dataset. Use a positive value for the option to indicate a sequential generation number. Use a value of 0 to get the base generation, the newest version, or the negative numbers –1, –2, and so on, for successively older historical generations. For actions that affect all the available generations, for example, to rename them all at once, use the option GENNUM=*.

Use the GENNUM statement option in the DELETE statement of the DATASETS proc to delete selected generations of the SAS dataset. Write the option after a slash at the end of the statement. In addition to the numeric options, you can use the option ALL to delete all generations, HIST to delete all historical generations while keeping the base generation, and REVERT to delete the base generation (which makes the most recent historical generation the base generation). For example, this step deletes all generations except the current one in the generation dataset CORP.EVENT:

```
PROC DATASETS LIBRARY=CORP NOLIST;
DELETE EVENT / GENNUM=HIST;
RUN;
```

The MODIFY statement of the DATASETS proc can set a new value for the GENMAX= option. To delete the oldest generations, reduce the GENMAX= value to something less than the number of generations that currently exist.

If you delete or rename the base generation, the newest historical generation automatically becomes the base generation again.

Passwords

Password protection for a file makes the file available for access only when a secret word, the password, is supplied. The SAS language provides a mechanism for password protection of SAS datasets and some other SAS files. But the SAS file password scheme requires the passwords to be written down in the SAS programs that access the files. That would tend to make the passwords not quite so secret as a password is supposed to be. If you use SAS file passwords, make sure you also have a way to protect the SAS programs that contain the passwords.

For password purposes, access to SAS files is divided into three levels: READ, for any kind of access that does not change the file; WRITE, for actions that may change the data values or add or remove data; and ALTER, for any other actions on a file. You can use different passwords for these three levels, use the same password for all levels, or protect the higher levels while leaving lower levels unprotected.

Password protection is implemented primarily with dataset options. The dataset options are READ=, WRITE=, and ALTER= for the three levels of password protection, or PW=, which applies to all three levels at once.

Use a password option to apply a password when you create a file, then use the same option later to access the file.

You can change, add, or remove passwords by using the password options in the MODIFY statement of the DATASETS proc. Generally, write the old password, then a slash, then the new password. Specifically, these are the ways the password options can be used:

Option Syntax	Effect
password option=*password* . . .	Assign passwords when creating a file
password option=*password*	Access a protected file
password option=/*password*	Add a password to an existing file
password option=*password*/	Remove a password
ALTER=*password* *password option*=/	
WRITE=*password* READ=*password*/ [a]	
password option=*old password*/*new password*	Change a password
ALTER=*password* *password option*=/*new password*	
WRITE=*password* READ=*old password*/*new password* [a]	

[a] Removing or changing the READ password for a file that is write-protected but not alter-protected

There are two other dataset options that affect the way passwords work. The ENCRYPT=YES dataset option encrypts a password-protected file, including any indexes, making it more difficult for any other program to read any data values in the file. Passwords of an encrypted SAS file cannot be changed, because the password is part of the encryption process. To give the file a new password, you would have to create a new copy of the file.

The PWREQ= dataset option lets you control the use of dialog boxes, called requestor windows, for passwords. By default, if you are running a SAS program in an interactive session, and the program attempts to access a protected SAS file, but does not supply a password or supplies an incorrect password, the SAS supervisor displays a requestor window in which you can enter the password. With the dataset option PWREQ=NO, the SAS supervisor does not display the password requestor window. It simply creates an error condition when the correct password does not appear in the program.

Password options can also be used to protect other types of SAS files, with the exception of catalogs. Password options can be used with SQL views, even though SQL views do not support any other kind of dataset option.

Audit Trails

In some applications it can be necessary to know not just the current state of the data in a SAS data file, but also how it got to be that way. Storing the details of changes to the data in a file can make it possible to correct errors and detect and deter fraud. Fraud is a special concern in accounting and security applications. But even when fraud is less of a concern, as in a file of customer information, a record of changes may still be essential to discover and correct errors that are made as data is updated.

An audit trail is a feature of a SAS data file that records all or selected changes to the observations of the file. The audit trail associates the SAS data file with a second SAS data file, the audit file, which records the details of changes to the SAS data file. The audit file duplicates all the variables of the SAS data file and has additional variables of its own.

When the SAS data file is updated, any number of changes in one observation are considered as a single event. Each changed observation is treated as a separate event. The audit file records one observation for each event in which a record is added or deleted. It records two observations for an event that changes an existing record — one observation that shows the values of variables before the changes, and another that shows the updated values.

Establishing an Audit Trail

The AUDIT and INITIATE statements of the DATASETS proc establish an audit trail and create the audit file. Indicate the library in the `PROC DATASETS` statement and the member name of the SAS data file in the AUDIT statement. For example, these statement establish an audit trail for a SAS data file named CORP.SALES:

```
PROC DATASETS LIBRARY=CORP NOLIST;
AUDIT SALES;
INITIATE;
```

If the SAS data file is password-protected, you must indicate the password as an option in the AUDIT statement.

When you establish an audit trail, you can use the LOG statement to limit the events logged in the audit file. The default is to log all events, which you could write as:

```
LOG BEFORE_IMAGE=YES DATA_IMAGE=YES ERROR_IMAGE=YES;
```

Use the value NO with one or two of these options to avoid logging before-update, successful, and unsuccessful events, respectively.

Audit Variables

The audit file contains all the variables of the SAS data file. It also contains six variables to describe the event that resulted in the observation in the audit file:

Audit Variable	Description
ATOBSNO	The observation number of the observation that was changed
ATDATETIME	A SAS datetime value indicating the date and time of the event
ATOPCODE	An opcode, indicating the type of event (see below)
ATUSERID	The user ID of the user or job
ATRETURNCODE	The return code from the event, with a zero value indicating that the event completed successfully
ATMESSAGE	The text of the log message, if any, generated by the event

The two letters of the opcode indicate the type of event that the audit observation logs. The first letter is D for a successful event or E for an event that fails with an error condition. The second letter is A, D, R, or W, indicating an added observation, a deleted observation, an observation logged before an update, and an observation logged after an update. This results in seven possible opcode values:

Action on observation	Before	Success	Failure
Add		DA	EA
Delete		DD	ED
Update	DR	DW	EW

An audit trail can also contain user variables, which are variables that are stored in the audit file, but are accessed as if they were in the SAS data file when you open the file to update it. Define the attributes of any user variables in the USER_VAR statement when you initiate the audit trail. For each variable, write:

- The name.
- For a character variable, $ to indicate the character data type.
- The length of the variable, if necessary.
- A LABEL= option for the label attribute, if necessary.

This is an example of the use of the USER_VAR statement to create a user variable:

```
USER_VAR REASON $ 27 LABEL='Reason for update';
```

Using and Managing the Audit Trail

The audit file has the same name as the SAS data file it logs, but it has the SAS dataset type AUDIT. Use the dataset option TYPE=AUDIT to access the audit file directly. You can print the audit file or use it as input to a program that analyzes the events of the audit trail.

The audit trail automatically adds observations to the audit file when changes are made to the SAS data file. This can slow down operations on the SAS data file, especially when a program makes a large number of changes in the file. You may want to suspend the audit trail before such a program and resume it after the program runs. To suspend an audit trail, use the AUDIT statement with the SUSPEND statement:

```
PROC DATASETS LIBRARY=libref NOLIST;
AUDIT member;
SUSPEND;
```

Later, use the RESUME statement to reactivate the audit trail:

```
PROC DATASETS LIBRARY=libref NOLIST;
AUDIT member;
RESUME;
```

To terminate the audit trail and delete the audit file, use the TERMINATE statement:

```
PROC DATASETS LIBRARY=libref NOLIST;
AUDIT member;
TERMINATE;
```

The audit trail of a SAS data file is automatically terminated if the SAS data file is deleted, replaced, sorted in place, or moved.

SAS Dataset Information

Chapter 5, "Variables and Values," mentioned the CONTENTS proc as a way to look at the attributes of variables in a SAS dataset. The CONTENTS proc also provides detailed information about the SAS dataset as a whole. This information may include:

- Dataset attributes, including the engine, member type, number of variables, observations, indexes, and integrity constraints, and degree of password protection
- Engine/Host Dependent Information, information items from the specific engine or operating system
- Alphabetical List of Variables and Attributes
- Alphabetical List of Integrity Constraints
- Alphabetical List of Indexes and Attributes
- Sort Information

The CONTENTS proc provides this information in a report and can also create SAS datasets of some of the information. In the interactive SAS environment, you can see similar information by looking at the properties of a SAS dataset.

All options for the proc are written in the `PROC CONTENTS` statement. The following table lists the available options.

Option	Value	Description
DATA=	*SAS dataset*	Describes one SAS dataset.
	libref._ALL_	Describes the SAS datasets of a library.
MEMTYPE=	DATA	Limits processing to SAS data files.
	VIEW	Limits processing to views.
	ALL	Processes all SAS datasets in the library.
DIRECTORY		Prints a directory of the library.
NODS		Prints only the directory of the library.
DETAILS		Includes number of observations, number of variables, and data set labels in the directory.
NODETAILS		Does not include number of observations, number of variables, and data set labels in the directory.
SHORT		Prints only variable names, sort information, and indexes.
NOPRINT		Does not print a report.
VARNUM POSITION		Prints a second list of variables in order of position.
OUT=	*SAS dataset*	Creates a SAS dataset of dataset and variable attributes.
OUT2=	*SAS dataset*	Creates a SAS dataset of indexes and integrity constraints.
CENTILES		Prints centiles of indexes.

This is an example of PROC CONTENTS output:

```
The CONTENTS Procedure

Data Set Name: CENTER.POINTS                          Observations:            1128
Member Type:   DATA                                   Variables:               6
Engine:        V8                                     Indexes:                 2
Created:       21:14 Monday, July 31, 2000            Integrity Constraints:   2
Last Modified: 21:14 Monday, July 31, 2000            Observation Length:      72
Protection:                                           Deleted Observations:    0
Data Set Type:                                        Compressed:              NO
Label:                                                Sorted:                  YES


                -----Engine/Host Dependent Information-----

...

-----Alphabetic List of Variables and Attributes-----

#    Variable    Type    Len    Pos
-----------------------------------
3    FG          Num       8      0
5    FT          Num       8     16
4    P3          Num       8      8
2    PLAYER      Char     20     50
6    PTS         Num       8     24
1    TEAM        Char     18     32
```

```
-----Alphabetic List of Integrity Constraints-----

     Integrity
#    Constraint    Type         Variables
---------------------------------------
1    ISPLAYER      Not Null     PLAYER
2    ISTEAM        Not Null     TEAM


-----Alphabetic List of Indexes and Attributes-----

               # of
               Unique
#    Index     Values    Variables
----------------------------------
1    MEMBER      1128    TEAM PLAYER
2    PLAYER      1128


 -----Sort Information-----

Sortedby:       DESCENDING PTS
Validated:      YES
Character Set: ANSI
```

System Options for SAS Files

Several system options affect the way SAS software works with SAS files. Unlike dataset options, these system options affect properties of the SAS session, or they affect all SAS datasets in the session.

One set of system options determines the SAS supervisor's reaction to certain unexpected events related to SAS dataset I/O. These options are summarized in the following table.

Option	Unexpected Event	Option Value	Action
DKRICOND=	A DROP=, KEEP=, or RENAME= dataset option for an input SAS dataset uses a variable name incorrectly	ERROR	Error condition
		WARNING WARN	A warning message
		NOWARNING NOWARN	No action
DKROCOND=	A DROP=, KEEP=, or RENAME= dataset option for an output SAS dataset uses a variable name incorrectly	ERROR	Error condition
		WARNING WARN	A warning message
		NOWARNING NOWARN	No action
BYERR	Attempting to sort a SAS dataset with no variables	On	Error condition
		Off	No action
DSNFERR	An input SAS dataset does not exist	On	Error condition
		Off	Uses a dummy input SAS dataset with no variables or observations

continued

Option	Unexpected Event	Option Value	Action
VNFERR	A variable is required in an input SAS dataset that does not exist	On	Error condition
		Off	The input SAS dataset is processed as if it contained the required variable
DLDMGACTION=	A catalog or SAS dataset is damaged	FAIL	Error condition
		ABORT	Terminate session
		REPAIR	Try to repair file
		PROMPT	Ask user

A related option, the DATASTMTCHK= option, determines what words the SAS supervisor will permit as one-level SAS dataset names in the DATA statement. With the option value NONE, any SAS name can be used. With the option value COREKEYWORDS, the keywords RETAIN, SET, MERGE, and UPDATE are not allowed. With the option value ALLKEYWORDS, any keyword that begins a data step statement is not allowed. The idea of this option is to catch coding errors in which the semicolon is omitted at the end of the DATA statement.

There are a few system options for managing SAS files. The ENGINE= option selects the library engine for new libraries created in the session. The TAPECLOSE= option controls the action of a tape drive after a library on tape is closed. The allowable values vary by operating system, but may include LEAVE, REREAD, REWIND, and FREE.

The WORKINIT and WORKTERM options initialize or erase the WORK library at the beginning and end of the SAS session, respectively. These options are on by default. Turn them off if you need to retain a WORK library from one session to the next.

The _LAST_= system option identifies the most recently created SAS dataset. This is the default input SAS dataset for procs and the SET statement. You can set the value of this option to have the SAS supervisor treat a certain SAS dataset as the most recently created.

Several system options are equivalent to dataset options. The system options FIRSTOBS=, OBS=, COMPRESS=, REUSE=, and REPLACE set defaults that apply to SAS datasets generally. You can override the system options by using the dataset options of the same name. The FIRSTOBS= and OBS= system option also apply to input text files.

Finally, there are a few options that might be used to fine-tune the performance of SAS files for a particular application. The BUFNO= and BUFSIZE= options control the number and size of buffers for SAS datasets. The CBUFNO= controls the number of buffers for each catalog, and the CATCACHE= option determines how many catalogs the SAS supervisor keeps open after they are used in case they might be used again.

12

Text File I/O

A data step can read and write virtually any kind of file. The syntax for text file I/O can be powerful and concise when used with ordinary text data files. At the same time, it is flexible enough to work with the most loosely structured text or binary files.

Statements

All text file I/O is accomplished with four data step statements: INFILE, INPUT, FILE, and PUT. The INFILE statement selects the input text file. Then, the INPUT statement locates and interprets fields from the input file. The FILE and PUT statements work in a similar way for output. The FILE statement selects the output text file. The PUT statement forms the output fields and records.

You can use options to control the way the program uses the file. Write options for an input text file in the INFILE statement and options for an output text file in the FILE statement. Many of the options are the same in the two statements.

The following example demonstrates all four of these statements. This program reads a file in one format and writes the same data to a file in another format.

```
DATA _NULL_;
  INFILE DLM TRUNCOVER;
  INPUT QUANTITY : COMMA11. SIZE 4.;
  FILE FIXED;
  PUT QUANTITY 8. SIZE BEST4.;
RUN;
```

In this program:

- The DATA statement with the special name _NULL_ indicates a data step that does not create a SAS dataset.
- The INFILE statement identifies the input file, using the fileref DLM. Options in the INFILE statement, such as the TRUNCOVER option, affect the way the input file is used.
- The INPUT statement reads two fields from the input record, creating values for the variables QUANTITY and SIZE.
- The FILE statement identifies the output file, using the fileref FIXED.

- The PUT statement writes two variables as fields in the output record.
- The automatic data step loop repeats the INFILE, INPUT, FILE, and PUT statements until the INPUT statement reaches the end of the input file.
- Earlier FILENAME statements might have defined the filerefs DLM and FIXED, or the filerefs could have been defined in some other way.

Selecting Files and Setting Options

Before you can read a record from or write a record to a text file, you have to select the file you want to use. Use the INFILE statement to select an input text file or the FILE statement to select an output text file. After the INFILE statement executes to select a file, subsequent INPUT statement read from that file. Similarly, after the FILE statement selects a file, subsequent PUT statements write to that file.

In the INFILE or FILE statement, you can use either a fileref or a physical file name to identify the file. If you use a physical file name, write it as a character constant, that is, enclosed in quotes. Identify the file with the first term in the statement, and write any options as subsequent terms.

```
INFILE fileref options;
INFILE 'physical file name' options;
FILE fileref options;
FILE 'physical file name' options;
```

Usually, a data step writes to only one text file, and the FILE statement can appear anywhere between the beginning of the step and the first PUT statement. However, a data step can write to multiple text files. Several FILE statements may be necessary to direct the PUT statements to the intended output files. Similarly, it is possible for a data step to have several INFILE statements and read from several different input text files.

Another reason you might have multiple FILE or INFILE statements is to use different options for working with different parts of a file. Each statement would indicate the same file, but with a change in options.

The INFILE and FILE statements are executable statements — statements that represent actions that take place when the data step executes. The statements can be conditional, which means you can use IF-THEN statements and other control flow statements to select different files or options in different conditions.

Default Files

The data step automatically resets the input and output files at the beginning of each repetition of the observation loop. If you omit the INFILE statement, the default input text file is the special fileref CARDS, which refers to data lines that can appear in the program file after a CARDS (or DATALINES) statement that marks the end of the data step. If you omit the FILE statement, the default output text file is the special fileref LOG, which refers to the log file.

If you want to read from a file other than the program file or write to a file other than the log, the INFILE or FILE statement must execute in every repetition of the observation loop.

Options for Files

There are many statement options available for use in the INFILE and FILE statements. These options change the way you access the file or give you more control over or information about the I/O process.

The most important category of options has to do with the way records in the file are considered. Other options have to do with the meanings of characters in the file, accessing the file as a whole, or the pointers used to identify locations inside the file.

Records

A *record* is a short segment of a file that is treated as a logical unit. When you display or edit a text file, each record is displayed as a separate line. In a text data file, each record usually contains one instance of the data being recorded in the file — that is, it can be considered as one observation. In a text file that contains a text document, each record may contain one paragraph.

A file system may have various ways of determining how the characters in a file are divided into records, which are identified as different *record formats*. Most commonly, special control characters contained in the file mark the end of each record. Different file systems use different characters as record delimiters. Among computers that use the ASCII character set, the file system may use the carriage return character, the line feed character, or a combination of both at the end of each record. These ASCII control characters ordinarily do not display as character forms. The carriage return (CR) and line feed (LF) characters were originally intended to control the actions of printer-based terminals. As an extension of that use, they came to be used to mark off records in text files.

In the very early days of computing, years before the invention of the electronic computer, the original digital text files were stacks of punch cards, large cards with holes punched in them to represent data values. In a stack of cards, of course, every card would be the same size. In the most common card format, each punch card had 80 columns, or character positions. If you were to copy data from punch cards to a computer file, it would make sense to make a computer file in which all the records had the same 80-character length. In a file with a fixed record length, there are no characters in the file itself to separate the records. If the record length is 80, as it usually is, each 80 characters in the file are treated as a record. The first 80 characters are the first record, the next 80 characters are the second record, and so on. Some file systems keep track of the record length of a file that has a fixed record format. In other file systems, those that do not keep track of the record length, you have to indicate the length in the program in order to access the file correctly.

Options for Use of Records

Option	Value	Use[a]	Description
RECFM=		I, O	Record format. The available values vary by operating system.
	F		Fixed
	V		Variable
	D		Data sensitive
	U		Undefined
	N		No format
	S		Stream (for TCP/IP and FTP)
	P		Print
LRECL=	*integer*	I, O	Logical record length. The number of characters in a record in the file.
BLKSIZE=	*integer*	I, O	Block size
LINESIZE= LS=	*integer*	I, O	Line size. The maximum number of characters in a record that are accessible to the program.
PAD		I, O	Pads short records with trailing spaces to make them as long as the logical record length.
NOPAD		I, O	Does not pad short records. Allows records of varying lengths.
N=	*integer*	I, O	The number of records accessed at a time.
FIRSTOBS=	*integer*	I	The number of the first record to read from the input file. This statement option overrides the system option.
OBS=	*integer*	I	The number of the last record to read from the input file. This statement option overrides the system option.
END=	*numeric variable*	I	The INPUT statement sets the value of this variable to 1 when it reads the last record in the file, or 0 when it reads all other records.
UNBUFFERED UNBUF		I	Unbuffered access. SAS does not read past the end of the current record.
OLD NEW[b]		O	Starts writing records at the beginning of the file.
MOD		O	Starts writing records at the end of the file.

[a] I=option for input file, in INFILE statement; O=option for output file, in FILE statement

[b] In some operating systems, the NEW option can be used as an equivalent to the OLD option.

In the INFILE or FILE statement, the RECFM= and LRECL= options indicate how records are defined in a file. These options are not necessary if the file system keeps track of these properties of the file. The RECFM= option uses one-letter codes to identify the record format of the file. Use the value F to indicate a fixed record format. Then, use the LRECL= option to indicate the length of the record. The table on the previous page lists these and other options that affect the way a data step accesses records.

There are several more options you can choose from to control the action of an INPUT or PUT statement when a record is not long enough or does not hold as much data as expected. When an INPUT statement is attempting to read a field, but there are no fields left in the record, the INPUT statement can stop reading at that point or continue with the next record, depending on the option you select in the INFILE statement. Similarly, when a PUT statement attempts to write a field that would extend beyond the length of the record, it can, if you prefer, continue writing on the next record. These options are summarized in the table below.

Options for Overflow in Text Records

Option	Use[a]	Description
MISSOVER	I	Assigns missing values to the remaining variables.
TRUNCOVER	I	Uses a short field at the end of the record if there is one, then assigns missing values to the remaining variables.
DROPOVER	O	Does not write the remaining fields.
FLOWOVER	I, O	Continues reading or writing at the beginning of the next record. This is the default.
SCANOVER	I	Continues scanning on the next record if it is the scanning pointer control that reaches the end of the record. If it is any other action that reaches the end of the record, this option works the same as the FLOWOVER option.
STOPOVER	I, O	Creates an error condition.

Use one of these options to determine the response when an INPUT or PUT statement reaches the end of a record unexpectedly. For an input file, this condition occurs when the INPUT statement does not find values for all its variables because an input record is too short. For an output file, this condition occurs when the line size is not sufficient for the PUT statement to write all of its fields.

[a] I=option for input file, in INFILE statement; O=option for output file, in FILE statement

Variables in Options

Many of the options for input and output files create data step variables, or designate variables for specific uses. For example, the END= option creates a variable that identifies the last record read from an input file. Variables used in any option in the INFILE or FILE statement, or in any other I/O statement, are treated differently from other data step variables. They are not automatically reset to missing in the observation loop, and they are not available to be written to any output SAS dataset.

Special Characters

A few options have to do with the meanings of characters. These special characters are not part of the data values in a record, but help to define the structure of the record.

DELIMITER= Delimiters in a record are characters that mark the end of one field and the beginning of the next. Usually, just one character is used as a delimiter. Use the DELIMITER= option to identify that character. The value of the option can be a character constant that indicates a character to use as a delimiter, or a character variable whose value is the delimiter character. For example, with the option `DELIMITER=','`, a comma is used as a delimiter.

Delimiters are used in list input and list output to read and write data values that vary in length. The default delimiter character for list input depends on the DSD option. With the DSD option, the default delimiter is a comma; otherwise, it is a space. The default for list output is not to write a delimiter, but to simply advance a column after each field, not writing any character in that column. If you did not already write a character in that column of the record, that makes the delimiter a space. But if you previously wrote a character in that column of that record, that character remains in that column.

For an input file, you can use more than one delimiter character. The various delimiter characters can be used interchangeably in the input data; they are equivalent to each other, as far as the list input processed is concerned. List output uses only one delimiter character. If the DELIMITER= option indicates more than one character, only the first character is used.

The alias for the DELIMITER= option is DLM.

DSD The DSD option changes the way delimiters work in list input and output. With the DSD option, an output field is written as a quoted string if it contains the delimiter character, and an input field written as a quoted string can be interpreted. Also, in input, consecutive delimiter characters are interpreted separately. This is useful for reading files that may contain null fields.

EXPANDTABS When tab characters appear in an input file, you can have them converted to spaces, based on tab stops at every 8 columns. Use the EXPANDTABS option for that behavior. Use the NOEXPANDTABS option to read tab characters without converting them.

Options for File Access

There are a few options that have to do with the physical file as a whole. These options, summarized in the table that follows, are especially useful when you want to treat a combination of several files as if they were a single file.

Options for File Access

Option	Value	Use[a]	Description
FILENAME=	*character variable*	I, O	A variable that shows the physical file name of the file.
FILEVAR=	*character variable*	I, O	You can change the file being used by assigning a different physical file name to the variable.
EOV=	*numeric variable*	I	A variable whose value is set to 1 for the first record read from a concatenated physical file.

[a] I=option for input file, in INFILE statement; O=option for output file, in FILE statement

A fileref can refer to two or more physical files; to define a fileref this way in the FILENAME statement, write a list of physical file names in parentheses. When you use the fileref for input, the files are concatenated. The FILENAME= and EOV= options can help you keep track of which physical file a record came from, in much the same way that the IN= dataset option can tell you which SAS dataset an observation came from.

The FILEVAR= option lets you use data values to change the file you are accessing. You could, for example, write each BY group in the data to a different physical file. With the FILEVAR= option, use a fileref that has not previously been defined. The INFILE or FILE statement checks the FILEVAR= variable each time it executes. The first time, it opens the file named in the variable. Subsequently, if the value of the variable has changed, it closes the file and opens the new file.

Pointer Options

The INPUT and PUT statements use pointers to keep track of locations in files. There is only one pointer for each file, but it is usually described as two separate integer values: a column pointer location and a line pointer location. The column pointer value tracks the character position within the current record. If you use the N= option to access several records at a time, the line pointer value keep track of which of those records is the current record.

Both pointers have positive integer values. The column pointer value can range from 1 to the line size. The line pointer value can range from 1 to the number of lines accessed at one time, which is the value of the N= option.

There are options that make the pointer values accessible as variables, so that you can use them in conditions. Use the COLUMN= option with the name of a numeric variable to create a variable that holds the column pointer value. The alias for this option is COL. Likewise, use the LINE= option for the line pointer value.

The following example demonstrates the use of a column pointer variable.

```
DATA _NULL_;
  INFILE FROM COLUMN=COLUMNPOINTER;
  INPUT @25 @;
  PUT COLUMNPOINTER=;
  STOP;
RUN;
```

```
COLUMNPOINTER=25
```

In this example, the INPUT statement places the pointer at column 25. The COLUMN= option in the INFILE statement creates the variable COLUMNPOINTER that contains the column pointer value.

You cannot assign new values to the COLUMN= or LINE= variables to change the pointer location. Instead, use the appropriate terms in the INPUT or PUT statement to move the pointer.

Treating a Record as a Variable

When you need to, you can treat an entire input or output record as a character variable. This can let you process the record with functions and other elements of data step logic that go beyond what the INPUT and PUT statements can do.

The options that turn a record into a variable are the _INFILE_= option in the INFILE statement and the _FILE_= option in the FILE statement. The value of the option is a name for the variable. The variable must not have been used in any earlier statements in the data step. The record variable is a character variable with a length of 32767. You can use it in expressions, and you can assign values to it in assignment statements.

The variable is the actual file buffer for the input or output file. It contains the current record of the INPUT or PUT statement. The current input record is the last record read. The current output record is the next record to be written. If the most recent PUT statement ends in a trailing @, the last record referred to in that statement is the current record. The buffer variable contains only one record, even if you use the N= option to access several records at a time.

These are examples of ways you can use record variables:

- If the same characters appear in every output record, first assign those characters to the output record variable. Then use a PUT statement to write the variable fields in the output record.
- Use the TRANSLATE function to translate certain characters to other characters in the input record variable before you read the fields from the record with an INPUT statement.
- Write the entire input record variable to an output file.
- Use the LENGTH function to find the length of the data in the input record variable.

There is also an automatic variable *called* _INFILE_ that refers to the input record. This automatic variable is a pseudo-variable that may refer to different physical variables at different times. It always refers to the current input record of the current input text file, which means that after an INFILE statement selects a different input text file, the variable name _INFILE_ refers to a different variable.

There are two options you can use in the INFILE statement that can affect the way the variable _INFILE_ is written by the PUT statement. You can use these options to write only part of the input record. The START= and LENGTH= options name numeric variables that identify the first and last column, or character position, to use. For example, to write characters 9–72 of an input record, assign a value of 9 to the START= variable and a value of 72 to the LENGTH= variable, then use the variable _INFILE_ in a PUT statement. The values of the START= and LENGTH= variables affect only the way PUT statements write the variable _INFILE_. They do not affect the use of the variable in an expression.

Other Options

Other options of the INFILE and FILE statements are designed for use with print and binary files, to edit files, and to use the special features of specific device types and file systems.

Terms for Processing Records

The INPUT and PUT statements process input and output records. The syntax rules of the two statements are similar. In fact, they have so much in common that you cannot always tell the statements apart by the terms they contain.

The INPUT and PUT statements are not like other executable statements of the data step. With most statements, one statement indicates one action. By contrast, an INPUT or PUT statement can contain any number of actions. The terms in the statement represent a sequence of actions, which involve moving to specific locations in the file and processing fields at those locations.

The purposes of the INPUT and PUT statements are not the same. The INPUT statement reads variables, and the PUT statement writes them. Still, they sometimes use the same terms to do these different actions. That is, sometimes, you can use the exact same terms that read data in an INPUT statement to write the same data in the same way in a PUT statement.

An example of this is not hard to come by. Consider the following fixed-field text data file, shown with a ruler to make it easier to measure its fields:

```
----+----1----+----2
 1January  Januar
 2February Februar
 3March    März
 4April    April
 5May      Mai
 6June     Juni
 7July     Juli
 8August   August
```

```
 9SeptemberSeptember
10October  Oktober
11November November
12December Dezember
```

The records in the file contain three fields, which you could read and write with the following statements:

```
INPUT
   @1 NUMBER 2.
   @3 NAME_ENG $CHAR9.
   @12 NAME_DEU $CHAR9.
   ;
PUT
   @1 NUMBER 2.
   @3 NAME_ENG $CHAR9.
   @12 NAME_DEU $CHAR9.
   ;
```

In these statements, each term indicates a separate action in the input or output file. The terms are the same in the two statements, and the actions, although not exactly the same, are similar. The following table categorizes and describes the terms.

Category	Terms	Description
Pointer control	@1 @3 @12	Moves to a specific location in the file
Variable	NUMBER NAME_ENG NAME_DEU	A variable to read a value for (in the INPUT statement) or to write the value of (in the PUT statement)
Informat/Format	2. $CHAR9.	Routine that reads a value from a field in the input file (informat, in the INPUT statement) or writes a value to a field in the output file (format, in the PUT statement)

Pointer Controls

Pointer controls are terms in the INPUT and PUT statements that move to locations in the file. Column pointer controls move to different column positions in the same record. Line pointer controls move to the beginning of another record. Pointer controls work the same way in both the INPUT and PUT statements.

Most of the time, one INPUT or PUT statement processes one record. The record is released at the end of the statement. You can use line pointer controls to process multiple records in one statement, or to stay on the same record at the end of the statement so that a subsequent statement can do further processing on the same record.

Pointer controls for use with text data files are listed in the following tables. There are additional line pointer controls for use in the PUT statement for writing print files.

Column Pointer Controls

Term	Use[a]	Description
@positive integer *@numeric variable* *@(numeric expression)*	I, O	Moves to the column indicated by the value of the positive integer constant, numeric variable, or numeric expression enclosed in parentheses
+positive integer *+numeric variable* *+(numeric expression)*	I, O	Advances the pointer by the number of columns indicated by the positive integer constant, numeric variable, or numeric expression enclosed in parentheses
@'character value' *@character variable* *@(character expression)*	I	Scans the input record for the character value, and moves the column pointer to the next column after the character value. With the SCANOVER option, this term scans as many records as necessary, until it finds the character value.

[a] I=term for input file, in INPUT statement; O=term for output file, in PUT statement

Line Pointer Controls

Term	Use[a]	Description
/	I, O	Moves to the beginning of the next record
#n *#numeric variable* *#(numeric expression)*	I, O	Moves to the beginning of the record number indicated by the positive integer constant, numeric variable, or numeric expression enclosed in parentheses. This form of line pointer control can be used only if the N= option is used in the INFILE or FILE statement to access multiple records at a time. The value in the pointer control can be no greater than the number of records indicated in the N= option.
@ *(the last term in the INPUT statement)*	I	*Trailing at-sign:* Holds the current input record for more processing in a subsequent INPUT statement. However, the record is released at the end of the current repetition of the observation loop.
@@ *(the last term in the INPUT statement)*	I	*Double trailing at-sign:* Holds the current input record for more processing in a subsequent INPUT statement in either the current repetition or the next repetition of the observation loop
@ *(the last term in the PUT statement)*	O	*Trailing at-sign:* Holds the current output record for more processing in a subsequent PUT statement

[a] I=term for input file, in INPUT statement; O=term for output file, in PUT statement

Several pointer controls use integer values. These values can be positive integer constants, numeric variables, or numeric expressions enclosed in parentheses. If the value is not an integer, it is truncated and the integer part of the value is used. In a column pointer control, if the value is too low, such that it would move past the beginning of the record, the pointer moves to column 1, the position of the first character in the record.

If a pointer control or other term would move the column pointer beyond the end of the record, the resulting action of the INPUT or PUT statement depends on the record overflow option in the INFILE or FILE statement, as described previously. The default action is that of the FLOWOVER option, which moves to the beginning of the next record and continues processing there.

Fields

After moving to a specific location in a file, the next action is to process a field at that location, usually by reading or writing a variable. Several terms may be required to process a field in a file. The first term is the variable name. In the simplest case, the variable name is followed by an informat or format. The width of the informat or format is the width of the field and the number of characters that are read or written. In more complicated styles of input and output, a variable name may appear by itself, or followed by any of various combinations of terms that determine how it is read or written. The details are described separately for the INPUT and PUT statements later in this chapter.

In a PUT statement, you can also write constant fields. When a character constant is used as a term in the PUT statement, the statement writes that value to the output file. You can use a repetition factor to write a character value repeatedly. Write a positive integer, an asterisk, and the character constant, for example:

```
PUT 5*'<> ';
```

```
<> <> <> <> <>
```

Processing a field also moves the pointer. After it processes a field, the INPUT or PUT statement leaves the pointer at the next column after the last character it processes.

Defining Variables

The INPUT and PUT statements can be used to define variables. Usually, when variables appear in an INPUT statement, it is their first appearance in the data step, and they are defined as numeric or character variables according to the way they are used in the INPUT statement. A variable used with a character informat is a character variable. If you use a new variable without writing an informat, you can write the special term $ after the variable name to make it a character variable. Any other variable defined in an INPUT statement is a numeric variable. The INPUT statement uses the width of a character informat to determine the length of a character variable it defines. If there is nothing in the statement to indicate a width, it defines the variable with a length of 8.

It is rare that a variable makes its first appearance in a PUT statement. However, when that occurs, the PUT statement defines the variable in the same manner as the INPUT statement does. If an INPUT or PUT statement does not define a variable the way you want, use a LENGTH statement, located earlier in the data step, to declare the variable's type and length.

Lists

When there are repetitive parts of a record, you can make INPUT and PUT statements shorter and easier to read by using a list notation. In the list notation, the variable names are separated from the other terms that process the record. A variable list in parentheses is followed by a list of other terms, also in parentheses. This second list is called an informat list in the INPUT statement and a format list in the PUT statement. It can include pointer controls, informats or formats, and other terms that modify the informats or formats. A format list can also contain character constant terms. The informat list or format list is repeated as many times as necessary to process all the variables in the list.

The syntax, in the INPUT statement, is summarized as:

```
(variable list) (informat list)
```

In the PUT statement, it is:

```
(variable list) (format list)
```

Any of the terms in an informat list or format list can be preceded by a number and asterisk indicating a repetition factor. For example, the term 5* means to use the term that follows 5 times. This format list:

```
(2*7. 5*11.2)
```

is equivalent to:

```
(7. 7. 11.2 11.2 11.2 11.2 11.2)
```

Often, an informat list or format list contains only a single term, usually an informat or format. The term is applied to each of the variables in the variable list. This can be especially useful for processing a related group of variables that can be written as an abbreviated variable list. For example, this list combination:

```
(RATE1-RATE6) (11.2)
```

is equivalent to:

```
RATE1 11.2  RATE2 11.2  RATE3 11.2  RATE4 11.2  RATE5 11.2  RATE6 11.2
```

Wherever a variable list can be used in an INPUT or PUT statement, you can use an array name with the special subscript * to indicate a variable list of the array's elements. This works only for arrays whose elements are named variables; it does not work for arrays that use temporary variables as elements. For example, if the array STAR is defined as

```
ARRAY STAR{15} $ 18 STAR1-STAR15;
```

you could then use this PUT statement to write all the elements of STAR:

```
PUT (STAR{*}) ($18. /);
```

This would be the same as writing the PUT statement with the variable list that defined the array:

```
PUT (STAR1-STAR15) ($18. /);
```

In the INPUT and PUT statements, abbreviated variable list keywords can be used only in variable lists, for example, (_CHAR_) (:).

When an informat list or format list contains multiple terms, it repeats only until the last variable in the variable list has been processed. Regardless of where the term or terms that process the last variable appear in the list, processing of the list stops at that point. In the preceding example, the pointer control at the end of the list does not execute after the last variable is written. Or, as another example, consider this statement:

```
PUT (TIME1-TIME3) (MMSS5. ',');
```

It produces this kind of output line:

```
14:37,11:05,33:24
```

The character constant term ',' in the format list writes commas after the first and second variables in the variable list, but it does not write a comma after the last variable in the variable list. That is because processing of the format list stops as soon as the last variable in the variable list is written.

The interpretation of the term $ in an informat list or format list can be confusing. Its effect depends on whether it is followed first by a pointer control or an informat or format. If an informat or format occurs first, the $ is treated as part of the informat or format name (which usually results in an error condition). But if, after the $, a pointer control occurs before an informat or format term, the $ is treated as a separate informat or format term, equivalent to the terms : $F.. To avoid confusion, do not use $ as a term in an informat or format list.

If a list does not contain an informat or format, all the formatting terms in the list are applied to every variable in the list. For example, one common format list is (=). This format list is equivalent to writing = after each variable name. For example,

```
PUT (A B C) (=);
```

is equivalent to

```
PUT A= B= C=;
```

Pointer controls are executed before each variable. This is true even if the pointer controls are not written at the beginning of the list. For example,

```
PUT (A B C) (= / +2);
```

is interpreted as

```
PUT / +2 A= / +2 B= / +2 C=;
```

An informat list or format list can contain commas between terms. However, the commas have no effect on the execution of the list.

If it is necessary to repeat a list of informats or formats in a way that is too complicated for the syntax of an informat list or format list, use a DO loop with IF-THEN statements. When you use the trailing @ in INPUT or PUT statements, you can use as many statements as necessary to process a record.

Skipping Records

There are times when you want to skip over a record in a file without doing anything with it. You can do this by writing an INPUT or PUT statement without any terms, that is,

```
INPUT;
```

or

```
PUT;
```

At the end of an INPUT or PUT statement, the statement releases the current record in the file, even if nothing has been done with it. Thus, the effect of an INPUT or PUT statement without any terms is to advance to the next record. Use the INPUT statement without any terms to skip over an input record without reading it. Use the PUT statement without terms to write a blank output record.

To skip several records, use the / pointer control. Use one less / than the number of records you want to skip. For example, to write three blank output records, use this statement:

```
PUT //;
```

Processing Input Records

Terms in an INPUT statement set forth the actions that process an input record from an input text file. Often, a single INPUT statement is all that is needed to process an input record. When you need to take different actions with different input records, depending on conditions of the data, use several INPUT statements, combined with control flow statements.

The actions of an INPUT statement move to places in the input file and read fields, assigning the resulting values to variables. The pointer control terms that move to specific locations in the file are described in the previous section. If the pointer is not already at the beginning of a field you want to read, first use pointer controls to move it there.

The terms that read a field begin with a variable name, usually followed by an informat. Various other terms can affect the way the field is read. The following table lists the terms that can be used for reading a field. The only term that is required is the variable; the other terms are optional. An informat term is usually necessary to indicate the informat to use to interpret the field. However, in one style of input, known as *list input*, even the informat term is omitted.

INPUT Statement Terms for Reading Variables

Term	Description
variable	The value from the field is assigned to this variable.
:	Reads a field marked off by spaces (or other delimiter characters, depending on the DELIMITER= option).
&	Reads a field marked off by multiple spaces (or other delimiter characters). This makes it possible to read fields that contain single spaces (or delimiter characters).
?	Prevents the data error messages that are normally written to the log when the value of the field is not valid for the informat. The automatic variable _ERROR_ is still set to 1 to indicate an error condition.
??	Prevents the error condition that would normally result when the value of the field is not valid for the informat. There are no log messages and the automatic variable _ERROR_ is not set to 1.
$	Declares the variable as a character variable. (Do not use this term and an informat term together. If you do, the $ is treated as part of the informat name, which results in an error condition with a message that the informat cannot be found.)
informat	The routine that interprets the data in the field.

The table below shows examples of terms in the INPUT statement to read variable fields. For each example, the table describes how the terms interpret the field and move the pointer.

Terms	Field Interpretation	Pointer Movement
RECTYPE $CHAR1.	Reads one character and assigns the resulting value to the character variable RECTYPE.	Advances the pointer to the next column after the field.
QTY ?? 7.	Reads a 7-character field, interpreting it as a number and assigning the value to the numeric variable QTY. Does not create an error condition if a number is not found.	Advances the pointer to the next column after the field.
VOLUME : 12. (*with* DELIMITER=' ')	Starts at the next nonblank character. Reads a field of up to 12 characters, stopping at the next space. Interprets the field as a number and assigns the value to the numeric variable VOLUME.	Advances the pointer to the next column after the space character that marks the end of the field.

Informat Terms

An informat reference starts with the informat name. This is followed by a whole number argument that indicates the width, then a period. For numeric informats, a second argument can follow the period to indicate the number of decimal places to read. This decimal argument is ignored if there is an actual decimal point in the field. The informat reference is a single token, so there are no spaces between the informat name, width argument, period, and decimal argument.

The data type of an informat must match the type of the variable named for the field. If you create a new variable, the type of the informat determines the type of the variable. For a character variable, the width of the informat determines the length of the variable. Names of character informats begin with $.

You can omit the width argument. If you write an informat without indicating a width, a numeric informat uses its default width. A character informat uses the length of the character variable to determine its width. If you create a new character variable without anything to indicate its length, the SAS supervisor gives it a length of 8. If you modify an informat with the : or & scanning modifier, without writing a width argument for the informat, the informat reads the actual width of the field.

The $CHAR informat is sufficient for reading most character fields. It assigns the value of the field, unchanged, to the character variable. For example, to read an entire 80-character record as a single variable, you could use this statement:

```
INPUT RECORD $CHAR80.;
```

The informat reads 80 characters from the record, which are then assigned to the variable RECORD.

Ordinary numeric fields can be read with the standard numeric informat. The name of this informat is F — although this name can be (and usually is) omitted, as long as the width argument is present. The informat F7. or 7. reads a seven-character numeric field.

Use the decimal argument with the standard numeric informat to read fields that contain an implied decimal point. For example, the informat 5.2 would read the field 10000 as 100.00.

Any informat can be used in the INPUT statement. Informats and their use are described in more detail in the next chapter.

The informat term in an INPUT statement must be written as the last term associated with a field. When you use other terms that modify the way the field is read, such as & or ??, write them between the variable name and the informat.

It is possible to omit the informat term altogether. Then, the field is read with the scanning process with the standard numeric or standard character informat. It is equivalent to writing the terms : F. for a numeric variable or : $F. for a character variable. The standard character informat, $F, differs from the $CHAR informat only in that it left-aligns values and treats a single period as representing a blank value.

Demonstration of INPUT Statement Actions

The following example demonstrates the actions of an INPUT statement. The example reads these input records:

```
ABCDEFGHIJKLMNO
123456789000000
```

The INPUT statement that reads these records is:

```
INPUT @8 A $CHAR2. @5 INITIAL $CHAR1.
  / NUMBER 10. @4 RATE 5. +(-5) PERCENT 5.2;
```

To run this example, put it into a data step like this:

```
DATA _NULL_;
  INFILE CARDS;
  INPUT @8 A $CHAR2. @5 INITIAL $CHAR1.
    / NUMBER 10. @4 RATE 5. +(-5) PERCENT 5.2;
  PUT _ALL_;
CARDS;
ABCDEFGHIJKLMNO
123456789000000
;
```

The following table shows the results of each action in the INPUT statement. The pointer control terms only move the pointer. Terms for fields read the fields and move the pointer. There is no trailing @ in the statement, so the last action of the statement is an automatic action to release the current record.

Terms	Variable	Value	Pointer
@8			column 8, at "H"
A $CHAR2.	A	HI	column 10, at "J"
@5			column 5, at "E"
INITIAL $CHAR1.	INITIAL	E	column 6, at "F"
/			the next line, column 1, at "1"
NUMBER 10.	NUMBER	1234567890	column 11, at the second "0"
@4			column 4, at "4"
RATE 5.	RATE	45678	column 9, at "9"
+(-5)			column 4, at "4"
PERCENT 5.2	PERCENT	456.78	column 9, at "9"
(automatic action)			The INPUT statement releases the input record. The pointer is located after the end of this record.

In this example, notice these possibilities of the INPUT statement:

- Column pointer controls can move from anywhere in a record to anywhere else in the record. This allows you to read the fields of a record in any order.
- Either the @ or + pointer control can move to any location in a record.
- The INPUT statement can read any subset of the fields in a record. It can also read the same field more than once.

- The / pointer control allows a single INPUT statement to read more than one record.

List Input

The easiest kind of record to read in an INPUT statement is a record that contains a single numeric value written in standard or scientific notation. The INPUT statement only needs to indicate the variable name, for example:

```
INPUT X;
```

A statement such as this one can read input records such as these:

```
14.04
            -1
      1389298454374654312987436583746 53.234574356438783456
```

The value can be written in any length, with any number of spaces before and after it.

It is not much more complicated to read a record that contains two or more numeric values, as long as the values are separated by at least one space. List the variable names in order in the INPUT statement, for example:

```
INPUT X1 X2 X3;
```

This statement can read records such as this one:

```
40505  64009  51316
```

This input process, and the INPUT statement syntax that makes it work, is called *list input* — an appropriately descriptive name for a process that uses a list of variables to read a list of values. List input is the easiest way to read data in which values are separated by spaces. With list input, you do not have to specify the width of each field or the amount of space between fields. The list input process automatically adjusts to whatever field sizes and spacing is present.

You can sometimes shorten the INPUT statement by using abbreviated variable lists. For example, the previous INPUT statement could also be written:

```
INPUT X1-X3;
```

List input also works with character variables. If any variable in list input has been defined as a character variable, the field is read as a character value. However, because spaces determine the beginnings and ends of fields in list input, the character value cannot contain spaces. If there are leading spaces (spaces at the beginning of the value), the scanning process skips over them and starts reading the value at the first nonblank character. If there are embedded spaces (spaces in the middle of the value), the scanning process would treat the value as more than one value, and it would read the record incorrectly. Consider this example:

```
* This example contains a logical error. ;
DATA _NULL_;
```

```
  INFILE CARDS;
  LENGTH CITY $ 32 STATE $ 2;
  INPUT CITY STATE;
  PUT (CITY STATE) (=);
CARDS;
DENVER CO
BALTIMORE MD
NEW YORK NY
LOS ANGELES CA
;
```

This INPUT statement is intended to read records that contain a place name and state. It would work correctly for some records, such as the first two records shown above. But many place names contain spaces, and the INPUT statement would read them incorrectly, one word at a time, as you can see in the log output from the example:

```
CITY=DENVER STATE=CO
CITY=BALTIMORE STATE=MD
CITY=NEW STATE=YO
CITY=LOS STATE=AN
```

With this kind of input data, there is no easy way to change the way the INPUT statement is written to make it correctly separate the place name and state. Instead, read the data as a single field, and if necessary, use other data step statements to separate the two values.

Scanning Modifiers

The scanning modifiers of the INPUT statement can make list input more versatile. With scanning modifiers, you can specify an informat to read a field, and you can read some fields that contain single embedded spaces.

To select an informat for a variable in a record that you are reading with list input, write a colon and the informat after the variable name. You need to specify an informat to read a date value, for example:

```
START : DATE9.
```

The colon acts as a scanning modifier; with the colon, the field is read with the scanning process of list input, even though an informat is present.

An ampersand as a scanning modifier, used either with or without an informat, means that it takes two consecutive spaces to mark the end of the value. This allows a character value to contain single spaces. Revising an earlier example, if there are two spaces between the place name and state in an input record, you can use the & scanning modifier to read the place name:

```
DATA _NULL_;
  INFILE CARDS;
  LENGTH CITY $ 32 STATE $ 2;
  INPUT CITY & STATE;
  PUT (CITY STATE) (=);
CARDS;
DENVER  CO
BALTIMORE  MD
NEW YORK  NY
```

```
LOS ANGELES CA
;
```

```
CITY=DENVER STATE=CO
CITY=BALTIMORE STATE=MD
CITY=NEW YORK STATE=NY
CITY=LOS ANGELES STATE=CA
```

It is a good idea to write a space after the & scanning modifier. Ordinarily, SAS syntax does not require spaces around special characters. However, the ampersand also serves as an indicator in macro language. If there is no space between an ampersand and a name that follows it, the macro processor treats the combination as a macro variable reference, which usually results in a warning message or error condition. By writing a space after the ampersand, you avoid the potential for a macro error.

A third scanning modifier, the tilde (~), changes the interpretation of the quoted strings that can be found in some comma-delimited files. Its use is described later in this chapter.

Not every informat can be used with list input. Most binary informats are incompatible with the list input style, because the delimiter character could occur as part of a data value. If you use a scanning modifier with one of these binary informats, the INPUT statement disregards the scanning modifier and reads the field using formatted input.

Even with the modifications that are possible with scanning modifiers, the list input style has limitations, especially for character data.

- It cannot correctly read a value that begins with a space.
- Fields must have spaces between them.
- It cannot read a value that contains two or more consecutive spaces.
- Every field must contain a nonblank value. Unlike other input styles, list input cannot read a blank field as a missing value. (It can, however, read a single period as a missing value.)

List input syntax can be used for files that are delimited by characters other than spaces. The DELIMITER= option of the INFILE statement determines the delimiter character or characters, and the DSD option changes the way consecutive delimiters are interpreted. Techniques for delimited files are described later in this chapter.

Variable-Width Input Fields

List input can read certain kinds of variable-width fields, but the $VARYING informat provides a more precise way to read fields that can vary in width. Unlike other informats, the $VARYING informat's width is determined by a separate variable, called a *length variable,* which serves as an extra argument to the informat. Because the $VARYING informat depends on this extra argument, it is useful only in the INPUT statement.

This is an example of the use of the $VARYING informat in a term in the INPUT statement:

```
NAME $VARYING40. LENGTH
```

NAME is the variable being read, a character variable. LENGTH is the length variable, a numeric variable whose value determines how many characters informat reads. It the value of LENGTH is 1, the informat reads 1 character. If the value of LENGTH is 5, the informat reads 5 characters.

Usually, as in this example, the $VARYING informat is used with a regular width argument — in this case, 40. With a width argument of 40, the $VARYING informat does not read more than 40 characters, even if the value of the length variable is greater than 40. If the value of the length variable is between 1 and 40, the $VARYING informat reads that many characters. If the value of the length variable is less than 1, or if it is a missing value, the $VARYING informat does not read anything, and the resulting value of the variable being read is blank.

In practical applications, it is another field in the input data that determines the width of the field that you read with the $VARYING informat. Usually, this other field is found earlier in the same record. Using the $VARYING informat might require several statements: first, an INPUT statement to read the field that determines the length; then, additional statements to assign the appropriate value to the length variable; and finally, another INPUT statement to read the variable-length field itself. In the following example, the width of the description field can be any of various lengths, depending on the value found in a code field.

```
INPUT @7 DESLCODE $CHAR1. @;
SELECT (DESLCODE);
  WHEN ('2') LENGTH = 32;
  WHEN ('3') LENGTH = 48;
  OTHERWISE LENGTH = 16;
  END;
INPUT @14 DESCRPTN $VARYING48. LENGTH;
```

The first INPUT statement ends in @ so that the second INPUT statement will read from the same record.

Lines

Usually, an INPUT statement reads a single input record. However, it is possible for a single INPUT statement to read any number of records. Line pointer controls make this possible. If it makes sense to read the records in sequence, simply use the / pointer control after processing each record to advance to the next record.

As an example, consider these input records:

```
http://www.eqmag.com
EQ Magazine
http://www.mtv.com
MTV Online
http://www.riaa.com
Recording Industry Association of America
http://alanwhite.net
Alan White
```

Each pair of records indicates the url (uniform resource locator) and title of a web site. To read the two records that form each observation, you could use two INPUT statements:

```
INPUT @1 URL $64.;
INPUT @1 TITLE $48.;
```

Or, you could write the same actions in just one INPUT statement, using the / pointer control:

```
INPUT @1 URL $64.
      / @1 TITLE $48.;
```

For situations in which it is not practical to read records in sequence, you can treat several records as a group and use the # pointer control to jump from one line to another within the group. Use terms such as #1, #2, and so on, to move among the different lines.

For example, the previous example could be rewritten as:

```
INPUT #1 @1 URL $64.
      #2 @1 TITLE $48.;
```

You can refer to the lines in any order within the INPUT statement. Or you can use multiple INPUT statements to read the lines in the group. The INPUT statement stays in the same group of records if you use a trailing @ at the end of the INPUT statement or if you leave the pointer on a line other than the last line in the group. The group of lines is automatically released at the end of the repetition of the observation loop. You can read only one group of lines in one repetition of the observation loop. If, after you release a group of lines, you use a # pointer control later in the same repetition of the observation loop, it results in an error condition.

When it releases a group of lines, the INPUT statement positions the pointer after the last line in the group. For example, the previous example could be rearranged this way:

```
INPUT #2 @1 TITLE $48.
      #1 @1 URL $64.;
```

At the end of the statement, the pointer is located after line 2, even though line 1 was read after line 2.

Every line you access is considered part of the group, whether you read a field from that line or simply move the pointer to it. Make sure you include the last line in the group by accessing it at some point. If you do not read any fields from the last line in the group, access that line by using a # pointer control with the line number of the last line. For example, if you only read fields from the line 2 of a 5-line group, the INPUT statement might look like this:

```
INPUT #2 PLACE REGION
      #5;
```

It is sometimes necessary to use the N= option in the INFILE statement to indicate the number of records that form a group. For example, the option N=2 in the INFILE statement indicates groups of 2 lines. The default for the N= option is the highest constant line number used with the # pointer control. If this number is not the correct number of lines in a block, use the N= option in the INFILE statement to set the correct number. If the value of the N= option is not large enough, the INPUT statement will generate an error condition with a message that a record is not available.

You can use the N= option and the # line pointer control even when different groups of records contain different numbers of lines. To make this work:

- Determine the maximum number of lines that a group can have and use this number as the value of the N= option in the INFILE statement.
- You will usually need to use multiple INPUT statements for each group of records. Use a trailing @, if necessary, to ensure that the INPUT statement does not release the group of lines.
- In each group, determine how many lines it has, and be sure to access the last line in the group without accessing any lines after that one. If necessary, determine the line number of the last line in the group, assign this value to a variable, and use this variable with the # pointer control to move the pointer to the last line. For example, if the variable LINES is the number of lines in the group, you could use the term `#LINES` to move the pointer to the last line.

It is not always necessary to specifically indicate line breaks in the INPUT statement. With the FLOWOVER option in the INFILE statement, the INPUT statement automatically advances to the beginning of next record if it reaches the end of a record while looking for a field or executing a column pointer control. This works with both formatted and list input, but it is especially useful with list input. When the INPUT statement advances to a new record in this manner, it generates a log message that says, "`NOTE: SAS went to a new line when INPUT statement reached past the end of a line.`"

If an INPUT statement reaches the end of the input file unexpectedly, it generates a log message that says "`NOTE: Lost card.`" This message is generated if the end of the file is found as the result of a pointer control, or if the INPUT statement is still looking for a field when it reaches the end of the file.

Column and Named Input

Column input and named input are two rarely used styles of input that the INPUT statement supports. Column input is an alternative way of writing terms to read fields at fixed locations in a record. Named input is a way of reading variable values that are labeled with variable names.

In the syntax for column input, the variable name is followed by column numbers that simultaneously indicate pointer movement and field size. For example, the terms `N 10-17` mean to move the pointer to column 10, read an 8-character field, and assign the resulting value to the variable N. The field is read with the standard numeric informat, F, or the standard character informat, $F, depending on the data type of the variable. The following table describes the terms you can use to read fields with column input.

Column input syntax offers very little flexibility. For example, there is no way to select a different informat to use. At the same time, it can be harder to work with, because it does not directly indicate the width of a field. To keep things simple, use the standard syntax instead.

INPUT Statement Terms for Column Input

Term	Description
variable	The value from the field is assigned to this variable.
$	Declares the variable as a character variable.
n	The column of a 1-character field.
n–n	The starting and ending columns of a field.
.n	The decimal argument for the standard numeric informat.

Named input provides a way to read files in which each variable is written with the variable name, an equals sign, and the value, for example:

```
TEMPERATURE=25  PRECIPITATION=0.7
```

Values in the input record can contain single spaces, as long as variables are separated by two spaces. That is, in the input record, there must be at least two spaces before the variable name and two spaces after the end of the value. If not all the values fit on one record, the record can end with a slash to tell the INPUT statement to continue reading on the next record.

A named input record does not have to contain every variable. Variables that are not present are given missing values.

If you have a data file in which records follow this format strictly, you can read them with the named input style. The variables do not have to appear in any specific order in the input record, but there should not be any other data among or after the variables in the record. In the INPUT statement, write an equals sign immediately after the variable name, for example:

```
INPUT TEMPERATURE= PRECIPITATION=;
```

You can write an informat for a variable after the variable name and equals sign in the INPUT statement. The INPUT statement does not actually use the informat you name to read the value for that variable. Named input, like column input, always reads values with the standard numeric informat or standard character informat. The INPUT statement does, however, consider the informat in determining the type of the variable and the length of a character variable. This is useful when you use an INPUT statement with the named input style to create character variables. For example, the terms `NAME= $F27.` in an INPUT statement ensure that the variable NAME is created as a character variable with a length of 27.

You can also indicate columns at which named input should start to look for specific variables, or a decimal term for a numeric value that is written with an implied decimal point. These terms are the same as described above for column input. However, with named input, the end column is only an indication of the length of the value. It is not the end of the field.

Named input does not give you much control over the process of reading the file. In any critical application, you would probably want to read this kind of file using list or formatted input, with separate data step logic to interpret the values and respond to any data errors that might occur.

Processing Output Records

Terms in a PUT statement are actions that put together records to write to an output text file. Pointer control terms, which are largely the same as in the INPUT statement, move to different places in a record or from one record to another. The other terms in the statement write constant and variable fields.

To write a constant value to an output file, write a character constant as a term in a PUT statement. The PUT statement is often used to write diagnostic notes, which may simply be constant text values, as in this example:

```
IF _N_ > 1 THEN PUT 'ERROR: Control file contains multiple observations.';
```

To write the value of a variable to an output file, write the variable name in a PUT statement, followed by any other terms that may be necessary. The following table describes the terms you can use in writing a variable.

PUT Statement Terms for Variable Fields

Term	Description
variable	The value of this variable is written to the field.
:	Removes leading or trailing spaces, then advances one column after the value (or writes the delimiter character, if the DELIMITER= option is used in the FILE statement). (This is the default action when no format term is present.)
=	Writes the variable name and an equals sign, then writes the value the same way as the : modifier does.
$	Declares the variable as a character variable.
	(Do not use this term and a format term together. If you do, the $ is treated as part of the format name, which results in an error condition with a message that the format cannot be found.)
format	The routine that produces the field text from the value of the variable.
	The following alignment options modify the text produced by the format, by moving any leading or trailing spaces. They do not change the width of the field.
-L	Left-aligns. Produces a field with no leading spaces.
-C	Centers. Produces a field with approximately equal numbers of leading and trailing spaces.
-R	Right-aligns. Produces a field with no trailing spaces.

The order of terms is important. The variable name comes first. Write a : or = modifier before the format or $. Write an alignment option after the format.

These are examples of terms that write variable fields:

```
RECTYPE $CHAR1.
NAME
PRICE : COMMA10.2
```

```
X=
GROUP $9. -C
```

Format Terms

A format reference is formed the same way as an informat reference. It is a single token made up of the format name, a width argument, a period, and possibly a decimal argument, with no spaces between them. For example, in the format reference F7.4, the format name is F, the width argument is 7, and the decimal argument is 7.

The width argument determines the width of the field — the number of characters that the format produces. Most numeric formats can have a decimal argument, which indicates the number of decimal places — the number of digits to write after the decimal point. If you omit the decimal argument, most formats do not write any decimal places. You can also omit the width argument. Then, numeric formats use their default width. Character formats use a width that is sufficient to write the entire character value.

The $CHAR format, like the $CHAR informat, does not change the character value in any way. The standard character format, $F, works the same way as the $CHAR format. You can write the name of the standard character format without the F, simply as $, as long as there is a width argument.

The default numeric format is the BEST format, which writes a number with about as much precision as it can in the width it uses. The BEST format determines how many decimal places to write based on the width of the field and the magnitude of the value. It does not use a decimal argument. If you supply a decimal argument for the BEST format, it ignores it.

To write numbers with a fixed number of decimal places, use the standard numeric format, F. You can omit the name of this format as long as you have a width argument. The format 7.2 or F7.2 formats a number in a seven-character field, with two decimal places, for example:

```
1234.56
```

If you use a character format with a width that is less than the length of the value, the format writes only the beginning of the value. If the format width is longer than the value, the format writes trailing spaces after the value to fill the rest of the field. When you consider the widths of numeric fields, remember to allow a column for the negative sign if the field will contain negative values. Formats are described in more detail in the next chapter.

Demonstration of PUT Statement Actions

The following example demonstrates the actions of terms in the PUT statement. This example uses the same terms that appeared in the earlier demonstration of the INPUT statement. However, different actions result when these terms appear in the PUT statement.

```
A = 'ACT';
INITIAL = 'E';
NUMBER = -22222222;
```

```
RATE = 80;
PERCENT = 68.08;
PUT @8 A $CHAR2. @5 INITIAL $CHAR1.
   / NUMBER 10. @4 RATE 5. +(-5) PERCENT 5.2;
```

```
    E  AC
-268.0822
```

Note these points about the way that the PUT produced its two records of output text:

- The @ pointer controls locate specific columns in the current record. They can move the pointer forward or backward.
- Only two characters of the variable A are written, because the format has a width argument of 2.
- The / pointer control advances to the next record.
- Think of the +(-5) pointer control as backspacing 5 columns. The result is that the PERCENT variable field writes over the entire RATE variable field. It also writes over the middle of the NUMBER variable field.

The following table shows the text and pointer location that result from each pointer control and field in the PUT statement. The gray shading indicates the pointer location.

Terms	Output Text
@8	▮
A $CHAR2.	AC▮
@5	▮ AC
INITIAL $CHAR1.	E▮ AC
/	E AC ▮
NUMBER 10.	E AC -22222222▮
@4	E AC -22222222
RATE 5.	E AC -2 8022
+(-5)	E AC -2▮ 8022
PERCENT 5.2	E AC -268.0822

List Output

If you list variable names in the PUT statement without any terms to indicate their formatting, the PUT statement writes the values of the variables to the output file. It leaves spaces between the variables and removes leading and trailing spaces from the values. This style of output is called *list output*.

The following is an example of a PUT statement that uses the list output style. All the terms in the statement are variable names.

```
PUT CLOCK BUS DISK RAM CPU YEAR;
```

This is an example of output records in the list output style:

```
4 8 1 1 A 1986
8 16 1 16 C 1990
180 40 615 108 F 1995
```

In list output, the width of a field tends to vary from one record to the next, depending on the value of the variable.

List output checks the format attribute of each variable to determine the format to use to write the variable. Usually, though, the format attributes for variables have not been set. In that case, list output uses default formats. For a character variable, it uses the standard character format, $F, with a width equal to the length of the variable. For a numeric variable, it uses the BEST12. format.

To select a different format for a variable in list output, write a colon and the format reference after the variable name, for example:

```
START : DATE9.
```

In order to leave a space between one value and the next, list output advances the pointer by one column after it writes each value. You can change this default behavior or list output by setting a delimiter character in the DELIMITER= option of the FILE statement. Then, list output writes the delimiter character after each field (except the last field in the record). For example, with the option DELIMITER=',', the output records above would instead appear as:

```
4,8,1,1,A,1986
8,16,1,16,C,1990
180,40,615,108,F,1995
```

With the DELIMITER= option, you can use the list output style to write various kinds of delimited files. This is described later in this chapter.

When a character value is blank, list output writes the value as a single space. Combined with the spaces that the list output style provides before and after the value, this appears as three consecutive spaces.

List Output for List Input

An output file produced by list output can subsequently be read by the list input style, provided that certain conditions are met. Those conditions are:

1. No value can be completely blank.
2. No value can contain an embedded space.

These conditions are not usually a problem with numeric variables. They can, however, create a problem for some character values. If necessary, use data step logic in statements before the PUT statement to remove embedded spaces from character variables and convert blank values to periods. These statements accomplish these effects:

```
ARRAY CHAR{*} _CHARACTER_;
DO I = 1 TO DIM(CHAR);
  IF CHAR{I} = '' THEN CHAR{I} = '.';
  ELSE CHAR{I} = COMPRESS(CHAR{I});
  END;
```

Named Output

List output lines are easy enough to read if they contain only a few values. However, lines that present more than five values can be easier to read if you add the variable names. That is the idea of named output.

To write a variable in named output, write an equals sign after the variable name in the PUT statement. You can also indicate a format if necessary. In the output line, the PUT statement writes the variable and an equals sign immediately before the value of the variable. Other than these differences, named output is the same as list output. This is an example of a PUT statement in the named output style:

```
PUT ACCOUNT= BALANCE=COMMA12.2 LAST=DATE9.;
```

These are examples of output lines:

```
ACCOUNT=001-563487609 BALANCE=588.90 LAST=05MAR2001
ACCOUNT=001-565994009 BALANCE=3,407.05 LAST=06MAR2001
ACCOUNT=001-566001021 BALANCE=1,160.00 LAST=02MAR2001
```

The format list syntax is sometimes a convenient way to write list output. For example, this statement:

```
PUT A1= A2= A3= A4= A5= B1= B2= B3= B4= B5=;
```

can be written more concisely as:

```
PUT (A1-A5 B1-B5) (=);
```

When the abbreviated variable list _ALL_ appears in the PUT statement, it implies named output. That is, the name _ALL_ by itself is approximately equivalent to `(_ALL_) (=)`. The list _ALL_ includes all the variables in the data step — at least, all the variables defined up to the point where it is used in the step. When it is used as a separate term in the PUT statement, _ALL_ also includes automatic variables — _ERROR_, _N_, and any other automatic variables the step might have. So, in a typical data step, the term _ALL_ is equivalent to the terms

```
(_ALL_ _ERROR_ _N_) (=)
```

This statement:

```
PUT _ALL_;
```

is often useful in debugging a data step. Looking at the values of all the variables at a particular point in a data step is often the fastest way to track down a logical error.

List Output as Prose

The list output style can be combined with character constants in a PUT statement in order to write sentences that contain variable values. The following example demonstrates how this can work.

```
FACT = 1;
DO N = 1 TO 5;
  FACT = FACT*N;
  PUT 'The factorial of ' N 'is ' FACT +(-1) '.';
  END;
```

```
The factorial of 1 is 1.
The factorial of 2 is 2.
The factorial of 3 is 6.
The factorial of 4 is 24.
The factorial of 5 is 120.
```

Most of the time, the space that list output skips after a value is useful in forming output in a prose style. On the other hand, there are times when you want to write a punctuation mark or other value immediately after the value of a variable. Use the term +(-1) after a variable to backspace one space, to go back to the column that list output automatically skips. In the example above, this technique is used to correctly position the period at the end of the sentence.

If you have logic in a program that checks for data errors, you can write descriptive messages in the log using this technique. This is an example:

```
FILE LOG;
IF NAME = '' THEN
   PUT 'WARNING: There is no name for customer ID ' CUSTOMER +(-1) '.';
```

Lines

A PUT statement can write any number of output lines. Use the / pointer control to advance from one line to the next. This example demonstrates the use of the / pointer control to write multiple lines.

```
PUT 'Line 1'
   / 'Line 2'
   / 'Line 3';
```

```
Line 1
Line 2
Line 3
```

You can accomplish the same result by writing a separate PUT statement for each output record, for example:

```
PUT 'Line 1';
PUT 'Line 2';
PUT 'Line 3';
```

It is not always necessary to indicate line breaks in a PUT statement. The PUT statement automatically advances to the next line when it runs out of room on the current line. When there is not enough room on the current record to write an entire field, or when a column pointer control would move it past the end of the line, the PUT statement moves the pointer to the beginning of the next record. This is especially useful in list output and named output, which are often used to write more variables than can fit on a single record. The PUT statement writes as many variables or fields as it can on one record, then continues on the next record. This default behavior of the PUT statement corresponds to the FLOWOVER option in the FILE statement. To change the way the PUT statement responds when it reaches the end of a line, use the DROPOVER or STOPOVER option in the FILE statement.

To write a blank line, write a PUT statement with no terms, or write an extra slash at the appropriate point in a PUT statement, as this example demonstrates:

```
PUT / 'Line 1'
   // 'Line 2'
   // 'Line 3';
```

```
Line 1

Line 2

Line 3
```

If it is not practical to write a group of lines in sequence, use the # pointer control to move among the various lines that form a group of lines. The # pointer control works the same way as in the INPUT statement. It may be necessary to use the N= option in the FILE statement to indicate the number of lines that form a group.

Sometimes it is necessary to use several PUT statements to write a single line of output. When the last term in the PUT statement is @, called a *trailing at-sign*, it suppresses the PUT statement's automatic action of releasing the output line. Instead, the PUT statement holds onto the line so that the next PUT statement can write to the same line. The approach is especially useful when you use control flow statements to form repetitive parts of records, as in this example:

```
DO I = 1 TO 25;
  PUT I @;
  END;
PUT;
```

```
1 2 3 4 5 6 7 8 9 10 11 12 13 14 15 16 17 18 19 20 21 22 23 24 25
```

Variable-Width Output Fields

The PUT statement can write fields that vary in width. To write a value without its leading and trailing spaces, use the list output style or the : format modifier. If the width of a variable field is determined in any other way, use the $VARYING format to write the field.

The $VARYING format works in the PUT statement in much the same way that the $VARYING informat works in the INPUT statement. Like the informat, the $VARYING format uses a length variable to determine the width of the field. Sometimes the length variable is a separate variable in the data. Suppose, for example, that serial numbers may have various lengths in a particular application. If SLENGTH is a variable that indicates the length of the serial number in the variable SERIAL, write the serial number with these terms in a PUT statement:

```
SERIAL $VARYING. SLENGTH
```

The length variable might also be computed from the value of the variable involved. This example differs from list output, in that it does not remove leading spaces from the value it writes:

```
TEXT = ' Text   ';
LENGTH = LENGTH(TEXT);
PUT '*' TEXT $VARYING. LENGTH +1 '*';
```

```
* Text *
```

Practical applications often involve writing part of a composite value. For example, if the variable PLACE sometimes contains only a place name and sometimes contains a place name, a comma, and a state abbreviation, you can use the $VARYING format to write only the part of the value that contains the place name:

```
DO PLACE = 'King of Prussia, PA', 'Philadelphia', 'Lindenwold, NJ';
  COMMAX = INDEXC(PLACE, ','); * Location of comma in value, or 0.;
  IF COMMAX > 0 THEN PL_LEN = COMMAX - 1;
  ELSE PL_LEN = LENGTH(PLACE);
  PUT PLACE $VARYING. PL_LEN;
  END;
PUT;
```

```
King of Prussia
Philadelphia
Lindenwold
```

If you write a width argument for the $VARYING format, then the format writes no more than that many characters. Usually, it is not necessary to write a width argument; the default for the width argument is the length of the variable. Use a width argument whenever you need to limit the output field to a width that is less than the length of the variable.

Column Output

Column output is an alternative output style for writing fields at fixed positions in a record. The results it produces are the same as those of the usual syntax of formatted output, but it is no easier to use and very limited in what it can do, so it is rarely used. Its syntax mirrors the syntax of column input.

As in column input, the variable name is followed by column numbers that indicate pointer movement and field size. A character variable is written with the standard character format, $F. A numeric variable is written with the standard numeric format, F. For a numeric variable, you can write an additional term to indicate decimal places. The following table describes the terms you can use to write fields with column output.

PUT Statement Terms for Column Output

Term	Description
variable	The value of this variable is written to the field.
$	Declares the variable as a character variable. (This term is not necessary, but you might include it in order to have the terms of a PUT statement exactly match those of an INPUT statement.)
n	The column of a 1-character field.
n–n	The starting and ending columns of a field.
.*n*	The decimal argument for the standard numeric format.

This is an example of column output.

```
PI = 3.1415926535897932;
PUT PI 4-15 .6;
```

```
   3.141593
```

Print Output

There are additional terms that are used in the PUT statement for producing print output and object-oriented (ODS) table output. These other uses of the PUT statement are described in chapter 14, "Print Output."

Creating Error Messages

The purpose of the ERROR statement is to create an error condition and write a log message about the error. Terms in the ERROR statement are the same as in the PUT statement. The ERROR statement differs from the PUT statement in these ways:

- The keyword is ERROR instead of PUT.
- The ERROR statement writes only in the log file. It is not related in any way to the FILE statement.
- The ERROR statement sets the value of the automatic variable _ERROR_ to 1 to indicate an error condition.

Usually, the ERROR statement is the action of an IF-THEN statement. The condition checks for a specific error condition, and if it is found, the ERROR statement is executed to write a descriptive note in the log:

```
IF error condition THEN ERROR message terms;
```

This is an example of the use of the ERROR statement:

```
IF AGE < 0 THEN ERROR 'Observation ' _N_ 'has incorrect value for age: ' AGE;
```

```
Observation 3 has incorrect value for age: -5
```

Delimited Files

In the records of a delimited file, a specific delimiter character is used to separate fields. The delimiter could be any character, but it is most often a comma or a tab. The tab is a control character (ASCII character '09'X) that was originally designed for essentially this purpose. (To be specific, the tab character was intended to separate values in tables. The word *tab* is derived from a form of the word *table*.) In a delimited file, the delimiter character is not part of the data values. Instead, it indicates the end of one value and the beginning of the next. Delimited files are commonly used as a way of delivering data, and especially as a way of exchanging data between database management systems and office automation applications. One advantage that delimited files have is that they can be easily created, viewed, and edited using an ordinary text editor.

This is an example of comma-delimited data:

```
Adaptec Toast,3.5.4,1998,Adaptec
America Online,4.9,1998,America Online
Anarchie,2.0.1,1996,Stairways
AOL Instant Messenger,2.0.531,1999,America Online
ClarisWorks,4.0v4,1996,Claris
clip2gif,0.7.2,1995,Yves Piguet
ColorIt!,3.0.9,1996,MicroFrontier
```

Reading a Delimited File

A program to read a delimited file such as this one must do the following:

1. Use a LENGTH statement to declare the character variables with appropriate lengths.
2. Set the DELIMITER= and DSD options in the INFILE statement.
3. List the variables in order in the INPUT statement.

For example, the preceding records could be read by this data step:

```
DATA WORK.APP;
  LENGTH APP $ 34 VERSION $ 10 YEAR 4 OWNER $ 25;
  INFILE TEXT DELIMITER=',' DSD TRUNCOVER;
  INPUT APP VERSION YEAR OWNER;
RUN;
```

The DSD option is essential for reading a delimited file that could contain empty fields. Without the DSD option, the INPUT statement interprets two or more consecutive delimiter characters as a single delimiter. The result is that it misses the empty field and reads the rest of the record incorrectly.

The TRUNCOVER option is usually advisable when reading a delimited file. If there is a field missing in any record in the file, the INPUT statement will usually read that record incorrectly, but with the TRUNCOVER option, it will, at least, read the next record correctly. The usual behavior of the INPUT statement when a field is not found (corresponding to the FLOWOVER option) is to treat the next record as a continuation of the current record, and delimited files are usually not meant to work that way.

Writing a Delimited File

There are a few different ways to write a delimited file, depending on how you want the file to appear. Sometimes, the easiest way to write a delimited file is as regular formatted output, writing the delimiters as constant fields. This is an example of a PUT statement that writes fields and delimiters:

```
PUT YEAR 4. ',' MAKE $CHAR. ',' MODEL $CHAR. ',' BODY $CHAR.;
```

```
1986,CHEVROLET      ,SPRINT             ,CP
1988,TOYOTA         ,COROLLA            ,SDN
1990,TOYOTA         ,COROLLA            ,SW
1996,FORD           ,ECONOLINE          ,VAN
```

The output records contain extra spaces whenever values are shorter than the width of the field. To write fields without any leading or trailing spaces, use the DSD and DELIMITER= options in the FILE statement and a list output style in the PUT statement, as in this example:

```
FILE OUT DELIMITER=',' DSD;
PUT YEAR MAKE MODEL BODY;
```

```
1986,CHEVROLET,SPRINT,CP
1988,TOYOTA,COROLLA,SDN
1990,TOYOTA,COROLLA,SW
1996,FORD,ECONOLINE,VAN
```

Quoted Strings

For the delimited file format to work as described above, the data values cannot contain the delimiter character. One common way to work around this limitation in some kinds of delimited files is to quote any data value that contains the delimiter character. That is, the value is enclosed in double quotes, and any occurrence of the delimiter character in the quoted value is part of the value, and is not a delimiter. As in SAS quoted strings, if the double quote character must appear in the quoted string, it is written as two double quote characters.

This treatment of quoted strings is an additional effect of the DSD option. With the DSD option in the FILE statement, the PUT statement quotes any field that contains the delimiter character. For example, if the delimiter is a comma, and the values to write are `America Online, Inc.`, and `Atlantic City, NJ`, each of which contains a comma, the PUT statement writes:

```
"America Online, Inc.","Atlantic City, NJ"
```

The DSD option in the INFILE statement enables an INPUT statement to read quoted strings such as these when it reads a delimited file. It interprets the value represented by the quoted string — that is, it removes the double quotes and reduces each sequence of two consecutive double quotes to the one double quote that it represents. For example, if the field is `"Atlantic City, NJ"`, enclosed in double quotes, the resulting value of the variable is `Atlantic City, NJ`, with the double quotes removed.

The INPUT statement can read quoted strings as fields without interpreting the quoted string. To read a field this way, write the ~ (tilde) scanning modifier after a variable name. When you use the ~ scanning modifier, the double quotes that form the quoted string become part of the value of the variable. For example, if the field is "Atlantic City, NJ", enclosed in double quotes, the resulting value of the variable is "Atlantic City, NJ", with the double quotes intact.

The ~ scanning modifier is effective only with the DSD option. Using it with an informat, without the DSD option, results in a parsing error.

Some delimited file types do not allow quoted strings. Do not use the DSD option in the FILE statement when you write those files. Instead, if necessary, use separate statements in the data step to ensure that the values you will write in the PUT statement do not contain the delimiter character.

Options for Specific Delimited File Types

A CSV (comma separated value) file is a file that uses a comma as a delimiter and that may contain quoted strings. Use the DELIMITER=',' and DSD options in the INFILE or FILE statement to read or write a CSV file. For a tab-delimited file on an ASCII computer, use the option DELIMITER='09'X in the INFILE or FILE statement. The default form of list input and list output is another kind of delimited file, with the space character as the delimiter.

Fixed-Field Files

A fixed-field text data file contains fields at fixed positions in each record. Fixed-field text data files are widely used as a "lowest common denominator" form of data, particularly in legacy data center applications and when exchanging files on tapes between mainframe computer systems. In the data center world, these files are commonly called "flat" files.

Fields in a fixed-field text data file are read and written with formatted input and output. The INPUT and PUT statements can be written with two different pointer control styles. The more common approach uses absolute column pointer controls. An alternative style uses relative column pointer controls, or sometimes, no pointer controls at all.

Record Layout

A document that describes the way data is organized in a fixed-field text data file is a *record layout*. A record layout must contain enough information for you to determine, for each field:

- The starting column
- The width of the field
- The nature of the data contained in the field
- The way the value in the field is formatted, or organized

Absolute Column Locations

Select the appropriate informat or format for each field to match the formatting and width of the field as indicated in the record layout. Then, you can translate the record layout into INPUT statement terms in this form:

```
@starting column  variable informat
```

or PUT statement terms in this form:

```
@starting column  variable format
```

Often, the name of the informat is the same as the name of the format, and the terms may be exactly the same in the INPUT and PUT statements.

The INPUT statement below, repeated from earlier in this chapter, is written this way. The exact same terms can be used in a PUT statement.

```
INPUT
   @1 NUMBER 2.
   @3 NAME_ENG $CHAR9.
   @12 NAME_DEU $CHAR9.
   ;
```

Relative Column Locations

In the previous example, you might notice that the column pointer controls do not actually do anything. The term @1 moves the pointer to column 1 — but it appears at the beginning of the statement, when the pointer is already at column 1, so no movement is required. The same is true of the terms @3 and @12 — the pointer is already at the correct location after processing the previous field.

Most fixed-field text data files are set up this way, with no extra space between fields. With few exceptions, every column is part of a field. When a record is defined this way and you read or write every field in sequence using formatted input or output, column pointers are not necessary. The previous example could be rewritten as:

```
INPUT
   NUMBER 2.
   NAME_ENG $CHAR9.
   NAME_DEU $CHAR9.
   ;
```

To skip over filler or fields that are not being read in an input record, use the + pointer control. For example, use the term +4 to advance the pointer past a 4-character field that you do not need to read. The same considerations apply to a PUT statement that creates an output record. Use a +4 pointer control, for example, to leave a 4-character field blank. In this way of writing an INPUT or PUT statement for a fixed-field record, there are no absolute pointer controls. After the beginning of the record, pointer movement is always indicated relative to the previous pointer location.

This style has advantages and disadvantages. With no direct mention of column positions, it can be a challenge to match the statement against a record layout that is written in terms of columns. With all positions being described in relative terms, you have to be very careful in writing the statement; a mistake in the width of any field makes the rest of the statement line up incorrectly. On the other hand, the lack of absolute column pointers has advantages. It makes it possible to use informat lists or format lists for repetitive parts of the record. It may also make the statement easier to maintain, especially for files whose record layouts tend to change frequently. For example, if the record layout is altered to make one field wider, you only have to change the format width for that one field. By contrast, if you were using absolute column pointer controls, you would also have to correct every column pointer from that point to the end of the record to make that change.

Both styles of processing a fixed-field record do the same things, so choose the style that is easier to read and maintain in a particular application.

Options

If an input record is shorter than the fixed-field record an INPUT statement expects, problems can result. The default response of the INPUT statement is to continue reading on the next line, as if it were a continuation of the current line, but that is never appropriate in a file of fixed-field records. Instead, specify the TRUNCOVER option, or perhaps the MISSOVER or STOPOVER option, in the INFILE statement. In some circumstance, the PAD and LINESIZE= options may be helpful for a fixed-field text data file.

Records that are too short can also be a problem in writing fixed-field records, if you are writing to a file that has a fixed record length that is too short for the data you are writing. If this happens to you frequently, you may want to use the STOPOVER option in the FILE statement. This option stops the program with an error condition if the record length of the output file is not long enough. Then you can correct the problem in the program or in the file and run the program again.

Binary Files

In the SAS language, binary input and output files are treated as text files, and the same concepts of fields, pointers, informats, and formats apply. The specific informats and formats will likely be different, however. SAS includes dozens of binary informats and formats for reading and writing binary files. You can also use the $CHAR informat and format to read and write character variables that have binary values. Informats and formats are described in the next chapter.

Fixed-Length Binary Records

Some binary files are defined as a sequence of fixed-length records. There may be fields in fixed positions in each record. Reading or writing these records is not fundamentally any different from reading or writing a text file that has a fixed record length. Use pointer controls in the same manner — @ and + to move to different columns on a record and / to advance to the next record.

Indicate the structure of the file in options in the INFILE or FILE statement. Use the RECFM=F option to indicate fixed-length records and the LRECL= option to indicate the record length.

Byte Stream

More often, a binary file is easier to understand as a sequence of bytes. A byte is the binary equivalent of a character, and a *byte stream*, or a sequence of bytes, is the basic nature of all files in modern file systems.

In the INFILE or FILE statement, use the RECFM= (record format) option to indicate a binary file. For most operating systems, the correct setting is RECFM=N (no format). In IBM mainframe operating systems, it is RECFM=U (undefined).

Use only column pointer controls to move around the file. Process objects in the file as fields. Line pointer controls are not of any use, because there really are no records in the file. For the same reason, the INPUT or PUT statement does not release the record at the end of the statement, so the use of trailing @ or @@ to hold onto the current record is not necessary. Think of the file as consisting of one long record.

To read or write objects in the file, use informats or formats with whatever length the objects might be, provided that they are no more than 32,767 bytes long. Use the $CHAR informat or format for character variables that contain the binary values or data structures of the file. The $VARYING informat or format may be helpful for objects whose size is indicated in a separate field in the file. Use the + pointer control to advance the pointer past objects you do not want to read, or use it with a negative value (in parentheses) to go back. You can use the COLUMN= option in the INFILE or FILE statement to create a variable that has the current location in the file. Assign the value of that variable to another variable to mark a location in the file. You can then use the @ pointer control with that variable to return to that location.

The programming logic required to interpret or create a binary file can be as complicated as the structure of the file is. That is to say, it is often more complicated than anything that is discussed in this book. However, SAS data step logic and I/O syntax give you all the flexibility you need to work with virtually any kind of file.

Updating a Text File

You can use a data step to update a text file in place, changing some of the characters in the file while retaining others. This involves the same statements that do ordinary text file I/O in the data step: the INFILE, FILE, INPUT, and PUT statements. Indicate the same file in the FILE statement as in the INFILE statement. Then use an INPUT statement to read a record from the file, and use the PUT statement to write the same record back to the file, possibly with changes.

Usually, you should use the SHAREBUFFERS option in the INFILE statement. With this option, the INPUT and PUT statements work with the same buffer. The practical result is that characters in the file remain unchanged unless you specifically change them in the PUT statement. If you prefer to rewrite the entire file, do not use the SHAREBUFFERS option.

Any file options that can ordinarily appear in the INFILE or FILE statement must be written in the INFILE statement. They are ignored if you write them in the FILE statement. The INFILE statement must appear before the FILE statement. To avoid conflicts, be sure not to change the length of a record or to add or delete records. If you need to do that kind of processing, use a separate output file.

The example below updates a text file, changing only the e-mail address field. It looks at the domain name, which is the part of the e-mail address that follows the @ character, and converts any capital letters to lowercase letters.

```
DATA _NULL_;
  INFILE LIST SHAREBUFFERS STOPOVER;
  FILE LIST;
  INPUT @25 EMAIL $54.;
  * Lowercase domain name of e-mail address.;
  ATSIGN = INDEXC(EMAIL, '@');
  IF ATSIGN THEN DO;
    I = ATSIGN + 1;
    IF SUBSTR(EMAIL, I) NE LOWCASE(SUBSTR(EMAIL, I)) THEN
      SUBSTR(EMAIL, I) = LOWCASE(SUBSTR(EMAIL, I));
    END;
  PUT @25 EMAIL $54.;
RUN;
```

Because of the SHAREBUFFERS option, the rest of the data in the file is not affected. The STOPOVER option ensures that, if a record is shorter than expected, the step stops running before it damages the file.

This same technique can be used to update print files and binary files. To change selected data objects in a binary file, use the SHAREBUFFERS option along with the binary file techniques described in the previous section.

Data Step Windows

The WINDOW statement defines a window using terms similar to those of the INPUT and PUT statements. The window can then be displayed with a DISPLAY statement. The window definition can use the #, /, @, and + pointer controls and constant and variable fields. A variable term followed by the PROTECT=YES option indicates a protected variable field, which only displays the value of the variable. Otherwise, the user can alter the value of the variable in a variable field. The format term used with the field is also used as an informat term, which means the format name and informat name must be the same.

The statement below is an example of a WINDOW statement. It defines the window Observation Count with two fields: a constant field and a variable field that displays the variable COUNT.

```
WINDOW "Observation Count"N #2 @4 'Number of observations written: '
  #4 @4 COUNT F12. PROTECT=YES;
```

The window COUNT can be displayed by the DISPLAY statement:

```
DISPLAY "Observation Count"N;
```

When a data step window is displayed, it shows the current values of variables in the variable fields. The name of the window appears as its title. If the user changes the text in an unprotected variable field, it changes the value of the variable. Execution pauses when the window is displayed, and resumes after the user presses the enter key. If the window contains multiple unprotected variable fields, the user might have to press the enter key more than once. If the user enters the END command or otherwise closes the data step window, execution of the data step stops immediately.

To display a window without waiting for a user response, use the NOINPUT option in the DISPLAY statement. A data step window remains open until the end of the data step, unless you use the DELETE option in the DISPLAY statement. With the DELETE option, the window closes as soon as the user responds. There are other options in the WINDOW and DISPLAY statements to modify the use and appearance of the window and its fields.

Two automatic variables are used with data step windows. Before displaying the window, assign a text value to the variable _MSG_ to show a message in the message area of the window. After the window is displayed, if the user enters a command that the SAS supervisor does not recognize as a SAS command, the command text is contained in the variable _CMD_.

These are possible uses of data step windows:

- *Data entry.* Use unprotected variable fields, with constant fields to provide labels. Prompt the user to enter values for the variables for each observation, and to enter the END command when there are no more observations to enter.
- *Results of a data step.* Use a combination of constant fields and protected variable fields.
- *Progress.* Periodically recalculate and display the relative completion of or amount of data processed by a data step. Use the NOINPUT option in the DISPLAY statement.
- *Dialog.* Display a question in a protected field, with an unprotected variable field in which the user can indicate an answer. Use the DELETE option in the DISPLAY statement.

The presence of a DISPLAY statement triggers the automatic loop of the data step. If necessary, use the STOP statement to ensure that the data step does not repeat indefinitely. This example shows the use of the STOP statement to stop a data step after it displays a window:

```
DATA _NULL_;
  CLOCK = DATETIME();
  WINDOW "The Current Date and Time"N
    #4 @11 CLOCK DATETIME. PROTECT=YES;
  DISPLAY "The Current Date and Time"N;
  STOP;
RUN;
```

13

Informats and Formats

Informats and formats are routines that are an essential part of text I/O in SAS software. Informats convert text to data values that can be assigned to variables. Formats convert the values of variables to text that can be displayed or written to a file.

Types

There are two types of variables, so there have to be two types of informats and two types of formats. Numeric informats and formats work with numeric variables. Character informats and formats work with character variables. The different actions of the different types are compared in the following table.

Types of Informats and Formats

Routine (Examples)	Action	Input	Result
Character informat ($CHAR, $F)	Interprets character data	Text data	Character value
Numeric informat (F, COMMA)	Interprets numeric data	Text data	Numeric value
Character format ($CHAR, $UPCASE)	Formats character data	Character value	Text data
Numeric format (BEST, F, E, COMMA)	Formats numeric data	Numeric value	Text data

You can tell the two data types apart at a glance. Names of character informats and formats begin with a dollar sign. By contrast, the names of numeric informats and formats do not include any special characters. On the other hand, it is not so easy to see the difference between an informat and a format. Often, you can tell an informat from a format only by seeing the context in which they are used. The informat that interprets a certain kind of text field often has the same name as the format that creates that kind of text data.

References

A reference to an informat or format follows a syntax form that is unique to the SAS language. The name of the informat or format is followed by two numeric arguments, with a period between them. Any of these components of an informat or format reference can be omitted, in various combinations, except for the period, which is always required. There are no spaces between any of the parts of the reference. The following diagram identifies the components of an informat or format reference.

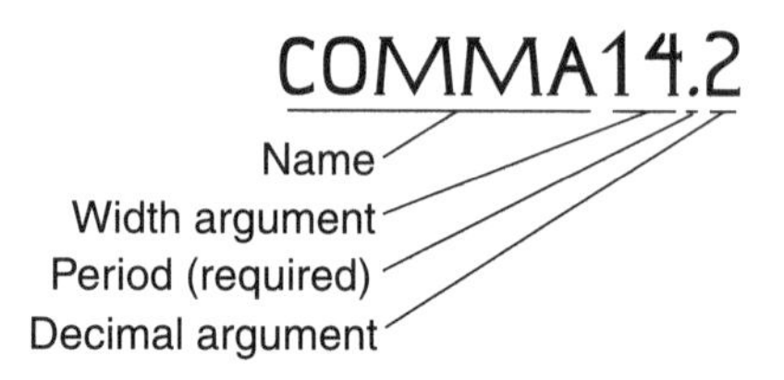

The name identifies the specific routine that does the work of interpreting or formatting the value. The $ character at the beginning of the name of a character informat or format is part of the name. The numeric suffix is never part of the name. Even though there is no space between the name and the number, the number is a separate component, the width argument. The width argument determines how many bytes or characters an informat reads or a format writes. The width argument is followed by a period, which is always a required part of the reference. After the period, there can be a second numeric argument, which is the decimal argument used by most numeric informats and formats. The decimal argument typically indicates the number of decimal places in the value.

The decimal argument is not always necessary. In addition, either the name or the width argument can be omitted. However, the width argument must be present if the decimal argument is used. This results in five possible ways of writing an informat or format reference, all of which are commonly used in SAS programs. The various combinations of components that can form a valid informat or format reference are listed in the table below.

Five Ways to Write an Informat or Format Reference

Examples	Components	Notes
$CHAR14. F5.	name, width argument, period	1
F5.2	name, width argument, period, decimal argument	2
$CHAR. F.	name, period	1, 3
5.	width argument, period	1, 4
5.2	width argument, period, decimal argument	4

1. When you do not write a decimal argument, the default decimal argument is 0.
2. Only numeric informats and formats use the decimal argument. However, it is not incorrect to write 0 as the decimal argument for a character informat or format.
3. The width argument is required when you use the $ alias of the $F informat or format.
4. When you omit the name, the default informat or format is F, the standard numeric informat or standard numeric format.

Arguments

The arguments to an informat or format determine specific details about the way the informat or format works. The width argument always determines the number of characters or bytes in the field — the number of bytes read by an informat or written by a format. Different formats allow different minimum and maximum widths, but the width must always be at least 1. You can usually leave out the width argument. If there is no width argument, the width is determined by the context in which the informat or format is used. These are examples of the sources for width values in some situations:

- The default width defined for the informat or format
- For a character informat in an INPUT statement, a width based on the length of the variable
- For a character format, a width sufficient to write the entire variable or value
- For a format in the REPORT proc, the width of the column
- The width of a field in an interactive application

The second argument, the decimal argument, can have a few different meanings, all having something to do with decimal places. Only numeric informats and formats use the decimal argument.

For a numeric informat, the decimal argument indicates the number of implied decimal places in the input field. For example, when the informat `4.2` reads the text `2325`, the resulting value is the number 23.25. However, the decimal argument is considered only when the input text does not contain a decimal point. For example, when the informat `4.2` reads `25.0`, the resulting value is the number 25. The decimal argument is not considered in this case because the field contains a decimal point.

The table below shows various possible combinations of width and decimal arguments for the standard numeric informat, along with the values the informat produces when it reads a sample input field containing all 8's.

Informat Value	1. 8	2. 88	3. 888	4. 8888	5. 88888	6. 888888
Informat Value	1.1 .8	2.1 8.8	3.1 88.8	4.1 888.8	5.1 8888.8	6.1 88888.8
Informat Value	1.2 .08	2.2 .88	3.2 8.88	4.2 88.88	5.2 888.88	6.2 8888.88
Informat Value	1.3 .008	2.3 .088	3.3 .888	4.3 8.888	5.3 88.888	6.3 888.888
Informat Value	1.4 .0008	2.4 .0088	3.4 .0888	4.4 .8888	5.4 8.8888	6.4 88.8888

Input text: `888888`

There are several informats that read numbers from binary fields. For these informats, the decimal argument represents a decimal scaling factor. It divides the value it reads from the field by the power of 10 indicated in the decimal argument. This is essentially the same thing that a decimal point does, but the difference is that the binary field does not actually contain decimal digits. For these informats, the decimal argument can range from 0 to 10. Consider, as an example, a one-character field whose value could be written in hexadecimal as '70'X. If you read this field with the informat PIB1., the value that results is 112. But if you read it with the informat PIB1.2, the value is 1.12.

For ordinary numeric formats, the decimal argument tells the format how many decimal places to write. When the decimal argument is 1 or more, the format writes a decimal point and the indicated number of digits after the decimal point. For example, the format 8.3 could write the following fields:

```
   1.250
 -57.108
   0.016
8192.000
```

As with binary informats, the decimal argument for binary formats is a decimal scaling factor that works like an implied decimal point. The format multiplies the value by the power of 10 indicated by the decimal argument before it writes the value. Thus, for example, the way the format IB4.2 writes the number 3.16 is the same as the way the format IB4. writes the number 316. As with binary informats, the decimal argument can range from 0 to 10.

Fields, Informats, and Formats

Informats and formats form the connections between the SAS data types and the various ways that data fields can be organized. Of all the categories of informats and formats, the ones that deal with character fields have the easiest job to do; there is no fundamental difference between the data in a character field and the data in a SAS character value. Informats and formats that deal with numbers have the more difficult task of dealing with the characters that are used to write a number. Those character representations of numbers are connected in a systematic but rather complicated way to the actual value of the number. There are also informats and formats that deal with the even more complicated notation with which dates and times of day are written, and other informats and formats that deal with various kinds of binary fields, each with its own way of representing a number or other object.

Informats and formats are two different kinds of routines, but they connect to the same kinds of fields; for almost every informat that reads a certain kind of field, there is a compatible format, with the same name, that writes that same kind of field. In this chapter, informats and formats are discussed together, but they also have to be considered separately. An informat and a format may have the same names, be written the same way, and deal with the same kinds of data, and in that connection, they can be considered together. On the other hand, their actions tend to be more opposite than alike, and are easier to describe when considered separately.

Many informats and a few formats have aliases — alternate names that can be used to refer to them. Part of the reason for aliases is that one informat may be compatible with more than one format. The informat may use several names, one to match each compatible format.

Informats and formats are designed to translate data from one form to another, each one doing so in a different way. On the other hand, you do not always need to change data in any way when you read it or write it. If you just want to copy data from a field to a variable or from a variable to a field without altering it, remember that the $CHAR informat and format are available for just that purpose.

Character Fields

The informats and formats in this category provide a connection between character variables and fields that contain character data. In the process or reading and writing the fields, they transform the data in various ways.

For all the character informats and formats listed in this section, the width argument *w* can range from 1 to 32,767 bytes — the same range that is allowed for the length of a character variable.

$CHAR — Character data The $CHAR informat and format merely copy data, without changing it or altering it in any way. They are the fastest informat and format, because they do not do any translation. They are often the best way to read and write fields of ID numbers, codes, and other identifying and categorical variables.

Informat: `$CHAR`*w*.	Format: `$CHAR`*w*.
Action: Transfers data unchanged.	Action: Transfers data unchanged.

$F — Words Most character values are words, names, codes, and similar text values — values that do not naturally contain leading or trailing spaces, even though the fields or variables that hold them might include spaces before or after the value. The standard character informat, $F, is designed to disregard any extra spaces that are present and look for words or something similar in a text field. The standard character format, $F, is meant to be compatible with the standard character informat. It acts the same as the $CHAR format, writing character values without changing or altering them.

To the $F informat, when a field contains a single period, the period is not a data value, but a placeholder that represents a missing value. The $F informat's ability to interpret a period as a missing, or blank, value, makes it especially useful for list input. List input uses spaces as delimiters, so every value in list input must be written in a nonblank form — even a blank value.

You can write this informat or format name simply as `$`, provided that the reference also has a width argument.

Informat: `$F`*w*., `$`*w*.	Format: `$F`*w*., `$`*w*.
Action: Drops leading spaces. Interprets a field that contains a single period as missing (that is, blank).	Action: Transfers a character value unchanged (the same as the $CHAR format).

Sometimes you want to drop leading spaces when you write a field, but the $F format does not do that the way the $F informat does when you read a field. However, you can use the -L alignment option after the $F format in the PUT statement to accomplish the same thing. You could also use the LEFT function in an assignment statement to remove leading spaces.

$UPCASE — Capitalize The $UPCASE informat and format convert lowercase letters to uppercase. The informat is handy when a variable should contain only capital letters but the input field could contain lowercase letters. Use the format when a variable contains lowercase and uppercase letters but you are creating a file that will be used in a context where only uppercase letters can be used. The effects of the informat and format are essentially the same as that of the UPCASE function. Characters that are not lowercase letters are not changed.

Informat: $UPCASE*w*. Action: Converts lowercase letters to uppercase.	Format: $UPCASE*w*. Action: Converts lowercase letters to uppercase.

$MSGCASE — Messages The $MSGCASE format is designed for writing messages. SAS System messages are ordinarily written with mixed-case letters, but if necessary, you can have them written in all uppercase letters. This is determined by the MSGCASE system option. When this option is on, the SAS supervisor converts all message text to uppercase letters. Any text you write with the $MSGCASE format is also affected by the MSGCASE option. So if you want to write messages that are in mixed case when they can be and in uppercase when they have to be, start with a mixed-case value and format it with the $MSGCASE format.

	Format: $MSGCASE*w*. Action: Converts lowercase letters to uppercase if the system option MSGCASE is on.

$QUOTE — Quoted string The $QUOTE informat and format work with fields that contain quoted strings. The informat interprets quoted strings the same way that the SAS supervisor interprets quoted strings in a SAS program. The strings can be quoted with either double quote marks or single quote marks.

The $QUOTE format writes quoted strings that can be interpreted by various programs that use quoted strings, such as spreadsheet programs, or by the $QUOTE informat. It is important to allow a sufficient width for the $QUOTE format. The quoted string that the format produces is longer than the value it contains. It is at least two characters longer, and on occasion may be four or more characters longer. If the field is not wide enough to write the entire quoted string, the $QUOTE format writes a blank field.

Informat: $QUOTE*w*.

Action: Removes double quote characters or single quote characters used to delimit a quoted string. Converts each pair of the active quote character inside the quoted string to one of the same quote character. Does not change the value if the input text is not a quoted string.

Format: $QUOTE*w*.

Action: Encloses the value, not including trailing spaces, in double quote characters. Converts each double quote character in the value to a pair of double quote characters.

$REVERJ — Reversed data The $REVERJ informat and format reverse the order of the bytes in character data or any other kind of data. Use them for fields in which bytes are stored in reverse order.

Informat: $REVERJ*w*.

Action: Reverses the order of the bytes in the value.

Format: $REVERJ*w*.

Action: Reverses the order of the bytes in the value.

$REVERS — Reversed text The $REVERS informat and format also reverse the order of characters, but they are designed to work with words and similar kinds of text values. They remove leading spaces from the result, similar to the way the $F informat does. This way, any trailing spaces in the input do not become leading spaces in the output. They are still trailing spaces.

Informat: $REVERS*w*.

Action: Reverses the order of characters, and removes leading spaces from the result.

Format: $REVERS*w*.

Action: Reverses the order of characters, and removes leading spaces from the result.

The width argument is important for both the $REVERS and $REVERJ formats. If the width of the $REVERS or $REVERJ format is less than the width of the value it writes, the format truncates the value before writing it. The result is that the format writes the reverse of the beginning of the value, as demonstrated in the example that follows. Both formats use a default width of 1 — a width that would defeat the purpose of the format.

```
X = 'abcdef';
PUT X $REVERJ2.
   / X $REVERS2.
   / X $REVERJ6.
   / X $REVERS6.
   ;
```

```
ba
ba
fedcba
fedcba
```

$ASCII — ASCII text The $ASCII informat and format work with fields that contain ASCII text. On ASCII computers, which includes most computers, the $ASCII informat and format transfer data without changing it in any way; they are the same as the $CHAR informat and format. But on EBCDIC computers, a category that includes IBM mainframe computers, they translate between the ASCII and EBCDIC character sets. Thus, the $ASCII informat and format provide a portable way to work with ASCII text fields. They handle these fields the same way whether the program is running on an ASCII computer or an EBCDIC computer.

Informat: `$ASCII`*w*.
Action on ASCII computers: Transfers characters unchanged (the same as the $CHAR informat).
Action on EBCDIC computers: Translates ASCII characters to EBCDIC characters.

Format: `$ASCII`*w*.
Action on ASCII computers: Transfers characters unchanged (the same as the $CHAR format).
Action on EBCDIC computers: Translates EBCDIC characters to ASCII characters.

$EBCDIC — EBCDIC text The same way that the $ASCII informat and format provide a portable way of working with ASCII text fields, the $EBCDIC informat and format provide a portable way of working with EBCDIC text fields.

Informat: `$EBCDIC`*w*.
Action on ASCII computers: Translates EBCDIC characters to ASCII characters.
Action on EBCDIC computers: Transfers characters unchanged (the same as the $CHAR informat).

Format: `$EBCDIC`*w*.
Action on ASCII computers: Translates ASCII characters to EBCDIC characters.
Action on EBCDIC computers: Transfers characters unchanged (the same as the $CHAR informat).

The $ASCII and $EBCDIC informats and formats let you translate between the ASCII and EBCDIC character sets, but to do the translation, you need to select the correct routine for the computer you are running on. The following table indicates the right routine to use.

Translating Between ASCII and EBCDIC Character Sets

Routine for translation	from ASCII text to EBCDIC text	from EBCDIC text to ASCII text
running on an ASCII computer system	$EBCDIC format	$EBCDIC informat
running on an EBCDIC computer system	$ASCII informat	$ASCII format

$VARYING — Variable-width field in INPUT/PUT statement The $VARYING informat and format work in the INPUT and PUT statements to read and write fields that vary in width. They depend on a length variable, which is a separate term in the statement that serves as an additional argument to the informat or format.

Informat:
$VARYING*w. length variable*
Action when used in the INPUT statement: Reads the number of characters indicated by the length variable.

Format: $VARYING*w. length variable*
Action when used in the PUT statement: Reads the number of characters indicated by the length variable.

If you use the $VARYING informat and format anywhere else, they act the same as the $CHAR informat and format.

$HEX, $OCTAL, and $BINARY — Byte values The value of a byte can be thought of as a number from 0 to 255. There are various ways of writing the number that represents a byte value. The $HEX format writes each byte value as two hexadecimal digits. The $OCTAL format writes each byte as three octal (base 8) digits. The $BINARY format writes each byte as eight binary digits, using the characters "0" and "1" to represent the binary digits. The corresponding informats read fields written in those formats. Character hexadecimal notation is often used when binary data must be converted to ordinary text characters so that it can be transferred accurately from one computer to another. The octal and binary notation could also be used for the same purpose.

The width of these informats is a multiple of the length of the values they represent. A character hexadecimal field must be twice as long as the value it contains; a character octal field, 3 times; and a character binary field, 8 times. If the length of an input field is not an exact multiple, the informat treats it as if it were padded with zeroes to complete a value that would be a whole number of bytes.

Informat: $HEX*w.*
Action: Reads each two hexadecimal digits as the value of 1 byte.

Format: $HEX*w.*
Action: Writes each byte value as two hexadecimal digits.

Informat: $OCTAL*w.*
Action: Reads each three octal digits as the value of 1 byte.

Format: $OCTAL*w.*
Action: Writes each byte value as three octal digits.

Informat: $BINARY*w.*
Action: Reads each eight binary digits (the characters "0" and "1") as the value of 1 byte.

Format: $BINARY*w.*
Action: Writes each byte value as eight binary digits (the characters "0" and "1").

DBCS Fields

Most character informats and formats treat each byte as a separate character, so they may not work correctly with DBCS fields. For most DBCS data, the $CHAR informat and format can read and write the fields. There are a few additional informats and formats that work with special situations involving DBCS data.

$KANJI, $KANJIX — Adding or removing shift codes Some DBCS systems use characters called shift codes to mark the beginning and end of DBCS data; others do not. Use the $KANJI and $KANJIX informats and formats to add and remove these shift codes.

Informat: $KANJI*w.*
Action: Removes shift codes from DBCS data.

Format: $KANJI*w.*
Action: Adds shift codes to DBCS data.

Informat: $KANJIX*w.*
Action: Adds shift codes to DBCS data.

Format: $KANJIX*w.*
Action: Removes shift codes from DBCS data.

The following table indicates the informats and formats to use for DBCS fields, based on whether shift codes are used in the field and variable.

Informats and Formats for DBCS fields

Variable:	With shift codes	Without shift codes
Field with shift codes	$CHAR	$KANJI
Field without shift codes	$KANJIX	$CHAR

There are two other DBCS informats. The JDATEYMD and JNENGO informats interpret two kinds of date values written in Japanese kanji DBCS text.

Numeric Fields

The biggest category of fields is made up of the fields that contain numerals — numbers written in a way that people can read them. Although they are designed primarily for people to read, they are also commonly used in files that transfer numeric data from one computer program to another.

Lexicon

In the data center environment, these fields are sometimes referred to as *visual* fields. The word "visual" emphasizes the fact that the fields are made of characters that are meant for people to read.

The width of most informats and formats that read and write numeric fields can range from 1 to 32 characters. Most can handle negative numbers, with the negative sign (the hyphen character) written immediately before the first digit of the number.

Most, also, can handle fractional numbers. The digits written after the decimal point (usually written as a period) indicate the fractional part of the number. When you determine the width necessary for a field, include character positions for the negative sign and decimal point, if necessary.

Informats can also read fields with implied decimal places; the decimal places are written in the field, but the actual decimal point is not present.

Instead, the decimal argument of the informat indicates the number of decimal places. The decimal argument is ignored, though, if there is an actual decimal point in the input text. The decimal argument for most formats must be less than the width argument. The decimal argument is optional; having no decimal argument is equivalent to having a decimal argument of 0.

F — Standard notation The standard numeric informat reads numbers written in the same way that standard numeric constants are written in a SAS program, without commas. This includes scientific notation, such as that produced by the E format, which is described next. The standard numeric format writes numbers in the standard notation, without commas. If there is a decimal argument, it writes a decimal point and a fixed number of decimal places, as indicated by the decimal argument.

This informat and format can be written with the name F, or they can be written without a name as long as the width argument is present.

Informat: F*w.d, w.d*
Width: $1 \leq w \leq 32$
Decimal: $d \leq 31$
Action: Interprets numbers in standard and scientific notation.
Aliases: BEST, D, E

Format: F*w.d, w.d*
Width: $1 \leq w \leq 32$
Decimal: $d < w$
Action: Writes numbers in standard notation.

E — Scientific notation Some numbers, especially in science, are too large or too small in magnitude to be written easily in standard notation. For example, the number known as Avogadro's number, which expresses the ratio between a gram and an atomic weight unit, is approximately this value:

```
602000000000000000000000
```

However, this number may be easier to appreciate when written in scientific notation, as

```
6.02E+23
```

The number after the `E`, the exponent, indicates a factor of a power of 10 and gives a quick indication of the magnitude of the number.

The E format writes numbers in this manner. It does not use a decimal argument; instead, it writes the number with as much precision as is possible in the space available. It uses the first column for a minus sign for negative values; this column is blank for positive values. It writes a mantissa, the part of the value before the `E`, that has one digit before the decimal point; that is, it is at least 1 but less than 10. It then writes the letter `E`, a sign for the exponent, and an exponent of at least two digits, using more digits if necessary. The minimum width it requires for this notation is 7 characters.

The standard numeric informat can read scientific notation, which does not have to strictly follow the format produced by the E format. It can have any number of digits in the mantissa, before or after the decimal point, or in the exponent; it can include or not include a plus sign before a positive mantissa or exponent; and the letter that connects the mantissa to the exponent can be an uppercase or lowercase E or D. The E informat is an alias for the standard numeric informat.

Informat: `E`*w.d*
An alias for the F informat.

Format: `E`*w.*
Width: $7 \le w \le 32$
Action: Writes numbers in scientific notation. The first character is a negative sign for a negative value, or a space for a positive value. This is followed by a mantissa, the letter E, a sign for the exponent, and an exponent indicating a power of 10. The format uses at least 2 digits for the exponent, and writes the value with as many decimal places as the width allows.

BEST — Maximum precision The BEST format, which is the default format for writing numeric values, writes numbers with as much precision as it can in the width you provide it. For whole numbers that fit in the field, the output is no different from the output produced by the F format. If a number has a fractional part, the BEST format writes as many decimal places as will fit, but without writing a 0 as the last decimal digit. The BEST format determines the number of decimal places to write based on the value it is writing, so it does not use a decimal argument. If the magnitude of a number is too great for it to fit in the field, the BEST format writes a form of scientific notation.

Informat: `BEST`*w.d*
An alias for the F informat.

Format: `BEST`*w.*
Width: $1 \le w \le 32$
Action: Writes numbers in standard notation with as much precision as possible in the width of the field. Writes in scientific notation if necessary.

Other numeric formats, such as F and COMMA, use the BEST format when a value does not fit in the field with the format indicated. This way, an indication of the value is written if it is possible to write it. For example, if you write the value 5000 with the `F3.` format, the value does not fit in the width, so the F format calls the BEST format, which writes the value as `5E3`.

When even the BEST format cannot write a value, it fills the field with asterisks. For example if you try to write the value –5000 in a three-character field, the resulting field is `***`.

The F, E, and BEST formats are all capable of writing scientific notation, but they do so in different ways. The F format uses the BEST format to create its scientific notation, but only for numbers that the F format itself cannot format. The following two examples show comparisons of the three formats. The first example writes a large number. When the F format calls the BEST format, it generates the log note that follows.

```
X = 5000000000;
PUT X F9. / X BEST9. / X E9.;
```

```
      5E9
      5E9
 5.00E+09
NOTE: At least one W.D format was too small for the number to be printed.
      The decimal may be shifted by the "BEST" format.
```

The second example writes a small number, which uses a negative exponent. The F format does not write this number in scientific notation, because it can write it as 0.

```
X = .000000009;
PUT X F9. / X BEST9. / X E9.;
```

```
        0
      9E-9
 9.00E-09
```

D — Significant digits When the BEST format writes a value, it can put the decimal point anywhere in the field. For some kinds of variables, this can make the values in a column difficult to compare to each other. The D format tries to address this problem by keeping the decimal point relatively fixed within ranges of values, but moving it when necessary. The second argument for the D format is not really a decimal argument. Instead, it indicates the minimum number of significant digits the format should maintain. Use a larger significant digits argument to get more precision; use a smaller value for the argument to keep the decimal point in more of a fixed position.

Informat: D*w.d*
An alias for the F informat.

Format: D*w.s*
Width: $1 \le w \le 32$
Significant digits: $s \le 16$; $s < w$
Action: Writes numbers with at least the indicated number of significant digits, while keeping the decimal point fixed within ranges of values.

If you like the idea of the D format, but are not completely satisfied with the results it produces, you may want to use the FORMAT proc to create a picture format. With a picture format, you can determine exactly how many decimal places to print in ranges that you specify. See chapter 24, "Creating Formats and Informats."

ROMAN, FRACT, WORDF, and WORDS — Alternative forms These four formats write numbers in ways that are not usually associated with computers, because they are not easy formats for computers to read. For that same reason, there are no informats to correspond to these formats.

The ROMAN format writes Roman numerals. It can write values from 1 to 9,999, which it writes as `MMMMMMMMMCMXCIX`. It writes values of 10,000 or more as ****************.

The FRACT format writes fractions, such as 1/25. It reduces fractions to their lowest terms before writing them. It writes numbers greater than 1 as whole number plus a fractional part. These are examples of the fractions that the FRACT format writes:

```
X1 = 20/100; X2 = 25/100; X3 = 250/100;
PUT (X1-X3) (FRACT8.)
```

```
     1/5       1/4    2+1/2
```

The WORDF and WORDS formats write numbers as words. The difference between these two formats is the way they present fractional parts of numbers, which they write as hundredths. The WORDF format writes them as a fraction, with 100 as the denominator. This is the way amounts are commonly written on checks. The WORDS format writes them as words. This example demonstrates the difference:

```
X = 350.17;
PUT X WORDF50.;
PUT X WORDS50.;
```

```
three hundred fifty and 17/100
three hundred fifty and seventeen hundredths
```

All of these formats can require large field widths, compared to other formats that write numbers.

Like most formats that write numeric values, the FRACT format right-aligns its results. That is, when the text it produces is shorter than the width of the field, it writes it at the end of the field and fills the rest of the field with leading spaces. The ROMAN, WORDF, and WORDS formats are different. They left-align their results. That is, they write the value starting at the beginning of the field and fill the rest of the field with trailing spaces.

The WORDS and WORDF formats, by default, write words in lowercase letters. However, they are affected by the MSGCASE system option. As described in the discussion of the $MSGCASE format, the MSGCASE option tells the SAS System to convert all messages to uppercase letters. When the MSGCASE option is on, the text produced by the WORDS and WORDF formats is also converted to uppercase. The MSGCASE option also affects the output of all other numeric formats that write lowercase letters.

Format: ROMAN*w*.
Width: $2 \leq w \leq 32$
Action: Writes Roman numerals in capital letters.

Format: FRACT*w*.
Width: $4 \leq w \leq 32$
Action: Writes fractions.

Format: WORDF*w.*
Width: $5 \leq w \leq 32767$
Action: Writes numbers as words with hundredths written as a fraction.

Format: WORDS*w.*
Width: $5 \leq w \leq 32767$
Action: Writes numbers as words with hundredths written as words.

HEX, OCTAL, and BINARY — Numbering systems The HEX, OCTAL, and BINARY informats and formats work with three numbering systems used frequently in computer work. The hexadecimal, octal, and binary number fields, respectively, can represent only whole numbers. The formats write negative numbers as positive and truncate any fractional parts of numbers to arrive at whole number values. They write leading zeroes to fill the width of the field.

The corresponding informats can read numbers with or without leading zeroes. They ignore leading and trailing spaces. If you use a decimal argument with the OCTAL and BINARY informats, they divide the value by that power of 10.

Informat: HEX*w.*
Width: $1 \leq w \leq 15$
Decimal: $d = 0$
Action: Interprets hexadecimal (base 16) integer values.

Format: HEX*w.*
Width: $1 \leq w \leq 15$
Decimal: $d = 0$
Action: Writes whole number hexadecimal (base 16) numerals with capital letters and leading zeroes.

Informat: OCTAL*w.d*
Width: $1 \leq w \leq 24$
Decimal: $d \leq 31$
Action: Interprets octal (base 8) integer values.

Format: OCTAL*w.*
Width: $1 \leq w \leq 24$
Decimal: $d \leq 31$
Action: Writes whole number octal (base 8) numerals with leading zeroes.

Informat: BINARY*w.d*
Width: $1 \leq w \leq 64$
Decimal: d ≤ 31
Action: Interprets binary integer values using the characters "0" and "1".

Format: BINARY*w.*
Width: $1 \leq w \leq 64$
Decimal: $d \leq 31$
Action: Writes whole number binary numerals, using the characters "0" and "1", with leading zeroes.

The following example demonstrates the output of these formats.

```
X = 45;
PUT X HEX8. / X OCTAL8. / X BINARY8.;
```

```
0000002D
00000055
00101101
```

When the field is not wide enough to write the entire binary, octal, or hexadecimal integer, the formats write the rightmost digits of the number, truncating the most significant digits. The following example demonstrates this truncation process. The two-digit version of the number contains the last two digits of the complete, ten-digit version of the number.

```
X = 425;
PUT X HEX10. +2 X HEX2.
   / X OCTAL10. +2 X OCTAL2.
   / X BINARY10. +2 X BINARY2.;
```

```
00000001A9   A9
0000000651   51
0110101001   01
```

With a width of 16, the HEX informat and format do something completely different, which is described later in this chapter.

BZ and Z — Zeroes When the F format, or most numeric formats, write small numbers, they use leading spaces to fill in the rest of the field, but sometimes you need to write a field that does not contain any spaces. Use the Z format to write numbers with leading zeroes.

Conversely, you may occasionally see an input field in which space characters actually mean zeroes, and the BZ informat reads those fields. The BZ informat is also useful for reading fields that have a negative sign (or a positive sign) in the first column of the field, followed by spaces before the first digit. The F informat cannot read such a field, because it contains embedded spaces. The BZ informat treats those intervening spaces as zeroes and reads the value correctly.

Use the F informat to read fields written with the Z format.

Format: Z*w.d*

Width: $1 \leq w \leq 32$

Decimal: $d < w$

Action: Writes the value with leading zeroes, so that the field does not contain any spaces.

Informat: BZ*w.d*

Width: $1 \leq w \leq 32$

Decimal: $d \leq 31$

Action: Converts spaces to zeroes, then interprets standard numeric data.

Use the F or BEST format to write fields that can be read with the BZ informat.

COMMA — Commas Long numbers are conventionally written with commas between every three digits. Use the COMMA format to write numbers this way. The COMMA informat can read numbers from fields that have commas and various other characters in them.

Informat: COMMA*w.d*	Format: COMMA*w.d*
Width: $1 \leq w \leq 32$	Width: $1 \leq w \leq 32$
Decimal: $d \leq 31$	Decimal: $d < w$
Action: Interprets numbers, disregarding commas, dollar signs, percent signs, spaces, hyphens, and parentheses. Interprets a left parenthesis or hyphen at the beginning of the value as a negative sign.	Action: Writes numbers with commas separating every three digits left of the decimal point.
Alias: DOLLAR, NEGPAREN	

NEGPAREN — Accounting values The NEGPAREN format is similar to the COMMA format, but it writes negative values differently. Instead of a negative sign, it writes parentheses around negative values, following the style usually used in accounting reports and financial statements. It uses the first and last characters of the field for the parentheses.

Informat: NEGPAREN*w.d*	Format: NEGPAREN*w.d*
An alias for the COMMA informat.	Width: $4 \leq w \leq 32$
	Decimal: $d < w$
	Action: Writes the value with commas separating every three digits. Writes negative values with parentheses in the first and last characters of the field.

PERCENT — Percents Relative frequencies, rates, relative changes, and similar numbers are often expressed as percents. The percent value is 100 times the relative value that it represents. The PERCENT format writes numbers as percents, with a percent sign (%) after the value. It uses the first and last characters in the field for parentheses that mark negative values. This means it takes a width of at least 4 to write a one-digit percent value.

The PERCENT informat reads percents. It also tolerates the same punctuation characters that the COMMA informat accepts.

Informat: PERCENT*w.d*	Format: PERCENT*w.d*
Width: $1 \leq w \leq 32$	Width: $4 \leq w \leq 32$
Decimal: $d \leq 31$	Decimal: $d < w - 3$
Action: The same actions as the COMMA informat. In addition, it looks for a percent sign after the value, and divides the value by 100 if the percent sign is present.	Action: Multiplies the value by 100 and write a percent sign after it. Writes negative values with parentheses in the first and last characters of the field.

DOLLAR — Currency The DOLLAR format is similar to the COMMA format, but it writes a dollar sign ($) immediately before the first digit of the value —

or in the case of a negative value, immediately before the negative sign. The dollar sign may be replaced with a different national currency symbol in the character sets in use in some countries.

Informat: DOLLAR*w.d*
An alias for the COMMA informat.

Format: DOLLAR*w.d*
Width: $2 \le w \le 32$
Decimal: $d < w$
Action: Writes numbers with commas separating every three digits left of the decimal point. Writes a dollar sign before the first character of the value.

The idea of the DOLLAR format is to have the dollar sign appear immediately before the value, with no intervening spaces. When you want the dollar sign to appear in a fixed position, write it as a character constant and use the COMMA format to write the number.

NUMX, COMMAX, and DOLLARX — National notation In some countries, it is common to use the comma character as a decimal point and a period to separate every three digits, especially when writing amounts of money. This is the reverse of the usual use of those characters in numbers. The NUMX, COMMAX, and DOLLARX informats and formats work with fields written in this notation.

Informat: NUMX*w.d*
Width: $1 \le w \le 32$
Decimal: $d \le 31$
Action: The same as the F informat, except that a comma, not a period, is read as a decimal point.

Format: NUMX*w.d*
Width: $1 \le w \le 32$
Decimal: $d < w$
Action: The same as the F format, except that a decimal point is written as a comma.

Informat: COMMAX*w.d*
Width: $1 \le w \le 32$
Decimal: $d \le 31$
Action: The same as the COMMA informat, except that commas represent decimal points and periods are disregarded.

Format: COMMAX*w.d*
Width: $2 \le w \le 32$
Decimal: $d < w$
Action: Writes numbers with periods separating every three digits left of the decimal point, which is written as a comma.

Informat: DOLLARX*w.d*
An alias for the COMMAX informat.

Format: DOLLARX*w.d*
Width: $2 \le w \le 32$
Decimal: $d < w$
Action: Writes numbers with periods separating every three digits left of the decimal point, which is written as a comma. Writes a dollar sign before the first character of the value.

YEN — Japanese currency The YEN informat and format work with fields that contain Japanese currency amounts. They work much like the DOLLAR informat and format, but with yen signs instead of dollar signs. The yen sign is character '5C'X in the Japanese ASCII character set and '5B'X in the Japanese EBCDIC character set.

Binary Fields

Some fields that are not really designed to be readable to people. These fields can be described collectively as binary fields. Some contain actual binary numbers; others contain multiple elements that, together, represent numbers or other objects.

There are matching informats and formats for these kinds of fields. The informat interprets the field, and the corresponding format organizes data in the exact same way. The informat and format have the same name and allow the same argument values. Most of these field types are very restricted in the field widths that are allowed.

Binary fields do not actually contain decimal points, so the decimal argument represents a scaling factor. The informat divides the value by the power of 10 indicated by the decimal argument; the format multiplies the value by the power of 10 before writing the field. The decimal argument does not necessarily have to be less than the width argument; it can range from 0 to 10.

A binary field may contain any byte value as part of the data it contains. Therefore, do not use binary fields in delimited files or in files that have variable-length records in which control characters mark the end of a record. If you created a file that way, some of the bytes in the binary fields would be mistaken for delimiter characters or control characters. Such a file would be impossible to interpret correctly. Use binary fields in binary files or in files that have fixed-length records.

Lexicon

> Binary fields are sometimes referred to as *mechanical* fields, especially in the data center environment. The word "mechanical" indicates that the fields are meant to be read by machines, that is, by computer programs, rather than by people.

Binary Data Issues

Working with binary data fields can be a complicated business. Different computer hardware and operating systems have different requirements and standards for binary data, which can create many different ways of organizing essentially the same kind of data value. Some of these issues were described for floating-point data in chapter 5, "Variables and Values," and the same issues apply for floating-point fields in files. The byte-ordering issue is the main issue with integer fields.

Binary numeric fields can be categorized as *big-endian* or *little-endian* according to the order of the bytes in the field. A field that starts with the most significant byte of a value, the byte that represents the greatest amount of magnitude, is big-endian. A field that starts with the least significant byte is little-endian.

In addition to informats and formats that deal with specific kinds of fields, the SAS System also includes native informats and formats. These informats and formats work with the field types that are native to the computer, so some details of the way they work may vary from one computer to another. This is especially significant when you transport or share data between different kinds of computers.

Any kind of binary data field is limited in what data values it can represent. This is an issue for formats, which must produce something for output no matter what input value is provided. They cannot simply create a blank field, because every byte value, even a space character, has a specific meaning in a binary field. Formats may write missing values as a value of 0. Formats that write only positive values have to handle negative values in some way; some write the absolute value of the number, while others write the largest value possible for the field. Values that are too large or too small in magnitude to fit in the field may be written as the closest available value or as the largest or smallest value that can be written in the field. Floating-point formats write these values as specific bit patterns that usually do not represent valid numbers. The way a format responds to values outside its range may be different on different kinds of computers. Test a format to find out what it will do if there is a chance that you will be using it with values that are outside its range.

Binary Integer Fields

A binary integer field can be signed, allowing it contain both negative and positive integer values, or unsigned, so that it can contain only positive integers and zero. It can be big-endian or little-endian, or it can follow the byte-ordering scheme that is native to the computer it is running on. This

Ranges of Binary Integer Fields

Width	Number of Different Values		Unsigned Maximum Value	Signed Maximum Value
1	2^{8}	256	255	127
2	2^{16}	65,536	65,535	32,767
3	2^{24}	16,777,216	16,777,215	8,388,607
4	2^{32}	4,294,967,296	4,294,967,295	2,147,483,647
5	2^{40}	1.0995E12	1.0995E12	549,755,813,887
6	2^{48}	2.8147E14	2.8147E14	1.4073E14
7	2^{56}	7.2057E16	7.2057E16	3.6028E16
8	2^{64}	1.8446E19	1.8446E19	9.2233E18

The minimum value for an unsigned field is 0. The minimum value for a signed field is 1 less than the negative of the maximum value.

leads to six combinations, so there are six informats and six corresponding formats for binary integer fields — along with one extra.

The width of a binary integer field can range from 1 to 8 bytes. The range of values of the field depends on its width and on whether it is signed or unsigned. The number of values in a field can be calculated from the width in bytes w, as

$$N = 2^{8w}$$

Then, for an unsigned field, the range of values is from 0 to $N - 1$. For a signed integer field, the range is from $-N/2$ to $N/2 - 1$. These results are summarized in the table at the bottom of the preceding page.

These are the informats and formats for binary integer fields:

Informat/Format: PIB*w.d*
Field type: Unsigned integer binary.
Byte ordering: Native

Informat/Format: S370FPIB*w.d*
Field type: Unsigned integer binary. The format writes negative values as the highest possible value for the field.
Byte ordering: Big-endian

Informat/Format: S370FIBU*w.d*
Field type: Unsigned integer binary. The format writes the absolute value of negative values. The informat is the same as the S370FPIB informat.
Byte ordering: Big-endian

Informat/Format: PIBR*w.d*
Field type: Unsigned integer binary.
Byte ordering: Little-endian

Informat/Format: IB*w.d*
Field type: Signed integer binary.
Byte ordering: Native

Informat/Format: S370FIB*w.d*
Field type: Signed integer binary.
Byte ordering: Big-endian

Informat/Format: IBR*w.d*
Field type: Signed integer binary.
Byte ordering: Little-endian

Floating-Point Fields

A floating-point field adds two more elements to a binary integer field in order to be able to represent fractional numbers and numbers of a wide range of magnitude. Details of the floating-point format were described in chapter 5, "Variables and Values." SAS software uses one kind in particular, native double-precision floating-point, for its own numeric values. There are many different kinds of floating-point fields, and there are informats and formats for several of these.

The maximum width of a binary floating-point format is the width that is required to write the entire value. Using a lesser width results in truncation, which reduces the precision of the value. The effects are the same as those described in chapter 5 for shorter lengths of SAS numeric variables.

Most of these informats and formats can use a decimal argument. As with binary integer informats and formats, the decimal argument is a decimal scaling factor with a range from 0 to 10.

The RB8. informat and format read and write native double-precision floating-point fields, a field type that corresponds to the internal format of SAS numeric values. This kind of field is the fastest way to store SAS numeric values in an external binary file, because the informat and format can simply transfer the data without translating or altering it in any way. However, this kind of field may not be portable; if you write a field with the RB format on one computer, you may not be able to read the field if you transport the file to a different kind of computer. If a file you create on one kind of computer might be used on a different kind of computer, use a portable format, such as IEEE, to create floating-point fields. The portable formats and informats work the same way on different kinds of computers.

Informat/Format: RB*w.d*
Width: $2 \le w \le 8$
Field type: Native double-precision floating point.

Informat/Format: FLOAT*w.d*
Width: $w = 4$
Field type: Native single-precision floating point. On IBM mainframe computers, FLOAT. is the same as RB4.

Informat/Format: IEEE*w.d*
Width: $3 \le w \le 4$
Field type: IEEE single-precision floating point.
Width: $5 \le w \le 8$
Field type: IEEE double-precision floating point.

Informat/Format: S370FRB*w.d*
Width: $2 \le w \le 8$
Field type: IBM mainframe floating point.

Informat/Format: MRB*w.d*
Width: $2 \le w \le 8$
Field type: Microsoft real binary floating point.

Informat/Format: VAXRB*w.d*
Width: $2 \le w \le 8$
Field type: VAX real binary floating point.

Informat/Format: `HEX`*w.d*
Width: $w = 16$
Field type: Hexadecimal expansion of native double-precision floating point. Each byte of the floating-point value is written as two hexadecimal digits in the field. The actions of the format are equivalent to applying the `RB8.` format, then the `$HEX16.` format. The actions of the informat are equivalent to applying the `$HEX16.` informat, then the `RB8.` informat.

Coded Decimal Fields

Some fields contain coded decimal digits of numbers. Such fields are primarily used in legacy accounting systems. They represent decimal numbers using their actual decimal digits, but to do so, they take up more space than binary integer fields.

Informat/Format: `S370FPD`*w.d*
Width: $1 \leq w \leq 16$
Field type: IBM mainframe packed decimal. Each half byte contains one decimal digit, except the last half byte, which contains an indication of the sign of the number: C, A, E, or F for a positive number, D or B for a negative number.

Informat/Format: `PD`*w.d*
Width: $1 \leq w \leq 16$
Field type: Native packed decimal.

Informat/Format: `S370FPDU`*w.d*
Width: $1 \leq w \leq 16$
Field type: IBM mainframe unsigned packed decimal. For positive values, this is the same as S370FPD, except that the format writes the sign value as F rather than C. For negative values, the format writes the absolute value; the informat creates an error condition.

Informat/Format: `PK`*w.d*
Width: $1 \leq w \leq 16$
Field type: Native unsigned packed decimal.

Informat/Format: `S370FZD`*w.d*, `ZD`*w.d*, `S370FZDL`*w.d*, `S370FZDS`*w.d*, `S370FZDT`*w.d*, `S370FZU`*w.d*
Informat only: `ZDV`*w.d*, `ZDB`*w.d*
Width: $1 \leq w \leq 32$
Field type: Zoned decimal (various different kinds).

Informat/Format: `S370FF`*w.d*
Width: $1 \leq w \leq 32$
Decimal: $d \leq 31$ (informat); $d < w$ (format)
Field type: EBCDIC numeric. On EBCDIC computers, this is the same as the F informat and format. On ASCII computers, the informat converts the EBCDIC text of the input field to ASCII text, then applies the F informat; the format creates a field with the F format, then converts it to EBCDIC text.

Bitfields

A bitfield contains various binary objects in individual bits or groups of bits. The BITS informat is designed to read a single element from a bitfield as an unsigned binary integer.

Informat only: `BITS`*w.d*
Width (in bits): $w \leq 64$
Displacement (in bits): $d \leq 64 - w$
Action: Skips *d* bits, then reads *w* bits
as an unsigned binary integer.

Binary Fields as Undifferentiated Data

It is not always necessary to use informats to interpret fields or to use formats to put fields together. Especially with certain kinds of binary data, it can be easier to deal with the field in other data step statements. A character variable can contain the entire contents of the field, and then you can work with the data using data step statements, operators, and functions.

To work with a field this way, create a character variable whose length is the same as the width of the field. Use the $CHAR informat to transfer the data from the field to the variable and the $CHAR format to transfer the data from the variable to a field in a file.

On occasion, a field may be stored in a file with its bytes in reverse order. You can use the $REVERJ informat and format for such a field. The $REVERJ informat and format reverse the order of bytes in a data value, but without making any other changes in the data.

Time Fields

Knowing when something happened can be essential for understanding a set of data. Points in time and durations of time can be more complicated to work with than other kinds of data values, though, because of the way they are written as multiple elements. A date, for example, is written as a year, month, and day — and these three elements can be written in various forms and in a few different orders. To deal with this complexity, SAS software includes many informats and formats for dealing with different kinds of time fields.

Elements of Time Fields and Values

You can use these kinds of elements in data values that represent points in time:

- *Year.* Years of the Gregorian calendar, from 1582 to 19999.
- *2-digit year.* Two-digit year numbers in input, year values from 0 to 99, fall into the century defined by the YEARCUTOFF= system option. The value of the YEARCUTOFF= option is the oldest year that can be represented in two digits. For example, with the version 8 default `YEARCUTOFF=1950`, two-digit years fall into the period from 1950 to 2049.

The year number 50 means 1950; the year number 0 means 2000; the year number 49 means 2049. In output, any year can be written with its last two digits.

- *Quarter.* Numbered sequentially in the year.
- *Month.* Numbered sequentially in the year.
- *Month name.* Written with an initial capital letter.
- *3-letter month abbreviation.* The first three letters of the month name. In output, this is written with an initial capital if it is followed by a space, or in all capital letters if it is concatenated to another element of the date.
- *Month initial.* The first letter of the name of the month.
- *Week.* Numbered sequentially in the year, from 0 or 1 to 52 or 53, starting week 1 on the first Sunday of the year.
- *Weekday number.* The day of the week, numbered sequentially from Sunday = 1 to Saturday = 7.
- *Weekday name.* Written with an initial capital letter.
- *3-letter weekday abbreviation.* The first three letters of the weekday name. In output, this is written with an initial capital.
- *Day.* Numbered sequentially in the month.
- *Day of year.* Numbered sequentially in the year.
- *SAS date.* Days elapsed since the beginning of 1960.
- *Day half.* The abbreviations AM and PM.
- *Hour.* The hour of the day, numbered from 0 to 23.
- *Hour of 12-hour clock.* The hour as displayed on the clock: 12, then 1 to 12, then 1 to 11.
- *Minute.* The minute of the hour, from 0 to 59.
- *Second.* The second of the minute, from 0 to 59.
- *SAS time.* Seconds elapsed since midnight.
- *SAS datetime.* Seconds elapsed since the beginning of 1960.
- *Punctuation.* The comma, slash, hyphen, period, and other punctuation symbols are used in output fields and can appear in input fields.

An interval of time is represented according to its starting point. For example, if you use a SAS date for the year 2003, it is the same SAS date value that represents January 1, 2003. Elapsed time can be measured in days, hours, minutes, and seconds. Elapsed time measured in seconds can be treated like a SAS time value.

Date Fields

There are many informats and formats to make the translation between SAS date values and various kinds of date fields. Because there are so many ways to write dates, there are more formats than informats. In fact, this is one of the largest categories of formats.

Although a range of widths are valid for each format, certain specific widths are the most useful for writing two- and four-digit years, and those widths are shown separately here. The formats write specific symbols as punctuation, if they use punctuation. The informats that interpret elements as numbers can accept almost any kind of punctuation you might find, or spaces in place of punctuation. If a field contains two digits for the year and the full width of all other numeric elements, no separation between the elements is necessary. For example, these are all fields that can be read by the YYMMDD informat:

```
2002 5 31        02-5-31        020531        2002/05/31        2002.05.31
```

The following table lists informats that read date fields and produce a SAS date value. There is a corresponding format for each of these informats.

Informat	Width	Field elements	Examples
DATE*w.*	$7 \leq w \leq 32$	day, 3-letter month abbreviation, year	01JUN00 01JUN2000
YYMMDD*w.*	$6 \leq w \leq 32$	year, month, day	1999.12.31
MMDDYY*w.*	$6 \leq w \leq 32$	month, day, year	5/27/05
DDMMYY*w.*	$6 \leq w \leq 32$	day, month, year	150708
MONYY*w.*	$5 \leq w \leq 32$	3-letter month abbreviation, year	mar2000
YYMMN*w.*	$4 \leq w \leq 32$	year, month (no punctuation)	9711
YYQ*w.*	$4 \leq w \leq 32$	year, "Q", quarter	2001Q1
JULIAN*w.*	$5 \leq w \leq 32$	year, day of year	2000 121

The table on the following pages lists formats that produce a date field from a SAS date value. The formats produce different output for each of the different widths shown.

There are another 33 formats that provide alternate punctuation for a few of the date formats. The names of these formats consist of the name of the base format and a one-letter suffix:

DDMMYYB	DDMMYYC	DDMMYYD	DDMMYYN	DDMMYYP	DDMMYYS
MMDDYYB	MMDDYYC	MMDDYYD	MMDDYYN	MMDDYYP	MMDDYYS
	YYMMC	YYMMD	YYMMN	YYMMP	YYMMS
YYMMDDB	YYMMDDC	YYMMDDD	YYMMDDN	YYMMDDP	YYMMDDS
	YYQC	YYQD	YYQN	YYQP	YYQS
	YYQRC	YYQRD	YYQRN	YYQRP	YYQRS

The last letter of the format name indicates the specific punctuation symbol that the format uses:

B	space	
C	colon	:
D	hyphen	-
N	none	
P	period	.
S	slash	/

The widths of these formats are the same as those of the formats they are based on, except for formats with the N suffix. The N suffix indicates no punctuation or space between elements. With punctuation removed, the format width can be 1 or 2 characters less.

Formats for SAS Date Values

Format	Field Elements	Examples
DATE5.	day, 3-letter month abbreviation	15JUN
DATE7.	day, 3-letter month abbreviation, 2-digit year	14JUN00 15AUG75
DATE9.	day, 3-letter month abbreviation, year	14JUN2000
MONYY5.	3-letter month abbreviation, 2-digit year	APR00
MONYY7.	3-letter month abbreviation, year	APR2000
YYMON5.	2-digit year, 3-letter month abbreviation	00APR
YYMON7.	year, 3-letter month abbreviation	2000APR
YYMM5.	2-digit year, "M", month	01M07
YYMM7.	year, "M", month	2001M07
YEAR2. YYMMDD2.	2-digit year	82 00
YEAR4.	year	1582
YYMMDD4.	2-digit year, month	0010
YYMMDD5.	2-digit year, "-", month	00-10
YYMMDD6.	2-digit year, month, day	001006
YYMMDD8.	2-digit year, "-", month, "-", day	00-10-06
YYMMDD10.	year, "-", month, "-", day	2000-10-06
MONTH2.	month	3
MMDDYY2.	month	08
MMDDYY4.	month, day	0818
MMDDYY5.	month, "/", day	08/18
MMDDYY6.	month, day, 2-digit year	081800
MMDDYY8.	month, "/", day, "/", 2-digit year	08/18/00
MMDDYY10.	month, "/", day, "/", year	08/18/2000
DAY2.	day	1

continued

Formats for SAS Date Values (continued)

Format	Field Elements	Examples
DDMMYY2.	day	01 21
DDMMYY4.	day, month	2103
DDMMYY5.	day, "/", month	21/03
DDMMYY6.	day, month, 2-digit year	210300
DDMMYY8.	day, "/", month, "/", 2-digit year	21/03/00
DDMMYY10.	day, "/", month, "/", year	21/03/2000
YYQ4.	year, "Q", quarter	01Q1
YYQ6.	2-digit year, "Q", quarter	2001Q1
YYQR6.	year, "Q", quarter as uppercase Roman numeral	01QI
YYQR8.	2-digit year, "Q", quarter as uppercase Roman numeral	2001QI
QTR1.	quarter	3
QTRR3.	quarter as uppercase Roman numeral	III
WEEKDAY1.	weekday number	6
DOWNAME1.	weekday name (first letter)	F
DOWNAME2.	weekday name (first 2 letters)	Fr
DOWNAME3. WEEKDATE3. WEEKDATX3.	3-letter weekday abbreviation	Fri Sat Sun
DOWNAME9. WEEKDATE9. WEEKDATX9.	weekday name	Monday Tuesday Wednesday
JULIAN5.	2-digit year, day of year	00121
JULIAN7.	year, day of year	2000121
JULDAY3.	day of year	366
MONNAME1.	month initial	F
MONNAME3. WORDDATE3. WORDDATX3.	3-letter month abbreviation	Feb Apr Oct

continued

Formats for SAS Date Values (continued)

Format	Field Elements	Examples
MONNAME9. WORDDATE9. WORDDATX9.	month name	`September` `October`
WORDDATE12.	3-letter month abbreviation, space, day, comma, space, year	`Dec 9, 1777`
WORDDATE18.	month name, space, day, comma, space, year	`December 9, 1777`
WEEKDATE15.	3-letter weekday abbreviation, comma, space, 3-letter month abbreviation, space, day, comma, space, 2-digit year	`Tue, Dec 9, 77`
WEEKDATE17.	3-letter weekday abbreviation, comma, space, 3-letter month abbreviation, space, day, comma, space, year	`Tue, Dec 9, 1777`
WEEKDATE23.	weekday name, comma, space, 3-letter month abbreviation, space, day, comma, space, year	†
WEEKDATE29.	weekday name, comma, space, month name, space, day, space, year	†
WORDDATX12.	day, space, 3-letter month abbreviation, space, year	`9 Dec 1777`
WORDDATX18.	day, space, month name, space, year	`9 December 1777`
WEEKDATX15.	3-letter weekday abbreviation, comma, space, day, space, 3-letter month abbreviation, space, 2-digit year	`Tue, 9 Dec 77`
WEEKDATX17.	3-letter weekday abbreviation, comma, space, day, space, 3-letter month abbreviation, space, year	`Tue, 9 Dec 1777`
WEEKDATX23.	weekday name, comma, space, day, space, 3-letter month abbreviation, space, year	†
WEEKDATX29.	weekday name, comma, space, day, space, month name, space, year	†

†Examples.

WEEKDATX23.	`Tuesday, Dec 9, 1777`
WEEKDATX29.	`Tuesday, December 9, 1777`
WEEKDATX23.	`Tuesday, 12 Oct 1777`
WEEKDATX29.	`Tuesday, 12 October 1777`

The various date formats that write words in mixed-case letters are affected by the MSGCASE system option. When that option is on, the text produced by the WORDDATE, WEEKDATE, DOWNAME, MONNAME, and other such formats is converted to all uppercase letters.

Other Time Fields

Other time fields involve the time of day, either in connection with the date or separately, or a duration of time, usually expressed in hours, minutes, and seconds.

There are a few informats, listed in the table below, to read fields that contain the time of day or the date combined with the time of day. There are corresponding formats for the TIME and DATETIME informats.

Informat	Width	Field Elements	Value
TIME*w*.	$5 \le w \le 32$	hour, minute, optional second	SAS time
DATETIME*w*.	$13 \le w \le 40$	day, 3-letter month abbreviation, year, hour, minute, optional second; *or* day, 3-letter month abbreviation, year, hour of 12-hour clock, minute, optional second, day half	SAS datetime
PDTIME*w*.	$w = 4$	hour, minute, second in IBM mainframe packed decimal format ('0*hhmmss*F'X)	SAS time

There are several formats that produce fields from SAS time and SAS datetime values. A few work with both kinds of values. These formats are for SAS time values only:

TIME HOUR

These work the same way for SAS time values and SAS datetime values:

TOD TIMEAMPM HHMM

These are formats for SAS datetime values only:

DATETIME DATEAMPM

The TIME and TOD formats both write hours, minute, and seconds, but they differ in two ways. First, the TOD format can write the time of day of a SAS datetime value, while the TIME format cannot. Instead, the TIME format can write values in which the number of hours is less than 0 or greater than 24. Second, the TOD format writes leading zeros for one-digit hours, while the TIME format writes leading spaces. Another format, the TIMEAMPM format, writes the time of day using the 12-hour clock.

The TIME and HOUR formats handle negative values in different ways. The TIME format writes a negative sign before the value. The HOUR format writes the absolute value of the negative value. Both the TIME and HOUR formats can write values of 24 hours of more.

The table on the next page describes the fields produced by these formats with various widths.

Format	Field Elements	Examples
HOUR2. TIME2. HHMM2.	hour	5 17 23
TOD2.	hour	05
TIME5.	hour, " : ", minute	0:00 23:59
TOD5.	hour, " : ", minute	00:00 23:59
TIME8.	hour, " : ", minute, " : ", second	0:00:00 15:45:00
TOD8.	hour, " : ", minute, " : ", second	00:00:00 15:45:00
TIME*w.d*	hour, " : ", minute, " : ", seconds ($10 \le w \le 20$, $d = w - 9$)	9:38:21.004 23:59:59.9
TOD*w.d*	hour, " : ", minute, " : ", seconds ($10 \le w \le 20$, $d = w - 9$)	09:38:21.004 23:59:59.9
HOUR*w.d*	hours ($2 \le w \le 20$, $d < w - 2$)	5 23.99999
TIMEAMPM2.	day half	AM PM
TIMEAMPM5.	hour of 12-hour clock, space, day half	11 AM 9 PM
TIMEAMPM8.	hour of 12-hour clock, " : ", minute, space, day half	11:00 AM
TIMEAMPM*w.d*	hour of 12-hour clock, " : ", minute, " : ", seconds, space, day half ($11 \le w \le 20$, $d = w - 12$)	10:32:07.420 AM
HHMM4.	hour, minute	9:31 1545
HHMM*w.d*	hour, " : ", minutes ($5 \le w \le 20$, $d = w - 6$)	09:31.25
DATETIME7. DATEAMPM7.	elements of a DATE7. field	11JAN00
DATETIME10. DATEAMPM10.	elements of a DATE7. field, " : ", hour	11JAN00:14
DATETIME*w.d*	elements of a DATE7. field, " : ", elements of a TOD field ($10 \le w \le 40$, $d = w - 17$)	11JAN00:14:30
DATEAMPM*w.d*	elements of a DATE7. field, " : ", elements of a TIMEAMPM field ($13 \le w \le 40$, $d = w - 20$)	11JAN00:02 PM 11JAN00:02:30 PM

The standard unit for measuring durations of time is the second, so a variable that indicates duration should usually be given a value that measures elapsed time as a number of seconds. You can use the TIME informat to read hours, minutes, and seconds or hours and minutes. The resulting value is a time duration value in seconds. This informat and the TIME format work correctly for duration values even when the number of hours is 24 or more. They also work correctly with negative values.

The TIME and HOUR formats can be used for writing time duration values measured in seconds in the same way that they write time of day. In addition, there is an MMSS format that writes time duration values in minutes and seconds.

Format	Field Elements	Examples
MMSS2.	minutes	7
MMSS4.	minutes, seconds	0709
MMSS*w.d*	minutes, ":", seconds ($5 \le w \le 20$, $d \le w - 6$)	7:09 7:09.1

To write a time duration value as days, hours, minutes, and seconds, first use the DATEPART function to extract the numbers of days from the value, assigning the result to another variable. Write the number of days with the F format or any other suitable format. Write the hours, minutes, and seconds with the TOD format.

International Date Informats and Formats

Most European countries use the same Gregorian calendar that is used in English-speaking countries, but they may use different names for the months and the days of the week and different ways of arranging the elements of a date or a time of day. The international date informats and formats adapt the output of several informats and formats to the languages and styles of more than 20 European countries.

The names of these informats and formats are constructed from a three-letter language prefix, the root name DF, and a two- to three-letter suffix that identifies the equivalent informat or format. The table on the following page shows the parts of the names.

Constructing the Names of International Date Informats and Formats

Prefix	Language[1]
AFR	Afrikaans
CAT	Catalan
CRO	Croatian
CSY	Czech
DAN	Danish
DES	Swiss_German
DEU	German
ENG	English
ESP	Spanish
FIN	Finnish
FRA	French
FRS	Swiss_French
HUN	Hungarian
ITA	Italian
MAC	Macedonian
NLD	Dutch
NOR	Norwegian
POL	Polish
PTG	Portuguese
RUS	Russian
SLO	Slovenian
SVE	Swedish
EUR	Selected in the DFLANG= system option[1]

\+

Root
DF

\+

Suffix	Informat Equivalent
DE	DATE
DT	DATETIME
MY	MONYY

Suffix	Format Equivalent
DD	DDMMYY
DE	DATE
DN	WEEKDAY
DT	DATETIME
DWN	DOWNAME
MN	MONNAME
MY	MONYY
WDX	WORDDATX
WKX	WEEKDATX

Examples:

Construct the name of the Finnish equivalent to the MONYY format
as FIN + DF + MY, resulting in the format name FINDFMY.

Construct the name of the French equivalent to the DATE informat
as FRA + DF + DE, resulting in the informat name FRADFDE.

[1] The values for the DFLANG= system option are the language names listed in the Language column.

The prefix EUR selects a language according to the DFLANG= system option. For example, with `DFLANG=English`, the EURDFDE informat works like the ENGDFDE informat to read date fields in English. Change the system option to `DFLANG=Spanish`, and the EURDFDE informat works like the ESPDFDE informat to read date fields in Spanish. Use the informats and formats that have the EUR prefix when you develop applications that may be localized to work in several different countries. The user will be able to set the national style of the informats and formats by changing the language selection of the DFLANG= system option.

The widths of the international date formats and informats are sometimes a few characters wider than the English-language versions of the formats and informats because of the different lengths of words and abbreviations in different languages.

Using Informats and Formats

Informats and formats are used throughout SAS software, anywhere there is a need to interpret or construct fields or other text values. Wherever they are put to use, they deal with data values and issues in the same way. The various uses of informats and formats use the same reference syntax, as shown at the beginning of this chapter.

In some places, the width argument of an informat or format reference determines the size of a field. In other places, the informat or format must use a particular width in order to fit the field or other object that it works with.

Using Informats

These are some of the ways informats can be used in SAS software:

- In the INPUT statement to read fields from an input text file.
- As an argument to a function, to apply the informat to a character variable as part of an expression.
- As the informat attribute of a variable, set in the INFORMAT or ATTRIBUTE statement or as an item option in an SQL statement.
- In the WINDOW statement, combined with a format, to define the behavior of a field.
- In interactive windows in the SAS environment that allow editing of data values.

Missing Values in Input Data

Most informats can read missing values in one way or another. For those that read decimal numbers, a blank field or a field that contains only a period is interpreted as a standard missing value. For ordinary character informats, a blank field results in a blank value. The standard character informat can also read a field that contains only a period as a missing value.

Many binary numeric informats cannot read missing values, because every possible data value is interpreted as a different number. For any other informat, some input data values are invalid, and reading one of those values results in a missing value and an error condition.

Ordinary numeric informats — those that read values written as decimal digits and that do not ordinarily allow letters as data values — can read letters as special missing values. Before an informat can recognize a character as a special missing value, you must declare the character in the MISSING statement.

The MISSING statement is a global statement that lists characters to be recognized as special missing values when they appear as the input text for an informat. The characters can be letters and the underscore character. Either an uppercase or lowercase letter can represent the corresponding special missing value, but you must list both the uppercase and lowercase letters in the MISSING statement if you want both to be recognized.

This is an example of the MISSING statement:

```
MISSING A;
```

This statement allows numeric informats to interpret the character A as the special missing value .A.

Invalid Input Data

When an informat finds input data that is not organized in the way it expects, it cannot interpret the data to produce a value. Instead, it produces a missing value and generates an error condition.

Not all informats have restrictions on the data values they can interpret. Most character informats and the binary integer informats, for example, can interpret all possible values of a field. But the standard numeric informat and similar informats are very restrictive about the data values they can use. If a field contains extra punctuation or embedded spaces, it is considered invalid data. Log notes describe the location and nature of the invalid data.

By default, the value that numeric informats generate for invalid data is a standard missing value. You can change the generated value is to a special missing value in the INVALIDDATA= system option. The value of the option is not the special missing value itself, but a character constant that contains the distinctive character that identifies the missing value. The default setting is INVALIDDATA='.', which generates the standard missing value . for invalid data. Changing this to INVALIDDATA='X', for example, would generate the special missing value .X for invalid data.

In the INPUT statement and INPUT function, you can use error controls to keep an informat from generating an error condition. With the error control ?? before the informat reference, the informat still produces a missing value when it finds invalid data, but it does not generate an error condition or log notes.

Using Formats

These are some of the ways in which formats are used in SAS software:

- In the PUT statement to write fields to an output text file or print file.
- In proc statements and options, to write variables in output.
- As an argument to a function, to apply the format to a value as part of an expression.
- As the format attribute of a variable, set in the FORMAT or ATTRIBUTE statement or as an item option in an SQL statement.
- In the WINDOW statement to define the appearance of a field.
- In interactive windows in the SAS environment that display data values.
- As an argument to the %SYSFUNC macro function, to format the value returned by a function.

Width Issues

The width of a format is not always the same as the width of the formatted data value that the format generates. The format adjusts its output to match the required width. This process works differently for character and numeric values.

A character format adjusts the width of its output the same way that character lengths are adjusted in SAS. If a character value is too short for a format width, the format treats the character value as if it were extended with spaces to make it the right length. If it is too long, the format uses the beginning of the value, as much of it as will fit in the field width.

Formatted numeric values are usually shorter than the width of the field. Most numeric formats right-align formatted values in the field — they add leading spaces to fill the rest of the field. Each format has its own default alignment. You can use alignment options after the informat reference in the PUT statement and PUT function. The alignment options `-L`, `-C`, and `-R` align the formatted value at the left, center, or right of the field, respectively.

If a formatted numeric value is too long for a field, many formats call the BEST format to produce a more compact version of the value. If that fails, the format fills the field with asterisks.

Writing Missing Values

By default, numeric formats write standard missing values as a period. The period is written the same way as a decimal point, so that missing values line up with other values in a column. You can change the character that formats write for a standard missing value in the MISSING= system option. For example, you might use the option `MISSING='*'` to call more attention to missing values, or the option `MISSING=' '` to leave the fields of missing values blank.

Formats write special missing values as capital letters, or the underscore character for the value `._`. The way special missing values are written is not affected by the MISSING= option.

Informats and Formats in Expressions

A special set of functions lets you use informats and formats in expressions. The PUT function lets you apply a format to an expression to create a character value that you can assign to a variable. The INPUT function lets you interpret a character value with an informat to create a numeric or character value. Four more functions let you do the same thing with informat and format references that are formed as expressions.

The PUT function has two arguments: a value and a format term. As always with formats, the type of the value should match the type of the format:

PUT(*character value*, *character format term*)
PUT(*numeric value*, *numeric format term*)

The value returned by the function is a character value whose length is the width of the format. For example, the function call PUT(5, Z4.) returns the character value '0005', a value with a length of 4.

You can write an alignment option, -L, -C, or -R, after the format reference, to change the way the result of the format is aligned, the same as in the PUT statement.

These are more examples of the use of the PUT function:

Function Call	Length Returned	Value Returned
PUT(480, HEX4.)	4	01E0
PUT(36525, DATE9.)	9	01JAN2060
PUT(1.25, 7.5)	7	1.25000
PUT(25, ROMAN10.)	10	XXV
PUT(397, HEX8.)	8	0000018D
PUT(24000, DOLLAR14.2)	14	24,000.00
PUT('lions', $UPCASE4.)	4	LION

The arguments to the INPUT function are a character value and an informat term. The character value acts as the input field for the informat. With a numeric informat, the function returns a numeric value. With a character informat, the function returns a character value whose length is determined by the informat.

INPUT(*character value*, *character informat term*)
INPUT(*character value*, *numeric informat term*)

The informat term is an informat reference, which can optionally be preceded by an error control. In the informat reference, it is important to supply a width argument for the informat; the informat cannot look at the text value to determine the width to use. The error controls ? and ?? work the same way as in the INPUT statement, to suppress the error condition and messages that an informat normally generates when the input data is invalid. These are examples of the use of the INPUT function:

Function Call	Data Type Returned	Value Returned
INPUT('480', F3.2)	Numeric	4.8
INPUT('1250000', 7.6)	Numeric	1.25
INPUT('01JAN1960', DATE.)	Numeric	0
INPUT('24:00:00', TIME.)	Numeric	86400
INPUT('01:00:00', TIME.)	Numeric	3600
INPUT('$64,000', DOLLAR7.)	Numeric	64000
INPUT('DX1070', $CHAR2.)	Character	DX
INPUT('lion', $UPCASE4.)	Character	LION
INPUT('FF'X, BITS3.5)	Numeric	7
INPUT('*****', ?? F.)	Numeric	.

A limitation of the INPUT and PUT functions is that their informat and format terms are a fixed part of the function call; they do not let you change the informat or format reference based on data values. The INPUTC, INPUTN, PUTC, and PUTN functions use a character argument for the informat or format reference. The informat or format argument could be a variable, letting you apply different informats or formats based on conditions of the data. In addition, you can supply the arguments to the informat or format as separate, numeric arguments if you wish.

This simple example uses a condition to set the decimal argument for the PUTN function:

```
IF 0 < PRICE < .1 THEN D = 4;
ELSE D = 2;
PRICEFIELD = PUTN(PRICE, 'F.', 6, D);
```

The IF-THEN statement sets a different value for the decimal argument based on the value to be formatted. If PRICE is 2.5, PRICEFIELD is ' 2.50', with 2 decimal places, but if PRICE is .048, PRICEFIELD is '0.0480', with 4 decimal places.

There are four different functions for the four different types of routines. INPUTC works with character informats; PUTC, with character formats; INPUTN, with numeric informats, and PUTN, with numeric formats.

```
INPUTC(character value, character informat, optional width argument)

INPUTN(character value, numeric informat,
       optional width argument, optional decimal argument)

PUTC(character value, character format, optional width argument)

PUTN(numeric value, numeric format,
       optional width argument, optional decimal argument)
```

The INPUTN function returns a numeric value. The INPUTC, PUTN, and PUTC functions return character values.

14

Print Output

Most proc steps produce print output. In some ways, print output from proc steps is the simplest kind of I/O in a SAS program. The proc produces the output, and then you can look at it. There are no file names or options in the proc step to direct the output to a specific file — it is automatically directed to the standard print file. Data steps can also create print output. A data step can direct its print output to the standard print file (using the fileref PRINT) or to any other file.

The difference between an ordinary text file and a print file is that a print file is organized to be printed on paper. The exact details of the form of a print file are different in different operating systems. In any environment, though, the SAS System includes features to break a print file into pages of a specific size and to arrange objects on the page.

Print Files and Pages

The rectangular area of a page is essential to the idea of a print file. When you print a document on paper, it has to fit within the width and height of the sheet of paper. Unless the document is small enough to fit on one page, this requires dividing the document into pages, with the idea that the reader looks at one page at a time. When you divide a document into pages, you often want to break it at logical points so that separate parts are on separate pages. These properties of a page and a printed document define the two essential properties that make a print file different from a regular text file. In a print file:

- The length of a line is no longer than the width of the printable area of the page.
- There is an indication of how to group the lines into pages.

Print files are based on the idea of a *monospaced* font, in which every character is printed with the same width. This allows the use of spaces to make columns of data line up. Most fonts are *proportional* fonts, in which some characters are wider than others. A print file can look almost unintelligible if it is printed or displayed with a proportional font.

Despite the name, print files are not necessarily printed. More often, they are only viewed on the computer screen. If you do plan to print a print file, you need to know some of the properties of the printer or print driver that will print it so that you can prepare the file to print correctly.

Print Control Characters

The print control characters that some operating systems use in print files are not control characters in the usual sense. Instead, they are ordinary visible characters used as codes to indicate the relative positions of lines on a page. The first column of each line is set aside to hold a print control character. On most lines, the print control character is a space, indicating to print the line right below the previous line. On lines that must be positioned in a certain way, the print control character may be any of various code letters, numbers, or symbols. There are various different systems of print control characters associated with different operating systems, printers, and programming environments. Depending on the system, a print control character may direct the printer to leave one blank line, leave two blank lines, or skip ahead to a specific point on the page.

The one essential print control character is the one that marks the beginning of a new page. Every system of print control characters designates a code for this purpose. SAS uses at least this one print control character when it creates a print file, and it usually uses a few other codes, but it does not use every code in any system of print control characters.

Other operating systems do not use print control characters. Instead, they use a regular ASCII control character, a nonprintable character, to indicate the start of a new page.

Title Lines

A SAS print file usually has a title line as the first line of every page. SAS provides a default title that appears in the title line. You can use TITLE statements to change or remove the title line or to add additional title lines, up to 10 in all.

The form of a TITLE statement is

```
TITLEn ' title text ';
```

The statement keyword is formed from the word TITLE and a numeric suffix from 1 to 10 indicating a specific title line. The second term in the statement is a character constant that indicates the text of the title line.

The TITLE1 statement is the statement that controls the first title line, which is the only one used on most print pages. You can write the keyword of this statement as `TITLE`, omitting the 1 suffix. The other title lines, if you use more than one, appear in order below the first title line.

A TITLE statement resets, or removes, all higher-numbered title lines. The TITLE1 statement resets all the other title lines. If you use several title lines, write the TITLE statements in order: TITLE1, then TITLE2, and so on. If you write a TITLE statement without a title, it resets that title line and all higher-numbered lines. The statement

```
TITLE1;
```

resets all the title lines, restoring the default SAS title to the first title line. To have a blank title instead of the default text in the first title line, use the statement

```
TITLE1 ' ';
```

Indicating a blank character value as a title line may also be necessary for the last title line you use. SAS prints additional title lines only as far as you have defined them. If you want to have a blank last title line, define the last title line that way, for example:

```
TITLE1 'Main Project';
TITLE5 ' ';
```

After you define title lines, they remain in effect for the rest of the SAS session or until you change them. To define title lines for just one step, write the TITLE statements before the step (after the RUN statement at the end of the previous step) and write new TITLE statements after the step (again, after the RUN statement).

Footnote Lines

The same way that SAS prints title lines at the top of the page, you can have it print footnote lines at the bottom of the page. Footnote lines are defined in FOOTNOTE statements, whose syntax is the same as for TITLE statements, except that the word FOOTNOTE replaces the word TITLE. If you use several footnote lines, the FOOTNOTE1 line appears first, with the FOOTNOTE2 line below it, and so on. There is no default footnote line; SAS writes footnote lines only if you define them.

System Options for Print Output

These are system options that affect the form of print output.

LINESIZE= The LINESIZE= option indicates the line size — the number of characters on a line. The value can range from 64 to 256. The alias for this option is LS.

PAGESIZE= The PAGESIZE= option indicates the page size — the number of lines on a page. There can be as few as 15 lines or as many as 32,767 lines on a page. The alias for this option is PS.

When you prepare a print file for printing, it is important that the line size and page size settings correspond to the actual paper size. If you set the line size too large, the right end of each line may not print. If you set the page size too large, the last lines of each page in the print file may print on a separate page of paper. If you are only viewing a print file on the computer screen, the line size and page size settings are less critical. You can scroll to view a page area that is taller or wider than the window it is displayed in.

CENTER With the option CENTER, which is the default, most print output that SAS produces is centered horizontally on the page, based on the current line size. Centering affects title lines, footnote lines, and most tables produced by procs. Use the option NOCENTER to have print output left-aligned. The alias for this option is CENTRE.

DATE With the DATE option, the SAS supervisor writes the date of the SAS session in the title lines. The date appears in the first title line if there is room; otherwise, it appears in the first title line that has room for it. The default is to include the date in the title lines. Use the NODATE option to remove the date from the title lines.

NUMBER Along with the date, the SAS supervisor, by default, writes a page number in the title lines, usually at the right side of the first title line. Use the NONUMBER option to remove the page number from the title lines.

PAGENO= The SAS supervisor counts the pages of print output and assigns page numbers to them. If you want the pages produced by a specific step to start at a specific number, use the PAGENO= option to set the page number. To reset page numbers at 1 for the next step, use this statement:

```
OPTIONS PAGENO=1;
```

The alias for this option is PAGNO.

OVP If you are preparing a print file for printing on a mechanical printer, the OVP option may be useful. With the OVP option in effect, you can overprint, that is, print more than one character at the same position on the page. With the OVP option, the SAS supervisor uses overprinting when it underlines program text in the log. Video displays and most current printers and printer drivers based on raster image processors do not support overprinting.

FORMDLIM= The FORMDLIM= option lets you simulate page breaks instead of creating actual page breaks in print output. The value of the option is a character that is printed across the page in a line that represents the place where a page break would be. For example, with `FORMDLIM='-'`, page breaks are replaced with lines of hyphens. Use `FORMDLIM=''` to restore the default action of actual page breaks.

SKIP= The SKIP= option leaves blank lines at the top of each page of print output, before the title lines. For example, use `SKIP=5` to have the title lines start on the sixth line of the page. With the default, `SKIP=0`, title lines start on the first line of the page.

Writing Print Files in the Data Step

You can write output print files in the data step using the same FILE and PUT statements that write output text files. There are additional options in the FILE statement and terms in the PUT statement that are designed for use with print files.

Options for Print Files

The FILE statement identifies the output file and sets options for it. You can write to the standard print file, fileref PRINT, or any other print file. Use the PRINT option to indicate that the file is a print file. The following table lists options of the FILE statement that are specifically for print files.

FILE Statement Options for Print Files

Option	Value	Description
PRINT		Identifies the file as a print file. This option is necessary for all print files except the fileref PRINT.
NOPRINT		Identifies the file as a nonprint file. This is the default for all files except the fileref PRINT.
TITLES TITLE		Writes title lines on each page of output. This is the default.
NOTITLES NOTITLE		Does not write title lines in the file.
FOOTNOTES FOOTNOTE		Writes footnote lines on each page of output.
NOFOOTNOTES NOFOOTNOTE		Does not write footnote lines in the file. This is the default.
N=	PS	Accesses all the lines on the page at one time, so that the # pointer control can be used. The only other value for the N= option that is valid for a print file is the default, 1.
LINESLEFT= LL=	*numeric variable*	A variable that indicates the number of lines left on the page, compared to the current pointer location. For example, if the value of the variable is 1, the pointer is on the last line of the page. Use the LINESLEFT= variable as part of a condition to determine when to advance to a new page.
HEADER=	*statement label*	The PUT statement branches to this statement label when it begins a new page. Execution returns to the PUT statement when a RETURN statement executes. You can use the HEADER= option to execute statements that write column headers. However, it is not a good idea to use those statements to write variables. The values the statements write are the values of variables at the time the HEADER= branch occurs, which are usually different from the values in the observations that appear on the page.

Line Pointer Controls for Print Files

When writing a print file, a PUT statement can use any of the terms it can use for writing a text file. It can also use the additional line pointer controls listed in the following table.

Line Pointer Controls for Print Files

Term	Description
PAGE	Starts a new page. Use this term at the beginning of a page, not at the end of a page.
BLANKPAGE	Starts a new page, even if nothing has been written on the current page. This term is especially useful if you need to start a report with a blank page.
OVERPRINT	Starts a new line that overprints the previous line. This term requires the system option OVP. Overprinting works only with certain kinds of printers.

Output Objects and ODS

Procs do not produce print output directly. Instead, they create output objects that are converted to print output and delivered to the standard print file by ODS, the Output Delivery System. The output objects can also be intercepted and directed to other forms of output, such as HTML pages and SAS datasets.

An output object is not merely text, but contains actual data values along with information about how they are logically organized. A proc only produces the data component of an output object, which has to be combined with a table definition, or template, to convert it to something visual. The table definition determines many of the details of the formatting of the print output that is produced. You can, for example, remove columns or change the width of columns of a table by using a different table definition with the same output object. After the data component and table definition are put together to form the output object, ODS can convert the output to print output or to any of the other destination formats that it supports.

The SAS System includes table definitions to form the output objects that you see from procs. There are also alternate table definitions you can select to change the form or appearance of the output, and you can create new or modified table definitions in the TEMPLATE proc.

There are some procs that produce visual output directly. In this case, the data component produced by the proc is the entire output object; no table definition is involved.

In addition to table definitions, ODS uses style definitions. Whereas a table definition affects the formatting of only a single table, a style definition affects details of the visual appearance of all elements of the output. Style definitions work only with destination formats such as HTML and RTF that allow for visual formatting. They affect such details as font selection, rules, colors, spacing, and justification.

None of these details need to concern you if you just want to run procs, apply the default table definitions, and get the resulting print output. But if you want to work with output objects in a different way — change the layout of the print output, or produce a web page or a SAS dataset — the various ODS features of the SAS language can accomplish that for you.

Data Step Output Objects

With special uses of the FILE and PUT statements, a data step can create an output table object. The FILE statement must indicate the PRINT fileref and the ODS or ODS= option:

```
FILE PRINT ODS;
FILE PRINT ODS=(ODS option list);
```

Of the FILE statement options that can be used with a print file, only the FLOWOVER, MISSOVER, STOPOVER, and N= options can be used for ODS output. As with a print file, the only values that are valid for the N= option are 1 and PS. In addition, the ODS= option can be used to indicate specific details of the way the ODS table object is constructed. The value of the option is a list, in parentheses, of options relating to the data step variables to use and the output that results.

In the PUT statement, you can use the keyword _ODS_ to write all the columns of the output table at once, that is:

```
PUT _ODS_;
```

Or you can construct the table by writing individual variables to individual table cells. In this form of the PUT statement, column pointers refer to ODS table columns rather than character positions. For example, the column pointer @1 refers to the first cell of a row in the table, not the first character of a record. You can indicate variables to write, but you cannot indicate formats for the variables, and you cannot write constant values.

By default, the data supplied by the PUT statement is combined with a default table definition for data step output, and ODS converts the resulting output object to print output. This default output presents the variables of the data step as separate columns, much like the output of the PRINT proc. The following example demonstrates ODS print output from the data step.

```
OPTIONS NODATE NONUMBER NOCENTER;
TITLE1 'Demonstration of Data Step ODS Print Output';
TITLE2 ' ';
DATA _NULL_;
  RETAIN ID 'Text' X1 1000 X2 2.005 X3 -50;
  FILE PRINT ODS;
  PUT _ODS_ / _ODS_;
RUN;
```

```
Demonstration of Data Step ODS Print Output

 ID          X1             X2             X3

Text            1000          2.005              -50
Text            1000          2.005              -50
```

Destination Formats

ODS can direct output to any combination of its destination formats at the same time. It supports these destination formats:

Listing The Listing destination format is the traditional print output, consisting of monospaced text divided into pages. The Listing destination format is open by default.

HTML The HTML destination format creates HTML (Hypertext Markup Language) files, files of text with formatting tags that can be displayed as web pages. HTML files can also be used as document files that can be imported by various applications.

Output The Output destination format creates SAS data files in which the table columns are used as variables and the rows as observations.

Printer The Printer destination format directs output to a printer for printing.

RTF The RTF destination format creates RTF (Rich Text Format) files that can be imported by Microsoft Word.

Other destination formats, though not yet available as of release 8.1, are expected. These are particularly noteworthy:

Document The document files created with the Document destination format can be stored and converted at a later time to any of the destination formats of ODS.

XML XML (Extensible Markup Language) files, like HTML files, are tagged text. An XML file, when combined with a style sheet, can be displayed as a web page. XML is also a standard format for exchanging data between applications.

PDF PDF (Portable Document Format) is a file format designed by Adobe to maintain consistent page layout and typography when electronic documents are exchanged between computers. Another use for PDF files is to prepare documents for printing. PDF files can be viewed in the free Acrobat Reader program.

PostScript Files in the PostScript (PS) language can be printed on some printers and can be converted to PDF files by Acrobat Distiller.

Lexicon

SAS Institute writers use the term *destination* for a *destination format*. In spite of the name, a *destination* is not a file or other container for output; it is only a data format.

Use the various forms of the ODS statement between steps to direct output to one or more of these destination formats. Use an ODS statement with a destination format before a step to open a destination format. Subsequent output objects are directed to that destination format. In the same statement, you can use other options to set other details of the actions of the destination format, such as a file name. Later, use the CLOSE option to close the destination format.

The following example directs the output of a PROC PRINT step to an HTML file instead of the default Listing destinatuion.

```
* Close the Listing destination format. ;
ODS LISTING CLOSE;
* Open the HTML destination format and create the file pyramid.html. ;
ODS HTML FILE='pyramid.html';
* Proc step produces an output object. ;
PROC PRINT DATA=MAIN.PYRAMID;
RUN;
* Close the HTML destination format. ;
ODS HTML CLOSE;
* Open the Listing destination format for subsequent steps. ;
ODS LISTING;
```

Style Definitions

Destination formats that allow for formatting are affected by style definitions. You can select a style definition with the STYLE= option in the ODS statement that opens a destination format, for example:

```
ODS HTML FILE='pyramid.html' STYLE=Beige;
```

Base SAS includes several style definitions, most notably Default, the standard style definition for HTML output, and Minimal, a streamlined style definition. Several other style definitions are variations on Default. Beige and Brick use different color schemes; Statdoc, D3d, and Brown change colors, spacing, and other style elements. It is possible to create other style definitions in the TEMPLATE proc.

Tracing Output Objects

The TRACE option in the ODS statement makes it possible to find out the names of the output objects from a proc. When tracing is on, ODS writes log notes that describe every output object that is created. Use this statement to turn tracing on:

```
ODS TRACE ON;
```

Log notes are not always enough to identify the various objects that some procs generate. For a more specific identification of output objects, use the LISTING option:

```
ODS TRACE ON / LISTING;
```

This option writes an identifying note for each output object immediately before the object text in the Listing destination format. This way, the objects and their names appear together.

To turn tracing off, use this statement:

```
ODS TRACE OFF;
```

The following example shows the kind of notes that the TRACE option generates.

```
ODS TRACE ON;
PROC CONTENTS DATA=MAIN.PYRAMIDS;
RUN;
ODS TRACE OFF;
```

```
Output Added:
-------------
Name:       Attributes
Label:      Attributes
Template:   Base.Contents.Attributes
Path:       Contents.DataSet.Attributes
-------------

Output Added:
-------------
Name:       EngineHost
Label:      Engine/Host Information
Template:   Base.Contents.EngineHost
Path:       Contents.DataSet.EngineHost
-------------
```

```
Output Added:
-------------
Name:       VariablesAlpha
Label:      Variables
Template:   Base.Contents.Variables
Path:       Contents.DataSet.VariablesAlpha
-------------

Output Added:
-------------
Name:       Sortedby
Label:      Sortedby
Template:   Base.Contents.Sortedby
Path:       Contents.DataSet.Sortedby
-------------
NOTE: PROCEDURE CONTENTS used:
      real time           0.79 seconds
```

Creating a SAS Data File

With the Output destination format, you can create a SAS data file from an output object. This can be useful when you want to do further analysis based on the output of a proc.

First, use tracing to determine the name of the output object. Then, in the ODS OUTPUT statement, write the object name followed by an equals sign and the SAS dataset name. The example below creates a SAS data file WORK.COUNT with an output object from the FREQ proc.

```
ODS LISTING CLOSE;
ODS OUTPUT LIST=WORK.COUNT;
PROC FREQ DATA=CORP.SOURCE;
  TABLES DEPT*VENDOR / LIST;
RUN;
ODS OUTPUT CLOSE;
ODS LISTING;
```

```
NOTE: The data set WORK.COUNT has 36 observations and 8 variables.
NOTE: There were 177 observations read from the dataset CORP.SOURCE.
NOTE: PROCEDURE FREQ used:
      real time           0.15 seconds
```

Selection and Exclusion Lists

You can use a selection list or an exclusion list to obtain a subset of the output objects of a proc step. To create or modify a selection list, use the ODS SELECT statement to list the objects to select:

```
ODS SELECT object . . . ;
```

It is possible for more than one output object to have the same name. In that situation, use a suffix such as #1, #2, etc., with an object name to select a particular occurrence of an object.

To create a selection list for one specific destination format, name the destination format before SELECT in the statement. This example creates a selection list for the Listing destination format:

```
ODS LISTING SELECT PRINT;
```

Use the keyword ALL or NONE to reset a selection list, for example:

```
ODS SELECT ALL;
```

Exclusion lists are formed with the word EXCLUDE in place of SELECT and result in the selection of all objects that are not in the list. ODS automatically resets selection and exclusion lists between steps unless they contain the keyword ALL or NONE. The default list is `SELECT ALL`.

Selection lists work slightly differently for the Output destination. ODS automatically removes a selected object from the list when it creates the output SAS dataset from the object. This way, the Output destination creates each indicated output SAS dataset only once.

Proc Step Output

Most procs produce some kind of print output. Every proc has a different purpose, so every proc's output is different. Even so, all proc step output has some qualities in common. Procs produce output objects that ODS, by default, converts to print output in the standard print file. There are system options that affect the appearance of the output objects and the resulting print output. There are also special features that allow the use of BY variables of proc steps in title lines.

System Options for Proc Output

Several system options affect the print output from procs:

LABEL The LABEL option affects the way procs identify variables. The LABEL option tells procs to use the label attribute of a variable to identify the variable in print output. Variables that do not have labels are still identified by their names. The NOLABEL option tells procs to use only names to identify variables.

FORMCHAR= Several procs use regular printable characters to form tables, boxes, rules, and other objects. The value of the FORMCHAR= option tells the procs which characters to use.

The default value for the option is `"|----|+|---+=|-/\<>*"`. Each character in this string is used as a different part of a table, box, or rule. The string can contain as many as 64 characters, but base SAS procs use only the first 20 characters, and only the first 11 characters are used to form the parts of a table. The following table describes each of the FORMCHAR characters.

FORMCHAR Characters

<table>
<tr><th>Character Position</th><th>Default Character</th><th>Use</th></tr>
<tr><td>1</td><td>|</td><td>vertical rule</td></tr>
<tr><td>2</td><td>-</td><td>horizontal rule</td></tr>
<tr><td>3</td><td>-</td><td>top left corner</td></tr>
<tr><td>4</td><td>-</td><td>top of vertical rule</td></tr>
<tr><td>5</td><td>-</td><td>top right corner</td></tr>
<tr><td>6</td><td>|</td><td>left of horizontal rule</td></tr>
<tr><td>7</td><td>+</td><td>horizontal rule crosses vertical rule</td></tr>
<tr><td>8</td><td>|</td><td>right of horizontal rule</td></tr>
<tr><td>9</td><td>-</td><td>bottom left corner</td></tr>
<tr><td>10</td><td>-</td><td>bottom of vertical rule</td></tr>
<tr><td>11</td><td>-</td><td>bottom right corner</td></tr>
<tr><td>12</td><td>+</td><td>start or end of event line</td></tr>
<tr><td>13</td><td>=</td><td>special event</td></tr>
<tr><td>14</td><td>|</td><td></td></tr>
<tr><td>15</td><td>-</td><td></td></tr>
<tr><td>16</td><td>/</td><td>separator in event line</td></tr>
<tr><td>17</td><td>\</td><td></td></tr>
<tr><td>18</td><td><</td><td>left arrow</td></tr>
<tr><td>19</td><td>></td><td>right arrow</td></tr>
<tr><td>20</td><td>*</td><td>highlighting</td></tr>
</table>

The following example shows a table produced by the TABULATE proc using the default FORMCHAR characters.

```
DATA TEST;
  RETAIN ID 'TEST';
  DO I = 1 TO 4; OUTPUT; END;
RUN;
TITLE1 'Default FORMCHAR Characters';
OPTIONS FORMCHAR='|----|+|---+=|-/\<>*' LS=72 NOCENTER NONUMBER NODATE;
PROC TABULATE DATA=TEST;
  CLASS ID;
  VAR I;
  TABLE ID, I;
RUN;
```

```
Default FORMCHAR Characters

--------------------------------
|                |      I      |
|                |-------------|
|                |     Sum     |
|----------------+-------------|
|ID              |             |
|----------------|             |
|TEST            |        10.00|
--------------------------------
```

In the following revision, the 11 FORMCHAR characters that the TABULATE proc uses are given distinct values so that you can see, in an example of actual print output, the use of each character.

```
TITLE1 'Distinct FORMCHAR Characters';
OPTIONS FORMCHAR='123456789AB+=|-/\<>*' LS=72 NOCENTER NONUMBER NODATE;
PROC TABULATE DATA=TEST;
  CLASS ID;
  VAR I;
  TABLE ID, I;
RUN;
```

```
Distinct FORMCHAR Characters

3222222222222222242222222222225
1                1      I      1
1                62222222222228
1                1     Sum     1
62222222222222227222222222222 8
1ID              1             1
62222222222222228             1
1TEST            1        10.001
9222222222222222A222222222222B
```

BYLINE When procs have a BY statement, they produce separate output for each of the BY groups. With the BYLINE option, which is the default, many procs identify each BY group with what is called a BY line — a line that indicates the name and value of each of the BY variables, such as:

```
STATE=IN COUNTY=LAKE
```

The NOBYLINE option tells procs not to write BY lines. In addition, it tells procs to put each BY group on a separate page, something many procs do in any case. The usual reason to suppress the BY line is because you are writing the BY variables in the title lines.

BY Variables in Title and Footnote Lines

You can include special code sequences in the text of title lines to write the names and values of BY variables in title lines. These title line BY variable codes work only in proc steps, and they usually work with only a specific SAS dataset, so using them usually requires separate TITLE statements for a specific proc step. The same codes also work in FOOTNOTE statements.

Use the BY variable codes listed below in the text in a TITLE or FOOTNOTE statement for a proc step that uses a BY statement. When the title or footnote line is formed, the proc step substitutes the names and values of the BY variables for the codes.

Title Line BY Variable Codes

Code	Substituted Text
#BYLINE	The text of the BY line
#BYVAR*n*	The label or name of the *n*th BY variable
#BYVAR(*variable name*)	The label or name of the BY variable
#BYVAL*n*	The value of the *n*th BY variable
#BYVAL(*variable name*)	The value of the BY variable

You can write a period after a BY variable code. The period is necessary if the code is followed by a character that could be part of a name: a letter, underscore, digit, or period.

The following example works with a proc step that uses three BY variables.

```
TITLE1 'Activity Details';
TITLE3 '#BYVAR1: #BYVAL1';
TITLE4 '#BYVAR2: #BYVAL2';
TITLE5 '#BYVAR3: #BYVAL3';
```

When used with an appropriate proc step, these lines produce title lines such as these:

```
Activity Details

CAMPUS: Main Campus
BUILDING: Pinnacle
ROOM: 311
```

The title lines are different for each BY group as the values of the BY variables change.

Redirecting the Standard Print File

You can redirect print output from the standard print file to any other file with the PRINTTO proc. This proc can also redirect the log. The syntax of the proc is:

```
PROC PRINTTO options;
RUN;
```

These are options you can use in the PROC PRINTTO statement:

Option	Action
PRINT=*fileref* PRINT='*physical file name*'	Redirects print output to the indicated print file.
NEW	Replaces the existing file.
PRINT=*entry*	Redirects print output to a catalog entry (of type OUTPUT or LOG).
LABEL='*description*'	A descriptive label for the catalog entry.
PRINT=PRINT	Returns print output to the standard print file.
LOG=*fileref* LOG='*physical file name*'	Redirects print output to the indicated print file.
LOG=*entry*	Redirects the log to a catalog entry (of type LOG or OUTPUT).
LOG=LOG	Returns the log to the log file.

15

Functions and CALL Routines

Functions and CALL routines are small, simple program units, or routines, that you can use in data step programming. The purpose of a function may be to compute a value based on one or more values that you supply as arguments to the function or to take some external action such as opening a file or executing an operating system command. A function returns a value that results from its actions. A CALL routine is essentially the same as a function. The difference is that a CALL routine does not return a value.

Lexicon

Functions and CALL routines are essentially the same as the functions of the C programming language. In fact, if you wanted to create functions and CALL routines to use in the SAS System, you would most likely code them as C functions and use SAS/TOOLKIT to link them to the SAS System. A CALL routine corresponds to a null function in C.

Functions

To write a function call, write the function name followed by parentheses. Inside the parentheses, list the arguments to the function, separated by commas. Different functions require different numbers and types of arguments.

Return values are essential to the way functions are used in a program. Using a function means using its return value. Each function returns a specific type of value; the function call is an expression of that type, and it is used in places where that type of expression can be used. Function calls are used in data step statements and in WHERE expressions.

The most common use of a function call is in an assignment statement that assigns the return value to a variable, as in this example:

```
TIME = HMS(HOURS, MINUTES, SECONDS);
```

The HMS function is a numeric function, which means it returns a numeric value. In this example, the return value is assigned to the numeric variable TIME.

Function calls are also commonly used in comparisons, as in this example:

```
IF ROUND(A) = ROUND(B) THEN DO;
```

The condition in this statement tests whether the variables A and B have the same value after they are rounded to the nearest integer.

Another common use of function calls that must be mentioned is as arguments to other functions, as in this example:

```
TRIM(LEFT(STRING))
```

The return value of the function call `LEFT(STRING)` is used as the argument to the TRIM function. The effect is that the two functions are called in succession: first the LEFT function, then the TRIM function. This frequently used combination of functions removes the leading and trailing spaces from a character string.

If you use a function call only for its external actions, and you have no use for the return value, write the function call in an assignment statement. If necessary, create a variable to assign the return value to.

CALL Routines

A CALL routine does not return a value, so it cannot be used in an expression. Instead, CALL routines are used in the CALL statement. The CALL statement is an executable statement of the data step. After the keyword CALL, the only term in the statement is the call to the CALL routine, which is written exactly the same way as a function call.

Arguments

Each function or CALL routine is designed to work with a certain number of arguments. Each argument is required to be a specific type. Most arguments are simply values that the function or CALL routine uses. These arguments are written as expressions, and they should be character expressions or numeric expressions, depending on what the routine requires.

Not all arguments are simply expressions. These are the different types of arguments that functions and CALL routines use:

- *Numeric expression.* A numeric expression argument provides a number that the routine uses in its computations or actions.
- *Character expression.* A character expression argument provides a character or character string that the routine uses.
- *Numeric variable.* When a routine requires an argument to be a numeric variable, it means the routine will do something to change the value of the variable, giving it a new numeric value. CALL routines, in particular, tend to do this. The routine might also use the original value of the variable as part of its processing.
- *Character variable.* Similarly, an argument that is required to be a character variable is used to hold a character value that is the result of the routine's processing. It is important that the character variable be defined with a length that is long enough to hold the value that the routine gives it.
- *Variable.* Sometimes, the argument has to be a variable, which can be either data type. The routine returns an attribute of the variable. Often, the argument can be either a variable or an array.
- *Array.* The array functions require an array as their first argument. The function returns a property of the array.

- *Informat term.* The INPUT function requires an informat term as its second argument.
- *Format term.* Similarly, the PUT function requires a format term as its second argument.

Mathematical Functions

The word *function* comes to computer programming from mathematics, and many mathematical functions have been implemented as SAS functions. This is the category of functions that gets the most use.

Rounding and Representational Functions

Rounding is a fact of life in computing. The double precision floating-point values that SAS software uses form a finite set of numbers that attempts to represent the infinite set of real numbers. Most of the time, a numeric value is only an approximation of the underlying number it represents. The purpose of the rounding and representational functions is to create various other more limited sets of numbers. For example, the CEIL, FLOOR, and INT functions produce only integer values. If the argument to the function is within the set of values that it produces, then the function returns the value of the argument. If not, the function returns a value that represents the argument value in some way.

FLOOR Use the FLOOR function to round down. Values that are not integers are rounded down to the next lower integer. Values that are within 10^{-12} of an integer are rounded to the integer.

Function call: `FLOOR(X)`

Argument: `X`, a real number.

Return value: For values that are within 10^{-12} of an integer, the integer value. For other values, the next lower integer.

CEIL The CEIL (ceiling) function rounds up. Noninteger values are rounded up to the next higher integer. Values that are within 10^{-12} of an integer are rounded to the integer.

Function call: `CEIL(X)`

Argument: `X`, a real number.

Return value: For values that are within 10^{-12} of an integer, the integer value. For other values, the next higher integer.

The following example demonstrates how the results of the CEIL function shift to the next higher integer at a point 1E-12 above an integer value.

```
A = CEIL(3 + .99E-12);
B = CEIL(3 + 1.0E-12);
PUT A= B=;
```

```
A=3 B=4
```

INT Like the FLOOR and CEIL functions, the INT function returns integers. The INT function removes the fractional part of a number to return the integer part. Values that are within 10^{-12} of an integer are rounded to the integer. For positive arguments, the result of the INT function is the same as the FLOOR function; for negative arguments, it is the same as the CEIL function.

Function call: `INT(X)`

Argument: `X`, a real number.

Return value: For values that are within 10^{-12} of an integer, the integer value. For other values, the integer part of the argument.

ROUND The ROUND function rounds to the nearest integer, or to the nearest multiple of a roundoff unit that you supply as a second argument. Values that are halfway between multiples of the roundoff unit are rounded up.

Function call: `ROUND(X)`

Argument: `X`, a real number.

Return value: The nearest integer to the argument. The next higher integer if the argument is halfway between two integers.

Function call: `ROUND(X, U)`

Arguments: `X`, a real number; `U`, a positive number used as a roundoff unit.

Return value: The multiple of `U` that is closest to `X`. The next higher multiple of `U` if the argument is halfway between two multiples of `U`.

FUZZ The FUZZ function rounds values that are very close to integer values to the integer value. The FUZZ function may be useful in removing small rounding errors that result from a succession of arithmetic operations.

Function call: `FUZZ(X)`

Argument: `X`, a real number.

Return value: For values that are within 10^{-12} of an integer, the integer value. Other values are returned unchanged.

TRUNC The TRUNC function duplicates the effect of storing a numeric value with a length less than 8. Specifically, it zeroes out the least significant bytes of a value. The TRUNC function is especially useful in comparisons when you do not want the truncation that results from storing a shortened numeric variable to affect the results of the comparison.

Function call: `TRUNC(X, N)`

Arguments: `X`, a numeric value; `N`, a counting number from 2 to 8, or from 3 to 8, depending on the kind of computer, representing the length of the value in bytes.

Return value: The numeric value `X` truncated to a length of `N` bytes.

The following short program demonstrates how the TRUNC function can be used for comparisons of a variable that has been stored with a length of less than 8. When the variable A is compared to the constant value .1, the comparison is false because of the truncation that was applied when A was stored. When the TRUNC function is used to apply the same truncation to the value .1, then the two values are equal.

```
DATA SHORT;
  A = .1;
  LENGTH A 4;
RUN;
DATA _NULL_;
  SET SHORT;
  IF A = .1 THEN PUT 'A is stored as exactly .1.';
  IF A = TRUNC(.1, 4) THEN PUT 'A is stored as approximately .1.';
RUN;
```

```
A is stored as approximately .1.
```

SIGN The SIGN function tells you the sign of a number — whether it is negative, 0, or positive.

Function call: SIGN(X)
Argument: X, a real number.
Return value: For a negative value, –1. For 0, 0. For a positive value, 1.

ABS The ABS function returns the absolute value of its argument. In mathematical notation, this function is indicated with two vertical bars around an expression:

$$|x|$$

The absolute value is always a positive number or 0. For a negative number, the absolute value is the negative of the number, which is a positive number.

Function call: ABS(X)
Argument: X, a real number.
Return value: The absolute value of the argument.

Algebraic and Transcendental Functions

Mathematicians call a function *algebraic* if it can be constructed by algebraic operations, which are certain uses of the addition, subtraction, multiplication, division, and exponentiation operators. Since those operators (+, -, *, /, **) are part of the SAS language, there is no special need for a collection of algebraic functions implemented as functions. However, there are two to note. The ABS function, mentioned above, is one, and the SQRT function is another.

SQRT The SQRT (square root) function calculates the square root of a number. This is a function that in mathematical notation is written as $\sqrt{x}$.

The square root of a number is a number that, when multiplied by itself, yields that number. For example, 2 is the square root of 4 because 2 times 2 equals 4. The argument to the square root function cannot be negative.

Function call: `SQRT(X)`
Argument: `X`, a nonnegative number.
Return value: The square root of the argument.

There are also many important mathematical functions that require an infinite number of algebraic operations to construct. These functions, called *transcendental* functions, are essential in branches of mathematics ranging from geometry to statistics.

Logarithmic and Exponential Functions Logarithms are calculations of the magnitudes of positive numbers. A logarithm is calculated with a particular base, so that the base raised to the power of the logarithm results in the original number. For example, the base 10 logarithm of 100, `LOG10(100)`, is 2, because 10^2 is 100.

The natural base for logarithms is found in the natural logarithm function, LOG, a function that is created by calculus as an integral of $1/x$. The base of this function is a number that mathematicians call e. The mathematical equation that defines e is $\ln e = 1$. The inverse of the natural logarithm function is the exponential function, EXP. You can describe the exponential function as raising e to a power. Another way to define e, then, is as `EXP(1)`, or with the mathematical equation $e = \exp 1$. The following SAS statements display the approximate value of e.

```
E = EXP(1);
PUT E=BEST15.;
```

```
E=2.718281828459
```

Function call: `LOG(X)`
Argument: `X`, a positive number.
Return value: The natural logarithm of the argument.

Function call: `LOG10(X)`
Argument: `X`, a positive number.
Return value: The base 10 logarithm (common logarithm) of the argument.

Function call: `LOG2(X)`
Argument: `X`, a positive number.
Return value: The base 2 logarithm of the argument.

Function call: `EXP(X)`
Argument: `X`, a real number.
Return value: The exponential of the argument. Approximately `2.718281828459045**X`.

Comparison: The exponential function is the inverse of the natural logarithm function.

Logarithms of any base are measures of the same essential quality, the magnitude of a number. The logarithms of one base are proportional to the logarithms of another base. In practice, all logarithms are computed from natural logarithms. The function `LOG10(X)` is computed as `LOG(X)/LOG(10)`.

Trigonometric functions Trigonometry is the study of angles. It provides a classic set of transcendental functions, the trigonometric functions that characterize angles, among their many other uses. Most of these functions are implemented as SAS functions; others can be computed easily from SAS functions. The table at the bottom of this page shows SAS expressions to calculate trigonometric functions.

SAS trigonometric functions are based on angles measured in radians, rather than the degrees that are the units more often used in physical measurements of angles. Degrees may be more familiar, but radians are mathematically more meaningful. A degree is just an arbitrary division of a circle, 1/360 of it, but a radian is a division of a circle that creates an arc whose length is equal to the radius of the circle. A complete circle is 360 degrees or 2π radians, resulting in the equation $360° = 2\pi$ rad. To convert degrees to radians, multiply by $\pi/180$, or by approximately .0174532925.

Trigonometric Functions

Mathematical Name and Symbol		Expression
sine	$\sin x$	`SIN(X)`
cosine	$\cos x$	`COS(X)`
tangent	$\tan x$	`TAN(X)`
secant	$\sec x$	`1/COS(X)`
cosecant	$\csc x$	`1/SIN(X)`
cotangent	$\cot x$	`1/TAN(X)`
arcsine	$\sin^{-1} x$	`ARSIN(X)`
arccosine	$\cos^{-1} x$	`ARCOS(X)`
arctangent	$\tan^{-1} x$	`ATAN(X)`
hyperbolic sine	$\sinh x$	`SINH(X)`
hyperbolic cosine	$\cosh x$	`COSH(X)`
hyperbolic tangent	$\tanh x$	`TANH(X)`
hyperbolic secant	$\text{sech}\, x$	`1/COSH(X)` *or* `2/(EXP(X) + EXP(-X))`
hyperbolic cosecant	$\text{csch}\, x$	`1/SINH(X)` *or* `2/(EXP(X) - EXP(-X))`
hyperbolic cotangent	$\coth x$	`1/TANH(X)`

These other transcendental functions are used in some mathematical models and statistical tests:

Function Call	Description
ERF(X)	The error function.
ERFC(X)	The complement of the error function. Equal to 1 - ERF(X).
GAMMA(X)	The gamma function.
DIGAMMA(X)	The derivative of the gamma function.
TRIGAMMA(X)	The second derivative of the gamma function.
AIRY(X)	The airy function.
DAIRY(X)	The derivative of the airy function.
JBESSEL(NU, X)	The bessel function of order NU.

Combinatorics Functions

In some problems, it is important to determine the number of different ways that objects in a model can be arranged or combined. These problems fall into the branch of mathematics called combinatorics. Combinatorics forms the foundation of some problems in probability. In many of the classic problems of probability, all outcomes are equally likely, so the probability of a specific outcome can be determined as soon as all the possible outcomes have been enumerated. For example, in the toss of a coin, there are 2 possible outcomes, each with a probability of 1/2; in the draw of a card from a standard deck of cards, there are 52 possible outcomes, each with a probability of 1/52.

The COMB function calculates the number of combinations of elements that can be selected from a set. For example, COMB(52, 2) is the number of ways that 2 cards can be selected from a deck of 52. When the order of elements is significant, the question is about permutations rather than combinations, and the PERM function calculates that answer. Both of these functions are based on factorials, which the FACT function computes.

Function call: FACT(N)
Argument: N, a counting number.
Return value: The factorial of N, defined as the product of all the counting numbers from 1 through N.

Function call: PERM(N, R)
Arguments: N, a counting number, the number of elements in the set; R, a counting number no larger than N, the number of elements selected from the set.
Return value: The number of permutations of N elements selected R at a time. This is equivalent to FACT(N)/FACT(N - R).

Function call: COMB(N, R)
Arguments: N, a counting number, the number of elements in the set; R, a

counting number no larger than N, the number of elements selected from the set.
Return value: The number of combinations of N elements selected R at a time. This is equivalent to PERM(N, R)/FACT(R).

Statistic Functions

A statistic is a value that is calculated from a set of numbers. The distinctive thing about statistic functions, compared to other mathematical functions, is that a statistic function treats all of its arguments the same way. That is, the order of the arguments does not affect the result.

The set of values for the SAS statistic functions is treated as a sample — as a set of measurements or objects of a certain kind. A sample is drawn from and is considered to represent a population, or the set of all such measurements or objects. The functions calculate sample statistics. Statistics of a sample are calculated differently from the way statistics are calculated for a population. Specifically, sample statistics use a divisor of $n - 1$ rather than n when calculating the variance and other statistics that are derived from the variance. This corresponds to the option VARDEF=DF in the PROC statement of procs that calculate these same statistics, such as the SUMMARY, REPORT, and UNIVARIATE procs. Sample statistics are calculated this way in order to provide the best estimates of the characteristics of the population.

The statistic functions work the same way regardless of the number of arguments. A sample can comprise any number of measurements, and the size of the sample does not affect the meaning of the statistics. Statistic functions are not affected by missing values used as arguments. They disregard missing arguments and compute the same results as if those arguments were not present. For situations when you need to determine the number of missing arguments, the NMISS function counts the missing values among its arguments. Each statistic function requires a certain minimum number of nonmissing arguments in order to be able to calculate the value of the statistic. If there are not enough nonmissing arguments, the function returns a missing value.

Some of the statistics are so familiar that you might not think of them as statistics. The sum, for example, is simply the result of adding the arguments. The maximum is the largest value among the arguments. Other statistics, such as the variance and skewness, are equally interesting measures of a set of data, but their meaning is less obvious, and interpreting them may require technical knowledge of the field of statistics.

Function call: N(X1, X2, . . .)
Arguments: X1, X2, . . . , a set of real numbers.
Return value: The sample size; the number of nonmissing values.

Function call: NMISS(X1, X2, . . .)
Arguments: X1, X2, . . . , a set of numeric values.
Return value: The number of missing values.

Function call: MIN(X1, X2, . . .)
Arguments: X1, X2, . . . , a set of real numbers.
Return value: The minimum; the lowest number.

Function call: MAX(X1, X2, . . .)
Arguments: X1, X2, . . . , a set of real numbers.
Return value: The maximum; the highest number.

Function call: RANGE(X1, X2, . . .)
Arguments: X1, X2, . . . , a set of at least 2 real numbers.
Return value: The range; the difference between the maximum and minimum.

Function call: SUM(X1, X2, . . .)
Arguments: X1, X2, . . . , a set of real numbers.
Return value: The sum; the result of adding the numbers.

Function call: MEAN(X1, X2, . . .)
Arguments: X1, X2, . . . , a set of real numbers.
Return value: The mean; the arithmetic average. Computed as SUM/N.

Function call: USS(X1, X2, . . .)
Arguments: X1, X2, . . . , a set of real numbers.
Return value: The uncorrected sum of squares; the sum of the squares of the values.

Function call: CSS(X1, X2, . . .)
Arguments: X1, X2, . . . , a set of real numbers.
Return value: The corrected sum of squares. Computed as USS – MEAN*SUM.

Function call: VAR(X1, X2, . . .)
Arguments: X1, X2, . . . , a set of at least 2 real numbers.
Return value: The variance, a measure of dispersion. Computed as CSS/(N – 1).

Function call: STD(X1, X2, . . .)
Arguments: X1, X2, . . . , a set of at least 2 real numbers.
Return value: The standard deviation; the square root of the variance.

Function call: SKEWNESS(X1, X2, . . .)
Arguments: X1, X2, . . . , a set of at least 3 real numbers in which not all values are equal.
Return value: The skewness, a measure of sidedness.

Function call: KURTOSIS(X1, X2, . . .)
Arguments: X1, X2, . . . , a set of at least 4 real numbers, not all equal.
Return value: The kurtosis, a measure of the importance of extreme values.

Function call: CV(X1, X2, . . .)
Arguments: X1, X2, . . . , a set of at least 2 real numbers with a nonzero mean.
Return value: The percent coefficient of variation.

Function call: STDERR(X1, X2, . . .)
Arguments: X1, X2, . . . , a set of at least 2 real numbers.
Return value: The standard error of the mean.

Function call: ORDINAL(N, X1, X2, . . .)
Arguments: N, a counting number no greater than the number of other arguments; X1, X2, . . . , a set of numeric values.
Return value: The Nth lowest value.

Statistic functions can use a set of variables as arguments, so they are often written with the OF keyword, which allows the arguments to be written as a variable list. The following example demonstrates the calculation of statistics and the use of the OF keyword.

```
DATA _NULL_;
  RETAIN X1 4 X2 . X3 0 X4 5 X5 6 X6 5;
  CSS = CSS(OF X1-X6);
  CV = CV(OF X1-X6);
  KURTOSIS = KURTOSIS(OF X1-X6);
  MAX = MAX(OF X1-X6);
  MEAN = MEAN(OF X1-X6);
  MIN = MIN(OF X1-X6);
  N = N(OF X1-X6);
  NMISS = NMISS(OF X1-X6);
  ORDINAL1 = ORDINAL(1, OF X1-X6);
  ORDINAL2 = ORDINAL(2, OF X1-X6);
  ORDINAL3 = ORDINAL(3, OF X1-X6);
  RANGE = RANGE(OF X1-X6);
  SKEWNESS = SKEWNESS(OF X1-X6);
  STD = STD(OF X1-X6);
  STDERR = STDERR(OF X1-X6);
  SUM = SUM(OF X1-X6);
  USS = USS(OF X1-X6);
  VAR = VAR(OF X1-X6);
  PUT _ALL_;
RUN;
```

```
X1=4 X2=. X3=0 X4=5 X5=6 X6=5 CSS=22 CV=58.630196998 KURTOSIS=3.3223140496
MAX=6 MEAN=4 MIN=0 N=5 NMISS=1 ORDINAL1=. ORDINAL2=0 ORDINAL3=4 RANGE=6
SKEWNESS=-1.744369497 STD=2.3452078799 STDERR=1.0488088482 SUM=20 USS=102
VAR=5.5 _ERROR_=0 _N_=1
```

Probability Distributions

Statistical models use probability distributions to describe random events. A probability distribution sets forth the relative frequencies with which a random variable takes on each of its possible values. There are several kinds of functions that can be associated with a specific probability distribution.

Functions for Probability Distributions

Distribution	Function Type	Function Call
Bernoulli	distribution	CDF('BERNOULLI', *X*, *P*)
	probability density	PDF('BERNOULLI', *X*, *P*)
beta	distribution	CDF('BETA', *X*, α, β, *left location*, *right location*)
		PROBBETA(*X*, α, β)
	probability density	PDF('BETA', *X*, α, β, *left location*, *right location*)
binomial	distribution	CDF('BINOMIAL', *X*, *P*, *n*)
		PROBBNML(*P*, *n*, *X*)
	probability density	PDF('BINOMIAL', *X*, *P*, *n*)
	random number	RANBIN(*seed*, *n*, *P*)
bivariate normal	distribution	PROBBNRM(*X*, *Y*, *correlation*)
Cauchy	distribution	CDF('CAUCHY', *X*, α, β)
	probability density	PDF('CAUCHY', *X*, α, β)
	random number[1]	RANCAU(*seed*)
chi-squared	distribution	CDF('CHISQUARED', *X*, *d.f.*)
		PROBCHI(*X*, *d.f.*)
	probability density	PDF('CHISQUARED', *X*, *d.f.*)
exponential	distribution	CDF('EXPONENTIAL', *X*, λ)
	probability density	PDF('EXPONENTIAL', *X*, λ)
	random number[1]	RANEXP(*seed*)
F	distribution	CDF('F', *X*, *n.d.f.*, *d.d.f.*)
		PROBF(*X*, *n.d.f.*, *d.d.f.*)
	probability density	PDF('F', *X*, *n.d.f.*, *d.d.f.*)
gamma	distribution	CDF('GAMMA', *X*, α , β)
		PROBGAM(*X*, α)
	probability density	PDF('GAMMA', *X*, α , β)
	random number[1]	RANGAM(*seed*, α)
geometric	distribution	CDF('GEOMETRIC', *X*, *P*)
	probability density	PDF('GEOMETRIC', *X*, *P*)
hypergeometric	distribution	CDF('HYPERGEOMETRIC', *X*, *population size*, *category size*, *sample size*)
		PROBHYPR(*population size*, *category size*, *sample size*, *X*)
	probability density	PDF('HYPERGEOMETRIC', *X*, *population size*, *category size*, *sample size*)
Laplace	distribution	CDF('LAPLACE', *X*, θ , λ)
	probability density	PDF('LAPLACE', *X*, θ , λ)

continued

Distribution	Function Type	Function Call
logistic	distribution	CDF('LOGISTIC', *X*, θ, λ)
	probability density	PDF('LOGISTIC', *X*, θ, λ)
lognormal	distribution	CDF('LOGNORMAL', *X*, θ, λ)
	probability density	PDF('LOGNORMAL', *X*, θ, λ)
negative binomial	distribution	CDF('NEGBINOMIAL', *X*, *P*, *n*)
		PROBNEGB(*P*, *n*, *X*)
	probability density	PDF('NEGBINOMIAL', *X*, *P*, *n*)
normal	distribution	CDF('NORMAL' *or* 'GAUSS', *X*, θ, λ)
		PROBNORM(*X*)
	inverse distribution[2]	PROBIT(*P*)
	probability density	PDF('NORMAL' *or* 'GAUSS', *X*, θ, λ)
	random number[2]	RANNOR(*seed*)
Pareto	distribution	CDF('PARETO', *X*, *a*, *k*)
	probability density	PDF('PARETO', *X*, *a*, *k*)
Poisson	distribution	CDF('POISSON', *X*, λ)
		POISSON(λ, *X*)
	probability density	PDF('POISSON', *X*, λ)
	random number	RANPOI(*seed*, λ)
T	distribution	CDF('T', *X*, *d.f.*)
		PROBT(*X*, *d.f.*)
	probability density	PDF('T', *X*, *d.f.*)
table	random number	RANTBL(*seed*, *f(1)*, *f(2)*, . . .)
triangular	random number[3]	RANTRI(*seed*, *hypotenuse*)
uniform	distribution	CDF('UNIFORM', *X*, *left location*, *right location*)
	probability density	PDF('UNIFORM', *X*, *left location*, *right location*)
	random number[3]	RANUNI(*seed*)
Wald	distribution	CDF('WALD' *or* 'IGAUSS', *X*, *d*)
	probability density	PDF('WALD' *or* 'IGAUSS', *X*, *d*)
Weibull	distribution	CDF('WEIBULL', *X*, *a*, *optional* λ)
	probability density	PDF('WEIBULL', *X*, *a*, *optional* λ)

For a survival function, use the SDF function with the same arguments that are indicated for the CDF function.

There is a random number CALL routine for every random number function listed. The seed argument for a random number CALL routine must be a variable. There is an additional, final argument for a random number CALL routine, also a variable, to which the CALL routine assigns the random number it generates.

[1] with parameter values of 1, for parameters not indicated as arguments

[2] with a mean of 0 and a standard deviation of 1

[3] on the interval from 0 to 1

- A *distribution function*, or *cumulative distribution function*, defines a probability distribution. For each value of a distribution, the value of the distribution function is the probability that a random variable from the distribution will be less than or equal to that value. The value of the function increases from 0 to 1.
- For a continuous distribution, a *probability density function*, or *probability mass function*, associates each value of a random variable with a probability density value. The probability density function is the derivative of the distribution function. Definite integrals of the probability density function determine the probability that a random variable from the distribution falls into a particular range.
- An *inverse distribution function*, or *probability function*, is the inverse of the distribution function. The argument to the function is a probability value, a value from 0 to 1.
- A *random number function* generates random numbers from the probability distribution.
- A *survival function* is the complement of the distribution function.

Base SAS includes one or more of these kinds of functions for more than 20 probability distributions. Most of these distributions have parameters that determine the location and shape of the distribution. The table on the two previous pages summarizes the available functions for probability distributions.

Random Numbers

The purpose of a random number function is to generate numbers that simulate the behavior of random variables — variables that come from the measurements of random events or randomly selected objects. Usually, function arguments indicate parameters of the probability distribution from which the random numbers are drawn. The first argument of a random number function is a seed argument, which provides a starting point for the random number generator. Usually, you call a random number function repeatedly, with the same arguments, to generate a set of random numbers. Each successive call to the function generates a different number.

The numbers generated by a random number generator do not have all the qualities of random numbers. When they are considered in blocks, they exhibit some measurable characteristics that true random numbers would not have. Still, they are sufficiently random to be used for most applications of random numbers, such as random sampling and stochastic simulations.

The most commonly used random number function is the RANUNI function, which generates random numbers that are uniformly distributed over the interval from 0 to 1. This is an example of random numbers generated by the RANUNI function:

```
DO I = 1 TO 100;
  X = RANUNI(25);
  PUT X 8.7 +1 @;
  END;
```

```
.6240642 .2522179 .9956077 .4849661 .0400644 .2319337 .5744793 .2922931
.2448507 .6641638 .4829643 .0967630 .3348578 .6796144 .9255964 .4298491
.8152947 .2482408 .4191216 .6626880 .2059138 .3190967 .9638528 .1806403
.8947326 .1246744 .8631030 .7591167 .5645382 .4253342 .2090390 .6650882
.2803399 .2280012 .7415084 .4305712 .7760608 .8107294 .9881437 .2022088
.1631526 .7776222 .4697383 .3660441 .5117708 .2195867 .2386414 .3897065
.7829471 .8301141 .4026445 .6258998 .1204666 .1137192 .3371688 .9888285
.3709293 .4618490 .9953051 .2341762 .0595542 .5841328 .5678080 .8221969
.4266151 .1498268 .6734306 .8224671 .5818186 .3793012 .7226466 .3714130
.5176897 .7934286 .9623459 .7858720 .7462475 .6991588 .0458881 .7527174
.1299779 .2294591 .0648610 .9802648 .9358785 .1460462 .2953988 .6305890
.4000938 .6985765 .2831491 .4718440 .0390784 .5166456 .7659830 .3302982
.5343133 .9494426 .7701571 .6933480
```

The first argument to a random number function is a seed argument. If the seed argument is an integer from 1 to 2,147,483,646, the random number generator uses the argument as its starting point for the stream of random numbers that it generates. This means that the function generates the same random numbers every time you run the program. If you use 0 or a negative value as the seed argument, the random number generator obtains a value from the computer clock to determine its starting point. This way, the function generates different random numbers every time you run the program.

A random number function call considers the seed argument only the first time it executes. If you want to use programming logic to evaluate or adjust the seed argument, use a random number CALL routine instead. The seed argument to the CALL routine must be a numeric variable. Initialize the random number stream by assigning a value to the seed variable before the first time you call the routine. The routine assigns a new value to the variable every time it generates a new random number.

The CALL routine returns a random number by assigning it to a variable you provide as the last argument to the routine. The example revises the previous example to use the `CALL RANUNI` routine instead of the RANUNI function. The results are the same.

```
SEED = 25;
DO I = 1 TO 100;
  CALL RANUNI(SEED, X);
  PUT X 8.7 +1 @;
  END;
```

```
.6240642 .2522179 .9956077 .4849661 .0400644 .2319337 .5744793 .2922931
.2448507 .6641638 .4829643 .0967630 .3348578 .6796144 .9255964 .4298491
.8152947 .2482408 .4191216 .6626880 .2059138 .3190967 .9638528 .1806403
.8947326 .1246744 .8631030 .7591167 .5645382 .4253342 .2090390 .6650882
...
```

It is important to initialize the seed variable only once, at the beginning of the loop in which the random numbers are generated. If you assign a value to the seed variable every time you call the random number routine, that determines the values that the routine generates. Compare the example below to the previous example. The seed value, assigned in every repetition of the loop, results in the same number being generated every time.

```
DO I = 1 TO 100;
  SEED = 25;
  CALL RANUNI(SEED, X);
  PUT X 8.7 +1 @;
  END;
```

```
.6240642 .6240642 .6240642 .6240642 .6240642 .6240642 .6240642 .6240642
.6240642 .6240642 .6240642 .6240642 .6240642 .6240642 .6240642 .6240642
.6240642 .6240642 .6240642 .6240642 .6240642 .6240642 .6240642 .6240642
...
```

To generate random numbers from a distribution for which the SAS System does not provide a random number function, start with the RANUNI function, then apply an appropriate transformation to the resulting values. To be specific, the transformation that is required is that of the inverse distribution function.

Often, the random numbers that are needed are random integers that are uniformly distributed over a particular range. You can accomplish this with the RANUNI function, the multiplication and addition operators, and the FLOOR function. The following example generates random integer values from 1 to 25:

```
DO I = 1 TO 100;
  X = FLOOR(1 + RANUNI(1)*25);
  PUT X 2. +1 @;
  END;
```

```
 5 25 10  7 24 25 14 14  2  2 21 14 22  2 24  8  7 18 25  6 18 11 14  8 12
22 16 15 15 10 19 13 24 24 15  8 10 12 17  5  5 22  8 24 23 15  2  4 13 11
 5 17 11  4 12  5 15 19 11  2 14  9  1 18 24 12 24 18  3  5  7 16 11  2  9
18  5  4 15  7  9 15  2 11 23 14 19 23 15  5  9 18  4  5  7 17 11  1  7 11
```

Character Functions

The SAS System includes a set of functions for working with character data. Some of these functions extract part of a character string or transform it in some other way. Other functions measure or test a character string, returning a numeric value.

Parts

The SUBSTR function is the most direct way to extract a part of a character string in the SAS language. You can obtain a single character or any segment of a string. The SCAN function is another function for extracting a part of a character string. It is useful when a character value is divided into meaningful units such as words, fields, or tokens.

SUBSTR — Substring The SUBSTR function returns a part of a character string. The character string, usually a variable, is the first argument. The part of the string that is extracted is indicated by its starting position and length in the second and third arguments. You must indicate a starting position that falls within the size of the character value; if that argument is invalid, the function creates an error condition.

Function call: SUBSTR(STRING, START, LENGTH)

Arguments: STRING, a character value; START, a counting number that indicates the starting position within the value of STRING; LENGTH, a counting number that indicates the number of characters to return.

Return value: The segment of STRING starting at the START position, and extending for the number of characters indicated by LENGTH, or to the end of the value if that comes first.

Function call: SUBSTR(STRING, START)

Arguments: STRING, a character value; START, a counting number that indicates the starting position within the value of STRING.

Return value: The segment of STRING starting at the START position, and extending to the end of the value.

Use 1 as the third argument to SUBSTR to extract a single character. For example, SUBSTR(STRING, 1, 1) is the first character in STRING; SUBSTR(STRING, 2, 1) is the second character. Use the function without the third argument to remove characters from the beginning of the string. For example, SUBSTR(STRING, 5) returns the value of STRING with the first 4 characters removed.

The SUBSTR function can also be used as the target of the assignment statement in order to replace selected characters of a character variable. When the function is used in this way, its first argument must be a character variable. In the following example, the SUBSTR function is used both as an expression and as a target.

```
DO I = 1 TO VLENGTH(CAT);
  IF SUBSTR(CAT, I, 1) = '+' THEN DO;
    IF I >= 2 THEN SUBSTR(CAT, I, 1) = SUBSTR(CAT, I - 1, 1);
    ELSE SUBSTR(CAT, I, 1) = ' ';
    END;
  END;
```

The purpose of this code fragment is to replace any plus signs in the variable CAT with a repetition of the preceding character. For example, the value '6E4+-++ +*' would be changed to '6E44--- *'. The VLENGTH function is used for the ending value of the loop to ensure that the values of the index variable I stay within the length of the variable CAT. The VLENGTH function returns the length attribute of the variable.

SCAN — Words The SCAN function divides a string into segments such as words, fields, or tokens and returns a selected segment. The arguments are the string, the number of the word or segment to return, and a set of characters to use as delimiters to divide the string.

Consider the simple case of a character variable that contains words with spaces between them. You can use the SCAN function to look at each word separately. The following example extracts each word and writes it as a separate line. When the SCAN function's number argument is greater than the number of words it finds, it returns a null value. The UNTIL condition on the DO loop stops execution of the loop at that point.

```
PHRASE = 'take chances pay attention';
LENGTH WORD $ 64;
DO I = 1 TO 16 UNTIL (WORD = '');
  WORD = SCAN(PHRASE, I, ' ');
  PUT WORD;
  END;
```

```
take
chances
pay
attention
```

The meaningful units in a character value are not always words, and they are not always separated by spaces. You might want to separate the libref from the member name of SAS file, using the period as a delimiter character, as in this example:

```
SASFILENAME = 'MAIN.PROJECT';
LENGTH LIBREF $ 8 MEMBER $32;
LIBREF = SCAN(SASFILENAME, 1, '. ');
MEMBER = SCAN(SASFILENAME, 2, '. ');
PUT LIBREF= MEMBER=;
```

```
LIBREF=MAIN MEMBER=PROJECT
```

Function call: SCAN(STRING, N, DELIMITERS)
Function call: SCAN(STRING, N)
Arguments: STRING, a character value; N, a positive or negative integer; DELIMITERS, a set of characters to use as delimiters.
Default delimiters: " !$%&()*+,-./;<|", plus at least one more special character, depending on the environment.
Return value: Word N of STRING, using the delimiter characters to divide the string into words. If there are fewer than N words in STRING, the function returns a null string. If N is negative, the function counts words backward from the end of the string.

Any character or characters can serve as the delimiters in the SCAN function. List any character that would not be part of a data value in the delimiter argument. The character strings or sequences that remain after the delimiters are removed are the values that the function returns. Because character variables often contain trailing spaces that are not actually part of the data value, it is usually useful to include the space character as one of the delimiters.

When you assign the result of the SCAN function to a character variable, it is usually a good idea to first declare the variable in a LENGTH statement, as in the preceding examples. Use the LENGTH statement to set an appropriate length for the variable. If you process the words you generate with loops and arrays, make the array dimensions large enough to handle the number of words that a variable might hold. The number of words that the SCAN function extracts from a variable can be no more than half the length of the variable, rounded up to the next whole number.

Character Effects

There are a dozen functions that modify character strings in some way. The return value is a modified version of the character string argument, modified with a different effect in each function.

LEFT, RIGHT — Alignment The LEFT and RIGHT functions left- and right-align text values. They do not actually change the length of a value. When a value has leading spaces, the LEFT function moves them to the end of the value. Similarly, the RIGHT function moves trailing spaces to the beginning of the value.

Function call: `LEFT(STRING)`
Argument: `STRING`, a character string.
Return value: The value of `STRING` left-aligned, with leading spaces moved to the end.

Function call: `RIGHT(STRING)`
Argument: `STRING`, a character string.
Return value: The value of `STRING` right-aligned, with trailing spaces moved to the beginning.

While the LEFT function can be used equally well with any character expression, the RIGHT function is usually applied to a variable. That way, the length of the variable determines the way the resulting value lines up. The value returned by the function may be assigned to the same variable or to another variable of the same length. The following example demonstrates the use of the RIGHT function to right-align the value of a variable.

```
LENGTH MONTH $ 9;
MONTH = 'May';
PUT '|' MONTH $CHAR9. '|';
MONTH = RIGHT(MONTH);
PUT '|' MONTH $CHAR9. '|';
```

```
|May      |
|      May|
```

The following example is shown for comparison, to demonstrate that the RIGHT function has no effect when its argument is a character expression that does not have any trailing spaces.

```
LENGTH MONTH $ 9;
MONTH = RIGHT('May'); * Function has no effect. ;
PUT '|' MONTH $CHAR9. '|';
```

```
|May      |
```

COMPRESS, COMPBL — Removing characters The COMPRESS function removes all occurrences of selected characters from a character value. The COMPBL removes characters in a different way. When the value contains consecutive spaces, the COMPBL function leaves the first space and removes the subsequent ones. To put it another way, it converts any sequence of consecutive spaces to a single space.

Function call: `COMPRESS(STRING, CHARACTERS)`
Arguments: `STRING`, a character value; `CHARACTERS`, a set of characters to remove.
Return value: The value of `STRING` with all occurrences of the indicated characters removed.

Function call: `COMPRESS(STRING)`
Argument: `STRING`, a character string.
Return value: The value of `STRING` with all spaces removed.

Function call: `COMPBL(STRING)`
Argument: `STRING`, a character string.
Return value: The value of `STRING` with any sequence of consecutive spaces converted to a single space.

A common use of the COMPRESS function is to remove punctuation from a serial number or other identifying code value. For example, the ISBN numbers that identify books are properly written with hyphens in specific places, but most book databases require that you supply the ISBN digits without any punctuation. You could use the COMPRESS function to remove hyphens, along with other punctuation marks, from ISBN values:

```
RETAIN PUNCTUATION '-., :;+_=()*#!/';
ISBN_KEY = COMPRESS(VISUAL_ISBN, PUNCTUATION);
```

Another use of the COMPRESS function might be to remove the commas from the written date values that the WORDDATE format creates:

```
DATE = '27AUG1977'D;
LENGTH DATEWORD1 $ 18 DATEWORD2 $ 17;
DATEWORD1 = PUT(DATE, WORDDATE18.);
DATEWORD2 = COMPRESS(DATEWORD1, ',');
PUT DATEWORD1 / DATEWORD2;
```

```
August 27, 1977
August 27 1977
```

TRIM, TRIMN — Trailing spaces The TRIM and TRIMN functions remove trailing spaces from character strings. They differ only in the way they respond to a completely blank value: the TRIM function returns a single space, while the TRIMN function returns a null string.

Function call: TRIM(STRING)

Argument: STRING, a character string.

Return value: The value of STRING with trailing spaces removed, except from the first character of a blank string.

Function call: TRIMN(STRING)

Argument: STRING, a character string.

Return value: The value of STRING with all trailing spaces removed.

The TRIM and TRIMN functions are especially useful when you combine strings with the concatenation operator. They are necessary, for example, in this code fragment, which constructs a sentence from an array of words:

```
LENGTH SENTENCE $ 160;
SENTENCE = '';
DO I = LBOUND(WORD) TO HBOUND(WORD) WHILE (WORD{I} NE '') ;
  SENTENCE = TRIMN(SENTENCE) || ' ' || WORD{I};
  END;
SENTENCE = TRIM(LEFT(SENTENCE)) || '.';
```

REVERSE — Reversing The REVERSE function reverses the order of characters in a character value.

Function call: REVERSE(STRING)

Argument: STRING, a character value.

Return value: The value of STRING with the order of characters reversed.

Use the TRIM function first if you want to reverse the letters in a word without moving any trailing spaces to the beginning of the value. Compare the effects of the REVERSE function with and without the TRIM function:

```
LENGTH WORD $ 8;
WORD = 'emit';
REVERSE1 = REVERSE(WORD);
PUT '|' REVERSE1 $CHAR8. '|';
REVERSE2 = REVERSE(TRIM(WORD));
PUT '|' REVERSE2 $CHAR8. '|';
```

```
|    time|
|time    |
```

The four trailing spaces that are part of the value of the variable WORD become leading spaces when WORD is reversed. The trailing spaces stay at the end of the value when the REVERSE function is applied to the value that results from the TRIM function.

REPEAT — Repeating Use the REPEAT function to repeat a character string a specific number of times.

Function call: REVERSE(STRING, N)

Arguments: STRING, a character string; N, a whole number indicating the number of times to repeat the string, in addition to the initial copy of it.

Return value: The character string repeated N + 1 times.

The REPEAT function can be used to make banners, rules, codes, and other strings of repeated characters, as shown in this example:

```
LENGTH REPEAT $ 25 CHAR $ 1;
DO CHAR = '*', '-', '=', '_';
  REPEAT = REPEAT(CHAR, 24);
  PUT REPEAT $CHAR25.;
  END;
```

```
*************************
-------------------------
=========================
_________________________
```

If the string argument is a variable with trailing spaces, you may want to combine the REPEAT function with the TRIM function to repeat the string without the trailing spaces, that is:

```
REPEAT(TRIM(variable), n)
```

TRANSLATE, TRANWRD — Translation The TRANSLATE function replaces specific characters in a string with specific other characters. The TRANWRD function does the same thing with substrings, that is, with short strings of characters that may be contained in the string.

Function call: TRANSLATE(STRING, TO, FROM)

Arguments: STRING, a character string; FROM, a set of characters to replace wherever they appear in the character string; TO, a corresponding set of replacement characters.

Return value: The character string with every occurrence of a FROM character replaced by the corresponding TO character.

Caution: Write the TO argument before the FROM argument.

Function call: TRANWRD(STRING, FROM, TO)

Arguments: STRING, a character string; FROM, a substring to replace if it is found in the character string; TO, a replacement substring.

Return value: The character string with every occurrence of the FROM substring replaced with the value of the TO string.

The effects of the TRANSLATE function are easiest to see when it is used to replace one punctuation mark with another. For example, the function call TRANSLATE('19301+0176', '-', '+') replaces plus signs with hyphens in the string value and returns '19301-0176'.

The same character may appear more than once in the set of replacement characters when you translate several different characters to the same character. You can see this in this example, which translates telephone numbers written with letters to digits:

```
TRANSLATE(PHONE,
      '22233344455566677778889990', 'ABCDEFGHIJKLMNOPQRSTUVWXYZ');
```

In this translation, the letters A, B, and C all translate to the digit 2, D, E, and F translate to 3, and so on.

The replacement substring in the TRANWRD function can be shorter or longer than the original substring, which can make the return value shorter or longer than the string argument. For example, `TRANWRD(STRING, '*', '**')` doubles every occurrence of an asterisk in the value of STRING. The return value could be as much as twice as long as the string argument.

If the replacement substring is a null string, the TRANWRD function removes all occurrences of the substring from the string value.

```
LENGTH STRING1 STRING2 $ 30;
STRING1 = '2001 (projected)';
STRING2 = TRANWRD(STRING1, '(projected)', '');
PUT STRING2;
```

```
2001
```

UPCASE, LOWCASE — Change case The UPCASE and LOWCASE functions do special kinds of character translation. The UPCASE function translates lowercase letters to uppercase letters; the LOWCASE function does the opposite.

Function call: `UPCASE(STRING)`
Argument: `STRING`, a character string.
Return value: The character string with all lowercase letters replaced with the corresponding uppercase letters.

Function call: `LOWCASE(STRING)`
Argument: `STRING`, a character string.
Return value: The character string with all uppercase letters replaced with the corresponding lowercase letters.

The UPCASE and LOWCASE functions are used mainly to standardize the capitalization of data values. Either function can also be useful in character comparisons. When you want to ignore case when comparing two character values, apply the function to both operands, for example:

```
UPCASE(STRING1) = UPCASE(STRING2)
```

The result of the comparison is true if the two values are made of the same letters, disregarding the case of the letters.

The table on the next page shows specific examples of the use of character functions.

Function Call	Return Value
SUBSTR('123456789', 5, 1)	'5'
SUBSTR('123456789', 2, 5)	'23456'
SCAN('05.jpg', -1, '.')	'jpg'
LEFT(' -- ')	'-- '
RIGHT(' -- ')	' --'
COMPRESS('10C', 'ABCDEF')	'10'
COMPRESS('CA 134840 8 ')	'CA1348408'
COMPBL('There I was ')	'There I was '
TRIM(' -- ')	' --'
TRIMN(' -- ')	' --'
REVERSE('123456789')	'987654321'
REPEAT('Duran ', 1)	'Duran Duran '
TRANSLATE('123456789', '**', '25')	'1*34*6789'
TRANSLATE(' $1,205.47', '*', ' ')	'****$1,205.47'
TRANSLATE('3451-0055', '5816209347', '0123456789')	'6208-5500'
TRANWRD('800 MHz', 'MHz', 'megahertz')	'800 megahertz'
TRANWRD('800 MHz', 'MHz', '')	'800'
UPCASE('Earth')	'EARTH'
LOWCASE('Earth')	'earth'
QUOTE('123456789')	'"123456789"'
HTMLENCODE('&')	'&'
HTMLDECODE('&')	'&'

A few other character effects can be accomplished with character formats and informats. Use the PUT or INPUT function to use a format or informat in an expression. If other kinds of character effects are required, you can code them using any of the available tools of data step logic. Use the SUBSTR function in a DO loop to consider each individual character of the character string.

Encoding and Decoding

The QUOTE function converts a character value to a quoted string. The URLENCODE and HTMLENCODE functions encode text in two different Internet standards, and there are corresponding functions for decoding. All of the encoding functions tend to return values that are longer than their character string arguments.

Function call: `QUOTE(STRING)`

Argument: `STRING`, a character string.

Return value: The string value written as a quoted string. Double quote marks are written before and after the value. Any double quote marks in the value are written twice.

Function call: `HTMLENCODE(STRING)`

Argument: `STRING`, a character string.

Return value: The character string encoded as HTML text. Any characters that cannot be written as part of an HTML text value are encoded as HTML escape sequences.

Function call: `HTMLDECODE(STRING)`

Argument: `STRING`, an HTML text value.

Return value: The decoded text. HTML escape sequences are converted to ordinary text characters.

Function call: `URLENCODE(ADDRESS)`

Argument: `ADDRESS`, an address of a file.

Return value: The address encoded as a uniform resource locator (URL). Any characters that are not in the URL character set are encoded as URL escape sequences.

Function call: `URLDECODE(URL)`

Argument: `URL`, a uniform resource locator.

Return value: The decoded address. Any URL escape sequences are converted to ordinary text characters.

Measuring Character Strings

There are several functions that provide information about a character string. The values returned by these functions can be used in conditions, to determine whether to take certain actions. They can also be used to determine the numeric arguments to some of the character functions described in the previous pages.

LENGTH — Text length The LENGTH function measures the length of a string, not counting trailing spaces.

Function call: `LENGTH(STRING)`

Argument: `STRING`, a character string.

Return value: The visible length of the string; the position of the last nonblank character in the string. For a blank or null string, the function returns 1.

VERIFY — Nonmatching characters The VERIFY function looks for characters in a string that are not part of a specific set of characters. If returns the position of the first character in the string that does not match one of the characters. If all the characters match, the function returns 0.

> Function call: VERIFY(STRING, CHARACTERS)
>
> Arguments: STRING, a character string; CHARACTERS, a set of characters.
>
> Return value: The position of the first character found in the string that is not one of the indicated characters. If all the characters in the string are in the set of characters, the function returns 0.

The VERIFY function returns 0 if it does not find a nonmatching character and a positive value if it does; therefore, it can be used as a condition. In some applications, a nonmatching character indicates an invalid data value that should be corrected or treated as a data error.

INDEXC, INDEX, INDEXW — Text search The INDEXC function searches a character string for any of a set of characters. When it finds one of the characters in the string, it returns the position of the character. It returns 0 if it does not find any of the characters in the string.

The INDEX and INDEXW functions search in the same way for a substring. The INDEXW function searches only at word boundaries.

> Function call: INDEXC(STRING, CHARACTERS)
>
> Arguments: STRING, a character string; CHARACTERS, a set of characters.
>
> Return value: The position of the first character found in the string that is one of the indicated characters. If none of the characters in the string are in the set of characters, the function returns 0.

> Function call: INDEX(STRING, SUBSTRING)
>
> Arguments: STRING, a character string; SUBSTRING, a short character string that might be found in the character string.
>
> Return value: The position of the first occurrence of the substring in the string. If the substring is not contained in the string, the function returns 0.

> Function call: INDEXW(STRING, SUBSTRING)
>
> Arguments: STRING, a character string; SUBSTRING, a short character string that might be found in the character string.
>
> Return value: The position of the first occurrence of the substring, starting at a word boundary, in the string. A word boundary is a space or special character or any character that follows a space or special character. If the substring is not found, the function returns 0.

These functions are often used as conditions. For example, to test whether the character variable TEXT contains a dollar sign, you can use the function call INDEXC(TEXT, '$') as the condition. To test whether it contains the word LIST, use INDEX(TEXT, 'LIST') as the condition.

Regular Expression

Regular expression is a text pattern matching syntax standard that has mostly been used in Perl programming. If you are familiar with regular expression syntax, you can use it in data step programming to find or change text patterns in character variables. The RXPARSE function parses a regular expression. Then, the RXMATCH function and the RXSUBSTR and RCHANGE CALL routines apply the actions indicated by the regular expression to a specific character value.

Spelling and Pronunciation

There are two functions that may be useful in situations where the spelling of words is uncertain. The SOUNDEX function applies Soundex encoding to a word. The SPEDIS calculates spelling distance from one word to another.

SOUNDEX — Pronunciation The Soundex algorithm is designed to identify words that sound similar. Words with similar pronunciations often have similar or identical Soundex encodings. The SOUNDEX function produces Soundex encodings for use in comparisons.

The Soundex encoding of a word is based on the first letter and any subsequent consonants. The first letter is capitalized. The subsequent consonants are categorized and encoded as digits. Vowels, other than the first letter, are disregarded.

Function call: `SOUNDEX(WORD)`
Argument: `WORD`, a word.
Return value: The Soundex encoding of the word.

SPEDIS — Spelling distance The SPEDIS function computes a score that measures the spelling distance between one word, a dictionary word, and another word, a user word. A higher score indicates more differences between the words and a lower likelihood that the dictionary word was accidentally misspelled as the user word.

Function call: `SPEDIS(USERWORD, DICTIONARYWORD)`
Arguments: `USERWORD`, a user word; `DICTIONARYWORD`, a dictionary word.
Return value: A score that measures the spelling distance, or the extent of the differences between the dictionary word and the user word.

DBCS Functions

The character functions discussed so far are designed to work with ordinary character values in which each character occupies one byte. In a DBCS value, a character can occupy two bytes, and the character functions may not work correctly. You cannot use the INDEXC function to find a two-byte character, for example, because the function treats each byte independently — and similar problems are possible with every one of the functions. To make it possible to do the same things with DBCS values, there is a set of DBCS functions. The names of these functions begin with the letter K.

Most of the DBCS functions duplicate the effects of the standard character functions:

KCOMPRESS	KSUBSTR	KINDEX
KLEFT	KTRANSLATE	KINDEXC
KLOWCASE	KTRIM	KLENGTH
KREVERSE	KUPCASE	KSCAN
KRIGHT		KVERIFY

The only differences are these:

- The names start with the letter K.
- The functions consider characters that do not necessarily occupy just one byte.
- They count positions as character positions, rather than byte positions.

There are three more functions that allow other actions on DBCS values. The KCOMPARE function compares two values. The KCOUNT function counts the characters in a value. The KUPDATE replaces a part of a DBCS character variable. Finally, the two functions KSUBSTRB and KUPDATEB do the same actions as the KSUBSTR and KUPDATE functions, but measure positions in bytes rather than characters.

Time Computation

Time computations can be complicated because of the way time is measured with so many different units of measurement and because some of the units are not a fixed size. Things are simplified somewhat in the SAS environment because SAS date, SAS datetime, and SAS time values use only days and seconds as their units. It is still necessary to convert between these and other ways of measuring and indicating time, and most of the SAS functions that have to do with time are about converting between one kind of measurement and another.

DATE, DATETIME, TIME — The current time The DATE, DATETIME, and TIME functions allow a program to measure the time during which it runs. They return the current time as a SAS date, SAS datetime, and SAS time value. These functions have no arguments, but the parentheses are still required as part of the syntax of the function call.

Function call: `DATE()`
Return value: The current date as a SAS date value.

Function call: `DATETIME()`
Return value: The current date and time as a SAS datetime value.

Function call: `TIME()`
Return value: The current time of day as a SAS time value.

MDY, YYQ, HMS, DHMS — Constructing a value These functions can be used to construct a SAS date, SAS time, or SAS datetime value from calendar and clock elements.

Function call: MDY(MONTH, DAY, YEAR)
Arguments: MONTH, a month number; DAY, a day of the month; YEAR, a year number.
Requirement: The arguments must correspond to a date of the Gregorian calendar.
Return value: The SAS date value of the day that is identified by the indicated year, month, and day.

Function call: YYQ(YEAR, QUARTER)
Arguments: YEAR, a year number; QUARTER, a quarter number.
Return value: The SAS date value of the first day of the indicated quarter.

Function call: HMS(HOUR, MINUTE, SECOND)
Arguments: HOUR, an hour number of the 24-hour clock; MINUTE, a minute number, from 0 to 59; SECOND, a second number, from 0 to 59.
Return value: The time of day indicated by the hour, minute, and second, as a SAS time value.

Function call: DHMS(DATE, HOUR, MINUTE, SECOND)
Arguments: DATE, a SAS date value; HOUR, an hour number of the 24-hour clock, from 0 to 23; MINUTE, a minute number, from 0 to 59; SECOND, a second number, from 0 to 59.
Return value: The SAS datetime value of the indicated date and time of day.

DATEPART, TIMEPART — Date and time of day The DATEPART and TIMEPART functions convert a SAS datetime value to a SAS date value and a SAS time value.

Function call: DATEPART(DATETIME)
Argument: DATETIME, a SAS datetime value.
Return value: The SAS date value of the day indicated in the SAS datetime value. The return value is a whole number.

Function call: TIMEPART(DATETIME)
Argument: DATETIME, a SAS datetime value.
Return value: The SAS time value of the time of day indicated in the SAS datetime value.

YEAR, QTR, MONTH, WEEKDAY, DAY — Calendar elements These functions provide individual calendar elements for a SAS date value.

Function call: YEAR(DATE)
Argument: DATE, a SAS date value.
Return value: The year of the SAS date value.

Function call: `QTR(DATE)`
Argument: `DATE`, a SAS date value.
Return value: The quarter of the SAS date value.

Function call: `MONTH(DATE)`
Argument: `DATE`, a SAS date value.
Return value: The month number of the SAS date value.

Function call: `WEEKDAY(DATE)`
Argument: `DATE`, a SAS date value.
Return value: The number of the day of the week of the SAS date value. Days of the week are counted starting with Sunday = 1.

Function call: `DAY(DATE)`
Argument: `DATE`, a SAS date value.
Return value: The day of the month of the SAS date value.

To get this same information for a SAS datetime value, use the DATEPART function to extract a SAS date value from the SAS datetime value, for example:

```
YEAR(DATEPART(DATETIME))
```

HOUR, MINUTE. SECOND — Clock elements The HOUR, MINUTE, and SECOND functions return the clock elements of a SAS time or SAS datetime value.

Function call: `HOUR(TIME)`
Argument: `TIME`, the time of day indicated by a SAS time value or SAS datetime value.
Return value: The hour of the 24-hour clock time. The return value is a whole number.

Function call: `MINUTE(TIME)`
Argument: `TIME`, the time of day indicated by a SAS time value or SAS datetime value.
Return value: The minute of the clock time. The return value is a whole number.

Function call: `SECOND(TIME)`
Argument: `TIME`, the time of day indicated by a SAS time value or SAS datetime value.
Return value: The second of the clock time.

Time Interval Arithmetic

Two functions, INTCK and INTNX, do arithmetic based on fixed time intervals. INTCK counts the number of months, days, or other intervals between two values. INTNX adds time intervals to a value to get another value.

Both functions use time interval codes to indicate the time unit of the intervals involved in the calculation. Examples of time interval codes are 'MONTH', 'DAY', and 'HOUR'. The time interval names can be modified by multiplier and shift arguments, resulting in many possible time interval codes. Some time interval codes work with SAS date values, others with SAS datetime values, and the rest with both SAS datetime and SAS time values.

INTCK — Elapsed time The INTCK function measures the elapsed time between two values that represent points in time. The value it returns is always a whole number because it only counts whole time intervals. For example, if the two values are the same interval, the return value is 0, because there are no time interval starting points between the two values.

Function call: INTCK(INTERVAL, VALUE1, VALUE2)
Arguments: INTERVAL, a time interval code; VALUE1 and VALUE2, two SAS date, SAS datetime, or SAS time values.
Return value: The number of time intervals from the time of VALUE1 to the time of VALUE2. The return value is always a whole number.

INTNX — Time offsets The INTNX function adjusts a value forward or backward by an indicated number of time intervals. The value it returns is the beginning of a time interval.

Function call: INTNX(INTERVAL, VALUE, N)
Arguments: INTERVAL, a time interval code; VALUE, a SAS date, SAS datetime, or SAS time value; N, a whole number of time intervals.
Return value: The starting point of the time interval adjusted N intervals from the time of VALUE. Use a positive value for N to return a value later than the value argument, a negative value for N to return a value earlier than the value argument, or a zero value for N to return the start of the interval that contains the value argument.

There is an optional fourth argument for INTNX that can make the function return a value from the middle or end of an interval, instead of the beginning. The value for this argument is:

- 'B' or 'BEGINNING' to return a value at the beginning of the interval, which is the default
- 'M' or 'MIDDLE' to return a value in the middle of the interval, a value one half of an interval later than the default value that is returned
- 'E' or 'END' to return a value at the end of the interval, a value that is one interval later than the default value that is returned.

The following table shows examples of the INTCK and INTNX functions.

Function Call	Return Value
INTCK('MONTH', '14MAR2001'D, '04JUN2001'D)	3
INTCK('MONTH', '04JUN2001'D, '14MAR2001'D)	-3
INTNX('YEAR', '25DEC2000'D, 1)	'01JAN2001'D
INTNX('QTR', '20JUN2000'D, 5)	'01JUL2001'D
INTNX('QTR', '20JUN2000'D, -5)	'01JAN1999'D
INTNX('QTR', '20JUN2000'D, 0)	'01APR2000'D

Time Interval Codes

Time interval codes are used the INTCK and INTNX functions and in some procs. The time interval codes are based on interval names, which indicate the interval and the kind of value. The table on the following page lists the interval names.

If an interval name by itself does not provide the time intervals you want to work with, you may be able to construct the appropriate interval by modifying an interval name with a multiplier, a shift value, or both. The form of a time interval code is one of the following:

NAME
NAME*m*
NAME.*s*
NAME*m*.*s*

where NAME is a time interval name, *m* is a counting number that indicates a multiplier, and *s* is a counting number that indicates a shift of the start of the interval. The multiplier indicates an interval that is a multiple of the base unit that the interval indicates. For example, MONTH2 indicates a two-month interval. The shift value determines the starting point of the interval.

The default is a shift value of 1, which sets the starting point of the interval at the point where it most naturally falls on the calendar or clock. For example, MONTH4 or MONTH4.1 intervals begin in January, May, and September. If the shift value is 2 or more, the starting point of the interval is set at the indicated time unit within the calendar or clock interval. For example, MONTH4.3 intervals start in March, July, and November, two months later than the MONTH4.1 intervals.

The units that the shift value counts are different for different time interval names. For most, such as MONTH, the unit is the base unit that the name indicates. For time intervals of a quarter or longer, the shift unit is a month. For these time interval names, it is possible to have a shift value without having a multiplier. For example, QUARTER.2 intervals are quarters that start in February, May, August, and November — at the second month of the calendar quarter.

The amount of the shift must be at least 1 and no greater than the total length of the interval. For most interval names, this restricts the shift value to be less than or equal to the multiplier value.

Time Interval Names

Interval Length	Starting Point	Interval Name	Kind of Value
Second	At clock second	SECOND	SAS datetime SAS time
Minute	At clock minute	MINUTE	SAS datetime SAS time
Hour	At clock hour	HOUR	SAS datetime SAS time
Day	Midnight	DAY	SAS date
		DTDAY	SAS datetime
Selected days of the week[1]	Each selected day	WEEKDAY WEEKDAY*n*...W[1]	SAS date
		DTWEEKDAY DTWEEKDAY*n*...W[1]	SAS datetime
Week	Sunday	WEEK	SAS date
		DTWEEK	SAS datetime
Third month (8–11 days)	Days 1, 11, 21 of month	TENDAY	SAS date
		DTTENDAY	SAS datetime
Half month (13–16 days)	Days 1, 16 of month	SEMIMONTH	SAS date
		DTSEMIMONTH	SAS datetime
Month (28–31 days)	Day 1 of month	MONTH	SAS date
		DTMONTH	SAS datetime
Quarter (90–92 days)	Jan. 1, Apr. 1, Jul. 1, Oct. 1	QTR	SAS date
		DTQTR	SAS datetime
Half year (181–184 days)	January 1, July 1	SEMIYEAR	SAS date
		DTSEMIYEAR	SAS datetime
Year (365–366 days)	First day of year	YEAR	SAS date
		DTYEAR	SAS datetime

[1] Select days of the week with digits between WEEKDAY and W in the interval name. The digits 1 through 7 represent the days Sunday through Saturday. For example, WEEKDAY246W or DTWEEKDAT246W selects Monday, Wednesday, and Friday. WEEKDAY is the same as WEEKDAY23456W, and DTWEEKDAY is the same as DTWEEKDAY23456W, to select the days of the week Monday through Friday.

When multipliers are used with interval names other than WEEK and DTWEEK, the starting point for the multi-period interval is determined by counting from 0, which is the SAS date or SAS datetime value of January 1, 1960, or the SAS time value of midnight. For example, YEAR20 intervals start in 1960, 1980, 2000, and so on. As another example, MINUTE15 intervals start at the quarter-hour points of the clock.

Multiple-week intervals are counted from December 27, 1959, which is the SAS date value –5 and is the Sunday of the week that contains January 1, 1960. This means, for example, that WEEK2 intervals start at SAS date values that are 5 less than a multiple of 14 — days such as January 10, 1960, which is SAS date value 9; January 24, 1960, which is SAS date value 23; and so on.

Other Categories of Functions

There are other functions that have more limited or specialized uses. These functions involve such subjects as finance, geographical identifying codes, queues, and bitwise logical operations.

Financial Functions

The SAS System includes several functions for depreciation, loan, and value calculations. These financial functions are used to describe and analyze the connections between payments or other amounts of money made at various different times, usually in a regular, periodic sequence. The periods in these functions are usually months or years, but they can be any time period you choose.

Interest rates in these functions are usually expressed as periodic fractions. For example, if the period you use with the function is a month, an interest rate value of .0120 means an interest rate of 1.20 percent per month.

Compound interest The COMPOUND function computes compound interest on a simple loan, bond, or savings account that does not have periodic payments.

> Function call: COMPOUND(A, F, R, N)
>
> Model: A simple loan with compound interest.
>
> Arguments: A, an initial value; F, a final value; R, a periodic interest rate; N, a number of periods. Provide values for three of these parameters, and provide a missing value for the remaining parameter.
>
> Return value: The function computes the parameter for which a missing value is supplied as an argument.

For example, COMPOUND(1000, 2000, ., 10) computes the effective annual interest rate that, with compound interest, would increase the value of a savings account from 1000 to 2000 in 10 years. The value it returns is approximately 0.0718, which would commonly be expressed as 7.18 percent.

Periodic payments Several functions analyze different properties of regular, periodic payments. In some functions the payments are of equal amounts; in others, each payment can be a different amount.

> Function call: SAVING(F, P, R, N)
>
> Model: A series of equal periodic payments with compound interest.
>
> Arguments: F, a final value, one period after the last payment; P, a periodic

payment amount; R, a periodic interest rate; N, a number of periods. Provide values for three of these parameters, and provide a missing value for the remaining parameter.
Return value: The function computes the parameter for which a missing value is supplied as an argument.

Function call: MORT(A, P, R, N)
Model: A loan that is repaid with fixed periodic payments.
Arguments: A, an initial value; P, a periodic payment amount; R, a periodic interest rate; N, a number of periods. Provide values for three of these parameters, and provide a missing value for the remaining parameter.
Return value: The function computes the parameter for which a missing value is supplied as an argument.

Function call: INTRR(M, P1, P2, P3, . . .)
Model: A series of cash flows representing payment and repayment.
Arguments: M, the number of periods per year; P1, P2, P3, . . . , a series of payment amounts, in which some are positive and some are negative.
Return value: The internal rate of return, expressed as an annual fraction.

Function call: IRR(M, P1, P2, P3, . . .)
Model: A series of cash flows representing payment and repayment.
Arguments: the same as for INTRR.
Return value: The internal rate of return, expressed as an annual percent.

Function call: NETPV(R, M, P1, P2, P3, . . .)
Model: A series of cash flows representing payment and repayment.
Arguments: R, an annual interest rate fraction; M, P1, P2, P3, . . . , the same as for INTRR.
Return value: The net present value.

Function call: NPV(R, M, P1, P2, P3, . . .)
Model: A series of cash flows representing payment and repayment.
Arguments: R, an annual interest rate percent; M, P1, P2, P3, . . . , the same as for INTRR.
Return value: The net present value.

Six more financial functions, PVP, YIELDP, DUR, DURP, CONVX, and CONVXP, provide more technical measures of models involving periodic payments.

Depreciation Depreciation is an accounting mechanism to estimate the decline in value of a piece of equipment or other capital asset over time. At the end of the depreciation process, the asset is estimated to have no value remaining. Separate functions compute the accumulated depreciation and the depreciation expense for a period for several common depreciation methods.

Function call: DACC*method*(AGE, VALUE, *parameters*) where *method* is a code that identifies a depreciation method and *parameters* are the parameters for that method, as indicated in the table below.

Arguments: AGE, the number of periods since depreciation began; VALUE, the initial value to depreciate.

Return value: The accumulated depreciation after the number of periods indicated.

Function call: DEP*method*(AGE, VALUE, *parameters*) where *method* is a code that identifies a depreciation method and *parameters* are the parameters for that method, as indicated in the table below.

Arguments: AGE, the number of periods since depreciation began; VALUE, the initial value to depreciate.

Return value: The depreciation expense in the period indicated.

If the AGE argument is not a whole number, the function does a linear interpolation of depreciation between the beginning and end of the period. The DBSL method requires the AGE argument to be a whole number. The following table lists the depreciation methods and their codes and parameters.

Depreciation Method	Code	Parameters
Declining balance	DB	recovery period, rate
Declining balance switching to straight line	DBSL	recovery period, rate
Straight line	SL	recovery period
Sum of years digits	SYD	recovery period
Table of rates	TAB	rate for period 1, rate for period 2, . . .

For example, DACCDBSL(3, 4898, 7, 2) calculates the depreciation of the first 3 years of an asset whose original value is 4,898, using the 200% declining balance switching to straight line method with a recovery period of 7 years.

Geographical Functions

The geographical functions may be useful for working with geographical data from the United States. They convert between various identifying codes and state names. The arguments and return values of the functions are these kinds of values:

- *ZIP codes.* These are the five-digit codes that identify postal zones, as character values. In the name of the function, this appears as the letters ZIP.
- *Postal codes.* These are the two-letter postal codes for the states, such as 'NY' and 'FL'. In the name of the function, this appears as the letters ST or STATE.
- *State names.* These are the names of the states, which can be in uppercase or mixed-case letters. This appears as NAME or NAMEL in the name of the function.

- *FIPS state codes.* These are numeric codes that identify states in some U.S. census data. This appears as FIPS in the name of the function.

The following table shows the argument and return value for each function.

Return value	Function Calls
Postal code	ZIPSTATE(*ZIP code*) FIPSTATE(*FIPS state code*)
State name	STNAMEL(*postal code*) ZIPNAMEL(*ZIP code*) FIPNAMEL(*FIPS state code*)
State name in uppercase letters	STNAME(*postal code*) ZIPNAME(*ZIP code*) FIPNAME(*FIPS state code*)
FIPS state code	STFIPS(*postal code*) ZIPFIPS(*ZIP code*)

These functions create error condition if you call them with invalid arguments. This occurs most often when a ZIP code in a data value is not in the SAS System's list of ZIP codes.

Bitwise Logical Functions

The bitwise logical functions apply logical operations to the individual bits of integer values. These bitwise logical operations can be an efficient way to deal with some operating system codes that are displayed as numbers, but actually represent bitfields, in which each bit has an independent meaning. For example, the code value 25, which in the binary numbering system is written as 11001, might actually represent the eight separate binary codes 0, 0, 0, 1, 1, 0, 0, 1. Arguments to the bitwise logical functions should be whole numbers less than 2^{32}. The functions use one or two arguments, depending on whether the logical operator works on one or two operands.

Function Call	Logical Operation
BAND(*n*, *n*)	AND
BOR(*n*, *n*)	OR
BNOT(*n*)	NOT
BXOR(*n*, *n*)	Exclusive OR
BLSHIFT(*n*, *distance*)	Shift left by the distance indicated
BRSHIFT(*n*, *distance*)	Shift right by the distance indicated

Queue

The LAG and DIF functions work by keeping a queue, or list, of the values that are supplied as arguments each time the function is called. The values in the queue determine the values that the functions return. The LAG functions simply return the value of a previous argument. The DIF functions require numeric arguments and return the different between the current argument and a previous argument.

The function name includes a numeric suffix from 1 to 100 that indicates the size of the queue. For example, LAG2 is the name of the LAG function with a queue of two items. The LAG1 and DIF1 functions can also be written as LAG and DIF, respectively.

Function call: `LAGn(VALUE)` where *n* is a counting number from 1 to 100.
Argument: `VALUE`, a value. It can be character or numeric.
Return value: The value supplied as an argument in the *n*th previous execution of the function call. The function returns a missing or blank value the first *n* times the function call is executed.

Function call: `DIFn(VALUE)` where *n* is a counting number from 1 to 100.
Argument: `VALUE`, a numeric value.
Return value: The difference of the value supplied as an argument in the current execution of the function call and the value supplied as an argument in the *n*th previous execution of the function call. The function returns a missing value the first *n* times the function call is executed.

If you use more than one call to a LAG or DIF function in the same data step, the separate function calls are completely independent. Each function call maintains its own queue and returns values that are based only on its own arguments.

Functions by Subject

In addition to the functions mentioned so far, there are several categories of functions that are covered elsewhere in this book.

- *Variable information functions.* These functions return information from the attributes of variables in the program. See chapter 5, "Variables and Values."
- *Informat and format functions.* These functions make it possible to use informats and formats in expressions. See the end of chapter 13, "Informats and Formats."
- *Environmental functions and CALL routines.* These functions provide interfaces to the operating system, objects in the SAS session, and the macro processor. See the end of chapter 4, "Control Flow."
- *Array functions.* These functions return the dimensions and subscript boundaries of an array. See chapter 8, "Loops and Arrays."
- *I/O functions and CALL routines.* These functions read and write files. See the next chapter.

16

Low-Level I/O Programming

SAS I/O statements are powerful and easy to use, but sometimes you need a more flexible way of reading and writing files. These are examples of things that cannot be done easily with I/O statements:

- Read a file whose name is determined by program logic.
- Use a data step to analyze data from a SAS data file in which the names or number of variables is not known in advance.
- Check for observations in a SAS data file and read data from the file only if it contains observations, executing an alternative action if it does not contain any observations.
- Read a directory and process the files in the directory.

For tasks such as these, SAS provides an extensive set of I/O functions.

The flexibility of file I/O programming with functions comes with a price: you have to manage more of the details of the I/O process. This is why I use the term *low-level* to describe I/O programming with functions. "Low-level" does not imply any judgement about the value or quality of this style or programming; it only means that the programmer is directly involved in more of the specific actions that the program takes when it executes.

In data step programming, which is what is described here, I/O functions can read and write text files and can read SAS datasets and directories, but their capabilities do not extend much beyond that. The functions described here have additional capabilities, and there are additional functions as well, in SAS Component Language (SCL). SCL is the programming language used to develop interactive applications with SAS/AF. In SCL, I/O functions can also create, update, and modify SAS data files, and they have some abilities to manage SAS files generally.

Comparison to C

The idea of functions that read and write files should be familiar to anyone who has used the C programming language. C was the first major language to use functions, rather than statements, for all I/O operations. The SAS I/O functions resemble the standard C functions that do similar things.

"Low-level" is a relative term. C has two sets of I/O functions, which are described as low-level and high-level. The SAS I/O functions are low-level when compared to the SAS I/O statements. However, they are more similar to the *high-level* I/O functions of C. They still provide some abstraction to protect the programmer from the finer details of file I/O processing.

Concepts

The low-level approach requires you to work directly with parts of the file I/O process that you otherwise might overlook. These are some of the concepts you will need:

Directory When you provide a name or path in order to access a file, it might look like the name or path takes you directly to the file. Actually, the name or path takes you only as far as the directory where the file is listed. The file system first has to look up the name of the file in the directory in order to find the actual location of the file on the storage volume. Only then can it access the file itself.

The directory matches up file names with physical locations, and it does more than that. Depending on the file system, the directory can contain various information items about a file. These can include such things as the size of the file, creation and modification dates, the ID of the user, job, or agent that created or modified the file, security or file sharing flags that determine who can access the file, and alternate names. Directories contain this kind of information to help you find and manage files.

Open You can access the data in a file only while the file is open. The computer keeps files closed most of the time and opens a file only when you specifically request it.

Why is this necessary? Think of the old-fashioned file folders that contain files in file cabinets. A closed file folder is protected; the file is not likely to be accidentally changed in any way. Also, there is a limit to how many file folders you can have open at once, based on how much work space you have available. And even if you had unlimited space, you would still not want to open too many folders at once, because it would be hard to keep track of them all.

Similar considerations apply to computer files. The operating system of a computer has to keep track of all the open files. The more files are open, the more memory space and processing time this takes. Closing a file when you are done with it in a program prevents the possibility that a subsequent logical error in the program will alter the file in an unintended way. A closed file is also much less likely to be damaged in the event that the SAS application or the operating system crashes.

One reason the operating system has to keep track of all open files is to prevent two programs from opening the same file simultaneously in a way that might conflict, resulting in possible damage to the file. If a file is already open, and you attempt to open it again, it can result in an error condition. This can happen even though it was another part of the same program that left the file open.

There is an even more important reason to close files when you are finished with them. There is a limit to the number of files an operating system can open at the same time. If you accidentally leave too many files open, the operating system can no longer open any more files. At that point, a computer cannot do much of anything, and you may have to restart the SAS application or reboot the computer.

When you access a file in low-level I/O programming, you always start by opening the file and end by closing it. Each action is a simple function call. For example, you can open a SAS data file with the OPEN function and close it with the CLOSE function. If you do not close files that you open in a data step, the SAS supervisor automatically closes them when the data step ends. If you open files in macro programming and do not close them, the SAS supervisor closes them when the SAS session ends.

Access modes When you open a file in a program, you can limit the way the program can access the file. The different sets of limitations you can choose are called *access modes* or *open modes*.

Among the available modes, the most important distinction to make is among input, update, and output modes. Use input mode to read a file without changing it. Use update mode when you will be changing some of the data in the file. Use output mode to create a new file or to completely replace an existing file. Another possibility is append mode, which lets you add records to the end of the file without reading the existing contents of the file. Other options in the access mode may limit access to sequential access, in which the records of the file are accessed in sequence.

File identifiers A function that opens a file establishes a connection to the file and returns a numeric code that identifies the open file. This identifier is a dataset ID for a SAS dataset, a file ID for a text file, or a directory ID for a directory. You must assign the identifier to a numeric variable. Then, use that variable to identify the file in all function calls that access the file, as long as the file is open.

If it fails to open the file, the function returns a value of 0. This value of 0 for a file identifier always indicates that the attempt to open the file failed. A zero value is never an identifier for a file.

Buffers You cannot access the records of a file directly. Instead, you can work with an image of the record in memory — a buffer. When you read a record from a file, you copy the record to the buffer, then read the data from the buffer. When you write a record to a file, you first organize the data in the buffer, then copy the buffer to the file.

A SAS dataset has a special kind of buffer called a *dataset data vector*, or DDV, which is organized as a set of variables. When you copy an observation from a SAS dataset to the DDV, the DDV takes on the values of the variables in that observation. Variables in the DDV also have all the attributes of the variables in the SAS dataset. It is important not to confuse the DDV with the PDV, the program data vector, which contains the variables of the data step program. Variables in the DDV and PDV can have the same names, but they are not the same variables, and they can different values at the same time. You can copy values from the DDV to the PDV, but it takes a specific action to do so.

Noting a record You can use functions to mark the current record in a file. The function returns a numeric identifier called a *note ID*. The note ID value lets you return to that same record later.

Return Codes

Most I/O functions return integer values. These *return codes* are generated by the file system to indicate the result of an action on a file. Each operating system has a different set of return codes. However, they are almost always error codes, which means that a value of 0 indicates the successful completion of an action, and different nonzero values indicate different reasons for failure or possible problems. In some systems, positive values are considered errors, and negative values are considered warnings. Sometimes, larger values represent more serious problems.

Some I/O functions generate return codes but do not return them because they have other values to return. After such a function call, you can use the SYSRC function to get the return code. The SYSRC function returns the most recent return code.

To find out the meaning of a return code, use the SYSMSG function, which returns the text of an error or warning message associated with the most recent return code. The SYSMSG function returns the message text for the most recent return code, but only once; if you call the function a second time, it returns a blank value.

Function call: `SYSRC()`

Return value: The most recent return code from the file system action of an I/O function.

Function call: `SYSMSG()`

Return value: The text of the error or warning message for the most recent return code from the file system action of an I/O function, occurring since the most recent call to the SYSMSG function.

Variables and Assignment Statements

In addition to return codes, I/O functions return such values as ID numbers, names, and attributes. Sometimes the return value of a function can be used directly as an argument to another function. Usually, though, the return value of an I/O function must be assigned to a variable so that it can be used in subsequent statements. Even when there is no reason to use the return code from a function call, the easiest way to call the function is still to assign the resulting value to a variable.

Low-level I/O programming, then, tends to take the form of a series of assignment statements, each of which is formed with a variable name, an equals sign, and a function call. These are examples of this kind of statement, using the functions described above:

```
RC = SYSRC();
MESSAGE = SYSMSG();
```

When you read this kind of code, remember that the action in a statement is really more than just the assignment of the value of an expression to a variable. The most important actions are in the function calls themselves.

SAS Dataset I/O Functions

The actions required to read data from a SAS dataset are done by various different functions. There are around 25 functions for SAS dataset I/O; about half of them are needed to read a SAS dataset.

Reading a SAS Dataset

By necessity, the actions involved in reading a SAS dataset follow almost a set sequence, which is reflected in a predictable pattern of function calls. These are the actions, in order, and the functions involved in each one:

Action	Functions
1. Define the libref (if necessary)	LIBNAME
2. Open the SAS dataset	OPEN
3. Find the variables	VARNUM, VARTYPE, ATTRN, VARNAME
4. Read an observation	FETCH
5. Retrieve the values of variables	GETVARC, GETVARN
6. Repeat (till end of file)	FETCH, GETVARC, GETVARN
7. Close the SAS dataset	CLOSE
8. Clear the libref (if no longer needed)	LIBNAME

The LIBNAME function defines librefs in much the same way as the LIBNAME statement. The OPEN function opens a SAS dataset, based on the SAS dataset name, and returns a dataset ID. If you know the name of a variable, the VARNUM function returns its variable number. Or, if appropriate, you can use a DO loop to check every variable. The ATTRN function can tell you the number of variables; variable numbers go from 1 to the number returned by ATTRN. With variable numbers, you can determine the name of a variable with the VARNAME function. It is always a good idea to check the type of a variable with the VARTYPE function.

The FETCH function reads an observation, which means it copies the values into the DDV. You can then retrieve the value of each variable using the variable number and the GETVARC or GETVARN function. Usually, you repeat the process of reading observations in a DO loop until you reach the end of the file, which is indicated by a return code of –1 from the FETCH function. When you are done reading a SAS dataset, close it with the CLOSE function. When you are done with a libref, you can clear it with the LIBNAME function. Details of each of these function calls are listed below.

This sequence approximately duplicates the effects of the SET statement. The purpose, of course, is not merely to have another, more complicated way of doing the same thing. Splitting the process of reading a SAS dataset into its component actions lets you apply any necessary programming logic to any of the actions separately.

Function call: `LIBNAME(LIBREF, FILE)`

Function call: `LIBNAME(LIBREF, FILE, ENGINE, OPTIONS)`

Arguments: `LIBREF`, a libref; `FILE`, the physical file name of a library; `ENGINE`, an optional engine name; `OPTIONS`, engine options written the same way as in the LIBNAME statement.

Action: Associates the libref with the physical file name.

Return value: A return code.

Function call: `OPEN(DATASET, MODE)`

Arguments: `DATASET`, a SAS dataset name; `MODE`, an optional code indicating an access mode. The SAS dataset can be a SAS data file or an SQL view. It can have dataset option in parentheses after the name. The access mode values are `'I'` for input, which is the default; `'IN'` for input limited to sequential access; and `'IS'` for input strictly limited to sequential access, with no possibility of returning to an observation.

Action: Opens the SAS dataset.

Return value: A dataset ID. It returns 0 if the attempt to open the file fails.

Function call: `VARNUM(DSID, NAME)`

Arguments: `DSID`, a dataset ID; `NAME`, the name of a variable in the SAS dataset.

Return value: The variable number of the variable. Variables in a SAS dataset are numbered sequentially according to their relative logical position in each observation. The function returns 0 if the variable name is not found in the SAS dataset.

Function call: `ATTRN(DSID, 'NVARS')`

Arguments: `DSID`, a dataset ID; `'NVARS'`, a code that requests the number of variables in the SAS dataset.

Return value: The number of variables in the SAS dataset.

Function call: `VARNAME(DSID, N)`

Arguments: `DSID`, a dataset ID; `N`, the variable number of a variable in the SAS dataset, which can be a number from 1 to the number of variables in the SAS dataset.

Return value: The name of the variable.

Function call: `VARTYPE(DSID, N)`

Arguments: `DSID`, a dataset ID; `N`, a variable number.

Return value: A code indicating the data type of the variable: `'C'` for a character variable, `'N'` for a numeric variable.

Function call: `FETCH(DSID)`

Argument: `DSID`, a dataset ID.

Action: Reads the next observation from the SAS dataset. It copies the values of variables from the observation to the DDV.

Return value: A return code. It returns –1 for an end of file condition.

Function call: `GETVARC(DSID, N)`

Arguments: `DSID`, a dataset ID; `N`, the variable number of a character variable.

Return value: The value of the variable in the DDV.

Function call: GETVARN(DSID, N)
Arguments: DSID, a dataset ID; N, the variable number of a numeric variable.
Return value: The value of the variable in the DDV.

Function call: CLOSE(DSID)
Argument: DSID, a dataset ID.
Action: Closes the SAS dataset.
Return value: A return code.

Function call: LIBNAME(LIBREF, '') or LIBNAME(LIBREF)
Arguments: LIBREF, a libref; '', a null string (in most operating systems, a blank value can also be used).
Action: Clears the libref.
Return value: A return code.

The following example puts these function calls together to form a data step that reads a SAS dataset.

```
*
  read1.sas
  Reads SAS dataset using I/O functions
  Rick Aster    May 2000
*;
DATA _NULL_;
  LENGTH NAME $ 60 MSG $ 75;
  FILE LOG;
*
  Define libref.
*;
  RC = LIBNAME('MAIN1', 'main');
  IF RC > 0 THEN GOTO E1;
*
  Open SAS dataset.
*;
  DSID = OPEN('MAIN1.LIST');
  IF NOT DSID THEN GOTO E2;
*
  Find variables NAME and NUMBER.
*;
  NAME_VAR = VARNUM(DSID, 'NAME');
  NUM_VAR = VARNUM(DSID, 'NUMBER');
  IF NOT (NAME_VAR AND NUM_VAR) THEN GOTO E3;
  IF NOT (VARTYPE(DSID, NAME_VAR) = 'C' AND VARTYPE(DSID, NUM_VAR) = 'N')
     THEN GOTO E4;

*
  Process observations.
*;
  DO WHILE (1);
    RC = FETCH(DSID);
    IF RC THEN LEAVE; * Check for end of file or error;
    NAME = GETVARC(DSID, NAME_VAR);
```

```
    NUMBER = GETVARN(DSID, NUM_VAR);
    PUT NUMBER NAME;
    END;

*
  Close SAS dataset and clear libref.
*;
CLOSE:
  RC = CLOSE(DSID);
  RC = LIBNAME('MAIN1');
  STOP;

*
  Error handling.
*;
E1:
  PUT 'Unable to find main library. Return code ' RC;
  MSG = SYSMSG();
  IF MSG NE '' THEN PUT MSG;
  STOP;

E2:
  RC = SYSRC();
  PUT 'Unable to open LIST member. Return code ' RC;
  MSG = SYSMSG();
  IF MSG NE '' THEN PUT MSG;
  STOP;

E3:
  PUT 'Variables NAME and NUMBER not found.';
  GOTO CLOSE;

E4:
  PUT 'Wrong data type for variable NAME or NUMBER.';
  GOTO CLOSE;
RUN;
```

This program shows a style of programming that is very different from what you might expect to see in a data step. Note these specific points about the program:

- The LENGTH statement sets the lengths of character variables whose values come from function calls.
- The variable RC is used for return codes from functions.
- IF-THEN statements check for problems in the values returned by functions. If they find problems, they branch to the statement labels E1, E2, . . . , at the end of the step.
- The program uses the libref MAIN1 rather than MAIN. This way, if the libref MAIN is already defined, the program still works correctly and does not interfere with the existing libref.
- The program looks only for the two variables NAME and NUMBER. It is not affected by any other variables that might be present in the SAS dataset. A program could read any number of variables in the same

way, or it could work with all the variables it finds in a SAS dataset.

- The program checks the data type of the variables. This is important because the GETVARC and GETVARN functions, which return the values of variables, work only with variables of their respective data types.
- Observations are processed in a DO loop. With the term `WHILE (1)`, the loop repeats indefinitely until the LEAVE statement stops it.
- After the FETCH function call, the IF-THEN statement checks the value of RC, the return code from the function. A nonzero (true) value for RC indicates that the function did not return an observation, either because an error occurred or the end of file was reached. When that happens, the LEAVE statement branches out of the loop. In some applications, it might be necessary to check separately for the end of file (indicated by return code –1) and other specific return code values.
- The PUT statement writes the values of the variables in the log. A typical program would use the variables in a more complicated way at this point in the program.
- Most of the logic of the program has to do with possible error conditions, and this is usually the case in this kind of program. The error handling and log messages make it much easier to identify problems that can occur in the files that a program works with. For some applications, more elaborate error-handling logic might be necessary. For other applications in which file problems are unlikely, error-handling logic might not be necessary.
- For error conditions that occur in a program after a file has been opened, the error-handling logic should include closing the file. That is the purpose of the CLOSE label in this program.

Actions on Observations

These functions and the `CALL SET` routine take other actions on observations in an open SAS dataset.

Function call: `CUROBS(DSID)`
Argument: `DSID`, a dataset ID.
Return value: The logical observation number of the current observation in the SAS dataset. If the library engine does not support logical observations for a SAS data file, the function returns a missing value. In a view, the function returns the sequential observation number generated by the view.

Function call: `NOTE(DSID)`
Argument: `DSID`, a dataset ID.
Action: Creates a note that marks the current observation.
Return value: A note ID.

Function call: `POINT(DSID, NOTE)`
Arguments: `DSID`, a dataset ID; `NOTE`, a note ID in that SAS dataset.
Action: Moves to the indicated observation.
Return value: A return code.

Function call: `DROPNOTE(DSID, NOTE)`
Arguments: `DSID`, a dataset ID or file ID; `NOTE`, a note ID in that SAS dataset or file.
Action: Removes the note from memory.
Return value: A return code.

Function call: `REWIND(DSID)`
Arguments: `DSID`, a dataset ID.
Action: Moves to the beginning of the SAS dataset.
Return value: A return code.

Statement: `CA:LL SET(DSID);`
Argument: `DSID`, a dataset ID.
Action: Creates an association between the DDV and any program variables of the same names, so that reading an observation into the DDV with the FETCH or FETCHOBS functions results in the values of the variables also being assigned to the variables in the program.
Return value: A return code.

Function call: `FETCH(DSID, 'NOSET')`
Arguments: `DSID`, a dataset ID; `'NOSET'`, an option code that overrides the effects of the CALL SET routine.
Action: Reads the next observation from the SAS dataset. It copies the values of variables from the observation to the DDV. It does not copy the values of variables to the program variables of the same names, even if the CALL SET routine has previously been called.
Return value: A return code. It returns –1 for an end of file condition.

Function call: `FETCHOBS(DSID, N, OPTIONS)`
Arguments: `DSID`, a dataset ID; `N`, an observation number; `OPTIONS`, a value that can contain either of the two options noted below.
Action: Reads observation `N` from the SAS dataset. It copies the values of variables from the observation to the DDV.
Options: `NOSET`, overrides the effects of the CALL SET routine. The function does not copy the values of variables to the program variables of the same names, even if the CALL SET routine has previously been called. `ABS`, uses physical rather than logical observation numbers.
Return value: A return code. It returns –1 for an end of file condition.

Variable Attributes

The VARNUM, VARNAME, and VARTYPE functions have already been mentioned. They belong to a set of functions that return attributes of variables in an open SAS dataset.

The names of these functions begin with VAR, followed by an abbreviation for the attribute. The arguments are the dataset ID and the variable number. The one exception is the VARNUM function, which uses the variable name and returns the variable number. The following list shows the functions with their arguments and return values.

Function Call[1]	Return Value
VARNUM(DSID, NAME)	The variable number.
VARNAME(DSID, N)	The name of the variable.
VARTYPE(DSID, N)	A code indicating the data type of the variable: 'C' for a character variable, 'N' for a numeric variable.
VARLEN(DSID, N)	The length of the variable.
VARLABEL(DSID, N)	The label of the variable.
VARFMT(DSID, N)	The format attribute of the variable.
VARINFMT(DSID, N)	The informat attribute of the variable.

[1] Arguments: DSID, a dataset ID; N, the variable number; NAME, the variable name.

SAS Dataset and Library Information

There are functions to return dozens of different items of information about a SAS dataset. There are also functions for information about a library.

Function call: EXIST(SASFILE) or EXIST(SASFILE, MTYPE)
Arguments: SASFILE, a SAS file name; MTYPE, the member type. The default member type is 'DATA'.
Return value: A Boolean value that indicates whether the SAS file exists: 1 if it exists, 0 if it does not exist.

Function call: DSNAME(DSID)
Argument: DSID, a dataset ID.
Return value: The two-level SAS dataset name.

Function call: ATTRC(DSID, ATTRIBUTE)
Function call: ATTRN(DSID, ATTRIBUTE)
Arguments: DSID, a dataset ID; ATTRIBUTE, a character value indicating one of the attribute names listed in the table on the following pages.
Return value: The value of the attribute of the SAS dataset. The ATTRC function returns character values; the ATTRN function returns numeric values.

Function call: LIBREF(LIBREF)
Argument: LIBREF, a libref.
Return value: A Boolean value that indicates whether the libref is assigned: 1 if the libref is assigned, 0 if it is not.

Function call: PATHNAME(LIBREF)
Argument: LIBREF, a libref, SAS file name, or fileref.
Return value: The physical file name of the library, SAS file, or fileref.

Names for SAS Dataset Attributes

Name	*	Return Value
ALTERPW	N	A Boolean value that indicates whether the SAS dataset is alter-protected: 1 if the SAS dataset has an alter-level password, 0 if it does not.
ANOBS	N	A Boolean value that indicates whether the number of observations is known: 1 if the engine can determine the number of observations, 0 if it cannot.
ANY	N	A code indicating the presence of variables and observations: –1 if there are no variables, 0 if there are no observations, 1 if there are observations and variables.
ARAND	N	A Boolean value that indicates whether random access is supported: 1 if the engine supports random access for the SAS dataset, 0 if it does not.
ARWU	N	A Boolean value that indicates whether the engine can write to files: 1 if the engine can write, 0 if the engine is read-only.
CHARSET	C	The character set of the sort order of the SAS dataset. The value can be ASCII, EBCDIC, OEM, or ANSI.[1]
CRDTE	N	Creation datetime.
ENCRYPT	C	A word that indicates whether the SAS dataset is encrypted: YES if the SAS dataset is encrypted, NO if it is not.
ENGINE	C	The name of the engine.
ICONST	N	A code indicating the kind of integrity constraints: 0 if there are no integrity constraints, 1 if there are general integrity constraints, 2 if there are referential integrity constraints, 3 if there are both kinds of integrity constraints.
INDEX	N	A Boolean value that indicates whether indexing is possible: 1 if the engine can index the SAS dataset, 0 if it cannot.
ISINDEX	N	A Boolean value that indicates whether indexes exist: 1 if the SAS dataset has an index, 0 if it does not.
ISSUBSET	N	A Boolean value that indicates whether a WHERE clause is in effect: 1 if the SAS dataset is being accessed with a WHERE clause, 0 if it is being accessed without a WHERE clause.
LABEL	C	The dataset label.
LIB	C	The libref.
LRECL	N	The logical record length of the SAS dataset.
LRID	N	The record ID length of the SAS dataset.
MAXGEN	N	Maximum number of generations.[2]
MODTE	N	The datetime when the SAS dataset was last modified.
MEM	C	The member name.
MODE	C	The access mode. The value is a code that could be used as the access mode option in the OPEN function.
MTYPE	C	The member type.

continued

Name	*	Return Value
NDEL	N	The number of deleted observations (that are still physically present in the SAS data file).
NEXTGEN	N	The next generation number.[2]
NLOBS	N	The number of logical observations.[3]
NLOBSF	N	The number of observations available after applying FIRSTOBS= and OBS= options and WHERE clauses. To determine this attribute value, the function may have to read the entire SAS dataset.
NOBS	N	The number of physical observations, which is the sum of the logical observations and deleted observations.[3]
NVARS	N	The number of variables.
PW	N	A Boolean value that indicates whether the SAS dataset is password-protected: 1 if the SAS dataset has a password, 0 if it does not.
RADIX	N	A Boolean value that indicates whether the SAS dataset is accessible by observation number: 1 if the engine can access the SAS dataset by observation number, 0 if it cannot.
RANDOM	N	Alias for the ARAND attribute.
READPW	N	A Boolean value that indicates whether the SAS dataset is read-protected: 1 if the SAS dataset has a read-level password, 0 if it does not.
SORTEDBY	C	The sort order clause, written the way it would appear in the BY statement of the SORT proc.[1]
SORTLVL	C	A word indicating how it was determined that the SAS dataset is sorted: WEAK if the sort order was claimed by a program, STRONG if the sort order was created by SAS processing.[1]
SORTSEQ	C	The name of the alternate collating sequence used to sort the file.[1]
TAPE	N	A Boolean value that indicates whether the SAS dataset is a sequential file: 1 if it is a sequential file, 0 if it is not.
TYPE	C	The dataset type.
VAROBS	N	Alias for the ANY attribute.
WHSTMT	N	A code that indicate the type of WHERE clause that is in effect: 0 if there is no WHERE clause, 1 if there is a permanent WHERE clause, 2 if there is a temporary WHERE clause, 3 if both kinds of WHERE clauses are in effect.[4]
WRITEPW	N	A Boolean value that indicates whether the SAS dataset is write-protected.

* Data type: C=character, N=numeric.

[1] The return value is blank if the SAS dataset is not sorted.

[2] The return value is 0 if the SAS dataset is not a generation dataset.

[3] The return value is –1 if the engine cannot determine the value.

[4] A permanent WHERE clause is applied when the SAS dataset is opened. A temporary WHERE clause is applied after the SAS dataset is open.

Text File I/O Functions

Reading a text file involves nearly the same process as reading a SAS data file, although a completely different set of functions is involved. You can also write and update text files. The same processes can be used for binary files.

Reading a Text File

The simplest way to access a text file is to read each record of the file, in sequence. These are the actions and the functions that are required:

Action	Functions
1. Define the fileref (if necessary)	FILENAME
2. Open the SAS dataset	FOPEN
3. Read a record	FREAD
4. Retrieve the text of the record	FGET
5. Repeat (till end of file)	FREAD, FGET
6. Close the file	FCLOSE
7. Clear the fileref (if it is no longer needed)	FILENAME

There are no variables in a text file, but other than that distinction, the actions for reading a text file match those described previously for reading a SAS dataset. The FILENAME function defines a fileref for the file. The FOPEN function opens the file, based on the fileref, and returns a file ID. The FREAD function reads a record; it copies data from the file to a buffer called the file data buffer, or FDB. The FGET function extracts data from the FDB.

A return code of –1 in the FREAD function indicates the end of the file. The FCLOSE function closes the file, and the FILENAME function clears the fileref. Details of the function calls are listed below.

Function call: FILENAME(FILEREF, FILE)
Function call: FILENAME(FILEREF, FILE, DEVICE, OPTIONS)
Arguments: FILEREF, a fileref; FILE, a physical file name; DEVICE, an optional device name; OPTIONS, host options.
Action: Associates the fileref with the physical file name. If the fileref argument is a blank character variable, the function generates a new fileref and assigns it to that variable. A fileref can be assigned even if the physical file does not exist.
Return value: A return code.

Function call: FOPEN(FILEREF)
Function call: FOPEN(FILEREF, MODE)
Function call: FOPEN(FILEREF, MODE, RECLEN, RECFM)
Arguments: FILEREF, the fileref of a text file; MODE, an optional code indicating an access mode; RECLEN, the record length; RECFM, an optional code indicating the record format. The access mode values are 'I' for input, which is the default, 'S' for sequential input, 'U' for update, 'W' for sequential update, 'O' for output, and 'A' for append. The record format codes are 'V', variable; 'F', fixed; 'P', print; 'B', binary; 'D', default; 'E', editable.
Action: Opens the file.
Return value: A file ID. It returns 0 if the attempt to open the file fails.

Function call: FREAD(FID)

Argument: FID, a file ID.

Action: Reads the next record from the file, copying the data to the FDB.

Return value: A return code. It returns –1 for an end of file condition.

Function call: FGET(FID, VAR, N)

Arguments: FID, a file ID; VAR, a character variable; N, the number of characters to retrieve.

Action: Retrieves the next N characters from the FDB and assigns them to the character variable.

Return value: A return code. It returns –1 for an end of buffer condition.

Function call: FCLOSE(FID)

Argument: FID, a file ID.

Action: Closes the file.

Return value: A return code.

Function call: FILENAME(FILEREF, '') or FILENAME(FILEREF)

Arguments: FILEREF, a fileref; '', a null string (in most operating systems, a blank value can also be used).

Action: Clears the fileref.

Return value: A return code.

The following program puts these function calls together to read a text file.

```
*
   READ2.SAS
   Reads text file using I/O functions
   Rick Aster    May 2000
*;
DATA _NULL_;
  LENGTH RECORD MSG $ 75 FILEREF $ 8;
  RETAIN RECORD FILEREF '';
  FILE LOG;
*
  Define fileref.
*;
  RC = FILENAME(FILEREF, 'any.txt');
  IF RC > 0 THEN GOTO E1;
*
  Open file.
*;
  FID = FOPEN(FILEREF, 'S');
  IF NOT FID THEN GOTO E1;

*
  Process records.
*;
  DO WHILE (1);
    RC = FREAD(FID);
    IF RC THEN LEAVE; * Check for end of file or error;
    RC = FGET(FID, RECORD, 75);
    PUT RECORD $CHAR75.;
```

```
    END;

*
  Close file and clear fileref.
*;
  RC = FCLOSE(FID);
  RC = FILENAME(FILEREF);
  STOP;

*
  Error handling.
*;
E1:
  RC = SYSRC();
  PUT 'Error defining fileref or opening file. Return code ' RC;
  MSG = SYSMSG();
  IF MSG NE '' THEN PUT MSG;
  STOP;
```

If you compare this program to the previous program, you can see how the logic involved in reading a text file is similar to that for reading a SAS dataset. Many of the issues concerning that program also apply to this one. In addition, notice these points about this program:

- The FILENAME function associates a fileref with a physical file name. This action works even if the physical file does not exist. However, it will fail if the physical file name is not a legal file name.
- The fileref argument to the FILENAME function is a blank character variable. The FILENAME function generates a new fileref and assigns it to this variable. The variable then can be used as the fileref argument for later function calls. When you have the FILENAME function generate a fileref, you can be sure that the fileref does not conflict with any filerefs that have already been defined.
- The FOPEN function opens the file indicated by the fileref argument. The S access mode opens the file for sequential input. This action fails if the file does not exist.
- The FREAD function works like the FETCH function, but for a text file. It copies the next input record to the FDB.
- The FGET function call obtains the first 75 characters of the record from the FDB. If the record length is 75 or less, the effect of the statement is to copy the entire text of the record to the variable RECORD. It is also possible to use the FGET function to extract the specific fields from an input record.
- The PUT statement writes the text of the input record in the log. A typical application would make a more substantial use of the input data at this point in the program.

Writing a Text File

Only two additional functions are needed to write to a text file: the FPUT function, which copies a character value to the FDB, and FWRITE, which writes the FDB as a record in the file.

Function call: FPUT(FID, DATA)
Arguments: FID, a file ID; DATA, a character value.
Action: Appends the character value to the FDB.
Return value: A return code.

Function call: FWRITE(FID)
Argument: FID, a file ID.
Action: Writes the FDB as a record in the file.
Return value: A return code.

To update a text file, use the logic shown earlier to read the text file, but open the file with the 'U' or 'W' access mode. After you read a record with the FREAD and FGET functions, use the FPUT function to make any necessary changes, then call the FWRITE function to write the modified record back to the file.

When you create a new file, first make sure that no file exists with the file name you want to use — or if such a file does exist, that it is okay to delete it. Open the file with the 'O' access mode, for output. If the file does exist, the existing contents of the file are replaced. If the file does not exist, a new file is created. Use the FPUT and FWRITE functions to build records and add them to the file.

The same program logic works for appending to a file, that is, adding records to the end of an existing file, but use the 'A' access mode. You can also use the FAPPEND function to write a record at the end of a file. This function is especially useful for adding records to a file that is opened for update with the 'U' access mode.

Function call: FAPPEND(FID)
Argument: FID, a file ID.
Action: Writes the FDB as a new record added to the end of the file.
Return value: A return code.

FDB Actions

There are a few other functions for working with the FDB. The FCOL and FPOS functions work with the pointer, which is used for moving around the FDB in much the same way that column pointer controls are used in the INPUT and PUT statements. The FSEP and FGET functions can be used together to read delimited data.

Function call: FCOL(FID)
Argument: FID, a file ID.
Return value: The current column position in the FDB for the indicated file.

Function call: FPOS(FID, N)
Arguments: FID, a file ID; N, a column position.
Action: Moves to column N in the FDB for the indicated file.
Return value: A return code.

Function call: `FSEP(FID, CHARS)`

Arguments: `FID`, a file ID; `CHARS`, a character value containing a character or characters that the FGET function will treat as delimiters.

Return value: A return code.

Function call: `FGET(FID, VAR)`

Arguments: `FID`, a file ID; `VAR`, a character variable.

Action: Retrieves the next field from the FDB and assigns it to the character variable. Any character indicated in the previous call to the FSEP function is treated as a delimiter character. Any sequence of consecutive delimiter characters is treated as one delimiter.

Return value: A return code. It returns –1 for an end of buffer condition.

Actions on Records

The same way that the NOTE, POINT, and REWIND functions work with the observations of a SAS dataset, the FNOTE, FPOINT, and FREWIND functions work with the records of a text file. The DROPNOTE function is the same for text files as it is for SAS datasets. The only differences in syntax are:

- Adding an "F" to the beginning of the function name (except for the DROPNOTE function).
- Using a file ID rather than a dataset ID as the first argument.

Opening a File in a Directory

A fileref can refer to a directory. To open a file in the directory, either use the FILENAME function to create a separate fileref for the file, or use the MOPEN function to open the file using the directory fileref.

Function call: `FILENAME(FILEREF, FILE, DEVICE, OPTIONS, DIRECTORY)`

Arguments: `FILEREF`, a fileref; `FILE`, a physical file name; `DEVICE`, an optional device name; `OPTIONS`, host options; `DIRECTORY`, the fileref of the directory that contains the file.

Action: Associates the fileref with the physical file name of the indicated file in the directory. If the fileref argument is a blank character variable, the function generates a new fileref and assigns it to that variable. A fileref can be assigned even if the physical file does not exist.

Return value: A return code.

Function call: `MOPEN(FILEREF, MEMBER, MODE, RECLEN, RECFM)`

Arguments: `FILEREF`, the fileref of a directory; `MEMBER`, the physical file name of a text file in the directory; `MODE`, `RECLEN`, `RECFM`, optional arguments, as described for the FOPEN function.

Action: Opens the file.

Return value: A file ID. It returns 0 if the attempt to open the file fails.

File Information

The FINFO function returns information items from the file system for an open text file. Different information items are available in different file systems. The information items are always returned as character values. FOPTNUM and FOPTNAME are supporting functions that tell you the number of available information items and the name of each information item.

Function call: FOPTNUM(FID)
Argument: FID, a file ID.
Return value: The number of information items available for the file.

Function call: FOPTNAME(FID, N)
Argument: FID, a file ID; N, a counting number.
Return value: The name of information item N for the file. The function returns a blank value if either argument is invalid.

Function call: FINFO(FID, NAME)
Arguments: FID, a file ID; NAME, the name of an information item.
Return value: The value of the information item for the file, a character value.

The following example shows how these functions can be used together. The statements write all available information items for a file in the log. The file first must be opened, with a file ID in the variable FID, before these statements can be executed.

```
LENGTH NAME $20 VALUE $75;
FILE LOG;
DO N = 1 TO FOPTNUM(FID);
  NAME = FOPTNAME(FID, N);
  VALUE = FINFO(FID, NAME);
  PUT NAME +(-1) ': ' VALUE;
  END;
```

The output lines from these statements might look like the lines below.

```
File Name: C:\My Documents\My SAS Files\V8\any.txt
RECFM: V
LRECL: 256
```

File Existence

The FILEEXIST, FEXIST, and FILEREF functions tell you about the existence of a file or fileref. You can use the FDELETE function to delete a file.

Function call: FILEEEXIST(FILE)
Argument: FILE, a physical file name.
Return value: A Boolean value that indicates whether the file exists: 1 if it exists, 0 if it does not exist.

Function call: FEXIST(FILEREF)
Argument: FILEREF, a fileref.
Return value: A Boolean value that indicates whether the physical file exists: 1 if it exists, 0 if it does not exist.

Function call: FILEREF(FILEREF)
Argument: FILEREF, a fileref.

Return value: A numeric code indicating the existence of the fileref and physical file. The value is 0 if the fileref is not defined. If the fileref is defined, the value is positive if the physical file exists, negative if it does not exist.

Function call: `FDELETE(FILEREF)`
Argument: `FILEREF`, a fileref.
Action: Deletes the physical file.
Return value: A return code.

After you delete a physical file with the FDELETE function, use the FILENAME function to clear the fileref if it is no longer needed.

Directory I/O Functions

A directory is a list of files, but it is also a file itself. The directory I/O functions open a directory in order to read the information it contains. Directories in different file systems contain different information items.

Function call: `DOPEN(FILEREF)`
Argument: `FILEREF`, the fileref of a directory.
Action: Opens the directory.
Return value: A directory ID. It returns 0 if the attempt to open the directory fails.

Function call: `DNUM(DID)`
Argument: `DID`, a directory ID.
Return value: The number of members in the directory.

Function call: `DREAD(DID, N)`
Arguments: `DID`, a directory ID; `N`, a counting number.
Return value: The name of member `N` of the directory.

Function call: `DOPTNUM(DID)`
Argument: `DID`, a directory ID.
Return value: The number of information items available in the directory.

Function call: `DOPTNAME(DID, N)`
Arguments: `DID`, a directory ID; `N`, a counting number.
Return value: The name of information item `N` in the directory. The function returns a blank value if either argument is invalid.

Function call: `DINFO(FID, NAME)`
Arguments: `DID`, a directory ID; `NAME`, the name of an information item.
Return value: The value of the information item for the directory.

Function call: `DCLOSE(DID)`
Argument: `DID`, a directory ID.
Action: Closes the directory.
Return value: A return code.

17

Execution

One way to look at the SAS programming language is as a set of rules of syntax — rules that define how components are combined correctly to form a SAS program. This analytic perspective lets you look at a SAS program at any level of detail. There are times, however, when it is necessary to understand the SAS language in a different way. The SAS language can also be understood in terms of the actions that the SAS supervisor takes in response to the words and symbols that it finds in a SAS program. That is, when a program executes, you can see its actions as an effect of the SAS supervisor rather than seeing them as an effect of the way the program is written.

When the SAS supervisor executes a SAS program, it starts at the beginning and works its way through the program, piece by piece. To think of a SAS program this way, you have to understand it in linear terms. In this perspective, a SAS program is a sequence of text objects, starting at the beginning of the program file and continuing in a linear manner to the end of the program file. There is a corresponding linear sequence of actions that the SAS supervisor takes when it executes the program.

This point of view often helps to point out logical errors in a program that are not easy to see in an analytic view of the program. It is the natural point of view to employ whenever you want to change the usual sequence of execution in any way, as when you execute views, compile data steps, or generate SAS code with macro language or the EXECUTE routine. And it is absolutely essential if you ever need to track down logical errors in the generation of SAS code.

The Execution Process

The way a program in any programming language turns into computer actions is a mysterious process that defies any easy description, and this is doubly so of SAS programs. Traditionally, programming languages are either compiled or interpreted. A compiler converts an entire program into machine language, which can then be executed separately. An interpreter reads a program and takes the actions required to carry out each statement it comes across (or sometimes, each token or each line). The details of either compiling or interpreting would be hard enough to describe, but SAS execution is even more complicated. The process the SAS supervisor follows to execute a SAS program involves some aspects of compiling and interpreting, but it is really something different from either of the two.

The SAS supervisor goes through five stages in processing a SAS program, but it does not go through each stage for the entire program at one time. Executing a SAS program is an iterative process, working through the program one piece at a time. In each stage of the execution process, the SAS supervisor identifies and acts on a different kind of program unit, which it constructs from the results of the previous stage. These five stages focus on five different program units:

1. Program lines
2. Characters
3. Tokens
4. Statements
5. Steps

Roughly speaking, the first three stages encompass preprocessing and parsing. Preprocessing, which only some programs require, implements directives to assemble the text of the program. Parsing is the algorithmic process that applies the rules of a language to arrive at the meaningful units in the text. Stage 4, which processes statements, can be compared to interpreting. Stage 5, which processes steps, can be compared to compiling. The actions in each stage of the execution process are described below.

1. Program Lines

The SAS supervisor starts by reading program lines from the program. In batch mode, the primary source for program lines is a program file. In interactive mode, it is a text editor window from which the user submits program lines. Several other sources can also pass program lines to the SAS supervisor, including the CALL EXECUTE routine and the SCL programs of AF applications. The program lines form an execution queue; the SAS supervisor keeps the lines in order and works with one line at a time.

Finding the program lines can involve more than merely reading each line that comes from the program file or text editor window. Some SAS statements, notably the CARDS and ENDSAS statements, tell the SAS supervisor that the lines that follow are not program lines. The CARDS statement indicates that the lines that follow are data lines, rather than program lines. The ENDSAS statement indicates the end of the SAS program. Any lines that follow the ENDSAS statement are ignored.

When it has processed the last program line and there are no more program lines to read, the SAS supervisor is done executing the SAS program. In batch mode, this means the SAS session is over, and the SAS supervisor cleans up session objects and ends the SAS session. If lines were submitted from a window in an interactive session, the SAS supervisor returns control to the user, so that the user can take other actions in the session.

2. Characters

The SAS supervisor takes one program line and considers it as a string of text characters. At this stage, it does preprocessing with the text, which means that when some character sequences appear in the text, it replaces them with other characters. This preprocessing falls into three separate categories: character code substitution, secondary program files, and macro language. Also at this stage, the SAS supervisor identifies any comments in the program text. The comments are not treated as part of the program and are not passed along to the next stage.

Character code substitution is designed to help users whose computers do not have a complete character set. You can use a sequence of two characters to substitute for a missing character. For example, if you cannot type the character { you can write the character code ?(in its place.

The %INCLUDE statement identifies a secondary program file. The text of that file is inserted in the program in place of the %INCLUDE statement.

There are various kinds of macro language references. The easiest kind to understand is a macro variable reference. Suppose the macro variable N is defined with the value 12. Then, if the characters &N or &N. appear in the program, those characters are removed and the characters 12 are substituted for them. Some macro objects do not resolve to any program text; instead, they take various actions of their own.

Macro objects and secondary program files can generate any number of program lines. These program lines are processed one at a time, in much the same manner that program lines from the primary program are processed. However, they are not added to the end of the queue of program lines. These generated program lines are processed in their entirety at the point where they are generated, even if this is in the middle of a program line of the primary program file.

The SAS supervisor looks for specific character patterns that require preprocessing. Each of these patterns indicates a preprocessing object:

- Specific two-character combinations consisting of a question mark followed by certain other special characters.
- An ampersand or percent sign followed by a letter or underscore.
- Consecutive ampersands.
- A percent sign followed by an asterisk.

When an ampersand or a percent sign is followed by a letter or an underscore, it indicates only the beginning of a preprocessor object. It is the macro processor that parses the object and determines how far it extends, and then processes it to create the appropriate substitute text.

After a preprocessing substitution, the SAS supervisor checks the resulting text again to see if more preprocessing is necessary. The SAS supervisor does only enough preprocessing at one time to generate one token. Preprocessing stops and the SAS supervisor moves on to the next stage of execution when it reaches a space, a special character, or the end of the program line and the program text up to that point does not contain a preprocessor object.

The SAS supervisor has to be aware of quoted strings and comments in order to do its preprocessing correctly. When a quoted string is enclosed in single quotes, the SAS supervisor treats an ampersand or percent sign in the text of the string as simply part of the data of the string. By contrast, when a quoted string is enclosed in double quotes, the SAS supervisor looks inside the string text for preprocessor objects that start with an ampersand or percent sign.

Comments are identified in this stage of processing. The SAS supervisor does not look for preprocessor objects or quoted strings inside comments, and it excludes comments from the subsequent stages of SAS execution. The SAS supervisor looks for the two kinds of comments that the SAS language has. A comment statement starts when an asterisk is found as the first nonblank character after the end of the previous statement. The comment statement ends at the first semicolon. A delimited comment starts with the character sequence /* anywhere in the program — except within a quoted string or comment statement. The comment ends when the characters */ are found.

The table below shows examples of the SAS supervisor's actions on program text.

Program Text	Actions
; * MODIFIED OCTOBER 31, 1997;	The comment statement is identified and excluded from execution.
FLOW?(X, I?)	Character code substitution results in FLOW{X, I}
%SQUARE(CORNER1, LENGTH)	The macro processor is called to substitute text for this macro object.
%INCLUDE CENSOR;	The text of the file identified by the CENSOR fileref is inserted at this point in the program.
TITLE1 'A&M Records';	The quoted string is inside single quote marks, so the ampersand does not indicate a macro object.
TITLE1 "&TITLETEXT";	The quoted string is inside double quote marks, so the macro processor is called to substitute text for the macro object.
/* "- - - - - - - -" */	The comment is excluded from execution. The quote marks do not indicate a quoted string.

In this stage, the SAS supervisor identifies the actual text of the SAS code — the characters it will execute. It next has to determine what the characters mean.

Lexicon

The word *source* is often used for a SAS program. The program lines are *source lines*. The program text is *source code*.

This use of the word *source* comes from the compilation process of other programming languages. The input file that a compiler reads, which contains the program as written by the programmer, is called a *source file*. The output file that a compiler produces, which contains some kind of executable code for the program, is called an *object file*.

3. *Tokens*

The SAS supervisor groups the characters of the program into the units that identify the objects and actions of the program. These meaningful units of a program are called tokens.

The tokens in a program line are usually not hard to pick out. For example, this line divides into seven tokens:

```
IF _N_ >= 5 THEN STOP;
```

Each token is written here on a separate line:

```
IF
_N_
>=
5
THEN
STOP
;
```

The SAS supervisor starts looking for the first token at the first character of the program, and it looks for each successive token immediately after the end of the previous token. It considers four kinds of characters as it identifies tokens. The first character it finds determines what kind of token it has and how it delineates the token. The process is described, in slightly simplified form, in the following table.

Token Identification

First character	Kind of Token	Extent of Token
Space, tab, or other whitespace	Not a token	Skips to the next character
Quote mark (" or ')	Quoted string	Continues until the same quote mark occurs again, not counting an occurrence of two consecutive quote marks
Letter, underscore, digit, or period	Word, numeric constant, etc.	Consecutive characters of this group of characters
Special character (all other characters)	Symbol	A single character

Lexicon

SAS Institute sometimes uses the words *word* and *symbol* as synonyms for *token*. Unfortunately, that leaves them with no distinct words to mean *word* and *symbol*. That usage makes explanations of parsing more difficult than they are already.

These are examples of the way tokens are formed in each of the categories:

Quoted string	Word, numeric constant, etc.	Symbol
"19301-0176"	TITLE1	=
"X"	MAIN.APPL.USER.KEYS	@
'Northwest Territories'	A	#
'*"*'	_IORC_	+
	.001	*
	100.	(
	0	)
	.Z	/
	025EX	
	25E6	
	COMMA14.6	

The process of identifying tokens is not quite as simple as this, because the process described so far splits some tokens into parts. The parts have to be put back together to form the whole token. These are some of the kinds of combinations that are necessary to form tokens:

- *Consecutive special characters.* Several symbol tokens consist of more than one character. Most of these are operators, such as >= and **.
- *Special quoted strings.* A quoted string by itself is a character literal constant, but if it is followed by certain letter codes, it can represent various other things. If the SAS supervisor finds a code such as N (name), T (SAS time), or X (character hexadecimal) after a quoted string, it treats that code as part of the same token.
- *Multilevel name.* A period between two words indicates that the words are part of a multilevel name, even if there are spaces before or after the period. A name literal, written as a quoted string followed by the letter N, can also be a part of a multilevel name.
- *Sign.* A numeric constant can be preceded by a - or + sign to indicate a negative or positive number. In most places, but not in an expression, the sign is treated as part of the constant. In expressions, the sign is kept as a separate token and treated as an operator.
- *Signed exponent.* A numeric constant in scientific notation can have a signed exponent. When this occurs, the letter E (or D) at the end of the constant value is followed by a - or + sign and one or more digits. The sign and digits are part of the constant value.
- *Special characters in names.* The dollar sign ($) is the first character of the name of a character informat or format, or it can serve as the entire name of the standard character format. When a dollar sign is followed by a token that could be an informat or format reference, the dollar sign is treated as part of that token. In some operating systems, filerefs can contain certain special characters, and those characters are treated as part of those names.
- *Compound keywords.* A few keywords are recognized when written as two words. Most notably, GO TO is recognized as the keyword GOTO.

These are examples of combinations that form tokens:

Quoted string	Word, numeric constant, etc.	Symbol
'31OCT2000'D	25.8E-29	||
'31OCT2000 00:00'DT	MAIN . APPL	^=
'0FFF'X	FIRST."Place Name"N	--
	$CHAR14.	
	-.06	

In general, it is permissible to write spaces before and after any special character in a program, even one that combines with other characters to form a token. There are only a few exceptions. In special quoted strings, there cannot be a space between the final quote mark and the letter code that indicates the special use of the quoted string. In scientific notation constants, it is possible for the exponent to have a sign, as, for example, the negative sign in 10E-12. A space is not permitted before or after this sign character.

Spaces are allowed after the $ that begins an informat or format name, between the special characters that form an operator, between the sign and magnitude of a numeric constant, and between levels of multilevel names. However, tokens are easier to recognize when you write them with no extra spaces in them.

Several different kinds of tokens can be formed out of letters, digits, underscores, and periods. They are all parsed in much the same way, but they are constructed differently and do very different things in statements.

- A *word* begins with a letter or underscore. All others characters are letters, underscores, or digits.
- A *standard numeric constant* consists mainly of digits. It can contain one period (used as a decimal point). The first character can be a sign character (- or +).
- A *numeric hexadecimal constant* uses digits and the letters A–F or a–f. The first character must be a digit. The last character is an X.
- A *standard missing value* is a period.
- A *special missing value* is a period followed by a letter or underscore.
- A *multilevel name* is made of two or more words (or name literals) with periods between them.
- A *numeric informat or format reference* consists of a word or one or more digits followed by a period, optionally followed by one or more digits.
- A *character informat or format reference* has the same elements as a numeric informat or format reference, and begins with a dollar sign.

The specific way a token is used depends on the sentence syntax. For example, a word may be a keyword of a statement or the name of a specific type of object, depending on where it appears.

Not everything that parses as a token is valid as a token in SAS syntax. However, the SAS supervisor does not check for the validity of a token until it forms a statement and checks its syntax. At that point, an invalid token generates a syntax error, often with an error message that suggests what would be a valid token at that point in the statement.

Some kinds of tokens contain parts that have separate, if not exactly independent, meanings. The most pervasive way this happens in SAS syntax is the numeric suffixes that many words have. For example, the keyword TITLE1 is formed of the root word TITLE and the numeric suffix 1. The word TITLE indicates a statement that defines a title line; the numeric suffix indicates which title line the statement defines. Numeric suffixes are found in a few keywords and in names of functions, options, and variables.

These are other ways that parts of tokens can have specific meanings:

- The parts of an informat or format reference can include a name, a width argument, and a decimal argument.
- Each level in a multi-level name has its own meaning. For example, in the four-level name of an entry, the four parts are a libref, member name, entry name, and entry type.
- A numeric constant in scientific notation has a mantissa part and an exponent part, with the letter E (or D) between them.
- SAS date, SAS time, and SAS datetime constants are written with specific calendar and clock parts — year, month, day, hour, minute, second — inside the quoted string.

4. Statements

After the tokens of a SAS program are identified, identifying the statements is a simple matter. Every statement ends with a semicolon. Every semicolon token marks the end of a statement.

The supervisor looks at the tokens of the statement to determine what kind of statement it is. Usually, the keyword that begins a statement identifies the statement, but there are a few special cases:

- A *null statement* has only one token, which is the semicolon.
- An *assignment statement* starts with a word and has an equals sign (=) as its second token, or as the next token after a pair of parentheses or braces that follow the first token.
- A *sum statement* starts with a word and has a plus sign (+) as its second token, or as the next token after a pair of parentheses or braces that follow the first token.

If a statement does not meet the rules for one of these special cases, then the statement must start with a word, a keyword that indicates a specific kind of statement in the SAS language.

The SAS language defines a limited set of statements, and most of these statements have limitations on where they can be used. If the keyword is not one that the SAS supervisor recognizes in the current context of the program, it considers the possibility that the word is a misspelled keyword. It can correct some minor spelling errors, as shown in this example:

```
RNU;
```

```
188  RNU;
     ---
     14
```

```
WARNING 14-169: Assuming the symbol RUN was misspelled as RNU.
```

If the SAS supervisor still does not recognize the statement, it generates an error condition, as demonstrated here:

```
ABRACADABRA;
```

```
1125  ABRACADABRA;
      -----------
      180

ERROR 180-322: Statement is not valid or it is used out of proper order.
```

Certain statements in the program can affect what statements the SAS supervisor subsequently recognizes. Data step statements are recognized only after a DATA statement. A PROC statement is required before the SAS supervisor can recognize the statements of a specific proc, and within a proc step, some statements may be recognized only when they come after specific other statements.

Even if the SAS supervisor does recognize a statement, the statement could still be in the wrong place. For example, an ELSE statement can come only after an IF-THEN statement. Otherwise, it generates a syntax error with the error message shown here:

```
DATA _NULL_;
  ELSE X = 8;
RUN;
```

```
126  DATA _NULL_;
127     ELSE X = 8;
ERROR: No matching IF-THEN clause.
128  RUN;
```

If the SAS supervisor does recognize and accept the statement, it can then check the syntax of the statement. Each statement of the SAS language has its own rules of syntax, and the SAS supervisor checks the program statement against those syntax rules. It is at this point in the execution of a SAS program that most of the syntax checking is done and most error messages are generated.

The SAS supervisor can sometimes correct spelling errors within a statement, as in this example:

```
PROC SUMMARY DATA=MAIN.ROUTE PRINT MAN MAX;
```

```
188   PROC SUMMARY DATA=MAIN.ROUTE PRINT MAN MAX;
                                         ---
                                         1
WARNING 1-322: Assuming the symbol MEAN was misspelled as MAN.
```

The SAS supervisor can use the statement in either of two ways, depending on the kind of statement. The statement can be a global statement, or it can belong to a step. If the statement is a global statement, the SAS supervisor carries out the actions of the statement at this point. If it belongs to a step, the SAS supervisor adds the statement to the step it is building.

Because the SAS supervisor executes global statements immediately, as soon as it reaches them, global statements execute between steps. Global statements that appear within a step are executed before that step executes. The SAS supervisor produces log notes that describe the execution of some global statements. Most notably, notes for a LIBNAME statement describe the library that is associated with a libref.

Control flow statements in the data step can contain other statements. For example, the statement

```
ELSE X = 8;
```

contains the statement

```
X = 8;
```

A few statements affect the way the SAS supervisor treats the text of the statement itself or the tokens or program lines that follow. For example, the CARDS statement indicates that the lines that follow are data lines, rather than program lines. Any text that follows the CARDS statement on the same line is not considered until after the data lines are processed. TITLE statements and statements of the FORMAT proc are parsed in a slightly different way in order to maintain compatibility with the different syntax rules of early versions of SAS software.

5. Steps

The SAS supervisor executes global statements as soon as it reaches them, but statements that belong to a step are merely collected as the step is built. A data step is not executed until the step boundary that indicates the end of the step is reached. The same is true of most proc steps, but the details of the way a proc step is executed depend on the proc.

A DATA or PROC statement marks the beginning of a step. The step continues until a subsequent statement specifically indicates a step boundary. The RUN statement is the usual indication of a step boundary. In a data step, the CARDS statement also indicates a step boundary. In a proc step, the QUIT or PARMCARDS statement is an indication of a step boundary. The beginning of the next step would also indicate the step boundary for the previous step. An ENDSAS statement or, in batch mode, the end of the program file can also serve as the step boundary for a step.

The SAS supervisor checks for errors as it builds a data step by checking the object names used in each statement against the properties of the objects. A mismatch is considered a semantic error. It is not a syntax error, because the form of the statement, considered by itself, is correct. The error is in the usage of a specific object. Using the same name for both a character and numeric variable, or as a variable and as an array, would be a semantic error. Using the wrong number of subscripts for an array, or the wrong number of arguments for a function or CALL routine, would also be a semantic error. The results of a semantic error are much the same as for a syntax error.

When it reaches the step boundary without errors, the SAS supervisor executes the step, taking the actions that are indicated by the various statements that make up the step.

After it executes a step, the SAS supervisor continues with the next token or the next line in the program. It continues the execution process in an iterative way until it reaches the end of the program.

In an interactive session, execution continues until the end of the submitted program lines. This can occur in the middle of a step, or even in the middle of a statement. If execution stops in the middle of a step, the SAS supervisor displays an indication of this. In recent releases, it has displayed this message in the title bar of the Program Editor window. When more program lines are submitted, the SAS supervisor continues to build the same step until a step boundary is reached.

Data Lines

A program file can contain certain kinds of input data. These data lines and parameter lines are written in the program file, but they are not actually part of the program. Data lines serve as an input file for a data step. Parameter lines contain parameters or directives that control a proc step. Write data lines and parameter lines at the end of the step they belong to.

To write data lines for a data step, use the CARDS statement, in place of the RUN statement, to mark the end of the step. DATALINES and LINES are aliases for CARDS. Write the data lines beginning on the next line. You can use a null statement to indicate the end of the data lines:

```
DATA . . .
  . . .
CARDS;
data lines
;
```

When it reads data lines, the SAS supervisor looks for a semicolon anywhere in the line. If it finds one, it assumes that the data lines are over and that the line that contains the semicolon is a program line.

For data lines that could contain a semicolon, write a 4 at the end of the keyword and use a line of four semicolons to mark the end of the data lines:

```
DATA . . .
  . . .
CARDS4;
data lines
;;;;
```

The line of four semicolons must appear exactly as shown here.

In the INFILE statement in the data step, use the special fileref CARDS to refer to the data lines in the program file. CARDS is the default fileref for input text data, so you might be able to omit the INFILE statement.

System Options for the Execution Process

Object	Option	Value	Action
Program lines	S=	Integer	Limits the length of lines from the program file
		MAX	Uses the entire line
	S2=	Integer	Limits the length of lines from secondary program files
		MAX	Uses the entire line
	CARDIMAGE	On	Uses 80 characters from each program line and allows tokens to be split between lines
		Off	Uses the actual length of each program line and treats the end of a line as a token boundary
Special characters	CHARCODE	On	Allows specific character sequences to substitute for special characters[1]
		Off	Does not recognize substitute character sequences
Letters	CAPS	On	Converts letters in the program to uppercase
		Off	Does not convert letters in the program
Names of variables	VALIDVARNAME=	ANY	Allows all characters in names[2]
		V7	Treats names as mixed-case
		UPCASE	Treats names as uppercase
		V6	Treats names as uppercase and limits their length to 8 characters
DATA statement: one-level names of SAS datasets	DATASTMTCHK=	NONE	Allows any valid name
		COREKEYWORDS	Creates an error condition if the names RETAIN, SET, MERGE, or UPDATE are used
		ALLKEYWORDS	Creates an error condition if any data step keyword is used
MERGE statement	MERGENOBY=	ERROR	Creates an error condition when a MERGE statement is not followed by a BY statement
		WARNING WARN	Writes a warning message when a MERGE statement is not followed by a BY statement
		NOWARNING NOWARN	Permits a MERGE statement to be used without a BY statement

[1] With the CHARCODE option, the SAS supervisor recognizes these character codes:

?: for `   ?, for \   ?(for {   ?) for }   ?= for ^ or ¬

?– for _   ?/ for | ?< for [?> for]

[2] If the name contains any characters that are not valid characters for SAS words, write the name as a name literal.

Parameter lines provide a way for procs to get parameters or options in text form rather than as statements. This feature is provided primarily to make it easier to develop add-on procs that can be used in a SAS program. There is only one proc in base SAS that uses parameter lines. The EXPLODE proc produces banner pages from text that you write in parameter lines. When a SAS program produces several long reports that are printed together, these banner pages can be useful in separating the reports. Parameter lines work the same way as data lines, but use the PARMCARDS statement to mark the end of the proc step and the beginning of the parameter lines. This is an example of the EXPLODE proc:

```
PROC EXPLODE;
PARMCARDS4;
 PART 1
;;;;
```

```
****      *    ****   *****          *
*   *    * *   *   *    *           **
*   *   *   *  *   *    *            *
****    *****  ****     *            *
*       *   *  * *      *            *
*       *   *  *  *     *            *
*       *   *  *   *    *          *****
```

System Options for SAS Execution

You can control many of the specific details of the execution process with system options. Each of these options affects the way the SAS supervisor recognizes or treats a certain kind of object in the program. The table on the previous page describes the options.

In addition to the options listed here, there are also system options that affect log messages, error handling, and macro processing.

Messages and Error Handling

Other actions of the SAS supervisor are closely related to its actions in executing a program. As it executes the program, it creates the log as a record of the program's actions and events. It handles any error conditions that come up, writing log messages about them and either stopping execution of the program at that point or continuing in some way.

Log Messages

As it processes the program lines, characters, statements, and steps of a program, the SAS supervisor writes lines in the log. The log serves as a record of the execution of the program. In batch mode, the log is simply a print file that SAS automatically creates and stores. In an interactive session, the log appears in the Log window. Typically, most of the lines in the log are the program lines that make up the program. These program lines are annotated with messages that describe the program and the events that occurred in its execution. The log can also contain lines written to it by data steps and procs.

Program lines As it executes each step and each global statement, the SAS supervisor numbers the program lines sequentially and writes the line numbers and program lines in the log. If a program line is too long to write on one line of the log, the SAS supervisor writes it on two or more lines, but still treats it as one line in the program. If a program line contains parts of two steps, the SAS supervisor writes the two parts of the program line at separate places in the log.

The line numbers in the log do not necessary correspond to the lines of the program file, because the SAS supervisor may also count program lines that come from secondary program files or are generated by the macro processor. In an interactive session, the SAS supervisor generates line numbers throughout the session, so the line numbers can start at 1 only for the first time in the session when statements are submitted for execution.

By default, only primary program lines appear in the log. In batch mode, these are the lines of the primary program file. In interactive mode, the primary program lines are lines submitted by the user or by an AF application. You can use system options to have program lines from secondary program files and those that result from macro processing appear in the log.

For certain kinds of error conditions, the SAS supervisor underlines specific tokens in the program lines. It writes an error code number under the underline. Then, a subsequent error message refers to that code number. The underlined token is the point where the SAS supervisor identified the error.

Notes Most of the messages in the log are considered notes, which means they are descriptive or informative in nature. Notes describe the execution of each step and of some global statements. Note lines begin with the label "`NOTE:`". In the Log window, they appear in a distinctive color.

There are a wide range of occurrences that result in log notes. Whenever it creates a SAS data file, the SAS supervisor writes a note that indicates the number of variables and observations. For every step, it generates a note that measures the system resources that the step used. Whenever a data step reads or writes a text file, there are notes to describe the extent of the data involved. Various unusual occurrences, such as automatically converting values from one data type to another or correcting the spelling of a word in the program, generate notes. When a session ends normally, the log ends with a note about the system resources used by the program and another note with SAS Institute's address. You can look in the log of any SAS program to see the kinds of notes that are produced.

Warning and error messages Warning messages and error messages appear in the log to indicate problems. Generally speaking, a warning message indicates a condition that might or might not be a problem and that does not prevent the SAS supervisor from executing the program. By contrast, an error message indicates a condition that appears to prevent the execution of the program or of some part of it. Warning lines begin with the label

"WARNING:". Error lines begin with the label "ERROR:". In the Log window in an interactive session, each kind of message appears in a distinctive color.

Program output Data steps, macro statements, and procs can write lines in the log. These lines can contain any kind of message or text. In a data step, use the LOG fileref to write to the log. In macro language, use the %PUT macro statement to write text in the log.

There are system options that you can use to control most of the lines of the log. Options determine whether the lines appear in the log or not. In some cases, options control details of the format of log messages. The following table lists system options for specific kinds of log lines.

Option	Log lines
SOURCE	Program lines from the primary program file
SOURCE2	Program lines from secondary program files (of %INCLUDE statements)
ECHOAUTO	Program lines from the autoexec (initialization) file
MPRINT, SYMBOLGEN, etc.	Program lines and tokens generated by the macro processor
ERRORS=*n*	Detailed messages for the first *n* occurrences of the same data error
NOTES	Notes
MSGLEVEL=N	The standard log messages for certain events
MSGLEVEL=I	Extra (informational) log messages for certain events
OPLIST	Command-line settings of system options
VERBOSE	System options set at initialization
STIMER	Operating system performance statistics for each step
FULLSTIMER	Detailed performance statistics for each step
PRINTMSGLIST	All the error messages from an event that generates a list of error messages

Several system options affect the way some log lines are formatted. The MSGCASE option affects the way messages appear in the log and elsewhere in the SAS environment. When this option is on, the SAS supervisor converts all messages to uppercase letters. When the option is off, messages appear in mixed case. MSGCASE is an initialization option, which can be set only at the beginning of the SAS session.

The LINESIZE= and PAGESIZE= options determine the width and height of the page in both the standard print file and the log.

The STIMEFMT= option determines the formatting of time measurements in the performance statistics for each step. Usually, time statistics should be measured in seconds; use the value SECONDS with this option. When a program works with large amounts of data, it may be more meaningful to write time measurements in minutes and seconds, or hours, minutes, and seconds; use the value MINUTES or HOURS, respectively. These option values can be abbreviated to the letters S, M, and H.

The UNBUFLOG option, an initialization option available only in some operating systems, tells the SAS supervisor to write the log without the use of a buffer. This takes slightly longer, but it ensures that the log is as complete as possible in the event that the SAS session ends abnormally.

Writing Note, Warning, and Error Lines

Programs can write the same kind of note, warning, and error messages that the SAS supervisor writes in the log. In an interactive session, the Log window displays these messages in the same distinctive colors that it uses for SAS System messages.

Simply write to the log a line that begins with the appropriate label: "NOTE:", "WARNING:", or "ERROR:". This is an example using the %PUT macro statement:

```
%PUT NOTE: Program version 1.2.;
```

```
NOTE: Program version 1.2.
```

If a message continues on a second line, write the label at the beginning of the line with a hyphen instead of a colon. The label does not appear in the Log window, but the line appears in the same distinctive color.

```
%PUT NOTE: Program version 1.2.;
%PUT NOTE- April 27, 2001.;
```

```
NOTE: Program version 1.2.
      April 27, 2001.
```

Error Conditions

Errors can occur during any stage of the execution process. The different kinds of errors the SAS supervisor handles include syntax errors, semantic errors, execution errors, password errors, and out-of-resource conditions. Syntax errors occur when a statement is formed incorrectly or is located at the wrong place in the program.

Semantic errors occur when a statement refers to an object in a way that is not valid for the object. These are examples of semantic errors:

- An input file that does not exist or is not available
- A reference to an array that has not been defined
- Defining the same variable name as two different data types
- A function call with a nonexistent function name or the wrong number of arguments
- The KEY= option referring to an index does not exist
- A GOTO statement referring to a statement label that does not exist

Execution errors are errors that occur in carrying out the actions of a statement or step — most often, in a data step. Some data errors merely result in log messages and missing values. Other errors can make a step stop running. Among the various kinds of error conditions can stop the execution of a data step, these are the two most common:

- Input observations that are not in the sequence indicated in the BY statement
- A subscript value that is outside the subscript range of an array

A password error occurs when the program attempts to access a password-protected SAS file and it either does not supply a password or supplies an incorrect password. If this occurs in an interactive session, the SAS supervisor prompts the user for the password. In a batch session, or if the user does not supply the password when prompted, the SAS supervisor generates an error condition, which is similar to the error condition that occurs when a program refers to an input file that does not exist.

An out-of-resource condition can occur when the SAS application needs more memory or storage space than is available. When this occurs in an interactive session, the SAS supervisor prompts the user with options for responding to the condition. This might include purging some resources, closing or deleting files, or stopping execution of the program.

In certain unusual circumstances, the SAS application may crash, or terminate abnormally. This can happen, for example, if the SAS software is incorrectly installed or if you attempt to run it with a very small amount of memory. The details of the way these kinds of errors can occur are different in different operating systems.

Whenever an error prevents the SAS supervisor from successfully executing a statement or step, the SAS supervisor writes an error message in the log. When an error occurs, the SAS supervisor's response depend on various factors, including:

- The type of error
- The type of statement
- The context of the statement
- The mode in which the program is executing
- System options that affect the responses to errors

The SAS supervisor's specific response to an error can be one or more of these actions:

Kind of Error	Action
Any	Write a log message.
Execution error in a data step	Set the automatic variable _ERROR_ to 1. This produces extra notes in the log.
Syntax error	Disregard the statement.
Any	Stop execution of the step, or skip the step entirely.
Invalid reference	Execute the step without the object.

With any kind of error, the SAS supervisor generates log messages, which can include both error message and notes. The log lines below are an example of log messages generated as the result of an execution error. The list of the values of all variables is generated when the automatic variable _ERROR_ has a value of 1.

```
ERROR: BY variables are not properly sorted on data set WORK.A.
A=2 FIRST.A=1 LAST.A=1 _ERROR_=1 _N_=1
NOTE: The SAS System stopped processing this step because of errors.
```

After it handles the error, the SAS supervisor's subsequent actions can be one of the following:

- Continue executing the program.
- Resume execution at the next step or global statement.
- Process the rest of the program in syntax check mode. In this mode, the SAS supervisor does not read any input records or observations, as if the system option `OBS=0` were in effect. It also does not replace existing SAS datasets, as if the system option `REPLACE=NO` were in effect.
- Stop executing the program.
- End the SAS session immediately, returning an error code to the operating system.

The idea of syntax check mode is to look for further errors in the program. This mode results from errors in steps that create SAS datasets in programs executing in batch mode. Log notes such as these indicate syntax check mode:

```
NOTE: The SAS System stopped processing this step because of errors.
NOTE: SAS set option OBS=0 and will continue to check statements. This may
cause NOTE: No observations in data set.
```

When there are errors in a program in batch mode, the end of the log contains an additional error message that indicates the presence of the errors and the locations of error messages in the log. This error message might look like this:

```
ERROR: Errors printed on page 1.
```

A few system options affect the way the SAS supervisor handles error conditions. The ERRORABEND option tells the SAS supervisor to end the SAS session immediately in most error conditions. This is useful in batch mode for some production jobs in which any error condition indicates a significant problem that has to be corrected.

The ERRORCHECK= option controls the way the SAS supervisor handles file errors that occur in the execution of LIBNAME, FILENAME, %INCLUDE, and LOCK statements in batch mode. Use the value NORMAL to have the SAS supervisor continue to process the program. Use the value STRICT to have the SAS supervisor stop processing the program. With `ERRORCHECK=STRICT`, the SAS session ends abnormally if a %INCLUDE statement refers to a file that does not exist.

The FMTERR option determines the SAS supervisor's response when a variable's format attribute refers to a format that the SAS supervisor cannot find. When this option is on, the SAS supervisor generates an error condition when it cannot find the format. When the option is off, the SAS supervisor uses the default format in place of the indicated format.

Several more system options determine the SAS supervisor's responses to certain kinds of errors in SAS file I/O operations. The BYERR, DKRICOND=, DKROCOND=, DLDMGACTION=, DSNFERR, and VNFERR options are described in chapter 11, "Options for SAS Datasets."

When people run SAS programs, they tend to check the log for possible problems. Therefore, when a well-behaved SAS program runs with the correct input files, its log should contain no error or warning messages and a minimum of cautionary notes. These are some of the things you can do in data step programming to produce a clean log in a program that runs correctly:

- Use error controls for input fields that may contain invalid data values.
- Before dividing, check the divisor to make sure it is not zero.
- Check that values are valid before using them as arguments to functions.
- If you use an expression as an array subscript, compare the resulting value against the subscript bounds. You can use the MAX and MIN operators and the LBOUND and HBOUND functions to do this.
- If you use a SET or MODIFY statement with the POINT= option, test the observation number value you use to make sure it is between 1 and the number of observations in the SAS dataset. You can determine the number of observations with the NOBS= option.
- After a SET or MODIFY statement with the KEY= option, test the value of the automatic variable _IORC_.
- Test error codes and ID values that I/O functions return.
- When necessary, reset the automatic variable _ERROR_ to 0.

Session

A SAS session, even in its simplest form, involves more than executing a single program. The SAS supervisor follows a sequence of actions to start and end the session. Those actions set up session objects at the beginning of the session and release computer resources at the end of the session. During a session, the SAS supervisor might execute several SAS programs and other kinds of programs. These programs can be essentially independent of each other, or they can be connected in various ways.

Startup

SAS software presents a universe of possibilities, in the form of modes, programs, windows, files, and much more. Any one SAS session can cover only a small extent of this territory, and the essential direction and character of a SAS session is determined in the initial moments of a SAS session, as part of the process of getting the session started.

System options are at the heart of the startup process. The SAS supervisor must set a value for every system option. Many system options, called initialization options, can be set only at startup. Some of these are options that control the startup process itself. Others are options that affect the operation of the core of the SAS System at a fundamental level, such that they cannot be altered once a session has started.

The first place where the SAS supervisor finds system options is in the operating system command line that starts the SAS application. Any system option can be set at this point.

The way the operating system command line is written depends on the operating system. There are just a few features in common across operating systems. The first part of the command line is a designation, often a file name, that identifies the SAS application. This part of the command varies from one computer to another according to the details of the way SAS software was installed, but it is usually some form of the word sas. The rest of the command line is a list of system options. The way system options are written in the command line varies by operating system, but most operating systems follow Unix conventions for writing options: option names and keywords are preceded by a hyphen, and a space separates an option name from a value. This might be an operating system command line to start the SAS application:

```
sas -config main.cfg -autoexec mainauto.sas
```

The most critical system option in the operating system command line is the CONFIG= option, which identifies a configuration file. Configuration files are the primary source for system option settings at SAS startup. A configuration file is a text file, and the name of the file often has a sas extension, but it is not a SAS program file. It can contain only system options, blank lines, and delimited comments that begin with /* and end with */. The syntax for writing system options in a configuration file varies by operating system. It is usually like the operating system command line, except that in a configuration file, each option can be written on a separate line.

A SAS session can use several configuration files. Typically, it makes sense to have a system configuration file for options related to the way SAS software is installed on the computer, project configuration files with options for each specific project, and user configuration files with options that reflect the preferences of each user. The SAS supervisor looks for configuration files according to specific rules that are different for each operating system, and it also uses any configuration files that are named in the CONFIG= option in the operating system command line.

To see what a configuration file looks like, you can look at the actual configuration files that a SAS session uses. Use this step to determine the value of the CONFIG= option:

```
PROC OPTIONS OPTION=CONFIG;
RUN;
```

The OPTIONS proc writes lines in the log that indicate the value of the option, which should show one or more file names. Look at those files to see the kind of lines that a configuration file contains in your operating system.

There must be at least one configuration file. In operating systems that use environment variables, specific environment variables can also be used to set system options at startup. When the same option is set in more than one place, the SAS supervisor uses the last value it comes to, except that options set on the operating system command line override options set elsewhere at startup. In general, the SAS supervisor applies system-level settings before user-level settings when it processes configuration files and environment variables, but the specific details vary. During the session, OPTIONS statements and the Options window can change the option values from the settings that were set at startup.

The SAS supervisor uses hundreds of system options, and typically, the configuration files mention fewer than half of them. When options are not given specific values in the startup process, the SAS supervisor gives them default values.

Startup Options

Many important system options can be set only at the beginning of a SAS session or SAS process. The table on the next page lists some of the more important startup options.

The ERRORCHECK= option, described earlier in the discussion of error handling, is another startup option.

Initialization options are system options that can be set only at the beginning of the SAS session. The VERBOSE, MSGCASE, and UNBUFLOG options, mentioned earlier in the discussion of log messages, are initialization options. There are many more initialization options that indicate locations of the essential files of the SAS System and details of the use of system resources. The available options and valid values vary by operating system.

Actions at Startup

The startup process of a SAS session follows this sequence of actions:

1. Set all system options to default values.
2. Set system options from configuration files and environment variables.
3. Set system options from the operating system command line.
4. Initialize log.
5. Write news text to log.
6. Execute the SAS statements of the autoexec file.
7. Execute the SAS statements of the INITSTMT= option.
8. Take the appropriate action for the session's execution mode, which can be one of the following:
    a. Execute the main program file of the SYSIN= option in batch mode.
    b. Execute the command of the INITCMD= option to open the window of an AF application.

Startup Options

Option	Value	Action
CONFIG=	File	Sets system options from the indicated configuration file.
NEWS=	File	Writes the text of the indicated news file to the log at startup.
AUTOEXEC= AE=	File	Executes the indicated autoexec file, a SAS program file, at the beginning of the SAS session.
INITSTMT= IS=	SAS code	Executes the SAS statement(s) before the primary program file, but after the autoexec file.
INITCMD=[1]	Command(s)	Executes the display manager command(s) to open an AF application instead of the usual windows of the interactive environment.
SYSIN=[1]	File	Executes the SAS program file in batch mode.
DMS[1]	On	Opens the programming (display manager) windows in the windowing environment.
DMSEXP[1]	On	Opens the programming and Explorer windows in the windowing environment.
EXPLORER[1]	On	Opens the Explorer window in the windowing environment.
LOG=	File	Use the file as the log file.
ALTLOG=	File or device	Writes a copy of the log to the indicated file.
PRINT=	File	Use the file as the standard print file.
ALTPRINT=	File or device	Writes a copy of the standard print file to the indicated file.
PRINTINIT	On	Initializes the standard print file immediately at startup.
	Off	Initializes the standard print file only when print output is generated, or not at all, depending on the operating system.
WORKINIT	On	Clears the WORK library at startup.[2]
	Off	Does not clear the WORK library.[2] This allows the SAS session to use any existing files in the WORK library.
RSASUSER	On	Allows only read access to the SASUSER library. The user profile and other user files can be used, but cannot be updated.
	Off	Allows read and write access to the SASUSER library. The user profile and other user files can be used and updated.

[1] Use only one of the DMS, DMSEXP, EXPLORER, INITCMD=, and SYSIN= options. Each of these execution-mode options overrides the others.

[2] In Unix, the WORKINIT option determines whether to create a new directory for the WORK library.

c. Open the windowing environment and open the windows indicated by the DMS, DMSEXP, or EXPLORER option.

The autoexec file is a SAS program file that is executed at the beginning of the SAS session. Any SAS statements that can appear in a SAS program file can be used in the autoexec file. Most often, the autoexec file contains LIBNAME, FILENAME, and TITLE statements, and other statements to define or create resources and objects that are used by the various SAS programs of a project. The autoexec file is identified by the AUTOEXEC= system option. If you do not use this option, the SAS supervisor looks for a file called `autoexec.sas`. The specific details of where it looks for this file vary by operating system. To prevent the SAS supervisor from using or looking for an autoexec file, you can use the NOAUTOEXEC option. There can be only one autoexec file, but the autoexec file can include a %INCLUDE statement to execute other SAS program files.

Lexicon

Previously, most writers used the term *initialization file* for the autoexec file. With version 7, *autoexec file*, or *SAS autoexec file*, became the officially preferred term.

Another way to execute SAS statements at the beginning of a SAS session is with the INITSTMT= option, whose value is a string of SAS code to execute. The INITSTMT= option makes it possible to write SAS code in the operating system command line or a configuration file. However, this can quickly become awkward, so the INITSTMT= option is usually limited to one or two short statements. To execute more extensive code at this point, use an autoexec file or use a %INCLUDE statement in the INITSTMT= option.

Several startup options are considered execution-mode options because they determine the kind of session that takes place. These options are mutually exclusive. If you indicate more than one execution-mode option, only the last such option is used. To start a session in batch mode, use the SYSIN= option. The option names the primary program file to execute in the session. In batch mode, the SAS session ends automatically when the end of the primary program file is reached. It is also possible to use the NOSYSIN option to start a session in batch mode with no primary program file. The session ends when the end of the autoexec file is reached.

The DMS, DMSEXP, and EXPLORER options start an interactive session. The only difference among these options is the selection of windows that are open at the beginning of the session.

The INITCMD= option makes it possible to hide most of the SAS interactive environment in a SAS session that runs an AF application. The value of the option is a display manager command, or a string of commands separated by semicolons. Usually, the first command is an AF command that executes the initial AF entry of the AF application. The first command must open an AF window. With the INITCMD= option, the SAS session displays that AF window at startup and does not open any of the usual programming or Explorer windows. When the AF application named in the INITCMD=

option ends, the SAS session ends. This way, the user can interact directly with the AF application and never has to interact with any other part of the SAS environment.

Programs and Session Objects

It is possible for a SAS session to run just one SAS program. It is also possible for a session to run several programs of various different kinds. The table on the following page lists various kinds of programs and tasks that a session might include, along with the options, statements, commands, and other actions that introduce them into the SAS session. When a session includes more than one program, the SAS supervisor keeps the programs straight and, to the extent possible, keeps them from interfering with each other.

Programs in a SAS Session

Kind of Program	Source
Autoexec file	AUTOEXEC= option
Initialization statements	INITSTMT= option
Primary program file (batch mode)	SYSIN= option
Submitted program lines (interactive mode)	SUBMIT command SUBMIT statement in SCL program of AF application CALL EXECUTE routine
Secondary program file	%INCLUDE statement
Macro programming	Included in SAS program lines Included in command lines RESOLVE function
AF application	AF command Other commands and user actions
Compiled data step	DATA statement with PGM= option
View	Use of input SAS dataset in SAS program or anywhere in SAS environment
Operating system command (synchronous)	X statement X command SYSTEM function CALL SYSTEM routine %SYSEXEC macro statement etc.
Operating system task (asynchronous)	SYSTASK statement
SAS process	STARTSAS statement

Running several programs in the same SAS session does not necessarily produce the same results as running each program in a separate session. A program can use objects created by or rely on session settings defined in an earlier program in the session. The autoexec file is the most obvious example of this. Usually, the entire purpose of the autoexec file is to define librefs, filerefs, title lines, macro variables, formats, and similar objects for use in the primary SAS program or in other programs in the SAS session.

There are other ways programs can interact. One program might call another, which might create a macro variable or a SAS data file to return its results to the first program. Any of the objects and properties of a session can potentially provide a way for two programs to work together — or to interfere with each other. These are some of the more important objects and properties a program might work with in a SAS session:

- Librefs
- Filerefs
- System options
- Title and footnote lines
- SAS datasets, catalogs, and other SAS files in the WORK library
- Macro variables
- Macros
- Formats and informats
- The log and standard print file
- Page numbers in the standard print file
- ODS settings, styles, objects, and files

When you submit one program after another in an interactive session, settings from a previous program can carry over and affect a subsequent program. Title lines can be one very visible indication of this. If a program has TITLE statements to define title lines for the pages it prints, those same title lines can appear on the pages printed by the next program unless the next program has TITLE statements of its own. A more serious problem occurs if a program tries to create a SAS data file with the same name as that of a view that a previous program creates. That results in an error condition, as the SAS supervisor cannot replace a view with a SAS data file, or vice versa. Logical errors can be created in macro programming if a program expects a macro variable to be uninitialized, but the same macro variable name was used by an earlier program in the session. Programs that should work together may have to be adjusted so that they do not create these kinds of conflicts when they are combined.

A more serious potential for conflict can occur between two programs that run at the same time. If there is any possibility that such programs could use the same macro variables, SAS files, or other objects, the programs must be designed to avoid any conflict. This is a particular concern in the SCL programs of AF applications, but it can also be an issue with views, macros, and SAS programs that are run as SAS processes.

Compiled Data Steps

With a compiled data step, it is possible to separate the process of defining a data step and executing it. The SAS supervisor initially compiles the data step program and stores it as a SAS file. Later, it can retrieve and execute the program separately from the data step statements that created it.

SAS Institute writers call a compiled data step a *stored compiled data step program*, a *stored compiled program*, or a *stored program*.

To compile a data step, use the PGM= option on the DATA statement:

```
DATA . . . / PGM=SAS file;
data step statements
RUN;
```

The PGM= option indicates the name of the SAS file in which the compiled data step will be stored. Later, to execute the compiled data step, use the PGM= option again, as the only term in a DATA statement, which is the only statement in a step:

```
DATA PGM=SAS file;
RUN;
```

The log of the following example demonstrates what it means for compilation and execution to be separate processes.

```
*
 This step compiles the data step program.
*;
DATA _NULL_ / PGM=WORK.RANDOM;
  RANDOM = FLOOR(RANUNI(5)*6) + 1;
  PUT RANDOM=;
RUN;

*
 This step executes the compiled data step.
*;
DATA PGM=WORK.RANDOM;
RUN;
```

```
233   *
234     This step compiles the data step program.
235   *;
236   DATA _NULL_ / PGM=WORK.RANDOM;
237      RANDOM = FLOOR(RANUNI(5)*6) + 1;
238      PUT RANDOM=;
239   RUN;

NOTE: DATA STEP program saved on file WORK.RANDOM.
NOTE: A stored DATA STEP program cannot run under a different operating
      system.
NOTE: DATA statement used:
      real time           0.15 seconds
```

```
240
241  *
242    This step executes the compiled data step.
243  *;
244  DATA PGM=WORK.RANDOM;
245  RUN;

NOTE: DATA STEP program loaded from file WORK.RANDOM.
RANDOM=6
NOTE: DATA statement used:
      real time           0.53 seconds
```

As the log indicates, the first step only compiles the data step program and saves it as a SAS file. The second step executes the compiled program, so it is at that point that the program's output appears in the log. Both compilation and execution use computer resources, as the performance statistic notes indicate.

In principle, compiling a data step that is run many times should save computer resources, because the step does not have to be compiled every time it executes. In practice, the compilation time for a typical data step may be insignificant compared to the programmer effort it takes to compile it separately. However, compiled data steps might be used to speed execution slightly in the most time-sensitive applications, mainly in processes that take place while the user is waiting in interactive applications.

When a compiled data step uses input SAS datasets, those SAS datasets must exist when the data step is compiled. In the input SAS datasets, the variables and attributes at execution time should be the same as at compilation time. If there are difference, the compiled data step might run incorrectly, or the SAS supervisor might not be able to run it at all.

However, the names of input and output SAS datasets can be changed at execution time. To do this, use the REDIRECT statement after the DATA PGM= statement. The form of the REDIRECT statement is:

REDIRECT INPUT *or* OUTPUT *logical name=actual name . . .*;

where INPUT or OUTPUT indicates an input or output SAS dataset, *logical name* is the name that was used for the SAS dataset when the program was compiled, and *actual name* is the name of the SAS dataset that the program actually uses when it executes. For example,

```
REDIRECT INPUT WORK.LIST=WORK.LIST1;
```

uses the SAS dataset WORK.LIST1 as the input SAS dataset that, in the data step program as it was compiled, was called WORK.LIST.

Views as Programs

A view organizes data that might be stored in various kinds of files so that it can be accessed as a SAS dataset. When you use a view in a SAS program or elsewhere in the SAS environment, you can use it in much the same way as any other SAS dataset. But a view is more than just a set of data. It is also the program that puts that data together. Every time you access the data of a view, you execute the program of the view.

The program logic of a view is easiest to see in a data step view, which is created as a data step. To create a data step view, write a data step that writes an output SAS dataset, and also name the SAS dataset in the VIEW= option of the DATA statement:

```
DATA SAS dataset / VIEW=SAS dataset;
data step statements
RUN;
```

When you run this data step, it creates the view, which it stores as a SAS file with a member type of VIEW. The program of the data step does not actually execute until you access the view and read the data in it. The view runs at the same time as the step that accesses it. The example below writes lines in the log in order to demonstrate the sequence of execution of a data step view when it is read by a data step.

First, this data step creates the data step view. Note that running this step does not result in any lines written to the log.

```
*
 This step creates the data step view.
*;
DATA WORK.COUNT / VIEW=WORK.COUNT;
  SET WORK.LIST;
  I = _N_;
  PUT 'Creating observation ' I=;
  OUTPUT;
RUN;
```

```
NOTE: DATA STEP view saved on file WORK.COUNT.
NOTE: A stored DATA STEP view cannot run under a different operating
      system.
NOTE: DATA statement used:
      real time           0.03 seconds
```

Later in the same SAS session, this data step reads the data step view. It is at this point that the PUT statements in the view write their lines in the log. In this case, the view executes 3 observation at a time, as you can see in the sequence in which the lines are written in the log. The performance and activity notes in the log also indicate that it is at this point that the view actually executes.

```
*
 This step accesses the data step view.
 The view runs at the same time that this step runs.
*;
DATA _NULL_;
  SET WORK.COUNT;
  PUT 'Reading observation ' I=;
RUN;
```

```
Creating observation I=1
Creating observation I=2
Creating observation I=3
Reading observation I=1
Reading observation I=2
Reading observation I=3
Creating observation I=4
Creating observation I=5
Creating observation I=6
Reading observation I=4
Reading observation I=5
Reading observation I=6
Creating observation I=7
Creating observation I=8
Creating observation I=9
Reading observation I=7
Reading observation I=8
Reading observation I=9
Creating observation I=10
Reading observation I=10
NOTE: View WORK.COUNT.VIEW used:
      real time           0.03 seconds

NOTE: There were 10 observations read from the dataset WORK.LIST.
NOTE: There were 10 observations read from the dataset WORK.COUNT.
NOTE: DATA statement used:
      real time           0.07 seconds
```

If another step reads from the view, the view executes all over again. The step below reads only 5 observations from the view. As a result, the view program does not execute completely. It stops executing when the step that reads it stops reading from it. The view, generating 3 observations at a time, generates 6 observations in this case, although only 5 of them are used.

```
*
 This step accesses the data step view again,
 reading only one observation.
*;
DATA _NULL_;
  SET WORK.COUNT (OBS=5);
  PUT 'Reading observation ' I=;
RUN;
```

```
Creating observation I=1
Creating observation I=2
Creating observation I=3
Reading observation I=1
Reading observation I=2
Reading observation I=3
Creating observation I=4
Creating observation I=5
Creating observation I=6
Reading observation I=4
Reading observation I=5
NOTE: View WORK.COUNT.VIEW used:
      real time           0.04 seconds

NOTE: There were 6 observations read from the dataset WORK.LIST.
NOTE: There were 5 observations read from the dataset WORK.COUNT.
NOTE: DATA statement used:
      real time           0.14 seconds
```

Ordinarily, a data step view does not execute one observation at a time. It executes only long enough to generate a block of output, which is a certain number of observations. In this example, WORK.COUNT generates 3 observations at a time, but more typically, a view generates hundreds of observations at once. The number depends on the length of the observation along with other factors.

The example above shows a SAS step reading a data step view, but it is also possible for a macro program or an interactive window to read a view. Whenever any kind of program reads a view, that program executes the view program. For any kind of view, there are a few implications to consider about the way view programs execute.

- The view program takes time to execute every time the view is accessed. This can represent a substantial amount of execution time if a view does a significant amount of processing on a large set of data and is accessed repeatedly.
- The data of the view is assembled again each time the view is accessed. As a result, the view data is always up to date, in the sense that it always reflects the current state of any files that the view reads from to assemble its data. On the other hand, the view data can potentially change from moment to moment, so a program cannot rely on data staying the same in a view. The data of a view could change even between steps in the same execution of a SAS program.
- The program that accesses the view must not conflict with the execution of the view. A particular problem can occur if both programs access the same data. If a view reads data from a file, the view usually should not be used as input to a program that updates or modifies that same file.
- Two or more views could be running at the same time, either because a program accesses multiple views, or because one view refers to another view. It is possible for the same view to execute more than once at the same time.

For data step views, there is greater potential for conflict. A data step, with its general programming capabilities, has access to many of the resources and objects of a SAS session, with the possibility of conflict with another program in each object it relies on or modifies. Other kinds of views access only the files that contain the data they use, and their potential for conflict is limited to conflicts over those files.

The program of a SQL view is a query expression written in SQL. It might also include libref definitions to use in identifying the tables used in the query. For a SAS/ACCESS view, the view program is little more than a list of parameters that identify the view's data.

Compilation and Source Code

Both compiled data steps and data step views are compiled programs. They are compiled from a source program — a program you can write. But they are executed from the compiled version of the program, without referring to the source statements that created the program.

With any kind of compiled program, it is important to save the program's source code. Even though the source code is not necessary to execute the program, it is valuable as a document that shows exactly what the program does, and it is essential if you ever want to modify the program. Save the SAS program file that creates the compiled data step or data step view, including any steps that are necessary to create the input SAS datasets the program uses.

The data step includes options for saving source code. To store the source code along with the compiled program, write the SOURCE=SAVE option in parentheses after the SAS file name in the PGM= or VIEW= option:

```
VIEW=SAS dataset (SOURCE=SAVE)

PGM=SAS file (SOURCE=SAVE)
```

Use the SOURCE=NOSAVE option to store the compiled program without the source code. Another option is SOURCE=ENCRYPT, to store the source code in encrypted form. This option works only with an alter-level password for the file. The alter-level password protects the file from modifications, but it allows the user to execute the compiled data step or access the data of the view without a password.

If the source code is saved in the file, you can use the DESCRIBE statement to retrieve it. Write a DATA statement with the VIEW= or PGM= option as the only term, followed by a DESCRIBE statement:

```
DATA PGM=SAS file or VIEW=SAS dataset;
  DESCRIBE;
RUN;
```

The SAS supervisor writes the source statements in the log. When you use the DESCRIBE statement, the SAS supervisor does not execute the program. To execute a compiled data step and write its statements in the log, use both the DESCRIBE and EXECUTE statements:

```
DATA PGM=SAS file;
  DESCRIBE;
  EXECUTE;
RUN;
```

Interactive Processes

For every window that displays more than static text or graphics, there is a process that determines what data to display in the window and how to respond to user actions in the window. There is no set limit to the number of windows that can be open in an interactive SAS session and the number of processes that can be executing at the same time. Some processes are of particular interest because they have the potential to affect or interfere with the execution of a SAS program.

Data The Viewtable window and various other windows display or edit the data of a SAS dataset. While one of these processes has a SAS dataset open, it limits the actions any other process can take on that SAS dataset. If the data is displayed in browse mode, no other process can update or modify the SAS dataset. If it is displayed in edit mode, no other process can read or open the SAS dataset.

Conflicts can occur easily when you edit a SAS dataset, then run a program that reads that SAS dataset. To avoid the possibility of a conflict between the interactive process and the SAS program, close the edit window before you submit the program.

Settings In the SAS System Options, Titles, and Footnotes windows, you can change the essential settings of the SAS session. If you open the window, make a change in the window, then close the window, saving the changes, it changes the settings. But if you leave these windows open, the effects are not consistent across recent SAS releases and environments.

Changes may take effect immediately and affect steps that execute after that, even if the windows are still open, or they may take effect only when you close the window. If you close the window without saving changes, or with the CANCEL command, it may restores the settings to what they were when the window was opened, or it may take not action. Conversely, changes you make in program statements may be reflected in the windows, or the window may not reflect such changes and may override them when you close the window. To avoid uncertainty about the effects of these windows, do not leave them open while anything else is executing.

AF applications can also change any of these settings. It is usual for an AF application to change some system option settings for its own purposes, and it may or may not change them back at the conclusion of its processing. The changed system options can cause unexpected behavior if you submit a SAS program while the AF application is open.

In one of the most common and disconcerting scenarios, the AF application can turn off the options SOURCE and NOTES. If you submit SAS program lines while these options are off, the log may give no indication that the program executed. You could easily get the impression that the program you submitted had not run, even though it did.

Conversely, if the program lines you submit contain an OPTIONS statement, the options you set in the program could adversely affect the AF application. If you can, close an AF application before you submit a separate SAS program. If an AF application does not reset system options when it closes, you might have to reset them yourself after you close the AF application.

Submitted statements In addition to the ability to open SAS datasets and change settings, AF applications have the ability to submit SAS statements for execution. These statements go into the same queue in which the SAS supervisor executes program lines you submit from the Program Editor window and elsewhere in the SAS session.

Most AF applications are executed with the AF command. There are a few AF applications that are included with specific SAS products, and they can be executed with other commands or with menu selections.

These are examples of AF applications in the SAS System:

Application	Software Product	Command
SQL Query	Base SAS	QUERY
SAS/ASSIST	SAS/ASSIST	ASSIST
Analyst	SAS/STAT	
SAS/PH-Clinical	SAS/PH-Clinical	PHCLINICAL
Projman	SAS/OR	PROJMAN
Time Series Forecasting	SAS/ETS	FORECAST

There are other SAS windows that can submit SAS statements, and it is possible for procs to submit SAS statements in the same way. With the CALL EXECUTE routine, data steps also have a limited ability to submit SAS statements. However, only one step can execute at a time, so statements submitted by a proc or by a data step do not execute until the proc or data step ends.

Some AF applications assemble and submit a program in pieces, a few statements at a time. These pieces are meant to go together, so do not submit any other SAS program lines in the middle of this process, and do not run other AF applications that might submit SAS statements at the same time.

When you run more than one interactive process at a time in a SAS session, and when you run SAS programs and interactive processes in the same session, be alert to the potential for conflict. If you discover that one interactive process can interfere with another, do not run both processes at the same time.

SAS Processes

SAS processes make it possible to run more than one program at the same time from within the same session. The SAS supervisor divides the available computer resources between the two programs. The most obvious use for a SAS process is to run a SAS program at the same time that you are working in an interactive SAS session. Ordinarily, when you submit a SAS program in an interactive SAS session, you have to wait for the program to finish before you can take any other action in the session. But when you run a SAS program as a separate process, you can do other things in the session while the program is still running.

Use the STARTSAS statement to start a new SAS process. In the STARTSAS statement, write system options for the process, using the same syntax for the options as you would use in the OPTIONS statement. You can also start a SAS process with the STARTSAS command, which uses a dialog box to select a program and set options.

The most important system options for a SAS process are SYSIN=, to identify the program file to run, LOG= and PRINT=, for the log and standard print files, and CONFIG=, for a configuration file. Any of the other options listed in the earlier "Startup Options" table can be used in the STARTSAS statement, as can most system options.

This is an example of a STARTSAS statement:

```
STARTSAS SYSIN="a.sas" LOG="a.log" PRINT="a.lst";
```

As in batch mode, when you use a SAS process to run a SAS program, the SAS supervisor writes the log to a file. After the program ends, you can read the log file to see how the program ran — and read the standard print file for the output of the program.

When the SAS supervisor starts a SAS process, it assigns it a process ID, which is a unique code name for the SAS process. Use the automatic macro variable SYSSTARTID to determine the process ID for a process you start. SYSSTARTID contains the process ID of the most recently started SAS process. The process ID can be used in subsequent statements to manage the SAS process. There is also a process name for each SAS process. The macro variable SYSSTARTNAME contains the name of the most recently started SAS process. The macro variables SYSPROCESSID and SYSPROCESSNAME contain the process ID and the process name of the current SAS process.

In a SAS process, the ENDSAS statement ends the SAS process, but it does not end the SAS session. To end a SAS process other than the current one, use the process ID as a term in the ENDSAS statement:

```
ENDSAS process ID . . . ;
```

To end the SAS session and all its SAS processes, use the ALL or _ALL_ keyword in the ENDSAS statement:

```
ENDSAS _ALL_;
```

```
ENDSAS ALL;
```

SAS processes can easily conflict with each other if they access the same files. If necessary, use the WAITSAS statement in one SAS process to have it wait for the completion of another SAS process:

```
WAITSAS process ID . . . ;
```

The WAITSAS and ENDSAS statements have no effect if you use them with a SAS process ID that is not valid or if the SAS process has already ended.

System Tasks

There are several ways a SAS program can ask the operating system to take a specific action in the form of an operating system command:

- SYSTEM function (in a data step)
- CALL SYSTEM routine (in a data step)
- %SYSEXEC macro statement
- X statement (a global statement)
- X command
- SYSTASK statement (in some environments)

The following table shows more details of each of these alternatives.

Element	Used In	Example
SYSTEM function	Data step	`RC = SYSTEM('copy a b');`
CALL SYSTEM routine	Data step	`CALL SYSTEM('copy a b');`
%SYSEXEC macro statement	Macro programming	`%SYSEXEC copy a b;`
X statement	Global statement	`X 'copy a b';`
X command	SAS command line	`x 'copy a b'`
SYSTASK statement	Global statement	`SYSTASK COMMAND 'copy a b' TASKNAME=COPY;`

The operating system command in any of these language elements is a command in the same form that you would use as an operating system command line. The language element executes the operating system command. Usually, the next statement does not execute until after the operating system command executes. All these language elements work that way except for the SYSTASK statement.

The `SYSTASK COMMAND` statement creates an asynchronous system task, one that executes separately, the same way that a SAS process executes separately. You can use the TASKNAME= option in the statement to give a name to the task. Then you can use that name to keep track of the task. The LISTTASK or `SYSTASK LIST` statement writes a log note with information about asynchronous system tasks:

```
SYSTASK LIST _ALL_;

LISTTASK _ALL_;

SYSTASK LIST task name;

LISTTASK task name;
```

You can use the `SYSTASK KILL` statement to end a system task before it completes:

```
SYSTASK KILL task name;
```

To have a SAS program wait for an asynchronous system task to complete, use the WAITFOR statement:

```
WAITFOR task name TIMEOUT=n;
```

The TIMEOUT= option indicates a maximum length of time to wait, in seconds.

To wait for any one of several asynchronous system tasks, use the _ANY_ keyword in the WAITFOR statement:

```
WAITFOR _ANY_ task name task name ... TIMEOUT=n;
```

To wait for all of a set of tasks to complete, use the _ALL_ keyword

```
WAITFOR _ALL_ task name task name ... TIMEOUT=n;
```

Multiple Users

Conflicts are nothing new in the file space of networks and multiuser computers. The conflicts that occur when two processes look for the same file at the same time also occur when two users or jobs attempt to access the same file. Usually, the file system gives access to the file to the first user who opens the file and returns an error code to the second user.

For an application in which it is important for several users to be able to update the same SAS files simultaneously, you can use SAS/SHARE to manage the libraries that contain those files. The SAS/SHARE server can allow different users to access different observations in the same SAS data file at the same time.

Conflicts and Errors

Conflicts between programs usually lead to errors — often, to predictable kinds of errors. The errors can be avoided by keeping the conflicting programs sufficiently separated.

One illustration of a conflict between two programs can be seen in the interactive environment. When you are writing a SAS program that uses a specific SAS dataset, you might open the Properties dialog box of the SAS dataset to see the names or other attributes of its variables. However, if you submit the program to execute while the Properties dialog is still open, the program may generate an error message like this one:

```
ERROR: You cannot open WORK.PRACTICE.DATA for input access with page-level
control because WORK.PRACTICE.DATA is in use by you in resource environment
DMSEXP.
```

The SAS program cannot open the SAS dataset it needs because the Properties dialog already has the SAS dataset open. The two programs — the SAS program and the Properties dialog — cannot both run at the same time. This problem is easy to fix. Close the Properties dialog, and then submit the SAS program to execute.

Whenever the SAS supervisor attempts to open a file that is not available because it is already being used, it generates an error condition. It could be another user or job or another process in the same SAS session that is using the file. A file can be open in two places at once only if both programs open the file in input mode, that is, only for the purpose of reading from the file. Any other attempt to open a file that is already open elsewhere results in an error. The SAS supervisor handles this kind of error in much the same way as the error that occurs when the file does not exist.

Some procs and AF applications that read the same SAS file more than once usually lock the file. This prevents any other process from opening the file they are working with. There are two main reasons to do this. First, it ensures that the file will continue to be available. Second, it ensures that the data in the file will not change in the middle of the action that the program is taking. If a SAS program depends on having the same data in a SAS file over a span of multiple steps, the program can lock the file with the LOCK statement. The LOCK statement can also lock an entire library or just a single entry:

```
LOCK libref;
LOCK SAS file;
LOCK entry;
```

Later, use the CLEAR option at the end of the LOCK statement to unlock the file:

```
LOCK libref CLEAR;
LOCK SAS file CLEAR;
LOCK entry CLEAR;
```

Other conflicts between programs in a SAS session are more likely to result in logical errors. For example, if one program requires a certain line size to print correctly, it may print incorrectly if another program sets the LINESIZE= system option to a different value. To avoid or minimize these kinds of conflicts, you might do one or more of the following:

- Run only one program at a time.
- For maximum isolation, run SAS programs in batch mode.
- Set any necessary system options and other settings at the beginning of each program that relies on them.
- Create and run a separate program that resets system options and other settings to the defaults for a project.
- If you create an AF application or other interactive application that sets system options, note the previous settings of those options and restore those settings afterward. If possible, restore system option settings before returning control to the user.

End

There are several ways a SAS session can end. In batch mode, the end of the program file leads to the end of the session. In interactive mode, the BYE or ENDSAS command or the appropriate menu selection or other user action ends the session. In any mode, you can use the ENDSAS statement to end the session.

By default, the SAS supervisor erases the WORK library at the end of the session. This removes temporary SAS files, macros, formats, and other objects. The process of clearing the WORK library is controlled by the WORKTERM system option. To avoid erasing the WORK library at the end of the session, when you have a reason to preserve the files of the WORK library, turn this option off. Use operating system commands to move the files from the WORK library to another location before you start the next SAS session.

Variability of the SAS Environment

A SAS program will not necessarily run the same way after you move it to a different place. A program that runs in one environment might run differently, or not at all, in a different environment. Conversely, you might have to change a program in order for it to produce the same results in a different context.

There are good reasons for the variations you will find in the SAS environment. Different SAS installations can have different products, versions, operating systems, and hardware. Regardless of the installation, there can be variations in modes, files, and system options. All of these are issues you might have to address when you develop a SAS program that is supposed to run in more than one place and when you want to work with SAS code that was written somewhere else.

Products

SAS Institute licenses SAS software as a set of products. Base SAS, the first product, includes support for the SAS language, SAS files, the windowing environment, macro language, SQL, and many other essential and generally useful features of the SAS System. Other products add specialized features. There are more than 25 SAS products, but many SAS sites have only base SAS, and others have various combinations of the available SAS products. The SAS products in a specific installation can change from time to time as products are added or taken away.

If a SAS program attempts to use a SAS product that is not available, the result is an error condition. The error message might say that the product is not licensed, that the necessary routines or files cannot be found, or that the statement or option is not recognized.

The programs in this book use the features of base SAS. When SAS programs refer to other SAS products, they most often use the procs or other routines in those products. A few products also have global statements and additional options for base SAS features. The following table lists selected SAS products that contain routines that can be used in SAS programs.

Product/Description	Routines and Language Elements
SAS/ACCESS Products for database and file access	Procs and engines for access to database files SQL components LIBNAME statement options Options for the IMPORT and EXPORT procs
SAS/CONNECT Connects to SAS software on another computer	Procs for server management and file transfer SIGNON, RSUBMIT, and 4 other statements for remote sessions LIBNAME statement options and SQL components for access to remote data

continued

Product/Description	Routines and Language Elements
SAS/ETS Time series analysis	15 procs PROBDF function
SAS/GIS Spatial data	GIS proc
SAS/GRAPH Graphics	Procs for creating and managing graphics GOPTIONS, AXIS, SYMBOL, LEGEND, and PATTERN statements Formatting options in TITLE and FOOTNOTE statements Data Step Graphics Interface (DSGI) functions for creating graphics
SAS/MDDB Server Multidimensional database	MDDB proc for building multidimensional databases
SAS/OR Operations research and project management	4 procs for project management 4 procs for network models 2 procs for linear and nonlinear programming
SAS/QC Quality	9 procs Functions
SAS/SHARE Data access for multiple users	SERVER and OPERATE procs for server management Engines and LIBNAME statement options for shared libraries
SAS/STAT Statistics	55 procs for statistical analysis, modeling, testing, and experimental design

In addition, it is possible for a SAS program to use add-on routines that are not part of the SAS System. Programmers develop custom engines, functions, procs, and other routines with SAS/TOOLKIT. Those routines can then be used in a SAS program in the same way that SAS routines can be used. A SAS program that relies on custom routines will work only where those routines are installed.

The SYSPROD function checks licensing information for a SAS product. In some programs, you may want to use the features of a SAS product only when the SYSPROD function indicates that the product is available.

Versions

The development of SAS software is an ongoing process. SAS Institute issues new releases at least once a year, adding new features and sometimes changing the existing features. On occasion, a program has to be revised slightly to work correctly in a new release. When a program relies on newer features of SAS, it is not always possible to run it in an older SAS release.

SAS software is identified according to three distinct levels of revisions. A whole number *version*, such as 5 or 6, represents a comprehensive change in SAS software. A *release* is a fractional number that begins with the version number, such as 6.09. It represents a significant change in SAS software, with new features. A *technical support level* identifies a specific image of a specific version. Technical support levels are whole numbers, often written with the letters TS, such as TS210. New technical support levels usually do not add new features. The distinction between *version* and *release* is especially significant in the titles of SAS Institute books.

The following table summarizes, briefly, the history of SAS software. When you work with an older SAS program, it is helpful to consider the context in which it was written.

Time Period	Version	Description	Key features
early 1980s	79, 82	Statistics package	SAS language/datasets Statistics tools IBM mainframe
late 1980s	4, 5	A tool for data center programmers	Analysis tools Macro language Multiuser platforms
1990s	6	Integrated software for working with data	Modular design Portable Introduced most SAS features
c. 2000	7, 8	The center of business intelligence solutions	Object-oriented design Integration features

Most SAS products use the same version numbers as the SAS System. However, some specialized interactive products have their own independent version numbers.

The automatic variables SYSVER and SYSVLONG provide SAS version information. SYSVER indicates the release number; SYSVLONG, the release number and technical support level. Use these macro variables in macro programming logic if it is necessary for a SAS program to have different statements for different SAS releases.

Operating Systems

SAS software runs on several operating systems and includes features to work with the unique capabilities of each environment. SAS programs that rely on the specific features of an operating system have to be modified if you move them to a different operating system. Operating system features may include these:

- Operating system commands
- The file system, including file names, the forms of directories, paths, libraries, etc., and support for specific file types

- A specific standard for the format of text files
- The character set
- Numeric data formats
- Special keys on the keyboard (such as function keys)
- Fonts
- Graphical user interface components
- Networking and communications standards

Within a particular operating system, there can be significant variations. There are different versions of each operating system. An operating system could have optional components that supply important functionality. It has become common for an operating system to support more than one file system.

The macro variables SYSSCP and SYSSCPL identify the operating system. If you want to write a single SAS program that runs in more than one operating system, you might need to use macro programming logic that tests the values of these macro variables in order to generate different SAS statements for different operating systems.

Other Possible Dependencies

Other properties of the SAS environment are more fluid, tending to vary from one computer to the next, or capable of changing from one day to the next. It is still possible for a SAS program to depend on these qualities of the SAS environment, so these are dependencies that should be looked for when a program is first moved from one place to another.

Hardware One responsibility of an operating system is to connect the computer system's software to its hardware. It does so with the use of routines called drivers. Drivers can be part of the operating system, or they can be routines that are added on to support a specific device, such as a modem, a sound card, or a printer. The drivers make the features of the hardware device available to the computer. Programs that rely on specific features of specific hardware devices or drivers may have to be rewritten if they are moved to a different computer with different hardware and drivers.

There are variations in the computer hardware that is connected to computers. For example, not every keyboard has the same set of keys; not every mouse has the same number of buttons; video displays vary in size and in the set of colors they can display. If you develop programs that use hardware features such as these, limit the program to the common features of hardware when you can.

Modes A SAS program can run in either batch mode or the windowing environment. In some operating systems, SAS also supports other execution modes for SAS programs. A few details of the way the SAS supervisor executes a program depend on the execution mode.

In particular, the SAS supervisor can prompt for the password of a SAS file only if a user is available; in batch mode, passwords must be written in the program. The way the SAS supervisor recovers from most errors differs between windowing and batch modes. In batch mode, it goes into syntax check mode. In an interactive session, it continues to execute the program as well as it can.

Files Most SAS programs work from specific input data files. Other kinds of files are often necessary to a program: secondary program files, libraries, directories, formats, macro libraries, and so on. To make a program easier to move and maintain, you can write it so that all specific file dependencies are in LIBNAME and FILENAME statements at the beginning of the program. Sometimes, all these file declarations can be moved to an autoexec file.

System options A program can depend on specific system option settings. Ideally, these system options should be set in the program itself or in the process of running the program. If a project uses a particular set of system options, it is best to write the options in a separate configuration file specifically for that project.

Be aware of a SAS program's use of computer resources, particularly its temporary use of hard disk space. Running out of hard disk space was once the most frequent reason for a correctly written SAS program to fail, and it still happens. Be alert to the possibility of running out of storage space when you when you move a program from one computer to another or when a program's resource requirements increase because it is working with a larger body of data.

18

Macro Language

Macro language is a language that is designed for generating SAS statements. A SAS program can contain references to macro variables and other macro objects. The macro processor resolves those references, converting them into the text of the SAS program that the SAS supervisor executes. By using macro language, you can have a SAS program file that does not execute the same SAS code every time it runs. This can give a SAS program more flexibility that it could have with the SAS language alone. Macro language can also be used as a programming language in its own right.

The Objects of Macro Language

The macro processor works with specific types of objects. Most macro objects generate program text when the macro process resolves them. Others do not generate any program text, but take specific actions of their own.

Macro Variables

A macro variable is a text value that is associated with a name. It can be used like a SAS character variable, but it is not the same; a macro variable does not have a fixed length, and it can hold a value of any length.

The %LET statement is the most direct way to give a value to a macro variable. This macro statement works much like the assignment statement of the data step, but the value on the right side of the equals sign is just text. This statement gives the macro variable PROJECT the value `MAIN`:

```
%LET PROJECT = MAIN;
```

If the macro variable does not already exist when you assign it a value, the macro processor creates it. If the macro variable does exist, the %LET statement gives it a new value.

There are various other ways a macro variable can obtain a value. In macro language, a macro variable can get a value as a parameter of a macro or an argument to the %SYSFUNC macro function, among other ways. A data step can use the `CALL SYMPUT` routine to give a value to a macro variable. A macro variable can also be defined in a %GLOBAL statement or in some other way that does not specifically give it a value; the new macro variable is given a null value, a value with a length of 0.

An existing macro variable can be used to generate characters of a SAS program. A macro variable reference can take either of the following forms:

```
&name
&name.
```

A reference to the PROJECT macro variable would be either of:

```
&PROJECT
&PROJECT.
```

The final period is necessary as a separator if the character that follows the macro variable reference is a letter, underscore, digit, or period.

This is an example of a SAS statement that contains a macro variable reference:

```
DATA &PROJECT..QUALIFY;
```

The macro processor substitutes the value of the macro variable for the macro variable reference. If the value of the macro variable PROJECT is MAIN, then the actual SAS statement that results from the program line above is:

```
DATA MAIN.QUALIFY;
```

Once a macro variable exists, there is no statement you can use to destroy it. You can, however, use the %LET statement to assign a null value to it, for example:

```
%LET PROBLEM = ;
```

When a macro variable has a null value, a reference to it has no effect. The macro processor removes the reference and does not substitute any text for it.

There are several automatic macro variables. These macro variables contain values that reflect the state of the SAS session. You do not have to define them to use them. These are examples of automatic macro variables:

SYSDATE9	The date when the SAS session started, written in DATE9. format
SYSDAY	The day of the week when the SAS session started
SYSMAXLONG	The maximum value of a long integer
SYSPARM	The parameter string defined in the SYSPARM= system option
SYSTIME	The time of day when the SAS session started
SYSUSERID	The user ID

To write the names and values of all automatic macro variables in the log, use this macro statement:

```
%PUT _AUTOMATIC_;
```

Macros

A macro, like a macro variable, is a text value that is associated with a name. In its simplest form, a macro could be used in much the same way as a macro variable. However, there are several qualities of macros that make them different from macro variables.

- A macro can have parameters. Parameters are macro variables that are given values in the macro reference.
- A macro can contain macro control flow statements to generate text repeatedly or to generate different text in different conditions.
- A macro can have local macro variables, which are available only while the macro is active. They disappear as soon as the macro is completely resolved.
- A macro is stored in a file.
- Any macro references that are contained in a macro definition are not resolved when the macro is defined. The macro processor resolves them when it resolves the macro. This allows a macro to generate different text each time it is called.

A macro is defined with the %MACRO and %MEND macro statements. In a simple macro with no parameters, the %MACRO statement just indicates the name of the macro. The %MEND statement marks the end of the text of the macro.

```
%MACRO name;
text
%MEND;
```

The following is an example of a simple macro definition. This macro determines the current month and year and uses them to generate a TITLE statement.

```
%MACRO MTITLE;
%LET MONTH = %SYSFUNC(DATE(), MONYY7.);
TITLE1 "&MONTH  Product Details by Region";
%MEND;
```

Note these points about this example:

- The macro definition starts with the %MACRO statement, which indicates the name of the macro. In this case, the macro name is MTITLE.
- The %SYSFUNC macro function uses the DATE function to determine the current date. It uses the MONYY7. format to convert the resulting value to text. The %LET statement assigns the resulting value to the macro variable MONTH.
- Because MONTH is created inside the macro, it is a local macro variable. It can be used only inside the macro.
- In the TITLE statement, &MONTH is a reference to the macro variable MONTH.

- The character constant in the TITLE statement is written with double quotes so that the macro processor will resolve the macro variable reference inside the quoted string.

A simple macro reference is a percent sign followed by the name of the macro:

```
%name
```

For example, after the MTITLE macro is defined, you can use it by writing this macro reference, or macro call, in the program:

```
%MTITLE
```

When the macro processor resolves, or executes, the MTITLE macro, it generates text to form a TITLE statement such as the following:

```
TITLE1 "AUG2001   Product Details by Region";
```

However, the title line contains the current month and year, as of the moment the macro obtains that information from the DATE function. This example illustrates the usefulness of macro programming. If you want a title line to contain the current month and year, you can use this kind of macro to automatically generate the appropriate TITLE statement. The effect is the same as if you changed the program every month to create a new TITLE statement, but the macro makes the change in the SAS code automatically, without requiring any changes in the program file.

The variable information of the MTITLE macro comes from the system clock, but most macros get their key information from parameters. The macro reference indicates values for the parameters, which the macro definition uses as local macro variables. The parameters are written in parentheses after the macro name, with commas between them, much like the arguments of a function or CALL routine.

There are two kinds of parameters. Most often, a macro uses positional parameters. The macro definition lists the names of the parameters, in order:

```
%MACRO macro(parameter 1 name, parameter 2 name, . . . );
text
%MEND;
```

The macro reference lists values in the same order:

```
%macro(parameter 1 value, parameter 2 value, . . . )
```

Some macros use keyword parameters, which are like options. The macro definition lists the names and default values for the parameters:

```
%MACRO macro(parameter=default, parameter=default, . . . );
text
%MEND;
```

The macro reference can list names and values for keyword parameters in any order:

```
%macro(parameter=value, parameter=value, . . . )
```

Keyword parameters can be omitted in the macro reference. Then the macro uses the default value for that parameter. If a macro uses both positional and keyword parameters, the positional parameters must be listed first.

The example below defines a macro that uses both kinds of parameters. Like the previous example, this macro generates a TITLE statement that contains the current month. However, it derives the text of the title line from a parameter.

```
%MACRO LTITLE(TEXT, LINE=1);
%LET MONTH = %SYSFUNC(DATE(), MONYY7.);
TITLE&LINE "&MONTH  &TEXT";
%MEND;
```

These are several different ways that the macro reference could be written for this macro, along with the program text that the macro generates as a result (assuming the program runs during the month of August 2001):

Macro call: `%LTITLE(Product Details by Region)`
Generates: `TITLE1 "AUG2001  Product Details by Region";`

Macro call: `%LTITLE(Product Details by Region, LINE=1)`
Generates: `TITLE1 "AUG2001  Product Details by Region";`

Macro call: `%LTITLE(Region Summary, LINE=3)`
Generates: `TITLE3 "AUG2001  Region Summary";`

Macro call: `%LTITLE(Table of Contents)`
Generates: `TITLE1 "AUG2001  Table of Contents";`

Note these points about the LTITLE macro:

- The parameters are TEXT and LINE. They are used as macro variables in the macro definition.
- TEXT is a positional parameter. A value must be provided for it whenever the macro is called.
- LINE is a positional parameter. Its default value is 1. To provide a different value for this parameter, it is necessary to write both the parameter name and the value in the macro reference.
- TEXT, LINE, and MONTH are local macro variables.
- The first macro call shown generates the same program text as the MTITLE macro of the previous example.
- The second macro call shown works the same way as the first. It demonstrates that a macro call can list a keyword parameter with its default value, and this is the same as not listing the keyword parameter.
- With different parameter values, the macro generates different program lines. This demonstrates the flexibility of a macro with parameters.

When you define a macro, the SAS supervisor stores it as an entry in a catalog in the WORK library. You can then use the macro any number of times in the same SAS session. The simplest way to work with a macro is to define it and use it in the same SAS program. The stored macro is automatically deleted when the SAS supervisor clears out the WORK library at the end of the SAS session.

It is possible for a macro to contain a reference to another macro. In this example, the %RTITLE macro consists only of a call to the LTITLE macro:

```
%MACRO RTITLE;
%LTITLE(Product Details by Region)
%MEND;
```

When the macro processor resolves the RTITLE macro, it comes to the LTITLE macro call. It has to stop resolving the RTITLE macro temporarily in order to resolve the LTITLE macro. For that period of time, both macros are active. When the LTITLE macro ends, the macro processor goes back to the RTITLE macro. It is common in some styles of macro programming to have a macro call another macro, which in turn calls another macro, and so on. There is no fixed limit to the number of macros that can be executing at the same time.

The purpose of most macros is to generate program text. The actions that result from the macro are the actions of the SAS statements that the macro generates. However, it is also possible to write a macro that takes actions in its own right. Such a macro contains macro statements, rather than SAS statements, and it is easier to think of the macro processor as executing the macro rather than resolving it.

Macro Statements

Use macro statements for actions that take place during the process of resolving macro references. Other macro objects tell the macro processor what program text to generate, but macro statements do not generate any program text. Instead, they tell the macro processor to take separate actions. Several macro statements have been described already: the %PUT statement, to write messages in the log; the %LET statement, to assign a value to a macro variable; and the %MACRO and %MEND statements, to define a macro.

All macro statements begin with a percent sign and a keyword. Like SAS statements, they end with a semicolon. However, they are not the same as SAS statements. It is possible for a macro statement to appear anywhere in a SAS program. A macro statement is usually written between two SAS statements, but it does not have to be. The SAS supervisor and macro processor can still find a macro statement in the middle of a SAS statement, and sometimes, particularly with the %DO statement, a macro statement has to be written that way.

Some macro statements, particularly the macro control flow statements, can be used only in macro definitions. Others can be used anywhere in a SAS program or anywhere a macro reference can appear.

%GLOBAL and %LOCAL — Declaring macro variables The %GLOBAL and %LOCAL statements declare macro variables, especially in macro definitions. The only way a macro can create a global macro variable is by declaring it in a %GLOBAL statement, for example:

```
%GLOBAL PAGENUMBER;
%LET PAGENUMBER = 1;
```

The global macro variable remains available even after the macro ends.

Use the %LOCAL statement in a macro to ensure that a macro variable used in the macro is a local macro variable, even if a global macro variable already exists with the same name. A macro's local macro variables do not interfere with any global macro variables that exist. Local macro variables are available only while the macro is active and disappear as soon as the macro ends. If any global macro variables have the same names as the local macro variables of a macro, the global macro variables are inaccessible to the macro.

%PUT — Log lines The %PUT statement writes a line in the log. Any kind of text can appear in the message the %PUT statement writes. If there are any macro language references in the text of the %PUT statement, they are resolved before the text is written to the log. This is an example of the use of the %PUT statement:

```
%PUT The current date is %SYSFUNC(DATE(), DATE9.).;
%PUT The SAS session began on &SYSDATE9..;
```

```
The current date is 22JUN2000.
The SAS session began on 20JUN2000.
```

In addition to its ability to write a text value, the %PUT statement can write lists of macro variables in the log. To do this, use one of the following keywords in the %PUT statement:

ALL	All macro variables
GLOBAL	Global macro variables
LOCAL	Local macro variables of the current macro
USER	Global macro variables and local macro variables of all macros that are currently executing
AUTOMATIC	Automatic macro variables

This example shows the kind of output that the %PUT statement produces with these keywords:

```
%LET A = 1.0;
%LET B = 2.0;
%PUT _GLOBAL_;
```

```
GLOBAL A 1.0
GLOBAL B 2.0
```

Macro control flow There are several macro control flow statements that can be used in macros to control the way the macro executes macro statements and generates text. The following table shows the syntax of these statements and describes their actions.

Macro Statement	Action
%IF *condition* %THEN *text*;	Generates *text* only if the condition is true.
%ELSE *text*;	Generates *text* only if the condition of the preceding %IF-%THEN statement is false.
%DO; *text* %END;	Treats *text* as a block, especially for use in %IF-%THEN and %ELSE statements.
%DO *macro variable name* = *start* %TO *stop*; *text* %END; %DO *macro variable name* = *start* %TO *stop* %BY *increment*; *text* %END; %DO %WHILE(*condition*) *text* %END; %DO %UNTIL(*condition*) *text* %END;	Generates *text* repeatedly. The iteration controls of the %DO statement are similar to those of the DO statement of the data step, but should be used only with integer values.
%GOTO *label* ;	Jumps to the macro label. The label must be in the same macro.
%*label* :	A macro label. The target of a %GOTO statement.

The following example shows the use of an iterative %DO statement to generate repetitive program text in a macro. This macro generates a data step that can divide the observations of an input SAS dataset among any number of output SAS datasets based on the value of the variable REGION.

```
%MACRO REGIONAL(N);
DATA %DO I = 1 %TO &N; WORK.ACT&I (DROP=REGION) %END;;
  SET MAIN.ACT;
  SELECT (REGION);
%DO I = 1 %TO &N;
    WHEN(&I) OUTPUT WORK.ACT&I;
%END;
    OTHERWISE ;
    END;
RUN;
%MEND;
```

The number of SAS dataset that the data step creates depends on the parameter N. For example, with this macro call:

```
%REGIONAL(5)
```

the macro generates this program text:

```
DATA WORK.ACT1 (DROP=REGION) WORK.ACT2 (DROP=REGION) WORK.ACT3
(DROP=REGION) WORK.ACT4 (DROP=REGION) WORK.ACT5 (DROP=REGION);
SET MAIN.ACT;
SELECT (REGION);
WHEN(1) OUTPUT WORK.ACT1;
WHEN(2) OUTPUT WORK.ACT2;
WHEN(3) OUTPUT WORK.ACT3;
WHEN(4) OUTPUT WORK.ACT4;
WHEN(5) OUTPUT WORK.ACT5;
OTHERWISE ;
END;
RUN;
```

If the number of distinct values of REGION is a known constant, it would be easier to code this data step directly. However, if the number varies, the macro coding makes it easy to adjust the program to work with a different number — only the macro parameter value has to be changed to change the program.

The conditions of macro control flow statements are based on macro arithmetic, which uses only integer values. Comparison operators compare integer values or text values.

Macro control flow statements can be used with macro statements to control the way the macro statements execute. The following example shows how this can work. The %PUT statement is executed only when specific values are found for the macro variables MON and DAY.

```
%IF (&MON = JAN) AND (&DAY = 1) %THEN %PUT Happy new year!;
```

Macro comment statement A macro comment statement begins with %* and ends with a semicolon. It does not result in any action. Usually, macro comment statements are used to describe macros and macro statements. When a macro comment statement is used in a macro definition, it does not generate any program text.

Macro Functions

Macro functions are used like macros — in fact, many of them are macros — but they are a part of macro language. Most macro functions are designed to apply simple effects to text values. Others are used for special purposes or for macro quoting, which affects the way the macro processor responds to special characters in text.

One of the largest categories of macro functions are those that apply character effects or measure character values in much the same way that SAS functions do. These macro functions work in essentially the same way as the SAS functions of the same name:

%INDEX	%LOWCASE	%SUBSTR	%UPCASE
%LEFT	%SCAN	%TRIM	%VERIFY

The %CMPRES macro function removes leading and trailing spaces and converts multiple spaces to single spaces, similar to the effects of the function call COMPBL(TRIM(LEFT(STRING))). The %LENGTH macro function measures the length of its argument including any leading and trailing spaces.

The %EVAL macro function evaluates an expression, using the integer arithmetic of macro language. The %SYSEVALF macro function evaluates an expression of the sort that would be used in data step programming to produce a numeric result. It makes it possible to do floating-point arithmetic, with fractional values, in macro programming. An optional second argument applies a mathematical effect to the value of the expression. The effect codes that you can use as the second argument to the %SYSEVALF macro function are FLOOR, CEIL, INT or INTEGER, TRUNC, and BOOLEAN.

The following table shows examples of the use of macro functions.

Macro Function Call	Return Value
%SUBSTR(EASTERN, 1, 4)	EAST
%SCAN(MAIN.POWER, 2, .)	POWER
%LENGTH(MAIN.POWER)	10
%CMPRES(Atlantic City NJ)	Atlantic City NJ
%LOWCASE(EASTERN)	eastern
%EVAL(2*25 + 15)	65
%EVAL(45/4)	11
%SYSEVALF(45/4)	11.25
%SYSEVALF(45/4, CEIL)	12
%SYSEVALF(45/4, BOOLEAN)	1

Other macro functions apply macro quoting effects, access environmental information, and allow the use of SAS functions in macro programming.

In comparing macro functions to SAS functions, the most important difference to note is that all arguments and return values for macro functions are text values. The text arguments are not enclosed in quote marks the way character values used as arguments to SAS functions would be.

Most often, macro functions are used with macro variables as arguments or with arguments formed from macro variables. Write the macro variable references as usual, with an ampersand before the macro variable name, for example:

```
%LOWCASE(&FILENAME)

%LENGTH(&TITLETEXT)

%EVAL((132 - %LENGTH(&TITLETEXT))/2)
```

Macro Expressions

As the macro processor is designed to work with text, a macro expression is simply text that might contain macro objects.

The Macro Processor

The macro processor resolves macro objects in program lines — and in the macro objects used in those macro objects. It takes these actions:

- Executes macro statements.
- Substitutes values for macro variables.
- Executes macros and macro functions and substitutes the resulting text for the macro call or macro function call.
- Converts each two consecutive ampersands to a single ampersand.

If any macro objects remain in the resulting text, the macro processor resolves them too, repeating the process until it produces text that does not contain any macro objects.

Errors can occur if the macro processor cannot find a macro object or if the syntax of a macro object is incorrect. The macro processor produces error messages that describe the error. Errors can also occur if the macro objects generate SAS statements that are incorrect. The SAS supervisor produces those error messages, the same way that it does with statements that it retrieves directly from the program lines.

Macro language is mainly used in SAS programs, but it can also be used in other places in SAS software. Macro objects can be used in commands in the windows of the interactive SAS environment and in the SCL programs of AF applications.

Macro Operators

The macro processor looks for operators in order to evaluate expressions used as arguments to macro functions and as conditions in macro control flow statements.

You can use most of the SAS operators in macro expressions, but the results may be different from the results you find in the data step. The arithmetic operators work only with integer values, truncating any fractional part of a number. Comparison operators do numeric comparisons only if both arguments are integers; otherwise, they do text comparisons.

Use the %EVAL function to force the evaluation of a macro operator elsewhere in a program, as in this example:

```
TEAM%EVAL(2*&I)
```

If the value of the macro variable I is 8, then the expression evaluates as `TEAM16`.

To prevent the evaluation of an operator that might occur in the text of an argument to a macro function, use a macro quoting function.

Macro Quoting

Macro quoting, which is done by macro functions, changes the way the macro processor treats certain special characters, especially ampersands, percent signs, semicolons, spaces, unmatched parentheses, and unmatched quote marks. There are several macro quoting functions because quoting can be done in different ways. Each function has one argument, which is the text that the function quotes.

The most basic macro quoting function is the %QUOTE function. Its only action is to quote text. The result of quoting is that the macro processor does not look for operators or the punctuation characters of macro language; instead, it treats the value simply as text. Quoting is sometimes necessary for the values in a macro call or macro function call. In order to use a macro parameter value that contains a special character such as a comma, a semicolon, or an equals sign, the macro parameter must be quoted, so that the special character does not appear to be part of the macro language syntax.

For example, you can use the %SCAN function to divide a value at the commas in the value. The value `Atlantic City, NJ` would be divided into `Atlantic City` and `NJ`. To use arguments that contain commas, the arguments have to be quoted. Otherwise, the value that contains the comma would appear to be two separate values. By quoting the value, you remove the special meaning from the comma, an effect called *masking*, to allow the macro processor to see the comma as a text value.

```
%PUT %SCAN(%QUOTE(Atlantic City, NJ), 1, %QUOTE(,));
%PUT %SCAN(%QUOTE(Atlantic City, NJ), 2, %QUOTE(,));
```

```
Atlantic City
NJ
```

Quoting the argument is just as important if the argument is a macro variable whose value contains a comma, as in this example:

```
%LET PLACE = Atlantic City, NJ;
%PUT %SCAN(%QUOTE(&PLACE), 1, %QUOTE(,));
%PUT %SCAN(%QUOTE(&PLACE), 2, %QUOTE(,));
```

```
Atlantic City
NJ
```

Quoting also makes it possible for a value to contain leading or trailing spaces, which the macro processor otherwise does not consider part of a value. This example shows how quoting affects the way the macro processor identifies a value.

```
%PUT %LENGTH( S );
%PUT %LENGTH(%QUOTE( S ));
```

```
1
3
```

Without quoting, the macro processor uses only the letter S as the argument, and the %LENGTH function measures a length of 1. When the argument is quoted, the macro processor treats the leading and trailing spaces as part of the value, and the %LENGTH function measures a length of 3.

Unmatched quote marks and parentheses can cause parsing problems for the macro processor, so quote marks and parentheses that do not occur in pairs in the argument to the %QUOTE function must be marked with a preceding percent sign, as in this example:

```
%PUT %SCAN(%QUOTE('Atlantic City', NJ), 1, %QUOTE(,%'));
```

```
Atlantic City
```

In the quoted text, two consecutive percent signs are interpreted as one percent sign. When a percent sign appears before certain special characters, it must be written as two percent signs.

There are three variations on macro quoting for different situations that require different kinds of quoting.

Unresolved macro expression Some functions quote a macro expression before it is resolved. When the macro expression is resolved, the resolved value is not quoted. This is indicated by the letters STR in the function name. The

%STR function is especially useful for assigning text that contain semicolons or other special characters to a macro variable, as shown in this example:

```
%LET READ = %STR(MERGE MAIN.CAPACITY MAIN.NET; BY LOCATION;);
```

No rescan Some macro quoting functions mask any ampersand and percent sign that occur in the text that results from resolving the values that are initially found in the quoted value. The result is that the macro processor does not rescan to look for macro objects in the resolved value. This is indicated by the letters NR at the beginning of the function name. The functions %NRQUOTE and %NRSTR add this quoting property to the %QUOTE and %STR functions, respectively.

Unbalanced parentheses and quote marks The %BQUOTE function differs from the %QUOTE in that it allows unmatched parentheses and quote marks in the quoted value. The %NRBQUOTE function adds the no-rescan property.

The example below revises an earlier example to use the %BQUOTE function instead of the %QUOTE function. By comparing this example to the earlier example, you can see how the %BQUOTE function differs from the %QUOTE function.

```
%PUT %SCAN(%QUOTE('Atlantic City', NJ), 1, %BQUOTE(,'));
```

```
Atlantic City
```

Functions and CALL Routines in Macro Language

Most SAS functions and CALL routines can be used in macro programming. The %SYSFUNC macro function makes it possible to call a SAS function. The %SYSCALL statement does the same for a SAS CALL routine.

The function call is an argument to the %SYSFUNC function. However, the function call is written as a macro function call. Write character values without quote marks. To use a macro variable as an argument, write a macro variable reference. If a function requires a variable for a certain argument, use a macro variable name for the argument.

This example uses the EXIST function to check for the existence of the SAS data file MAIN.NET.

```
%SYSFUNC(EXIST(MAIN.NET, DATA))
```

An optional second argument to the %SYSFUNC function is a format to convert the return value to text. This is especially useful for functions that return SAS date values. This example writes the current date formatted with the WORDDATE format:

```
%PUT %SYSFUNC(DATE(), WORDDATE.);
```

The %SYSCALL statement works about the same way for CALL routines. However, in the %SYSCALL statement, all arguments must be written as macro variable names:

```
%SYSCALL CALL routine name(macro variable name, macro variable name, . . . );
```

To use a value as an argument, assign the value to the macro variable before the %SYSCALL statement.

Most functions and CALL routines can be used in macro programming, but there are a few exceptions. These kinds of functions and CALL routines cannot be used in macro programming:

- Functions and CALL routines that refer to macro objects
- Variable information functions
- CALL routines that refer to data step variables
- Array functions
- Queue functions

System Options for Macro Language

There are several system options that control the macro processor's actions and, especially, the log messages it generates. The following table lists the more essential system options that relate to the macro processor.

System Option	Action When Option Is On	Action When Option Is Off
MACRO	The macro processor resolves macro language references.	The macro processor is inactive.
SERROR	The macro processor generates a warning message when it finds an ampersand followed by a word and cannot resolve the apparent macro variable reference.	The macro processor does not generate these warning messages.
MERROR	The macro processor generates a warning message when it finds a percent sign followed by a word and cannot resolve the object as a macro, macro statement, or macro function.	The macro processor does not generate these warning messages.
SYMBOLGEN	The macro processor generates log notes that show the results of resolving macro variables.	The macro processor does not generate these log notes.
MPRINT	The macro processor writes in the log the program text that results from a macro.	The macro processor does not write the program text that results from macros.
MFILE	The output from the MPRINT option is redirected to the MPRINT fileref.	The output from the MPRINT option is written to the log.
MLOGIC	The macro processor writes log notes that describe the execution of macro statements and macros.	The macro processor does not write these log notes.

User Parameters

A macro program can obtain user input either from the operating system command line or from a window that it displays.

The Parameter String

The SYSPARM= system option creates a parameter string, which you can use to pass parameters from the operating system command line or a configuration file to the SAS program. You can use the automatic macro variable SYSPARM to get the value of the parameter string.

The purpose of parameters for a program is the same as the purpose of parameters for a macro. It lets you call the program to do something slightly different each time you run it. A parameter could be title text, the name of an input file, a libref, or any number of other parts of a program.

If there is only one parameter, you can simply indicate it in the SYSPARM= system option and use the SYSPARM macro variable itself as the parameter in the program. In this example, the parameter is the libref MAIN. In the operating system command line, the parameter might be indicated as:

```
-sysparm "MAIN"
```

In the program, the parameter appears as the macro variable SYSPARM, such as:

```
PROC SORT DATA=&SYSPARM..LIST;
```

This statement would resolve as:

```
PROC SORT DATA=MAIN.LIST;
```

If there are two or more parameters, separate them with commas or any other appropriate delimiter character in the parameter string. Use the %SCAN function to extract the values and assign them to separate macro variables.

In this example, the parameters are a libref, year, and letter:

```
-sysparm "MAIN, 1999, a"
```

In the program, use macro statements such as these to extract the individual parameters:

```
%LET LIB = %SCAN(%QUOTE(&SYSPARM), 1, %QUOTE(,));
%LET YEAR = %SCAN(%QUOTE(&SYSPARM), 2, %QUOTE(,));
%LET LETTER = %SCAN(%QUOTE(&SYSPARM), 3, %QUOTE(,));
```

The %QUOTE function is necessary for macro quoting when the delimiter is a comma or any of various other special characters.

It is also possible to use the parameter string in a data step. Use the SYSPARM function to return the value of the parameter string.

Macro Windows

Macro windows are similar to the data step windows described in chapter 12, "Text File I/O," but they are used in the macro language environment. Macro windows are typically used to prompt the user for parameter values that are used in macro programming to construct SAS program lines.

A macro window is defined in the %WINDOW statement and displayed by the %DISPLAY statement. These macro statements work much like the WINDOW and DISPLAY statements of the data step, but with a few differences. The variable fields of a macro window display macro variables. The format/informat terms of the WINDOW statement are replaced by numbers that indicate the lengths of the fields. The automatic macro variables SYSMSG and SYSCMD take the place of the automatic variables _MSG_ and _CMD_.

This is an example of defining and displaying a macro window:

```
%WINDOW SELECT #3 @5 'Enter the date to process:' +1 USERDATE 9;
%LET USERDATE = %SYSFUNC(DATE(), DATE9.);
%DISPLAY SELECT;
```

The term `USERDATE 9` indicates to display the macro variable USERDATE in an unprotected variable field with a length of 9. The %LET statement assigns an initial value to the macro variable before the window is displayed. When the %DISPLAY statement displays the window, the user can enter a different value for the macro variable.

If a macro displays a macro window, the window remains visible until the macro ends. If a macro window is displayed outside a macro, it disappears as soon as the user responds.

19

Proc Steps

A typical SAS program includes both data steps and proc steps. Although the data steps and proc steps might take up a similar number of program lines, most of the action takes place in the proc steps. The reason it can happen that way is that a proc is a program that is already written. To use the full power of a proc, the proc step only has to identify the proc and fill in any specific details of the actions the proc will take.

For the most part, procs work with data in SAS datasets. Different procs do different things. For example, the PRINT proc prints a SAS dataset; the SORT proc sorts it. The statements and options of the proc step indicate what SAS dataset to use, which of its variables to use, and what to do with those variables. For example, the PRINT proc uses statements to indicate which variables to print in a report and other statements and options that dictate specific details of the way the report is constructed.

Proc Step Syntax

The SAS language imposes only a few restrictions on the syntax of a proc step. Each proc defines its own statements and options. However, there are common elements and patterns of syntax that you can find in many procs. Some statements have essentially the same effect in every proc that uses them.

In a SAS program, a step appears as a sequence of statements. The first statement of a proc step is the PROC statement, which identifies the proc and the input SAS dataset and sets other options for the proc. Other statements can follow, depending on the proc. When the statements define related objects or a sequence of actions, the order of statements is significant. Otherwise, the statements usually can be written in any order.

The PROC Statement

Every proc step begins with a PROC statement. The general form of the statement is the same for every proc:

```
PROC proc terms/options;
```

The statement begins the keyword PROC and the name of the proc. The rest of the statement includes other terms and options that apply to the execution of the proc step as a whole. These other terms can be written in any order.

Most procs work with an input SAS dataset, identified in the DATA= option in the PROC statement. The default for this option is _LAST_, that is, the most recently created SAS dataset. This option has no effect in procs that do not use an input SAS dataset.

There are other common options that appear in multiple procs with the same or similar meanings. These options are used in the PROC statement of several procs, and sometimes in other statements.

Option	Description
OUT=*SAS dataset*	The output SAS dataset
PRINT	Produces print output
NOPRINT	Does not produce the usual print output
CATALOG=*catalog*	The catalog to use
LIBRARY=*libref*	The library to use
IN=*libref*	The input library
OUT=*libref*	The output library
MEMTYPE=*member type* MTYPE=*member type* MT=*member type*	Processes this member type
ENTRYTYPE=*entry type* ETYPE=*entry type* ET=*entry type*	Processes this entry type
MISSING	Uses observations that have missing values for class variables
ORDER=	The order of class values in output:
INTERNAL	Sorted order
FORMATTED	Sorted order of formatted values
DATA	Order of appearance in the input data
FREQ	Order of frequency (usually, descending order)
FORCE	Carries out the actions even if there are discrepancies or if data is destroyed
LABEL	Uses labels, when present, to identify variables in print output
SPLIT='*character*'	Splits lines of labels where this character appears
FORMCHAR='*characters*' FORMCHAR(*n*)= FORMCHAR(*n*, . . .)=	Overrides all or selected form characters of the FORMCHAR= system option
UNIFORM	Uses the same layout for every page or for every BY group

Variable List Statements and Categories of Variables

A proc might use several statements to define the way the variables of the input SAS dataset are used in its processing. The variables are assigned to categories, with each category of variables used in a different way.

The syntax of these statements is always the same, at least in their simplest form. The statement keyword followed by a variable name or a variable list:

```
keyword variable;
keyword variable list;
```

The primary set of variables that a proc uses is a set of analysis variables, which are listed in the VAR statement. In the PRINT proc, for example, the analysis variables are printed as regular columns of the report. This is an example of a VAR statement:

```
VAR EARTH AIR FIRE WATER;
```

The following table lists statement keywords that are used in many different procs to define categories of variables. Individual procs have other statements of this kind.

Statement Keyword	Kind of Variable	Description
VAR	Analysis	The primary set of variables
CLASS	Class	Variables that identify groups of observations
ID	ID	Other identifying variables
FREQ	Frequency	A variable that indicates the repetition of an observation
WEIGHT	Weight	A variable that indicates the relative importance of an observation in calculating statistics

There can be more than one VAR, CLASS, or ID statement, and the same is true for most statements in which you list variables for use in a proc step. Using several statements to list variables has the same effect as listing them all in one statement. For example, the statements

```
CLASS SCOPE;
CLASS SIZE;
CLASS SUBJECT;
```

are equivalent to the single statement

```
CLASS SCOPE SIZE SUBJECT;
```

The BY Statement and BY Group Processing

In the proc step, as in the data step, the BY statement declares the sequence of observations in the input data and divides the data into groups. The BY statement also has another more significant effect in the proc step. Most procs go through their processing separately for each BY group, treating each BY group as a separate set of data. A proc that prints a page of statistics, for example, prints a separate page of statistics for each BY group. When a proc that produces a chart is used with a BY statement, it produces a separate chart for each BY group. BY group processing provides an efficient way to do this kind of repetitive processing on all the groups that are found in a set of data.

The sort order clause in the BY statement is the same as it is in the data step: a list of variables, with the possibility of the DESCENDING option before each variable. The BY statement can use the NOTSORTED option, which indicates that the observations form groups, but the groups are not in sorted order.

```
BY sort order;
BY sort order NOTSORTED;
```

The BY statement forms groups in a slightly different way in the proc step. When BY variables have format attributes, their groups are based on the formatted values of the variables. That is, the format of the BY variable is used to convert the BY values to text, and those formatted text values are considered when forming BY groups. When the BY values of two consecutive BY groups are formatted as the same text value, they are combined into a single BY group.

These are examples of ways the use of formatted values can affect the formation of BY groups:

- If the BY variable includes fractional values, and you format it as an integer, different fractional values do not form different BY groups.
- A SAS date value with the YYMON format forms one BY group per month.
- With a small format width, a character variable forms BY groups based only on the beginning characters of the values.
- A value format that applies labels to ranges of values, used with a BY variable, can form a specific set of BY groups.

The use of formatted values to form BY groups does not necessarily mean that each formatted value forms only one BY group. Only consecutive BY groups can be combined because their formatted BY values are identical. When there are multiple BY variables, this tends to happen only for the last BY variable.

Variable Attribute Statements

In a proc that uses variables from an input SAS dataset, you can use statements to change the attributes of those variables for the proc step. The FORMAT, LABEL, INFORMAT, and ATTRIB statements have the same syntax as in the data step, except that you cannot set the length or type of a variable.

Most often, it is the FORMAT and LABEL statements that are used to control the appearance of print output produced by the proc. In some procs, you must use the LABEL option or another option in the PROC statement in order for the proc to use the label attributes of variables. The format attributes of BY variables may also affect the way BY groups are formed.

Other Common Statements

There are several more statements that can be used in many different procs.

RUN and QUIT The RUN and QUIT statements mark execution boundaries, points where the SAS supervisor can execute the preceding statements. In most procs, either statement marks the end of the step. In some procs, the RUN statement only marks the end of a run group, a set of statements that are executed as a unit. The QUIT statement marks the end of any proc step. The RUN statement can be used in any proc except the SQL proc.

WHERE As in the data step, the WHERE statement has the same effect as the WHERE= dataset option on the input SAS dataset. It indicates a condition for selecting observations from the SAS dataset. In some procs, if you use the statement rather than the dataset option, different run groups can use different WHERE clauses.

OUTPUT The OUTPUT statement, used in some analytic procs, tells the proc to create an output SAS dataset. The statement includes the OUT= option, which names the new SAS dataset, and other options, which depend on the proc, to control the output variables that are created.

SELECT and EXCLUDE The SELECT or EXCLUDE statement in some procs lets you select members, entries, or other files from a group of files for an action such as copying. Use only one of these two statements. The proc processes the files listed in the SELECT statement or all files that are not listed in the EXCLUDE statement. If neither statement is present, the proc processes all the available files.

In some procs, you can use a type option in the SELECT or EXCLUDE statement to limit the selection of files to one specific member type or entry type. The option is MEMTYPE= or ENTRYTYPE=, as described above for the PROC statement. Write the option after a slash at the end of the statement.

Proc Step Execution

Actions that can be taken in a specified sequence to complete a task are commonly called steps. This description of a step, as one of a series of actions, also describes the steps of a SAS program. It especially fits those proc steps that appear as a single action in the program. However, not all proc steps work this way. Some procs can execute several statements at a time, and others execute each statement separately. There are also procs that display windows in order to interact with the user.

Single Execution

Most proc steps represent a single action. That is, all the statements of the step serve to define or modify the same action. This does not necessarily mean that the proc does something simple, only that the proc step describes the work of the proc as a single action rather than as a sequence of actions. With all the statements working together, the proc cannot execute until after it considers all the statements of the step. This means the proc executes all at once. As soon as the SAS supervisor finds a RUN or QUIT statement or any other indication of the step boundary, the proc executes.

Years ago, all procs worked this way, and most still do. PRINT, SORT, REPORT, APPEND, and TABULATE are examples of procs that execute singly. The order of statements in the proc step usually does not matter, except in a few cases where one statement depends on another in some way.

This is the general form of this kind of proc step:

```
PROC proc  options;
  other statements
RUN; or QUIT;
```

Run Groups

Not every proc step is over when you get to the RUN statement. A few procs that use statements to represent actions, such as DATASETS and CATALOG, can continue with more statements after the RUN statement. The group of statements that runs at one time is called a run group. One reason to use run groups is to change title lines, system options, or other properties of the SAS environment for different statements in the same step. You can execute global statements to change title lines or system options between run groups. The QUIT statement was invented in case you need to definitively end the proc step of a run-group proc. This is the general form of a run-group proc step:

```
PROC proc  options;
  statements of first run group
RUN;
  statements of second run group
RUN;
  , , ,
  statements of last run group
RUN;
QUIT;
```

Each run group executes almost as if it were a separate proc step. You could get the same results by repeating the PROC statement at the beginning of every run group to create separate steps. In some procs, each run group defines a separate action or object. In the PMENU proc, for example, each run group defines a separate menu bar. In other procs, the various statements of a run group may have no particular connection to each other, and the significance of the run group is just that the statements of the run group execute at one time. The DATASETS proc is an example of this kind of proc. Related statements must appear in the same run group. For example, in the

DATASETS proc, you might use the MODIFY statement along with several secondary statements to modify a SAS dataset. These statements must be in the same run group so that the proc can make the connection between the secondary statements and the MODIFY statement.

A step that is divided into run groups is still a single step, and the most obvious indication of this is found in the performance statistics in the log. They appear only once, at the end of the step. The SAS supervisor waits until it reaches a QUIT statement or another indication of a step boundary before it calculates the performance statistics and writes them in the log. Another way to recognize a run-group proc is found in the way the SAS interactive environment responds when you submit a proc step that ends in a RUN statement. If it is a run-group proc, the interactive environment provides an indication that the step is still running. Use a QUIT statement when it is necessary to specifically mark the end of a proc step for a run-group proc.

If you use ODS selection or exclusion lists to determine which output objects appear in the output from a run-group proc, it is important to write a QUIT statement at the end of the proc step. This ensures that the lists are cleared correctly between steps.

In run-group procs that use input SAS datasets, you can use different WHERE and BY statements in each run group to process different subsets and groups of the input observations. The PLOT proc is the only base SAS proc that fits this description. In run-group procs, WHERE and BY statements persist from one run group to the next. To make a WHERE or BY statement apply to the entire step, you need to write it only in the first run group of the step. Subsequent run groups that do not have their own WHERE and BY statements use the previous WHERE and BY statements. If, after using a WHERE statement in a run group, you want a subsequent run group to run without a WHERE condition, use a WHERE CLEAR statement or a WHERE statement with no terms to clear the previous WHERE statement:

```
WHERE ;
WHERE CLEAR;
```

Similarly, to clear a previous BY statement, write a BY statement without a sort order clause:

```
BY ;
```

Global statements can execute between run groups. Any global statements that are present in a run group execute before the run group executes. You might use TITLE statements, for example, to have different title lines in the output each run group.

Interpretive Procs

An interpretive proc executes each statement separately. Each statement represents a separate action, and there is little or no connection between one statement and the next. In these proc steps, all statements execute in order. This includes global statements, which can appear between any two proc step statements. The two interpretive procs of base SAS are the FORMAT and

SQL procs. In the FORMAT proc, each statement creates an informat or format. In the SQL proc, each statement represents an action of database access or management.

The use of the RUN statement differs between these two procs. For the FORMAT proc, either the RUN or QUIT statement marks the end of the step. In the SQL proc, the RUN statement is disregarded, with this log note:

```
NOTE: PROC SQL statements are executed immediately; The RUN statement has
      no effect.
```

Use the QUIT statement to mark the end of the `PROC SQL` step.

Interactive Procs

Some procs display a window in order to interact with the user. Most or all of their processing results from the user selections and actions. The FSLIST proc, which displays a text file for browsing, is one example of this kind of proc. For these procs, the proc step is essentially like a command that opens a windowing application. In fact, there usually is a command available to access the same application. For example, the FSLIST command opens the same application as the FSLIST proc.

Some of these applications are capable of submitting SAS statements to execute. Using the command rather than the proc to open the application makes it possible for the submitted statements to execute immediately, while the windowing application is open. Only one step can execute at a time, so when a proc step calls a windowing application, and the application submits statements to execute, the submitted statements cannot execute until the proc step is over.

System Options for Procs

A few system options affect the processing of proc steps in general. The PROBSIG= option, which can have a value of 0, 1, or 2, affects the formatting of *p*-values in certain statistical procs. The PROC option, in some operating systems, makes it possible to call an external program in the PROC statement. The name of the external program appears in place of the name of the proc in the proc step. There are no PROC statement options or additional statements. The form of the step is:

```
PROC program;
RUN;
```

Common Routines of Procs

Not all the processing that procs do is unique to a single proc. There are several procs that do variations and combinations of the same kind of processing. This is especially significant for sorting and summary statistics. The SORT, REPORT, SQL, and other procs sort SAS datasets. SUMMARY, TABULATE, and REPORT are some of the procs that calculate summary statistics. Regardless of which of these procs you use to sort or summarize data, the actual processing is done by the same internal routines of the SAS System. This ensures that the different procs produce consistent results.

The commonalities between procs also extend to some of the syntax that controls the sorting process and the computation of summary statistics. These common elements of syntax are described in the sections that follow.

Summary Statistics

Statistics can be used to study a population, which is a complete set that is the subject of an inquiry, or a sample, which is a subset of a particular population. Procs can calculate statistics for either a population or a sample.

A statistic is a value that is derived from the values of a sample or population that is used as a measure of the sample or the population. The statistics in base SAS procs are considered descriptive or summary statistics; they are simple and general measures of a set of values. Most of these statistics are calculated algebraically and are also implemented as functions, as described in chapter 15, "Functions and CALL Routines." Several procs calculate this common set of statistics:

CSS	Corrected sum of squares
CV	Percent coefficient of variation
KURTOSIS	Kurtosis
MAX	Maximum
MEAN	Mean
MIN	Minimum
N	Number of values
NMISS	Number of missing values
RANGE	Range
SKEWNESS	Skewness
STD	Standard deviation
STDERR	Standard error of the mean
SUM	Sum
SUMWGT	Sum of weights
USS	Uncorrected sum of squares
VAR	Variance

Statistics are calculated on the analysis variables of an input SAS dataset, which in most procs are listed in the VAR statement. The variables must be numeric. Missing values are disregarded in calculating statistics; the statistic NMISS is available to let you know how many observations had missing values for a variable.

Quantiles

Quantiles, which are often called percentiles, are a different kind of statistic that cannot be calculated algebraically. Instead, they are computed by ranking the values and finding specific relative positions within the ranked values. The definition of a percentile is that a certain percent of the values in the set are less than or equal to the percentile value. For example, the 50th percentile, which is also called the median, is the middle value. Half, or 50 percent, of the values are lower than the median, and half of the values are higher. These are the common quantiles, from lowest to highest:

P1		1st percentile
P5		5th percentile
P10		10th percentile
Q1	or P25	1st quartile (25th percentile)
MEDIAN	or P50	Median (50th percentile)
Q3	or P75	3rd quartile (75th percentile)
P90		90th percentile
P95		95th percentile
P99		99th percentile

Statements and Options

Two statements in the proc step and three options in the PROC statement can affect the way statistics are computed. The FREQ and WEIGHT statements indicate variables that change the way observations are considered.

The FREQ statement identifies a frequency variable. The value of the frequency variable is a counting number. A value of 2 or more indicates that the observation represents multiple items, each with identical measurements. It is the same as if the identical observation appeared multiple times in the SAS dataset, the number of times indicated by the value of the frequency variable. There can be only one frequency variable.

```
FREQ variable;
```

The WEIGHT statement identifies a weight variable. The weight variable indicates the relative importance, or weight, of each observation, when some observations are more important than others. The weight value should be a positive value. A negative value for the weight variable is treated as a zero weight. Observations with a zero weight have no effect on any weighted statistics, and procs may exclude such observations from their analysis entirely.

```
WEIGHT variable;
```

The VARDEF=, EXCLNPWGT, and PCTLDEF= options in the PROC statement of some procs affect the way statistics are computed. The VARDEF= option can have any of four values in order to calculate population or sample statistics, weighted or unweighted. The values affect the variance divisor, which is used to calculate the variance and any other statistics that are based on the variance. The following table lists the option values and their effects.

VARDEF=	Type of statistics	Variance divisor
N	Population	N
DF	Sample	N – 1
WGT	Weighted population	SUMWGT
WDF	Weighted sample	SUMWGT – 1

If there is no weight variable, WGT is equivalent to N and WDF is equivalent to DF.

The EXCLNPWGT option in some procs affects the way weighted values are used. With this option, observations whose weight values are zero or negative are completely disregarded. In other procs that calculate statistics, observations with zero or negative weights are always disregarded.

The PCTLDEF= option affects the way quantiles are determined, especially for breaking ties when multiple observations have the same values. This table describes the effects of the possible values for the option.

PCTLDEF=	Rules for Quantile Value
1	Linear interpolation: When a quantile falls between two observations, it is determined by linear interpolation between the two closest observations.
2	Actual value: When a quantile falls between two observations, it uses the value of the closest observation. When a quantile falls halfway between two observations, it uses the value of the even-numbered observation.
3	Next higher value: When a quantile falls between two observations, it uses the value of the next higher observation.
4	Linear interpolation, based on one more observation than is actually present.
5	Higher value: When a quantile falls on an observation, it uses the average of the value of that observation and the value of the next higher observation. When a quantile falls between two observations, it uses the value of the next higher observation.

Sorting

Sorting means arranging objects in a specified order. SAS software uses the same sort program wherever it sorts observations. The SORT proc is the primary way to sort a SAS dataset. Sorting is also part of the processing of the REPORT and SQL procs and in some interactive windows.

The following example shows the effect of sorting. The same data is shown unsorted, sorted in ascending order, and sorted in descending order.

Before Sorting

Title	Length
Going for the One	5:30
Turn of the Century	7:58
Parallels	5:52
Wonderous Stories	3:45
Awaken	15:38

Sorted
BY TITLE;

Title	Length
Awaken	15:38
Going for the One	5:30
Parallels	5:52
Turn of the Century	7:58
Wonderous Stories	3:45

Sorted	Title	Length
`BY DESCENDING TITLE;`	Wonderous Stories	3:45
	Turn of the Century	7:58
	Parallels	5:52
	Going for the One	5:30
	Awaken	15:38

Sort Order Clause

A sort order clause determines the order of observations. The sort order clause indicates the variables to sort by and indicates if any of them should be put in descending order. In the SORT proc, the sort order clause appears in the BY statement, as in this example:

```
PROC SORT DATA=MAIN.STAFF;
  BY DESCENDING SALARY ID;
RUN;
```

After the keyword BY, the rest of the BY statement is a sort order clause, which lists the sort key variables in order of priority. In this example, the variables are SALARY and ID. The option DESCENDING before a variable name indicates descending order for that variable. In this example, the variable SALARY is sorted in descending order.

The BY statement is also used in other steps to indicate the order in which a SAS dataset was previously sorted. The use of the BY statement in the data step and proc step is described in chapter 10, "SAS Dataset I/O."

Sort order clauses also appear in the SORTEDBY= dataset option, which you can use to indicate that a SAS dataset is already in sorted order. If you create a new SAS dataset that is already in sorted order, the SORTEDBY= option might save some processing used to check its sort order. The SORTEDBY= option is not necessary if the SAS dataset is created as a result of a BY statement or by any other SAS processing that generates data in sorted order. In those situations, SAS automatically saves an indication of the sort order in the SAS dataset.

System Options for Sorting

Among the handful of system options that affect sorting, the most important are SORTPGM= and SORTSEQ=. The SORTPGM= option selects the sort program that SAS uses. If the value is SAS, SAS uses its own internal sort program. If the value is HOST, SAS uses an external sort program that is installed in the operating system. If the value is BEST, SAS selects between these two sort programs, based on the size of the SAS dataset. If no external sort program is available, SAS always uses its own sort program.

The SORTSEQ= option affects the way character variables are sorted. By default, character values are sorted in the same order that the comparison operators use, but you can select from various other collating sequences for sorting character values, including these:

```
ASCII
EBCDIC
DANISH
FINNISH
ITALIAN
NORWEGIAN
SPANISH
SWEDISH
REVERSE
```

It is possible to create custom collating sequences. These are entries of type TRANTAB stored in the SASUSER.PROFILE or SASHELP.HOST catalog. When these entries exist, you can use them as values for the SORTSEQ= option.

The SORTSEQ= option can also appear as a statement option in the `PROC SORT` statement.

The next few chapters describe the use of a few selected procs.

20
Reporting

There are two procs that are designed specifically to produce table reports of the data in SAS datasets. The PRINT proc is quick and simple. The REPORT proc provides more detailed control over the way the data appears in the report.

PRINT

The PRINT proc prints tables of SAS datasets with variables as columns and observations as rows. It can be used with very little coding to produce a standard table of all the data in a SAS dataset, or you can add statements and options to control specific details of the report.

The Simple PRINT Proc

To produce a report with the PRINT proc, it is only necessary to identify the SAS dataset to print. As in other procs, the input SAS dataset is identified in the DATA= option in the PROC statement. This is the most common way for the `PROC PRINT` step to be written:

```
PROC PRINT DATA=SAS dataset;
RUN;
```

As always, the DATA= option can be omitted if you want to use the most recently created SAS dataset.

With no statements or options for formatting, the PRINT proc produces a report that contains a column of observation numbers and one column for each variable. The variable name appears as the column header for each variable. There is a blank line between the header line and the first line of data. The variables are ordered in the same order in which they are stored in the SAS dataset. Each observation appears as a row in the table. Like the variables, the observations appear in order.

The format attribute of each variable is used to convert the values of the variable to the text that appears in the report. If a variable does not have a format attribute, the default formats are used. For numeric variables, the default format is `BEST12.`. For character variables, it is the $F format with a width equal to the length of the variable. Whenever it can, the PRINT proc reduces the width of a column on a page by eliminating spaces. Ordinarily, though, it does not reduce the width of a column that is defined with a format that has an explicit width argument.

The following output shows how the PRINT proc reduces the widths of columns on a page by eliminating blank areas. The first output page shows the full column widths (produced using the option `WIDTH=FULL`). The second output page shows the PRINT proc's default output, with the column widths reduced. The DATE column is not affected, because it uses its full width, but both the COUNTRY and COUNT columns are made narrower.

```
PRINT Table With Full Column Widths      10:35 Friday, November 3, 2000  16

Obs          DATE     COUNTRY                            COUNT

 1     2000-10-31     Argentina                             18
 2     2000-10-31     Australia                          4,301
 3     2000-10-31     Brazil                               515
 4     2000-10-31     Canada                            10,262
 5     2000-10-31     Egypt                                250
```

```
PRINT Table With Default Column Widths   10:35 Friday, November 3, 2000  17

Obs          DATE     COUNTRY          COUNT

 1     2000-10-31     Argentina           18
 2     2000-10-31     Australia        4,301
 3     2000-10-31     Brazil             515
 4     2000-10-31     Canada          10,262
 5     2000-10-31     Egypt              250
```

The proc has other ways of conserving horizontal space. If necessary, it will reduce the space between columns to as little as one character. It can also make some columns narrower by rearranging their headers. If several variables form columns that are narrower than the column header text, the PRINT proc may use several lines for the headers, splitting the header text or rotating the headers so that they are arranged vertically in order to fit more columns across the page.

The PRINT proc tends to rotate headers when there are many variables with short values, and the variable names are longer than the values, but not too long, as in this example of output:

```
Vertical Headers

           D                          V   V V V V V   V V V V V
O          A V   V V V   V V V   V V  X   X X X X X   X X X X X
b          T X   X X X   X X X   X X  1   1 1 1 1 1   1 1 1 1 2
s          E 1   2 3 4   5 6 7   8 9  0   1 2 3 4 5   6 7 8 9 0

1 2000-07-31 0   0 0 0 220 0 0   0 0  0   0 0 0 0 0   0 0 0 0 0
2 2000-10-31 0 180 0 0   0 0 0   0 0  0 804 0 0 0 0   0 0 0 0 0
3 2001-01-31 0 220 0 0   0 0 0   0 0  0   0 0 0 0 0 145 0 0 0 0
4 2001-04-30 0   0 0 0 100 0 0 100 0  0   0 0 0 0 0   0 0 0 0 0
5 2001-07-31 0   0 0 0 100 0 0   0 0 25   0 0 0 0 0   0 0 0 0 0
```

The PRINT proc is more likely to orient headers horizontally and split them when they are longer than the width of the column when the headers contain spaces or mixed-case letters. In the following example, the variable name VideoSampleDate is in mixed case, and the PRINT proc splits it before one of its capital letters.

```
Split Header

      Video
Obs SampleDate VX1 VX2 VX3 VX4 VX5 VX6 VX7 VX8 VX9 VX10 VX11 VX12 VX13 VX16

 1  2000-07-31  0    0  0   0  220  0   0    0  0    0    0    0    0     .
 2  2000-10-31  0  180  0   0    0  0   0    0  0    0  804    0    0     .
 3  2001-01-31  0  220  0   0    0  0   0    0  0    0    0    0    0   145
 4  2001-04-30  0    0  0   0  100  0   0  100  0    0    0    0    0     0
 5  2001-07-31  0    0  0   0  100  0   0    0  0   25    0    0    0     0
```

Pages and Panels

If there are more observations than can fit on one page, the PRINT proc prints as many observations as it can on the first page, then continues at the next observation on the next page. It formats each page independently — the column widths on each page may be different, depending on the lengths of the values that appear on that page, and this can result in different pages being arranged very differently.

If the total width of all the columns is too large to fit across the page, the proc divides the columns into panels. It might print two or three panels on a page in order to fit all the variables. The observation number column appears in each panel, so that you can put the different columns together for each observation. If all the variables still do not fit on one page, the PRINT proc prints as many pages as it takes to print the first set of observations, before it goes on to the next observation.

In the following example, the first page of the report is printed in two panels. The values for the remaining observations are smaller and are able to be printed in one panel on the second page of the report. Notice how the observation number column makes it easy to see that the first two panels represent the same set of observations and the next page is a different set of observations.

```
Areas                                           7:08 Monday, June 5, 2000 124

Obs     AREA                   CT1          CT2           CT3

  1     Central Valley    17699718     11174877        200981
  2     Stadium             255506      1700388           220
  3     Expressway             972         1629       1024431
  4     Northeast              123        32255          6709
  5     Downtown             32708           22            13
  6     University            1521         5651        172406

Obs       CT4        CT5       CT6          CT7         CT8

  1     33321     254364    213974     31759632       15354
  2     11619     242628     45388        22104         119
  3    914358        882      6617        89864        7478
  4        42     443633       335            6       15791
  5       156         28       117        76816      180837
  6       297        108        17           44         300
```

```
Areas                                       7:08 Monday, June 5, 2000 125

Obs  AREA          CT1    CT2   CT3    CT4   CT5    CT6    CT7    CT8

  7  Airport         9  17455  7045  47959    57  21914    273  46400
  8  W2             37     31  1291   1921  2549    724  20608  36247
  9  Beach          22    743    10  40057   702    367     64     35
 10  Hills         446   5689    16    159  2269   7381   5552     31
 11  Remote         17   9659   940    544    24   1168   1039   3175
 12  North        1327     32  1984    496   335   6265     82      9
 13  Industrial      6   2999    12    409   697   4718      9    100
 14  Satellite      22    112     9     68    98   2356      9   3231
 15  Wall           14   2596    93    112    13     22      5     72
 16  South         873    460   179     14    17    118    204      8
 17  Metro           8    184    23     35   485     84     41     68
 18  Media           7     11    22     18    39    230     90    125
 19  W1             10     25    15     34    22     25     10     17
```

Variables

The first thing you might want to do to control the appearance of the report that the PRINT proc produces is to select the variables that appear as columns in the report. The ID and VAR statements select variables. Variables listed in the ID statement are ID variables; those listed in the VAR statement are analysis variables. If there is no VAR statement, the proc uses all variables other than ID and BY variables as analysis variables.

These statements also indicate the order of the columns. The ID variables, if there are any, appear first, in the order indicated in the ID statement. They are followed by the analysis variables. You can list the same variable more than once to have it appear as more than one column in the report.

Both ID variables and analysis variables appear as columns, but there are two differences between an ID variable and an analysis variable. First, if there is an ID variable, the PRINT proc does not print a column of observation numbers. Therefore, the ID variables should be sufficiently distinct to identify the observations. Second, if the PRINT proc prints the report as panels, because the total width of the columns is more than can fit across the page, the ID variables appear in every panel. Because of this, the PRINT proc limits the total width of the ID variables. Combined, their width should be no more than half the width of the page. If the width of ID variables is too large, the proc treats some of them as analysis variables.

The Input Data

Standard features of the proc step and the SAS language can modify the input data in order to change many of the details of the report. These features work essentially the same way as for other procs.

FORMAT Use the FORMAT statement to change the formats of variables. Use the width argument of the format to change the width of a column.

BY Use a BY statement to have the proc print a separate table for each BY group. The BY statement can also change the order in which the proc reads the observations from an indexed SAS data file, which changes the order of rows in the table.

By default, the proc prints more than one BY group on a page when it can. To put BY groups on separate pages, use the PAGEBY statement. Indicate one BY variable in the PAGEBY statement; the proc goes to a new page for each BY group of that BY variable and the higher-level BY variables. To put every BY group on a separate page, indicate the last BY variable in the PAGEBY statement.

By default, the PRINT proc prints BY lines at the beginning of every BY group. The BY line shows the BY variables and their values. This is an example of a BY line for a PROC PRINT step that uses one BY variable, X:

```
X=2
```

If a BY group extends to more than one page, the BY line is repeated at the beginning of each additional page, with the note "`(continued)`", as shown here:

```
X=2
(continued)
```

You can use the NOBYLINE system option to suppress BY lines. Then the BY variables appear only if you include them in the title lines or if you also list them in the VAR statement as analysis variables.

WHERE If you do not want the report to include all observations, use a WHERE= dataset option or WHERE statement to define a condition to select observations that appear in the report. Alternatively, you can use the FIRSTOBS= and OBS= dataset options to select a range of observations by observation number.

KEEP/DROP You can use the KEEP= or DROP= dataset option instead of a VAR statement to select the variables that appear in the report. With this approach, you can select variables without determining the order of columns. The DROP= option may be more convenient than a VAR statement if you want to exclude only a few variables.

LABEL The proc can use labels as column headers. Use the LABEL statement to change the labels of variables. The proc uses labels as column headers if you use the LABEL or SPLIT= option in the PROC statement.

Formatting Options

There are several options you can use on the `PROC PRINT` statement to affect the way the report is formatted. These options are described in the following table.

Formatting Options in the PROC PRINT Statement

Option	Value	Effect
HEADING=	HORIZONTAL H	Horizontal column headers.
	VERTICAL V	Vertical column headers. (The default is to orient headers horizontally unless the proc estimates that it can save space by orienting them vertically.)
NOOBS		No observation number column.
OBS=	'*label*'	Header text for the observation number column.
LABEL L		The proc uses variable labels as column headers.
SPLIT= S=	'*character*'	The proc uses variable labels as column headers, splitting them wherever this character appears to form multiple lines of headers.
WIDTH=		The width of a column is:
	FULL	The full formatted width.
	UNIFORM U	The formatted width, if the format has an explicit width argument; otherwise, the length of the longest formatted value.
	UNIFORMBY UBY	The formatted width, if the format has an explicit width argument; otherwise, the length of the longest formatted value in the BY group.
	MINIMUM MIN	The length of the longest formatted value on the page. This is the default.
ROWS=	PAGE	Only one panel per page.
DOUBLE D		Double-spacing between observations.

This example shows multiple-line column headers defined with a split character:

```
PROC PRINT DATA=FUELEFF SPLIT='\';
  ID MODEL;
  VAR INJECT TRANS TREAD TIRESIZE CITY HIGHWAY;
  LABEL MODEL='Model' TREAD='Tire\Tread' TIRESIZE='Tire\Size'
    CITY='MPG\City' HIGHWAY='MPG\Hwy'
    INJECT='Fuel\Injection' TRANS='Transmission';
RUN;
```

```
                     Fuel                        Tire       Tire      MPG  MPG
      Model        Injection  Transmission      Tread       Size      City Hwy

XXXXXXXXXXXXXXX XXXXXXXXXXX XXXXXXXXXXX  XXXXXXXXXX XXX/XXXXXX 55.5 55.5
XXXXXXXXXXXXXXX XXXXXXXXXXX XXXXXXXXXXX  XXXXXXXXXX XXX/XXXXXX 55.5 55.5
```

Summary Lines

Two statements and two options can be used to generate summary lines in the report. The SUM statement lists variables whose sums appear in summary lines. Variables listed in the SUM statement are treated as analysis variables even if they are omitted from the VAR statement. If there is a SUM statement, a summary line of totals appears at the end of the report.

If there are BY groups, you can use the SUMBY statement with one of the BY variables to have summary lines appear for BY groups. Summary lines present subtotals for BY groups of the indicated BY variable and any higher-level BY variables.

The ROUND option in the PROC PRINT statement rounds values to two decimal places for summing purposes.

Another PROC statement option, the N option, generates a line at the end of the report or at the end of each BY group that indicates the number of observations used. If there are BY groups and a SUM statement, the number of observations appears both for the BY groups and for the report as a whole.

You can control the text that explains the observation count by writing the option as N= followed by a character constant. If there is a SUM statement and a BY statement, write two character constants. The first one is the label for the BY group; the second one, the label for the report total.

REPORT

Both the PRINT and REPORT procs produce table reports from SAS datasets, but where the PRINT proc is designed to be easy to program, the REPORT proc is designed to give you detailed control over the structure of the report. The syntax of the REPORT proc is built around columns rather than variables. With this approach, the REPORT proc usually requires quite a few statements to define even a simple report. But then, it is easy to adjust the report by adding and changing options.

Core Statements

In a PROC REPORT step, the COLUMN and DEFINE statements form the heart of a report definition. The COLUMN statement lists the columns that appear in the report. A separate DEFINE statement defines each column.

The essential statements of the REPORT proc tend to follow this pattern:

```
PROC REPORT DATA=SAS dataset NOWD;
  COLUMN column 1  column 2  . . . ;
  DEFINE column 1 / 'header' . . . options;
  DEFINE column 2 / 'header' . . . options;
  . . .
```

For many reports, these statements are sufficient to define the report.

The PROC REPORT statement sets options that apply to the report as a whole. The two options that are usually required are the DATA= option and the NOWD option. The DATA= identifies the input SAS dataset. The NOWD (or NOWINDOWS) option overrides the proc's default behavior of opening a window in which you can modify the report definition interactively.

The COLUMN statement lists the items used as columns of the report in order. In the simplest case, this is a list of variables from the input SAS dataset.

The DEFINE statement for each item sets options for the column. The first term in the DEFINE statement is the name of the item, as it appears in the COLUMN statement. The item name is followed by a slash and item options.

Among the many available options, the two most important are the header text and the FORMAT= option. Write the header text as a character constant. To display a header that takes up two or more lines, write two or more character constants.

The value of the FORMAT= option is the format used for the values. The width argument of the format determines the width of the column.

The following example demonstrates these features of a simple PROC REPORT step.

```
PROC REPORT DATA=LOCALGRP NOWD;
   COLUMN NAME MAGNITUDE DISTANCE MASS DIAMETER;
   DEFINE NAME / 'Name' FORMAT=$F12.;
   DEFINE MAGNITUDE / 'Magnitude' FORMAT=9.1;
   DEFINE DISTANCE / 'Distance' FORMAT=COMMA8.;
   DEFINE MASS / 'Mass' FORMAT=COMMA10.;
   DEFINE DIAMETER / 'Diameter' FORMAT=COMMA8.;
RUN;
```

```
Local Group                              20:21 Wednesday, November 29, 2000   1

  Name          Magnitude  Distance        Mass  Diameter
  M31                 3.4     2,900     400,000       195
  Milky Way             .         0     750,000       100
  M33                 5.7     3,000      25,000        60
  LMC                 0.1       179      20,000        30
  SMC                 2.3       210       6,000        25
  M110                8.5     2,900      10,000        17
  M32                 8.1     2,900       3,000         8
  Leo I               9.8       880           .         .
```

More details of the PROC REPORT, COLUMN, and DEFINE statements follow.

Report Formatting Options

In addition to the DATA= and NOWD options, the PROC REPORT statement can include formatting options that apply to the report as a whole. The table on the following page lists the available options.

Formatting Options in the PROC REPORT Statement

Option	Value	Effect
CENTER		Centers the report.
NOCENTER		Left-aligns the report.
LS=	*counting number*	The line size: the number of characters in a line.
PS=	*counting number*	The page size: the number of lines in a page.
PANELS=	*counting number*	The number of panels. The proc divides the width of the page into this many panels, if possible.
PSPACE=	*counting number*	The horizontal spacing between panels. The default is 4.
NOHEADER		Omits headers.
NAMED		Omits headers and writes the column name and an equals sign before each value.
HEADLINE		Draws a horizontal line below the header lines.
HEADSKIP		Leaves a blank line below the header lines.
SPLIT=	'*character*'	The split character. The proc splits column headers between lines where the split character appears. It also splits values at the split character, if FLOW is used as an item option. The default split character is '/'.
BOX		Draws table rules around each page and around and between the rows and columns. This option is not compatible with the WRAP option or the FLOW item option.
FORMCHAR= FORMCHAR(*n*, . . .)=	'*characters*'	Overrides all or selected form characters of the FORMCHAR= system option.
COLWIDTH=	*counting number*	The default column width. The default is 9.
SPACING=	*counting number*	The default column spacing. The default is 2.
SHOWALL		Includes all defined columns, including hidden columns, in the report.
WRAP		Wraps, if necessary, to display all columns together. With this option, if the columns do not all fit across the page, a second line of columns appears on the next line after the first line of columns for a table row.

Column Definitions

Most of the time, a column definition is just a variable name. However, there are several other ways a column definition can appear in the COLUMN statement.

Alias If you use the same variable in more than one column, define an alias for each column so that you can set header text and options for each column in DEFINE statements. The alias is just a distinct name for the column. Write an equals sign and the alias at the end of the column definition. Then use the alias as the item name in the DEFINE statement.

The following example uses aliases in order to show the variable AMOUNT with two different formats.

```
PROC REPORT DATA=MAIN.MONEY NOWD NOCENTER HEADSKIP;
  COLUMN SERIAL AMOUNT=NUMBER AMOUNT=WORDS;
  DEFINE SERIAL / '#';
  DEFINE NUMBER / 'Amount (Number)' FORMAT=COMMA15.2;
  DEFINE WORDS / 'Amount (Words)' FORMAT=WORDF50. LEFT;
RUN;
```

```
 #     Amount (Number)  Amount (Words)

 0001            39.08  thirty-nine and 08/100
 0002            81.00  eighty-one and 00/100
 0003             0.44  zero and 44/100
 0004           607.02  six hundred seven and 02/100
 0005           248.78  two hundred forty-eight and 78/100
```

In this example, NUMBER and WORDS are aliases for specific columns that contain the variable AMOUNT. This example also demonstrates the HEADSKIP option, which creates the blank line below the headers, and the LEFT option, which is used to left-align the last column.

Spanning headers You can define headers that extend across multiple columns. In the COLUMN statement, in parentheses, write the header text as one or more character constants, followed by the definitions for the columns that appear under the header.

The example below creates a header that spans two related columns.

```
COLUMN ('Effective Dates' START END);
DEFINE START / 'Start' FORMAT=DATE7.;
DEFINE END / 'End' FORMAT=DATE7.;
```

```
Effective Dates
  Start      End
```

Statistics It is possible for a column to display a summary statistic rather than the actual value of a variable. The variable displayed in the column is considered an analysis variable. In a detail row, the statistic is often the same as the value of the variable. Usually, it is simplest to indicate the statistic in the DEFINE statement; however, it is also possible to indicate a statistic as part of the column definition. Write the column definition with the variable name and the statistic name, with a comma between them:

```
variable, statistic
statistic, variable
```

Or define several columns at once by using lists, in parentheses, of variables and statistics:

```
item, (list)
(list), (list)
```

A variable and a statistic can combine to define a single column, but they are treated as separate items. The column has headers for both items. The order of items in the column definition determines the order of the headers.

The available statistics are the common statistics described in chapter 19, "Proc Steps," except for skewness and kurtosis, plus two additional statistics: PCTN, relative frequency, and PCTSUM, fraction of total. The statistic that is used most often is SUM. The statistics N and PCTN can appear by themselves as column definitions. When used this way, N represents the absolute frequency.

Header Text

Indicate header text for a specific column as an item option in the DEFINE statement. If you do not specific header text for a column, the item name — usually, the variable name — is used as the column header. Indicate header text that spans several columns in the COLUMN statement, in parentheses that also enclose the definitions of the columns involved.

The REPORT proc automatically expands headers to fill the width of the column if the first and last characters of the header text are one of the following:

```
-   *   .   :   \   _   +   =   < and >   > and <
```

For example, if the header text is `'- Telephone -'` and the column width is 17, the header appears as `--- Telephone ---`. This header expansion occurs only in the default Listing destination format of ODS.

Usually, it is easiest to write multiple character constants to define multiple lines of a header. However, if necessary, you can define a multiple-line header in a single constant value with the use of a split character. When the split character appears in the header text, the REPORT proc does not show the split character as part of the header. Instead, it uses it as an indication of a line break. It displays the text after the split character on the next line of the header. The default split character is /. Use the SPLIT= option in the `PROC REPORT` statement to select a different split character. There is no way to show the split character in a header, so select a split character that does not appear in the text of any header. If you use the slash character in the headers of a report, you must use the SPLIT= option to indicate a different split character.

Item Options for Basic Column Formatting

The DEFINE statement can list various item options for a column in the report. The header text and FORMAT= item options have already been mentioned. Several more item options can affect the basic formatting of a column.

The SPACING= option sets the spacing to the left of the column. The default is set by the SPACING= option in the PROC FORMAT statement. Use the item option `SPACING=0` for the first column of the report to have that column appear at the left margin with no space before it.

The WIDTH= option sets the width of the column. The default is set by the WIDTH= option in the `PROC FORMAT` statement. This item option is not usually necessary because the width argument in the FORMAT= option is sufficient to set the column width.

You can use the NOPRINT or NOZERO item option to hide a column. A column defined with the NOPRINT item option does not appear in the report. A column defined with the NOZERO option does not appear if all its values are zero or missing. These options have no effect when the SHOWALL option is used in the `PROC REPORT` statement. The SHOWALL option lets you produce the report with all columns visible so that you can see the hidden columns.

Usage

In the REPORT proc, the *usage* of a variable determines how the variable relates to the structure of the report. The usage can be set as the item option DISPLAY, ORDER, COMPUTED, ANALYSIS, GROUP, or ACROSS in the DEFINE statement.

DISPLAY A variable with no special properties is a display variable. The formatted values of the variable appear in its column in the report. This usage is the default for character variables.

ORDER If there are order variables, they determines the order of the rows in the report. The REPORT proc sorts the data in the order of the order variables so that the rows appear in that order. Order variables typically have the same value in several consecutive rows, but the value is not repeated on each row. Instead, the REPORT proc writes the value of an order variable only on the first row of the group, that is, when the value of the variable changes. This makes it easier to pick out the groups in the report. If a group continues into another page, the values of the group variables are repeated on the first row of the new page.

Order variables must be the first columns in the report. That is, if an order variable is not the first column listed in the COLUMN statement, then all columns listed before it must also be order variables.

The item options ORDER= and DESCENDING determine the way the order variable is sorted. Values for the ORDER= option can be INTERNAL, FORMATTED, FREQ, or DATA to determine the order in which values appear. The DESCENDING option reverses the order of the values.

COMPUTED A report can include new variables that are not present in the input SAS dataset. These computed variables get their values from formulas or programming logic in the REPORT proc itself.

ANALYSIS Analysis variables are numeric variables for which statistics are computed and displayed. Analysis is the default usage for numeric variables. An analysis variable with the statistic MEAN, MIN, MAX, or SUM usually appears the same as a display variable in the detail lines of a report. The main difference is that analysis variables also appear in any summary lines that you define for the report. Display variables do not appear in summary lines.

Another difference is that analysis variables are affected by frequency and weight variables if those variables are defined in the FREQ and WEIGHT statements.

Indicate the statistic as item option for the analysis variable. If you do not indicate a statistic, the default statistic is SUM.

The PCTSUM statistic lets you show the fraction of the total that a value represents. Multiply this statistical value by 100 or display it with the PERCENT format to show a percent of the total. You can display both the value of a variable and the percent by using the variable in two different columns, as indicated in the example below. This example also shows the use of the RBREAK statement to create a summary line at the end of the report.

```
PROC REPORT DATA=MAIN.MONEY NOWD NOCENTER HEADSKIP;
  COLUMN SERIAL AMOUNT=NUMBER AMOUNT=PCT;
  DEFINE SERIAL / '#';
  DEFINE NUMBER / 'Amount' SUM FORMAT=COMMA15.2;
  DEFINE PCT / '% of Total' PCTSUM FORMAT=PERCENT11.4;
  RBREAK AFTER / OL SUMMARIZE;
RUN;
```

```
 #               Amount   % of Total

 0001             39.08      4.0028%
 0002             81.00      8.2965%
 0003              0.44      0.0451%
 0004            607.02     62.1743%
 0005            248.78     25.4814%
       ---------------  -----------
                 976.32    100.0000%
```

GROUP and ACROSS The two other usages, GROUP and ACROSS, define two kinds of class variables. They are used only in summary reports, as described in the next chapter. When a report definition contains a group or across variable, it cannot contain any order or display variables.

The default usage for character variables is display; for numeric variables, the default usage is analysis. If you indicate the display usage for a numeric variable and also indicate a statistic for the variable, the proc makes it an analysis variable.

There must be at least one order or display variable for the REPORT proc to produce a detail report with one row per observation. If the report includes only analysis and computed variables, the proc generates a summary report with only one row. If necessary, add a hidden column with a display variable to force the proc to produce a detail report.

Tiling and Wrapping

If the table is too wide to fit on the output page, the REPORT proc can make it fit by either tiling or wrapping. The default action of tiling uses multiple pages to fit the columns for each set of rows. The proc places as many columns as it can on the first page and places additional columns on the second page, continuing with more pages if necessary. The item options PAGE and ID can affect the details of the way this works.

Use the ID option to indicate ID variables. If the first column or the first few columns contain variables that identify a row, you can make those variables ID variables. The ID variables appear at the left side of each page. This can make it easier to put the various columns together for a specific row. When you use the ID item option for a variable, that variable and all preceding variables are ID variables.

The PAGE item option forces a page break to the left of a specific column. Use this option to control the locations of page breaks.

To avoid the need for tiling, you might reduce the widths of some of the longer columns. The FLOW item option makes it possible to display an entire value when the value is longer than the width of the column. The proc uses two or more lines to display the value. With the FLOW option, the proc looks for the split character in the value and uses that character as an indication of a line break. If you use the FLOW option, use the WIDTH= option to set the width of the column to something less than the format width. The FLOW option is especially useful when most values of a variable are short, but you want to display the entire value when it is longer.

The alternative to tiling is wrapping. The second set of columns appears on the next line after the first set of columns. This way, all the columns for one row appear together, on the same page. The column headers are also wrapped; the headers for the first set of columns are followed immediately by the headers for the second set of columns.

For some kind of reports, this is an effective way to present data. For other reports, it may make it too difficult to match up the data values with the column headers. In that situation, you can use the NAMED option in the `PROC REPORT` statement to have the column names appear immediately before the values in each column.

Computed Variables

The value of a computed variable is determined in programming statements in the `PROC REPORT` step. The statements appear as a code segment in a COMPUTE block.

The COMPUTE statement at the beginning of the COMPUTE block indicates the name of the variable for which the block computes a value. For a numeric variable, the form of the statement is:

```
COMPUTE variable;
```

For a character variable, the statement must also include options to indicate the type and length of the variable. The form of the statement is:

```
COMPUTE variable / CHARACTER LENGTH=length;
```

CHAR is an alias for the CHARACTER option.

An ENDCOMP statement marks the end of the COMPUTE block. The code segment of a compute block consists of programming statements that assign a value to the computed variable. The programming of a code segment is much like that of a data step. These data step statements are allowed:

- Assignment
- Sum
- LENGTH
- CALL
- Comment
- Null

In addition, you can use these control flow statements of the data step:

- DO, END
- GOTO
- IF-THEN, ELSE
- LINK, RETURN
- SELECT, WHEN, OTHERWISE, END

The following example works with U.S. address data. The SAS dataset contains a variable with a ZIP code. The COMPUTE block uses that value to determine the state.

```
PROC REPORT DATA=MAIN.METRO NOWD NOCENTER;
  COLUMN ZIP STATE;
  DEFINE ZIP / 'ZIP Code' DISPLAY FORMAT=$F10.;
  DEFINE STATE / 'State' COMPUTED FORMAT=$F2. WIDTH=5;
  COMPUTE STATE / CHARACTER LENGTH=2;
    STATE = ZIPSTATE(ZIP);
    ENDCOMP;
RUN;
```

Variables in Code Segments

Code segments are very limited in what variables they can use in their computations. The limitations are based on the order of execution of a report. The REPORT proc builds the cells of a table from left to right within each row, and it builds the rows in order. Table variables from one row are not available to be used in the next row. The only table variables that can be used in a code segment that computes a variable are those that are used in cells to the left of the computed variable.

If often happens that you want to compute a variable using variables that are not shown in the table at all. To do so, you must include those variables in the table definition. Position them to the left of the computed variable and define them with the NOPRINT option so that they are hidden.

Similarly, to refer to a variable in a column that appears to the right of the computed variable, also add the same variable as a hidden column to the left of the computed variable. Assign an alias to the column so that you can use a DEFINE statement to hide the column with the NOPRINT option. Use the alias to refer to the column value in the code segment.

If a column is defined with an alias, you must use the alias to refer to that column in the code segment. To refer to an analysis variable that does not have an alias, you must use a two-level name that indicates the variable and the statistic, for example, PRICE.SUM. Remember that numeric variables are analysis variables with the statistic SUM unless you specifically define them differently. If you attempt to refer to an analysis variable without the statistic as part of the name, the proc does not recognize the variable and generates log messages such as these:

```
NOTE: Variable I is uninitialized.
NOTE: Missing values were generated as a result of performing an operation
      on missing values.
```

The two-level names are not used for analysis variable columns that have aliases. Use only the alias as the variable name in the code segment.

There are automatic variables you can use in a code segment to address the columns of the current row in the report. The automatic variable _COL_ contains the column number of the computed variable. Automatic variables of the form `_Cn_`, where *n* is a column number, contain the values of the columns in the row.

Code segments can also create variables other than the variables that the code segment is computing. These code segment variables survive from one code segment to the next and from one row to the next. You can use these variables to pass values from one code segment to another or to calculate sums or counts across the rows of the table. Use the LENGTH statement to set the length of character variables that you create in a code segment.

Break Lines

The BREAK and RBREAK statements define break lines in a report. You can add break lines at the beginning or end of the report and at the beginning or end of each value of an order variable. The statement to create break lines at the beginning or end of the report is:

```
RBREAK BEFORE or AFTER / break options;
```

The statement to create break lines before or after each value of an order variable is:

```
BREAK BEFORE or AFTER break variable / break options;
```

Each of the break options in the RBREAK or BREAK statement creates a different object or effect in the break lines. The options and effects are listed in the following table.

Break Options

Option	Effect
OL	Overline
DOL	Double overline
PAGE	Page break
SKIP	A blank line
SUMMARIZE	A summary line
SUPPRESS	Hides the break variable in the summary line
UL	Underline
DUL	Double underline

The summary line contains analysis variables and computed variables. The summary line for a break variable also includes the break variable unless you use the SUPPRESS option to hide the break variable. To keep a numeric variable from appearing in summary lines, make it a display variable.

If the report contains computed variables and break lines, compute the variables in a way that makes sense for the break lines. If the computed variables do not make sense in the break line, assign them missing values when they are computed in a break line. The automatic variable _BREAK_ can help you do this. This variable is blank in a detail row, but in a break line, it contains the name of the break variable (or at least, the first 8 characters of the variable name). In a break line for the report, it contains the value "RBREAK".

You can also use a COMPUTE block to write break lines. Instead of a computed variable, the COMPUTE statement indicates a location, which can be any of those shown in the table below.

COMPUTE Locations

Statement	Location of Break Lines
COMPUTE BEFORE;	At the beginning of the report
COMPUTE AFTER;	At the end of the report
COMPUTE BEFORE *break variable*;	Before each value of the break variable
COMPUTE AFTER *break variable*;	After each value of the break variable
COMPUTE BEFORE _PAGE_;	At the top of each page
COMPUTE AFTER _PAGE_;	At the bottom of each page

With a _PAGE_ location, you can control the alignment of the break lines with an alignment option in the COMPUTE statement. Write the option LEFT, CENTER, or RIGHT after a slash at the end of the statement.

In the code segment, use LINE statements to write lines in the report. The LINE statement looks like the PUT statement of the data step, but it has only a small subset of the features of the PUT statement. To write a field, use a variable followed by a format, or a character constant. A character constant can be preceded by an integer and * to write the constant value multiple times. The only pointer controls are @ or + followed by a positive integer or a numeric variable. All these terms work the same way as in the PUT statement. The following list summarizes the terms that can be used in a LINE statement.

```
variable format
character constant
n*character constant
@n
@variable
+n
+variable
```

A single LINE statement can write only one line. To write several lines, use several LINE statements in the same code segment. To write a blank line, use a LINE statement with a blank character constant:

```
LINE ' ';
```

The character constant must contain at least one space; a LINE statement with a null constant has no effect. A LINE statement with no terms would be a syntax error.

The automatic variable _BREAK_ is blank when it is used with the _PAGE_ location.

The REPORT proc can also be used to generate reports of summary statistics. This use of the proc is described in the next chapter.

The PRINT and REPORT procs are just two of the many alternatives that SAS software offers for producing reports. SQL and data step programming can also create table reports, and data step programming can generate text reports in any kind of layout. The CALENDAR and FORMS procs produce other kinds or reports. You can use the PLOT, CHART, and TIMEPLOT procs and the many procs of SAS/GRAPH to present data in a graphical format. Last but not least, most SAS procs analyze data in some way and generate reports that present a summary, measurement, test, or analysis of the data. This is the subject of the next chapter.

21

Analyzing Data

The alternatives that SAS software provides for analyzing data are too numerous to count. This chapter may provide a starting point for SAS data analysis by describing the use of procs that analyze data in some of the more common, familiar ways. These procs use SAS datasets as input and produce SAS datasets and reports as output.

SUMMARY

The SUMMARY proc calculates summary statistics for variables in a SAS dataset. It calculates them for the entire set of observations and for groups for observations formed by class variables. It can deliver its results in a simple report or as a SAS dataset.

Frequency

The SUMMARY proc reads an input SAS dataset, analyzes it, and writes the results in an output SAS dataset that it creates. The DATA= option in the PROC SUMMARY statement identifies the input SAS dataset. The OUT= option in the OUTPUT statement names the output SAS dataset. These terms are sufficient to run the proc and generate a frequency count, as in this example:

```
PROC SUMMARY DATA=WORK.STAFF;
  OUTPUT OUT=WORK.COUNT;
RUN;
PROC PRINT DATA=WORK.COUNT;
RUN;
```

```
Obs     _TYPE_     _FREQ_

 1         0         38
```

When you run it this way, the SUMMARY proc generates one output observation with two variables. These two variables are always present in the output SAS dataset of the SUMMARY proc. The variable _TYPE_ has the value 0; this variable is of interest only when you use class variables. The variable _FREQ_ contains the absolute frequency, a count of the observations in the input SAS dataset.

Statistics

Usually, the SUMMARY proc is used to calculate statistics. This requires the addition of a VAR statement and statistic terms in the OUTPUT statement. The VAR statement identifies analysis variables — the numeric variables of the input SAS dataset for which the proc calculates statistics. Terms in the OUTPUT statement apply specific statistics to specific analysis variables and provide the names for the output variables that contain the statistics.

The most detailed form of a statistic term in the OUTPUT statement is

statistic(analysis variable)=output variable

If you give the output variable the same name as the analysis variable, this can be abbreviated to

statistic(analysis variable)=

To apply one statistic to all of the analysis variables and give the output variables the same names as the analysis variables, write the term as

statistic=

A statistic term can also be formed with variable lists:

statistic(analysis variable list)=output variable list
statistic(analysis variable list)=
statistic=output variable list

The statistics that can be used are the standard statistics and quantiles described in chapter 19, "Proc Steps"; the confidence limit statistics CLM, UCLM, and LCLM; and QRANGE, which is the difference between the quartiles Q3 and Q1. A few common statistics are SUM, MIN, MAX, MEAN, and STD. For confidence level statistics, use the ALPHA= option in the PROC statement to set the confidence level.

The following example calculates the sums of the variables WORDCT and PAGECT from the SAS dataset WORK.NEWS. The results are stored as variables of the same names in the one observation of a new SAS dataset called WORK.NEWSSUM.

```
PROC SUMMARY DATA=WORK.NEWS;
  VAR WORDS PICTURES;
  OUTPUT OUT=WORK.NEWSSUM SUM=;
RUN;
PROC PRINT DATA=WORK.NEWSSUM;
RUN;
```

```
Obs    _TYPE_    _FREQ_    WORDS    PICTURES

 1        0        145     42794       28
```

Statistical Report

With the PRINT option, the SUMMARY proc can produce a simple report of summary statistics. The report is a table with each statistic in a separate column and each variable represented as a separate row. Write the PRINT option and a list of statistics in the PROC SUMMARY statement. The OUTPUT statement is not required when the PRINT option is used. This example shows the use of the PRINT option and the style of the resulting report:

```
PROC SUMMARY DATA=WORK.NEWS PRINT
    NMISS MIN Q1 MEDIAN Q3 MAX;
  VAR WORDS PICTURES;
RUN;
```

```
The SUMMARY Procedure

                 N                       Lower                          Upper
Variable      Miss       Minimum       Quartile          Median       Quartile
------------------------------------------------------------------------------
WORDS            0     4.0000000    131.0000000     300.0000000    456.0000000
PICTURES         0             0              0               0              0
------------------------------------------------------------------------------

Variable         Maximum
------------------------
WORDS        596.0000000
PICTURES       1.0000000
------------------------
```

If the system option LABEL is on and at least one analysis variable has a label, the proc also shows the variable labels as a column in the report. If there is only one analysis variable, the proc shows its name and label, if applicable, in a separate line at the top of the page.

Two other options, FW= and MAXDEC=, control the way the report shows statistical values. FW= sets the width of the printed statistics. MAXDEC= sets the maximum number of decimal places.

Class Variables

Class variables classify the input observations, dividing them into categories. The SUMMARY proc calculates statistics for each group of observations that the class variables form. Class variables appear in the output data along with the statistics.

If there is one class variable, the proc calculates statistics separately for the observations that belong to each value of the class variable. It creates a separate output observation for each class value. It also calculates the statistics for all the observations together, the same as it does when there is no class variable. This creates an output observation in which the class variable has a missing value.

In the following example, there is one class variable, TEAM, which has two distinct values. Specifically, the example uses data from the 2000 World Bowl game, and the class variable identifies the two competing teams. The SUMMARY proc produces three output observations — one observation for the entire set of data, plus two observations for the two teams.

```
TITLE1 'World Bowl 2000 Pass Receiving';
PROC SUMMARY DATA=WORLDRCP CHARTYPE;
  CLASS TEAM;
  VAR RCP YDS TD;
```

```
  OUTPUT OUT=TEAMRCP SUM=;
RUN;
PROC PRINT;

RUN;
```

```
World Bowl 2000 Pass Receiving

Obs    TEAM         _TYPE_    _FREQ_    RCP    YDS    TD

 1                    0         12       31    296     0
 2     Claymores      1          5       16    177     0
 3     Fire           1          7       15    119     0
```

You can use the output of this example to see how the two teams, the Claymores and the Fire, compared in pass receiving in this particular game. The Claymores had more receptions and more receiving yards than the Fire, and neither team scored a touchdown on a pass reception.

The _TYPE_ variable indicates the *type* of each output observation, which has to do with the use of the class variable in that observation. Each type is a different way of segmenting the data. Type 0, in the first observation, indicates that the class variable is not a part of that observation. Type 1, in the other observations, indicates that the class variable is a part of those observations. Each type 1 observation has a different value for the class variable.

By default, _TYPE_ is a numeric variable, but you can use the CHARTYPE option to make it a character variable, as in this example. Either way, when there is one class variable, _TYPE_ shows values of 0 and 1.

The _FREQ_ variable counts the number of input observations that are included in each output observation. In this example, the SUMMARY proc used 12 observations from the input SAS dataset, so the values of _FREQ_ total 12 for each different type. Similarly, with the sum statistic used in this example, the totals of the analysis variables for type 0 match the totals shown in type 1.

If there are multiple class variables, the proc considers all possible combinations of the class variables and creates several sets of output observations that summarize different parts of the data. For example, if the class variables are FLAVOR and FORM, there are four sets of output observations to summarize the data with:

- all observations together
- all values of FORM
- all values of FLAVOR
- all combinations of values of FLAVOR and FORM

The _TYPE_ variable distinguishes the different groups of output observations. The meaning of the type value is easiest to see if _TYPE_ is a character variable, the result of the CHARTYPE option in the PROC statement. Reading _TYPE_ from left to right, each character is a 1 or 0 to indicate the presence or absence of the corresponding class variable. For example, when the first character of _TYPE_ is 1, it indicates that the output observation is formed using the first class variable. If all the characters of _TYPE_ are 1s, it indicates that the output observation is based on a combination of all the class variables.

If _TYPE_ is numeric, its value is the value that would be formed by reading the character value as a binary integer with the BINARY informat. _TYPE_ can be a numeric variable only if there are no more than 32 class variables. To show the numeric _TYPE_ value as 1s and 0s, write it using the BINARY format; it appears the same as the character _TYPE_ value.

The proc writes the output observation groups in increasing order of _TYPE_. Type 0, which summarizes the entire set of observations, comes first. Consider the example of the class variables FLAVOR and FORM, as defined by the CLASS statement:

```
CLASS FLAVOR FORM;
```

The following table shows the four different type values that are formed by the different combinations of class variables.

TYPE (numeric)	_TYPE_ (character)	Class variables
0	00	none
1	01	FORM
2	10	FLAVOR
3	11	FLAVOR and FORM

The output from the following program shows the way the _TYPE_ variable and class variables appear in the output observations.

```
TITLE1 'Coffee Totals';
PROC SUMMARY DATA=CAFE CHARTYPE;
  CLASS FLAVOR FORM;
  VAR MTD STOCK;
  OUTPUT OUT=CAFETYPE SUM=;
RUN;
PROC PRINT;
RUN;
```

```
Coffee Totals

Obs  FLAVOR                   FORM               _TYPE_  _FREQ_  MTD  STOCK

  1                                                00       60   149   112
  2                           Espresso             01        5     3     2
  3                           Universal grind      01       25    15    16
  4                           Whole Bean           01       30   131    94
  5  Arabian Mocha Sanani                          10       10    23    23
  6  Caffë Verona®                                 10       10     8    18
  7  Espresso Roast                                10       15    24    23
  8  French Roast                                  10        5     9     8
  9  Panama La Florentina™                         10       10    48    19
 10  Yukon Blend®                                  10       10    37    21
 11  Arabian Mocha Sanani     Universal grind      11        5     5     6
 12  Arabian Mocha Sanani     Whole Bean           11        5    18    17
 13  Caffë Verona®            Universal grind      11        5     3     4
 14  Caffë Verona®            Whole Bean           11        5     5    14
 15  Espresso Roast           Espresso             11        5     3     2
```

```
16  Espresso Roast             Universal grind    11      5     2    5
17  Espresso Roast             Whole Bean         11      5    19   16
18  French Roast               Whole Bean         11      5     9    8
19  Panama La Florentina™      Universal grind    11      5     2    1
20  Panama La Florentina™      Whole Bean         11      5    46   18
21  Yukon Blend®               Universal grind    11      5     3    0
22  Yukon Blend®               Whole Bean         11      5    34   21
```

When you use the output SAS dataset from the SUMMARY proc, you may need to work with only one of the summary levels. Write a WHERE condition that tests the value of _TYPE_, such as:

```
SET CAFETYPE (WHERE=(_TYPE_ = '01'));
```

There are several PROC statement options that affect the use of class variables. Two additional statements, the WAYS and TYPES statements, provide more control over the class combinations that appear in the output. Options in the CLASS statement also affect the way class variables are used.

In the PROC SUMMARY statement, the CHARTYPE option, as previously mentioned, makes _TYPE_ a character variable. The MISSING and ORDER= options work the same as in other procs, as described in chapter 19, "Proc Steps." The MISSING option allows the proc to use observations in which class values are missing. The ORDER= option determines the order of class values in the output SAS dataset.

The DESCENDING option orders the _TYPE_ values in descending order in the output SAS dataset. This can be useful if you want to have the totals for the entire set of data as the last output observation. The aliases for this option are DESCEND and DESCENDTYPES.

The NWAY option uses only the highest _TYPE_ value, which forms categories using all the class variables. Use this option to summarize a SAS dataset at only one level of detail.

With the COMPLETETYPES option and multiple class variables, the proc generates all combinations of values of the class variables, including those that are not present in the input data. The example below modifies the previous example to demonstrate the effects of NWAY and COMPLETETYPES options.

```
TITLE1 'Coffee Totals With the NWAY and COMPLETETYPES Options';
PROC SUMMARY DATA=CAFE CHARTYPE NWAY COMPLETETYPES;
  CLASS FLAVOR FORM;
  VAR MTD STOCK;
  OUTPUT OUT=CAFENWAY SUM=;
RUN;
PROC PRINT;
RUN;
```

```
Coffee Totals With the NWAY and COMPLETETYPES Options

Obs  FLAVOR                 FORM               _TYPE_  _FREQ_  MTD  STOCK

  1  Arabian Mocha Sanani   Espresso             11       0      .      .
  2  Arabian Mocha Sanani   Universal grind      11       5      5      6
  3  Arabian Mocha Sanani   Whole Bean           11       5     18     17
  4  Caffè Verona®          Espresso             11       0      .      .
```

```
 5  Caffè Verona®          Universal grind   11    5     3     4
 6  Caffè Verona®          Whole Bean        11    5     5    14
 7  Espresso Roast         Espresso          11    5     3     2
 8  Espresso Roast         Universal grind   11    5     2     5
 9  Espresso Roast         Whole Bean        11    5    19    16
10  French Roast           Espresso          11    0     .     .
11  French Roast           Universal grind   11    0     .     .
12  French Roast           Whole Bean        11    5     9     8
13  Panama La Florentina™  Espresso          11    0     .     .
14  Panama La Florentina™  Universal grind   11    5     2     1
15  Panama La Florentina™  Whole Bean        11    5    46    18
16  Yukon Blend®           Espresso          11    0     .     .
17  Yukon Blend®           Universal grind   11    5     3     0
18  Yukon Blend®           Whole Bean        11    5    34    21
```

The SUMMARY proc can use a class combination dataset, a second input SAS dataset that contains values of the class variables. All combinations of class values that are found in the class combination dataset are written in the output even if they are not present in the input data. Use the CLASSDATA= option to identify the class combination dataset. If you also use the EXCLUSIVE option, the class combinations from the CLASSDATA= dataset are the only ones to appear in the output; any others that might be present in the input data are disregarded. With the CLASSDATA= and EXCLUSIVE options, the output SAS dataset always contains the same set of class values, regardless of the data values that are found in the input SAS dataset.

Use the WAYS or TYPES statement to select types; that is, to indicate how to combine class variables. A term in the WAYS statement indicates how many class variables to combine at one time. The numbers you can use as terms in the WAYS statement are the whole numbers from 0 to the number of class variables. The WAYS statement is most commonly written in these three ways:

```
WAYS 1;
```

forms categories with each individual class variable, but does not combine class variables;

```
WAYS 0 1;
```

forms categories with each individual class variable, and also creates the overall total; and

```
WAYS 0 n;
```

where n is the number of class variables, forms categories by combining all the class variables, and also creates the overall total.

The TYPES statement requests specific combinations of class variables. Terms in the TYPES statement can be individual class variables, two or more class variables joined by asterisks to request a specific class variable combination, and () for the overall total.

The following table shows all the different combinations of numbers that are possible in the WAYS statement when there are two class variables, then shows the equivalent TYPES statement syntax using the example of the class variables FLAVOR and FORM. The table also shows the NWAY option as equivalent to one form of the WAYS statement.

WAYS statement	TYPES statement	Option
WAYS 0;	TYPES ();	
WAYS 1;	TYPES FLAVOR FORM;	
WAYS 2;	TYPES FLAVOR*FORM;	NWAY
WAYS 0 1;	TYPES () FLAVOR FORM;	
WAYS 0 2;	TYPES () FLAVOR*FORM;	
WAYS 1 2;	TYPES FLAVOR FORM FLAVOR*FORM;	
WAYS 0 1 2;	TYPES () FLAVOR FORM FLAVOR*FORM;	*(default)*

The TYPES statement is usually used to request specific sets of class variable combinations that cannot be indicated in the WAYS statement, for example:

```
TYPES () FLAVOR FLAVOR*FORM;
```

If you use both the WAYS and TYPES statements, the proc creates the combinations of the TYPES statement in addition to those formed by the WAYS statement.

When there are three or more class variables, using parentheses for grouping can simplify the TYPES statement. For example, this term in the TYPES statement

```
REGIMEN*(VISIT CENTER DIAGNOSIS)
```

is equivalent to the terms

```
REGIMEN*VISIT REGIMEN*CENTER REGIMEN*DIAGNOSIS
```

Options in the CLASS statement affect the way the class variables are used. Write the options after a slash that follows the list of variables. To use different options with different class variables, write multiple CLASS statements.

The MISSING option, as in the PROC statement, allows missing values as valid values for the class variable. By using the option in the CLASS statement rather than the PROC statement, you can allow missing values for some class variables, but not others. For example, with these statements:

```
CLASS FLAVOR;
CLASS FORM / MISSING;
```

the proc uses input observations that have a missing value for FORM, but disregards those that have a missing value for FLAVOR.

The GROUPINTERNAL option forms class groups based on the internal values of the class variables rather than their formatted values.

The ASCENDING, DESCENDING, and ORDER= options determine the order of class values in the output SAS dataset. The ORDER= option is the same as the option in the PROC statement. The DESCENDING option, however, should not be confused with the DESCENDING option of the PROC statement. In the PROC statement, the DESCENDING option affects the _TYPE_ variable; in the CLASS statement, the DESCENDING option affects the class variables. Several other CLASS statement options let you use formats to control the values of class variables.

ID Variables

The ID statement lists ID variables. Usually, these are other variables that are associated with the class variables. Values of these variables are copied to the output SAS dataset. When different input observations have different values for the ID variables, the proc takes the values from the observation that has the highest ID value. You can use the IDMIN option in the PROC statement to take values from the observation that has the lowest ID value.

The OUTPUT statement can contain terms to select ID values. These terms have a more complex syntax that gives them more flexibility in the rules they use to select the observations from which they take ID values.

BY Groups

If you use a BY statement to divide the input SAS dataset into groups of observations, the SUMMARY proc processes each group separately. It writes a separate set of output observations for each BY group, which it computes as if the BY group were a separate SAS dataset. Thus, for example, the overall total that the proc calculates is the total for the BY group, not the total for the entire SAS dataset.

Variables in the Statistical Report

BY, class, and ID variables affect the way the SUMMARY proc produces its statistical report. If there is a BY statement, the proc produces a separate statistical summary of each BY group.

Class values appear as separate groups of rows in the report; you can use the PRINTALLTYPES or PRINTALL option to print all types, that is, all combinations of class variables, or use the WAYS or TYPES statement to select specific types. A column with the heading `N Obs` shows the number of observations in each class group. You can use the NONOBS option to suppress this column.

ID variables appear only if you use the PRINTIDVARS or PRINTIDS option. If the LABEL option is in effect, labels of class and ID variables appear in place of their names.

Other Options

As in many other procs, you can use the WEIGHT and FREQ statements to identify a weight variable and frequency variable.

The SUMMARY proc has additional options to control details of statistical computations and details of variables in the output SAS dataset.

MEANS

The MEANS proc is the same as the SUMMARY proc, but with different defaults. In the MEANS proc, the PRINT option is the default, and you can use the NOPRINT option to suppress the print output. If there is no VAR statement, the MEANS proc uses all the numeric variables of the input SAS dataset as analysis variables. This contrasts with the SUMMARY proc, which does not use any analysis variables if there is no VAR statement.

TABULATE

The TABULATE proc has much in common with the SUMMARY proc. It calculates summary statistics in the same way using nearly the same syntax. However, its output is completely different. The TABULATE proc generates tabular reports that are defined as dimensional hierarchies.

Input Data and Statistics

The TABULATE proc accesses its input data in nearly the same way as the SUMMARY proc. This requires the PROC TABULATE statement and usually both of the CLASS and VAR statements.

Identify the input SAS dataset in the DATA= option of the PROC TABULATE statement. The PROC TABULATE statement can also use the MISSING, CLASSDATA=, EXCLUSIVE, and ORDER= options, which are the same as in the PROC SUMMARY statement.

List the class variables in the CLASS statement and the analysis variables in the VAR statement. These statements can use the same options as in the SUMMARY proc. In the TABULATE proc, you cannot use the same variable both as a class variable and as an analysis variable. If applicable, use a WEIGHT statement to indicate a weight variable and a FREQ statement to indicate a frequency variable.

The statements in the example below are sufficient to access the data in a SAS dataset for the TABULATE proc. These statements are changed only slightly from statements used in an example of the SUMMARY proc earlier in the chapter. A TABLE statement must be added to form a complete PROC TABULATE step.

```
PROC TABULATE DATA=CAFE;
  CLASS FLAVOR FORM;
  VAR MTD STOCK;
```

Statistics

The TABULATE proc uses a slightly different set of statistics. It does not support SKEWNESS or KURTOSIS or confidence level statistics. It adds a set of percent statistics. All of the percent statistics calculate the percent of either a frequency or a sum, but they differ according to how the base, or denominator, of the percent is defined:

Statistic Keyword		Denominator (Base)
Percent of Frequency	Percent of Sum	
ROWPCTN	ROWPCTSUM	Report row
COLPCTN	COLPCTSUM	Report column
PAGEPCTN	PAGEPCTSUM	Report page
REPPCTN	REPPCTSUM	Entire report
PCTN	PCTSUM	All cells of statistic
PCTN<*class combination*>	PCTSUM<*class combination*>	Class group

Most percents are based on a class group. Write the denominator expression that defines the class group in angle brackets after the PCTN or PCTSUM statistic. The denominator must be a class variable or combination of class variables that appears in the table definition as part of the definition for the cell.

TABLE Statement

The TABLE statement defines the form, or shape, of the table that the TABULATE proc produces. The form of the TABLE statement is:

TABLE *table expression / table options*;

The first part of the statement is a table expression, which defines the way variables, statistics, formats, and other elements are combined to form the table. The table expression can be followed by a slash and table options, which modify the formatting of the table.

The TABLE statement must appear after the CLASS and VAR statements that identify the variables it uses. There can be multiple TABLE statements to create multiple tables from the same input data.

Dimensions

A table is usually defined in two dimensions, although it can have as many as three or as few as one. The three available dimensions are pages, rows, and columns. Each dimension is defined in its own expression within the table expression, with commas separating the dimension expressions. These are the combinations of dimensions that are possible in a table expression:

column dimension
row dimension, column dimension
page dimension, row dimension, column dimension

Each dimension expression is defined as a combination of variables, statistics, and other elements. The simplest way to define a dimension is as a class variable. If FLAVOR and FORM are class variables, then this statement defines a table in which the values of FLAVOR appear as rows and the values of FORM appear as columns:

TABLE FLAVOR, FORM;

	FORM		
	Espresso	Universal grind	Whole Bean
	N	N	N
FLAVOR			
Arabian Mocha Sanani	.	5.00	5.00
Caffë Verona®	.	5.00	5.00
Espresso Roast	5.00	5.00	5.00

```
|French Roast    |           .|           .|        5.00|
|----------------+------------+------------+------------|
|Panama La       |            |            |            |
|Florentina™     |           .|        5.00|        5.00|
|----------------+------------+------------+------------|
|Yukon Blend®    |           .|        5.00|        5.00|
-------------------------------------------------------
```

When table cells are defined by class variables only, as in this example, the values that appear in the table cells are frequencies, indicating the number of observations contained in the cell.

Within a dimension, elements can be joined in two different ways. If the elements have only a space between them, they form a list. These elements appear next to each other in the table, but do not interact with each other. If elements have an asterisk between them, they form a hierarchy. The first element is at the top level of the hierarchy, and the element that follows the asterisk is on the next level down. The example below demonstrates both ways of joining elements in a dimension.

```
PROC TABULATE DATA=CAFE;
  CLASS FLAVOR FORM;
  VAR MTD STOCK;
  TABLE FLAVOR*FORM, N MTD STOCK / RTSPACE=35;
RUN;
```

```
-----------------------------------------------------------------------
|                                 |            |    MTD     |   STOCK    |
|                                 |            |------------+------------|
|                                 |     N      |    Sum     |    Sum     |
|---------------------------------+------------+------------+------------|
|FLAVOR          |FORM            |            |            |            |
|----------------+----------------|            |            |            |
|Arabian Mocha   |Universal grind |        5.00|        5.00|        6.00|
|Sanani          |----------------+------------+------------+------------|
|                |Whole Bean      |        5.00|       18.00|       17.00|
|----------------+----------------+------------+------------+------------|
|Caffë Verona®   |Universal grind |        5.00|        3.00|        4.00|
|                |----------------+------------+------------+------------|
|                |Whole Bean      |        5.00|        5.00|       14.00|
|----------------+----------------+------------+------------+------------|
|Espresso Roast  |Espresso        |        5.00|        3.00|        2.00|
|                |----------------+------------+------------+------------|
|                |Universal grind |        5.00|        2.00|        5.00|
|                |----------------+------------+------------+------------|
|                |Whole Bean      |        5.00|       19.00|       16.00|
|----------------+----------------+------------+------------+------------|
|French Roast    |Whole Bean      |        5.00|        9.00|        8.00|
|----------------+----------------+------------+------------+------------|
|Panama La       |Universal grind |        5.00|        2.00|        1.00|
|Florentina™     |----------------+------------+------------+------------|
|                |Whole Bean      |        5.00|       46.00|       18.00|
|----------------+----------------+------------+------------+------------|
|Yukon Blend®    |Universal grind |        5.00|        3.00|        0.00|
|                |----------------+------------+------------+------------|
|                |Whole Bean      |        5.00|       34.00|       21.00|
-----------------------------------------------------------------------
```

When an analysis variable is used without a statistic, as in this example, the table cells show the default SUM statistic. The N statistic shows frequency counts, the same values that appeared in the previous example.

This example also demonstrates the RTSPACE= option, which sets the width of the space used for row titles. The space is divided evenly between the two row variables.

Dimension Elements

The elements that make up a dimension can include these:

Element	Description
ALL	Unclassified summary. ALL acts like a class variable, but it does not segment the data.
class variable	Class variable. Each value of the class variable is shown as a separate page, row, or column.
analysis variable	Analysis variable. A statistic is computed and displayed in the table cell. The default statistic is SUM.
statistic	Statistic. This statistic is computed for the analysis variable. The N statistic and the related percent statistics can be computed with or without an analysis variable.
element='*label*'	Apples a label to a variable or statistic element.
F=*format*	The format used for cell values.
(*element element*...)	Consecutive elements treated as a group.

Within a dimension, all the levels of a hierarchy (joined with asterisks) are put together, and then combined with the other dimensions, if any, to determine the content of each cell in the table. Each cell can be defined with any number of class variables, but with no more than one format, statistic, or analysis variable.

Typically, a cell is defined with an analysis variable, a statistic, and at least one class variable. Formats are usually used as part of the column dimension, where the format width can determine the column width. This is shown in this example:

```
PROC TABULATE DATA=CAFE;
  CLASS FORM;
  VAR MTD STOCK;
  TABLE FORM, N (MTD STOCK)*(SUM MIN MAX)*F=F5. / RTSPACE=21;
RUN;
```

		MTD			STOCK		
	N	Sum	Min	Max	Sum	Min	Max
FORM							
Espresso	5.00	3	0	1	2	0	1
Universal grind	25.00	15	0	4	16	0	5
Whole Bean	30.00	131	0	16	94	0	11

Formatting

The TABULATE proc has many options for formatting the report it produces. A few options can appear in the PROC TABULATE statement. Others are table options that appear in the TABLE statement.

The most important formatting option in the PROC TABULATE statement is the FORMAT= option, which sets a default format for table cells. The default format if this option is not used is F12.2.

The FORMCHAR= option sets form characters. This option in the PROC TABULATE statement overrides the system option. The FORMCHAR= option is described in chapter 15, "Print Output," and chapter 19, "Proc Steps." The NOSEPS option eliminates the horizontal rules between the rows of the table.

The TABLE statement can contain these table options:

Option	Option Value	Description
RTSPACE= RTS=	*integer*	The width of the row title space.
ROW=	CONSTANT CONST	Divides the row title space among all row elements.
	FLOAT	Divides the row title space among nonblank row elements.
INDENT=	*integer*	Writes all row titles in one column, with lower-level elements indented by the indicated number of characters.
BOX=		The contents of the box in the upper left corner of the table:
	'text'	The indicated text.
	PAGE	The page dimension text.
	variable	The label or name of the indicated variable
CONDENSE		Pages are combined, when possible.
NOCONTINUED		Does not print a continuation message at the bottom of the page in a multi-page table.
PRINTMISS		Forms all combinations of class values, including those that do not appear in the data.
MISSTEXT=	*'text'*	The text to write in cells that contain missing values.
FUZZ=	*small positive number*	Values that are closer to 0 than this are treated as 0 values.

Labels can appear with the elements in the table definition, or you can use the LABEL statement to provide labels for variables and the KEYLABEL statement to provide labels for statistics. This is an example of the KEYLABEL statement:

```
KEYLABEL SUM='Total' MEAN='Average';
```

The KEYLABEL statement must precede the TABLE statement.

Other Options

The TABULATE proc has other statements and options to format ODS output and to create an output SAS dataset.

UNIVARIATE

The UNIVARIATE proc provides more detailed statistical descriptions of individual variables than are possible with the other procs of base SAS. Use the DATA= option in the PROC UNIVARIATE statement to identify the input SAS dataset. List the analysis variables in the VAR statement, or omit the VAR statement to have the proc analyze all numeric variables.

This example shows the default report that the UNIVARIATE proc generates:

```
TITLE1 'World Bowl 2000 Pass Receiving';
PROC UNIVARIATE DATA=WORLDRCP;
  VAR RCP;
RUN;
```

```
World Bowl 2000 Pass Receiving

The UNIVARIATE Procedure
Variable:  RCP

                             Moments

N                           12    Sum Weights                 12
Mean                2.58333333    Sum Observations            31
Std Deviation       1.08362467    Variance            1.17424242
Skewness            0.00119075    Kurtosis            -1.1526993
Uncorrected SS              93    Corrected SS        12.9166667
Coeff Variation     41.9467614    Std Error Mean       0.3128155


              Basic Statistical Measures

    Location                    Variability

Mean     2.583333     Std Deviation            1.08362
Median   2.500000     Variance                 1.17424
Mode     2.000000     Range                    3.00000
                      Interquartile Range      1.50000


            Tests for Location: Mu0=0

Test             -Statistic-    -----p Value------

Student's t     t  8.258329    Pr > |t|     <.0001
Sign            M         6    Pr >= |M|    0.0005
Signed Rank     S        39    Pr >= |S|    0.0005


Quantiles (Definition 5)

Quantile       Estimate

100% Max            4.0
99%                 4.0
95%                 4.0
90%                 4.0
75% Q3              3.5
50% Median          2.5
25% Q1              2.0
```

```
10%                  1.0
5%                   1.0
1%                   1.0
0% Min               1.0


        Extreme Observations

----Lowest----          ----Highest---

Value      Obs          Value      Obs

    1       12              3        6
    1        5              3        7
    2       11              4        1
    2       10              4        2
    2        9              4        3
```

The ALL option produces a much longer report, which also includes modes, tests for normality, trimmed means, robust measures of scale, confidence limit statistics for quantiles, and a normal probability plot, among other things. These are selected highlights of the additional output:

```
    Modes

Mode    Count

  2        4


     Basic Confidence Limits Assuming Normality

Parameter          Estimate     95% Confidence Limits

Mean                2.58333       1.89483     3.27184
Std Deviation       1.08362       0.76763     1.83986
Variance            1.17424       0.58926     3.38509

...
                    Tests for Normality

Test                  --Statistic---      -----p Value------

Shapiro-Wilk          W      0.890109     Pr < W        0.1182
Kolmogorov-Smirnov    D      0.204821     Pr > D       >0.1500
Cramer-von Mises      W-Sq    0.08343     Pr > W-Sq     0.1741
Anderson-Darling      A-Sq   0.536151     Pr > A-Sq     0.1378

...

          Robust Measures of Scale

                                          Estimate
Measure                         Value     of Sigma

Interquartile Range          1.500000     1.111951
Gini's Mean Difference       1.257576     1.114497
MAD                          0.500000     0.741300
Sn                           1.192600     1.192600
Qn                           2.221900     1.687519

...
```

```
                         Frequency Counts

            Percents                                 Percents
Value Count  Cell    Cum            Value Count  Cell    Cum

    1     2  16.7   16.7                3     3  25.0   75.0
    2     4  33.3   50.0                4     3  25.0  100.0

...
                      Normal Probability Plot
   4.25+                                        *   *   ++*+
       |                                          +++++
       |                                    *  *+*+++
   2.75+                                 ++++
       |                           *   *+*+*
       |                         ++++
   1.25+                 *  +++*+
        +----+----+----+----+----+----+----+----+----+----+
            -2        -1         0        +1        +2
```

The UNIVARIATE proc has other statements and options to control details of the statistical calculations and the report, create an output SAS dataset, and generate several other kinds of graphs.

REPORT

The use of the REPORT proc to generate detail reports was described in the previous chapter. The REPORT proc can also generate summary reports. These summary reports have the same appearance and structure as the detail reports, and they are produced with nearly the same syntax.

Summary Row

Sometimes, the default action of the REPORT proc is to generate a summary report with only one row that shows the sums of variables. The default usage of a numeric variable is as an analysis variable. Also, variables that are defined as display variables become analysis variables if they are associated with a statistic. When a report is defined with only analysis variables or only analysis and computed variables, the REPORT proc generates a one-row summary report.

The following example shows a program that generates a summary row report. The only difference between this program and an example in the previous chapter is that the display variable was removed. With only analysis variables remaining, the program generates a report with only one row. The variables use the default statistic, SUM.

```
PROC REPORT DATA=LOCALGRP NOWD;
   COLUMN MAGNITUDE DISTANCE MASS DIAMETER;
   DEFINE MAGNITUDE / 'Magnitude' FORMAT=9.1;
   DEFINE DISTANCE / 'Distance' FORMAT=COMMA8.;
   DEFINE MASS / 'Mass' FORMAT=COMMA10.;
   DEFINE DIAMETER / 'Diameter' FORMAT=COMMA8.;
RUN;
```

```
Local Group                              20:21 Wednesday, November 29, 2000   1

  Magnitude  Distance        Mass  Diameter
       37.9    12,969   1,214,000       435
```

A summary row report cannot have break lines at the end of the report. The only break lines it allows are break lines defined at the beginning of the report and at the top and bottom of the page. However, the statement `RBREAK BEFORE / SUMMARIZE;` merely produces a repeat of the summary row.

Summary Tables

The more common kind of summary report is a summary table that uses class variables and shows one row for each combination of class variable values. The class variables are defined as group variables in the report definition. The summary report is nearly the same as a detail report, with only these differences:

- Group variables are used instead of order variables.
- There are no display variables. The report can be include group, analysis, and computed variables.
- There is only one row per group.

A summary report uses break lines the same way as a detail report. The difference is that group variables are the break variables.

When there is only one observation per group, a summary report defined with group variables can be exactly the same as a detail report defined with order variables.

Statistical Options

The EXCLNPWGT, VARDEF=, and MISSING options in the `PROC REPORT` statement affect the way the proc calculates statistics. These options work the same way as for other procs, as described in chapter 19, "Proc Steps."

Across Variables

Across variables are class variables that are displayed as columns rather than rows. The across usage is most useful for class variables that have only a few values. Each value of the across variable is displayed in its own column, so using a variable that has more than a few values can result in a very wide table. Across variables are usually used in combination with group variables to create a table that has multiple rows and columns. When an across variable is defined with its own column, the column contains frequency counts. A report that is defined with one group variable and one across variable produces a simple two-way frequency table, as this example demonstrates:

```
TITLE1 'Transfers';
PROC REPORT DATA=TRANSFER NOWD HEADSKIP;
  COLUMN SOURCE DEST;
  DEFINE SOURCE / 'Source' '--' GROUP;
  DEFINE DEST / 'Destination' '--' ACROSS;
RUN;
```

```
Transfers                                     7:02 Thursday, June 1, 2000  10

                                 Destination
  Source       ------------------------------------------
  ------------ Daytona        Springfield   Valley Forge

  Daytona              .              66           136
  Springfield          1               .          1013
  Valley Forge       415             140             .
```

Cells for which there are no observations show a period, indicating a missing value. This example also demonstrates the use of the label '--' to generate horizontal lines in the headers.

It is possible for two across variables to be combined in the same column. This is generally only useful when two class variables each have only two or three values. As always in the REPORT proc, combine items in a column by writing them in the COLUMN statement with a comma between them.

More often, an across variable appears in the same column as an analysis variable. The table cells show the statistic computed for the analysis variable for the group of observations defined by the group and across variables.

Columns that contain across variables are not addressable by name or alias in code segments. Use column numbers to identify the columns in code segment expressions.

FREQ

The FREQ proc is designed specifically for frequency tables. It shows absolute and relative frequencies in the same table. Tables are defined in the TABLES statement using a syntax much like that of the TYPES statement of the SUMMARY proc. You can list individual variables, use asterisks to indicate combinations of variables, and use parentheses for grouping. This example shows one-way and two-way frequency tables.

```
TITLE1 'Coffee Frequencies';
PROC FREQ DATA=CAFE;
  TABLES FLAVOR FORM FLAVOR*FORM;
RUN;
```

```
Coffee Frequencies                            7:02 Thursday, June 1, 2000   39

The FREQ Procedure

                                                    Cumulative    Cumulative
FLAVOR                    Frequency     Percent     Frequency      Percent
-----------------------------------------------------------------------------
Arabian Mocha Sanani            10       16.67            10        16.67
Caffè Verona®                   10       16.67            20        33.33
Espresso Roast                  15       25.00            35        58.33
French Roast                     5        8.33            40        66.67
Panama La Florentina™           10       16.67            50        83.33
Yukon Blend®                    10       16.67            60       100.00
```

```
                                                     Cumulative    Cumulative
FORM                 Frequency     Percent           Frequency       Percent
-------------------------------------------------------------------------------
Espresso                     5        8.33                   5          8.33
Universal grind             25       41.67                  30         50.00
Whole Bean                  30       50.00                  60        100.00
```

```
Coffee Frequencies                                7:02 Thursday, June 1, 2000  40

The FREQ Procedure

Table of FLAVOR by FORM

FLAVOR            FORM

Frequency        |
Percent          |
Row Pct          |
Col Pct          |Espresso|Universa|Whole Be|  Total
                 |        |l grind |an      |
-----------------+--------+--------+--------+
Arabian Mocha Sa |      0 |      5 |      5 |     10
nani             |   0.00 |   8.33 |   8.33 |  16.67
                 |   0.00 |  50.00 |  50.00 |
                 |   0.00 |  20.00 |  16.67 |
-----------------+--------+--------+--------+
Caffë Verona®    |      0 |      5 |      5 |     10
                 |   0.00 |   8.33 |   8.33 |  16.67
                 |   0.00 |  50.00 |  50.00 |
                 |   0.00 |  20.00 |  16.67 |
-----------------+--------+--------+--------+
Espresso Roast   |      5 |      5 |      5 |     15
                 |   8.33 |   8.33 |   8.33 |  25.00
                 |  33.33 |  33.33 |  33.33 |
                 | 100.00 |  20.00 |  16.67 |
-----------------+--------+--------+--------+
French Roast     |      0 |      0 |      5 |      5
                 |   0.00 |   0.00 |   8.33 |   8.33
                 |   0.00 |   0.00 | 100.00 |
                 |   0.00 |   0.00 |  16.67 |
-----------------+--------+--------+--------+
Panama La Floren |      0 |      5 |      5 |     10
tina™            |   0.00 |   8.33 |   8.33 |  16.67
                 |   0.00 |  50.00 |  50.00 |
                 |   0.00 |  20.00 |  16.67 |
-----------------+--------+--------+--------+
Yukon Blend®     |      0 |      5 |      5 |     10
                 |   0.00 |   8.33 |   8.33 |  16.67
                 |   0.00 |  50.00 |  50.00 |
                 |   0.00 |  20.00 |  16.67 |
-----------------+--------+--------+--------+
Total                   5       25       30       60
                     8.33    41.67    50.00   100.00
```

If you combine three or more variables in a frequency table, the last two variables form the rows and columns and the other variables form pages.

There are several options that can be used in the TABLES statement to control details of the frequency table. Write options after a slash at the end of the list of table requests. The most important option is LIST, which writes a frequency table of two or more variables in the form of a list rather than a crosstabulation. The example below demonstrates the LIST option. The list format shows the same cell frequencies but does not have the row and column frequencies that are found in a two-way table.

```
TITLE1 'Coffee Frequencies';
PROC FREQ DATA=CAFE;
  TABLES FLAVOR*FORM / LIST;
RUN;
```

```
Coffee Frequencies

The FREQ Procedure

                                                              Cumulative  Cumulative
FLAVOR                  FORM             Frequency  Percent    Frequency    Percent
-----------------------------------------------------------------------------------
Arabian Mocha Sanani    Universal grind          5     8.33            5       8.33
Arabian Mocha Sanani    Whole Bean               5     8.33           10      16.67
Caffè Verona®           Universal grind          5     8.33           15      25.00
Caffè Verona®           Whole Bean               5     8.33           20      33.33
Espresso Roast          Espresso                 5     8.33           25      41.67
Espresso Roast          Universal grind          5     8.33           30      50.00
Espresso Roast          Whole Bean               5     8.33           35      58.33
French Roast            Whole Bean               5     8.33           40      66.67
Panama La Florentina™   Universal grind          5     8.33           45      75.00
Panama La Florentina™   Whole Bean               5     8.33           50      83.33
Yukon Blend®            Universal grind          5     8.33           55      91.67
Yukon Blend®            Whole Bean               5     8.33           60     100.00
```

The NOPERCENT, NOCUM, NOROW, NOCOL, and NOFREQ options suppress parts of the frequency table. The TOTPCT and MISSPRINT options add additional elements to the table. With the LIST option, the SPARSE option includes all combinations in the output, even when some have no data.

The FREQ proc has additional statements and options to apply statistical tests based on frequencies and to create output SAS datasets with frequencies and statistics.

CORR

The statistics described so far in this chapter consider just one variable at a time, but there are times when you need to analyze the connections between one variable and another. Is there a connection, for example, between moon phases and the severity of earthquakes? Or between the sound pressure levels of a concert and its ticket prices? The CORR proc can provide part of the answer to questions such as these by looking for linear relationships between variables in a SAS dataset. It computes statistics from pairs and combinations of variables.

These statistics, such as the covariance and Pearson correlation coefficient, are different from the statistics that are computed on a single variable. The covariance is a measure of the magnitude of the linear relationship between two variables. Correlation coefficients measure the strength of the linear relationship between two variables; the Pearson correlation coefficient is computed from the values of those two variables only.

The CORR proc produces a statistical report that can show various statistics that may have to do with linear relationships between variables. The default report shows several descriptive statistics for the individual analysis variables and the Pearson correlation coefficients for all pairs of analysis variables. This example of PROC CORR output shows the correlation table:

```
PROC CORR DATA=RIAA.YEAREND;
RUN;
```

```
RIAA Yearend Statistics (Millions Shipped)

The CORR Procedure
...
              Pearson Correlation Coefficients, N = 10
                      Prob > |r| under H0: Rho=0

                YEAR            CD      CASSETTE         LP_EP        SINGLE

YEAR         1.00000       0.97989      -0.97701      -0.49855      -0.74528
                            <.0001        <.0001        0.1425        0.0134

CD           0.97989       1.00000      -0.92759      -0.51980      -0.75079
              <.0001                      0.0001        0.1236        0.0123

CASSETTE    -0.97701      -0.92759       1.00000       0.46010       0.76238
              <.0001        0.0001                      0.1809        0.0104

LP_EP       -0.49855      -0.51980       0.46010       1.00000       0.59123
              0.1425        0.1236        0.1809                      0.0718

SINGLE      -0.74528      -0.75079       0.76238       0.59123       1.00000
              0.0134        0.0123        0.0104        0.0718
```

The NOSIMPLE, NOCORR, and NOPROB options suppress parts of the report. The COV option adds a covariance matrix to the report. The CORR proc has other statements and options to calculate other statistics, to control the details of the statistical computations, and to create output SAS datasets of statistics.

A correlation is not the same as a cause-and-effect relationship between variables. If there is a reason to hypothesize cause and effect, several procs of SAS/STAT can apply regression models and other statistical models that can estimate the form and magnitude of the causal relationships among a set of variables.

There are many other procs in base SAS and other SAS products that analyze data. The PRINT proc, described in the previous chapter, has options for calculating sums and counting observations. The SQL proc can summarize a data set and calculate descriptive statistics. The RANK and STANDARD procs analyze variables in order to apply transformations to them. The COMPARE proc measures the differences between two variables or SAS datasets. The various procs of SAS/STAT and several other SAS products apply statistical tests to data.

Graphs can also be useful for analyzing data. The PLOT, CHART, TIMEPLOT, and UNIVARIATE procs of base SAS and the various procs of SAS/GRAPH produce graphs.

22

Managing Data

The uses of the DATASETS and CONTENTS procs for managing SAS files have already been described in chapter 11, "Options for SAS Datasets." This chapter describes several more procs whose actions can be useful in managing SAS datasets and catalogs.

SORT

The SORT proc is the primary means of sorting SAS datasets. The proc uses the PROC SORT statement to set options and the BY statement to set the sort order.

The two essential options are DATA=, to identify the input SAS dataset, and OUT=, to name the output SAS dataset when it is different from the input SAS dataset. If you omit the OUT= option, the proc sorts the SAS dataset in place; this is a valid operation only for a SAS data file whose engine supports direct access.

The sort order clause in the BY statement is the same as anywhere else in a SAS program. List the sort key variables in order. Use the DESCENDING option before a variable to sort that variable in reverse order.

This is an example of sorting a SAS dataset in place:

```
PROC SORT DATA=MAIN.MEMBERS;
  BY NAME;
RUN;
```

This is an example of creating a new SAS dataset as a sorted copy of a SAS dataset:

```
PROC SORT DATA=MAIN.LIST OUT=WORK.LISTSORT;
  BY ZIP ADDRESS1 ADDRESS2 NAME;
RUN;
```

Other useful PROC SORT options are EQUALS, NODUPKEY, FORCE, and TAGSORT. The EQUALS and NODUPKEY options are significant when multiple observations may have the same sort key values. With the EQUALS option, observations that have the same key values are kept in the same order as in the input SAS dataset. With the default NOEQUALS option, observations that have the same key values could appear in any order in the output SAS dataset. The NODUPKEY option treats observations that have the same key values as duplicates and keeps only one of them. If you use the

NODUPKEY and EQUALS options together, the SORT proc keeps only the first input observation that it finds with a particular key value.

The FORCE option is necessary if sorting a SAS dataset in place will destroy data. Use this option to verify that you want to discard observations if you use the NODUPKEY option, the WHERE= dataset option, or the FIRSTOBS= or OBS= option while sorting in place. The FORCE option also has to do with indexes. The SAS supervisor deletes most indexes when you sort a SAS dataset in place. Use the FORCE option to verify that you want to sort an indexed SAS dataset and delete its indexes. Any indexes that are owned by integrity constraints are rebuilt after the sort. You can also use the FORCE option to force the SORT proc to sort a SAS dataset whose attributes indicate that it is already in sorted order.

The TAGSORT option indicates a *tag sort*. The key variables are sorted first, before the rest of the variables. A tag sort uses less temporary disk space. Depending on the data and the computer hardware, a tag sort might be slower or faster than the default sort method, which sorts all the variables at the same time.

COPY

The COPY proc copies SAS files. It can copy an entire library, or it can copy selected members from one library to another. To copy an entire library, only the PROC COPY statement is required. Use the IN= option to identify the source libref and the OUT= option to identify the destination libref. For example, this step copies all the members of the MAIN library to the OTHER library:

```
PROC COPY IN=MAIN OUT=OTHER;
RUN;
```

Use the SELECT statement to copy only selected members. For example, this step copies MAIN.GEO to OTHER.GEO:

```
PROC COPY IN=MAIN OUT=OTHER;
  SELECT GEO;
RUN;
```

Alternatively, use the EXCLUDE statement to copy all members except for those you specifically leave out.

To limit the copy to certain member types, use the MTYPE= option in the PROC COPY statement. Write one member type or a list of member types in parentheses. For example, the option MTYPE=DATA copies only SAS data files; MTYPE=CATALOG copies only catalogs; and MTYPE=(DATA VIEW) copies only SAS datasets. This option can be abbreviated as MT.

Other options in the PROC COPY statement affect details of the way SAS data files are copied. Use the option INDEX=YES to copy indexes of SAS data files, or INDEX=NO to copy SAS data files without indexes. With the INDEX=YES option, the option CONSTRAINT=YES copies the integrity constraints of SAS data files. Use the CONSTRAINT=NO option to copy SAS data files without their integrity constraints. Integrity constraints cannot be copied unless indexes are copied. The CLONE option reproduces the BUFSIZE=, COMPRESS=, and

REUSE= options of a SAS data file. With the NOCLONE option, the COPY proc sets those options for the copies of SAS data files based on the current system option settings.

If you need to change the name of a SAS file you are copying, use the COPY proc to copy it, then use the DATASETS proc to change the name of the copy.

APPEND

The APPEND proc combines two SAS datasets by adding the observations of the input SAS dataset to the end of the output SAS dataset. The input SAS dataset is unchanged by this process. The only change to the output SAS dataset is that more observations are added to it. This is an example of the APPEND proc:

```
PROC APPEND DATA=WORK.DAYTRANS OUT=MAIN.TRANS;
RUN;
```

The observations of the input SAS dataset, WORK.DAYTRANS, are added to the end of the output SAS dataset, MAIN.TRANS.

In the `PROC APPEND` statement, NEW is an alias for the DATA= option. BASE is an alias for the OUT= option.

For the append process to work correctly, the input SAS dataset should have the same variables with the same essential attributes as the output SAS dataset. However, this is not a strict requirement; the proc can negotiate some differences in variable definitions between the two SAS datasets.

- If the input SAS dataset lacks some of the variables of the output SAS dataset, those variables are given missing values in the new observations.
- If variable names differ, use the RENAME= dataset option on the input SAS dataset to make the names match those of the output SAS dataset.
- If a character variable in the input SAS dataset is shorter than in the output SAS dataset, the values are padded with spaces.
- If a variable in the input SAS dataset is longer than in the output SAS dataset, the values are truncated. This requires the FORCE option.
- If the input SAS dataset contains variables that are not in the output SAS dataset, the variables are discarded. This requires the FORCE option, or the KEEP= or DROP= dataset option on the input SAS dataset.

If the output SAS dataset does not already exist, the APPEND proc creates a new SAS dataset and copies the entire input SAS dataset to it. This may be helpful if you periodically accumulate data in a particular SAS dataset. The SAS dataset does not have to exist before the first time you append to it.

CATALOG

The CATALOG proc lets you manage the entries in a catalog. You can copy, rename, and delete entries. Much of the syntax of the CATALOG proc mirrors the syntax that the DATASETS proc uses for managing SAS files.

A PROC CATALOG step works with the entries of one catalog. Name the catalog in the CATALOG= term in the PROC CATALOG statement. CAT and C are aliases for this term. The proc ordinarily opens the catalog with member control, which gives it exclusive access to the catalog. To open the catalog with record control, use the FORCE option. This makes it possible to take actions on a shared catalog.

The CONTENTS statement prints a list of the catalog's entries. The list includes the name, type, descriptions, and creation and modification dates of each entry. To create a SAS dataset of the entries, use the OUT= option in the CONTENTS statement with a SAS dataset name.

Other statements in the step take actions on specific entries.

Rename The CHANGE and EXCHANGE statements change the names of entries. Use the CHANGE statement to rename individual entries. In the CHANGE statement, write the two-level name of the entry, then write an equals sign and the new name. This step changes the name of MAIN.APPL.SPLASH.FMENU to MAIN.APPL.DOC.FMENU:

```
PROC CATALOG CAT=MAIN.APPL;
  CHANGE SPLASH.PMENU=DOC.PMENU;
RUN;
```

Use the EXCHANGE statement, which has the same syntax, to swap the names of two entries. You cannot use the CHANGE or EXCHANGE statement to change the entry type of an entry.

The MODIFY statement changes the description of an entry. The only significance of the description of an entry is that it appears in lists of the entries in the catalog. The form of the statement is:

```
MODIFY entry (DESCRIPTION='description');
```

Delete The DELETE and SAVE statements delete entries from a catalog. Use the DELETE statement to list specific entries to delete. The SAVE statement lists specific entries to save; it deletes all other entries. This statement removes all the entries from the catalog except for the four that are listed in the statement:

```
SAVE MORNING.OUTPUT AFTERNOON.OUTPUT EVENING.OUTPUT MIDNIGHT.OUTPUT;
```

To delete all the entries in a catalog, use the KILL option in the PROC CATALOG statement.

Copy The COPY statement copies entries to another catalog. Name the destination catalog in the OUT= option. By itself, the COPY statement copies all the entries in the source catalog. To copy only selected entries, list the entries to copy in the SELECT statement after the COPY statement. To copy all but a few excluded entries, list the entries to exclude in the EXCLUDE statement after the COPY statement.

COMPARE

The COMPARE proc compares two SAS datasets and makes a list of the differences it finds. It can also compare two variables in the same SAS dataset. It prints a report or creates a SAS dataset of the differences in values between the two SAS datasets or variables.

These options in the PROC COMPARE statement identify the SAS datasets that the proc uses:

Option	Description	Default
DATA=*SAS dataset* BASE=*SAS dataset*	Base dataset: the SAS dataset to compare to	_LAST_
COMPARE=*SAS dataset* COMP=*SAS dataset* C=*SAS dataset*	Comparison dataset: the SAS dataset to compare	The same as the base dataset
OUT=*SAS dataset*	The output SAS dataset	None
OUTSTATS=*SAS dataset*	The output SAS dataset of summary statistics	None

The other statements in the COMPARE proc indicate the use of variables in the comparison:

Statement	Description	Default
VAR *variables*;	Base variables: variables in the base dataset to use in the comparison	All variables of the base dataset that match variables in the comparison dataset, other than BY and ID variables
WITH *variables*;	Comparison variables: variables to compare against the base variables, listed in the same order	Variables with the same names as the base variables
ID *sort order clause*;	ID variables: variables used to match observations between the two datasets	Observations are matched by observation number
BY *sort order clause*;	BY variables: form BY groups in the base dataset, each of which is processed separately	None

The BY statement refers to the base dataset. The BY variables define groups of observations in and indicate the sort order of the base dataset. If the BY variables are also present in the comparison dataset, that dataset must also be in sorted order, and each BY group of the comparison dataset is checked against the matching BY group of the base dataset. If none of the BY variables are in the comparison dataset, the proc compares the entire comparison dataset to each BY group of the base dataset. Usually, BY variables are not necessary for the COMPARE proc, and key variables should be indicated in the ID statement. Use the BY statement only when you want to perform a separate comparison with a separate report for each BY group.

The ID variables should form a unique key, so that no two observations have the same key values. If there are multiple observations for a key value in either dataset, the proc uses only the first one and treats the rest as duplicates. The proc indicates the number of duplicate observations in its report.

The comparison variables in the WITH statement are compared to the base variables in the VAR statement. If there is no WITH statement, the comparison variables are assumed to have the same names as the base variables. If there are fewer variables in the WITH statement than in the VAR statement, the remaining comparison variables are assumed to have the same names as the base variables.

The COMPARE proc has other options that affect the contents of the report and the output SAS dataset and determine the equality criterion. By default, the proc treats numeric values as unequal unless they are exactly equal. With the equality criterion options, you can have the proc treat values that are only slightly different as equal.

TRANSPOSE

The TRANSPOSE proc reshapes data by turning variables into observations and observations into variables. Usually, it works with a BY statement and changes the input variables into observations for each output BY group. In the TRANSPOSE proc:

- The DATA= and OUT= options in the `PROC TRANSPOSE` statement identify the input and output SAS datasets.
- The BY and COPY statements identify input variables that also appear as variables in the output SAS dataset.
- The VAR statement identifies input variables that are transposed to form output observations.
- The PREFIX=, LET, NAME=, and LABEL= options and ID and IDLABEL statements control the way output variables are formed.

Input Variables

The BY variables are the same in the input and output SAS datasets. The output SAS dataset has the same BY groups in the same order. If there is no BY statement, the proc transposes the same as if there were one BY group.

The input variables listed in the COPY statement also appear as variables in the output SAS dataset. With a COPY statement, the output BY group always has at least as many observations as the input BY group, in order to hold the COPY variables. If there are more output observations than input observations in the BY group, the COPY variables have missing values for the remaining output observations.

The analysis variables listed in the VAR statement are transposed to form observations in the output SAS dataset. This means that each input observation in a BY group forms an output variable in that BY group. The number of output variables is at least as great as the number of observations in the largest input BY group. If there is no VAR statement, the proc uses all numeric variables as analysis variables.

Output Variables

The TRANSPOSE proc creates an output SAS dataset that contains these variables:

- BY variables, which form the same BY groups as in the input SAS dataset.
- COPY variables, which are copied from the input SAS dataset within each BY group.
- Transposed variables: variables that are formed by transposing the input observations. These are numeric variables if all the analysis variables are numeric. Otherwise, they are character variables. The length of the transposed variables is the longest length of any analysis variable.
- A variable that contains the names of the analysis variables. The default name of this variable is _NAME_. You can specify a different name in the NAME= option.
- A variable that contains the labels of the analysis variables. The default name of this variable is _LABEL_. You can specify a different name in the LABEL= option.

The TRANSPOSE proc can form the names of the transposed variables in a few different ways. These components may be part of the names:

1. *The values of an ID variable.* Use the ID statement to indicate this input variable. If there is no ID statement, the proc uses an input variable called _NAME_, if it exists. The proc uses the values of this variable to form the names of the transposed variables. If the values of the ID variable do not form SAS names, the proc transforms them to form valid names it can use. For example, it removes spaces and special characters.
2. *A prefix.* Use the PREFIX= option in the `PROC TRANSPOSE` statement to indicate a prefix that begins the names of all the transposed variables. If there is no PREFIX= option and one is required for numeric ID values, the proc uses an underscore as the prefix. If there is no PREFIX= option or ID variable, the proc uses COL as its prefix.
3. *Numeric suffixes.* If there is no ID variable, the proc generates the counting number suffixes 1, 2, 3, . . . The suffix corresponds to the sequence of the input observation within the BY group.

The LET option determines how the proc responds to duplicate observations, that is, multiple observations within a BY group that have the same value for the ID variable. With the LET option, the proc uses the value from the last observation that has that particular value. That is, when there are duplicates observations, it disregards all but the last one it finds. Without the LET option, the proc generates an error condition and stops processing if it finds duplicate observations.

The IDLABEL or IDL statement names an input variable whose values are used as the labels of the output variables.

The following example shows the effect of transposing. The SAS dataset RIAA.YEAREND1, shown in the first output, is transposed to create the SAS dataset RIAA.YEAREND2, shown in the second output.

```
TITLE1 'RIAA Yearend Statistics (Millions Shipped) Before Transposing';
PROC PRINT DATA=RIAA.YEAREND1;
RUN;
PROC TRANSPOSE DATA=RIAA.YEAREND1 OUT=RIAA.YEAREND2
   NAME=CONFIGURATION PREFIX=UNIT;
  ID YEAR;
  IDLABEL YEAR;
  VAR CD CASSETTE LP_EP SINGLE;
RUN;
TITLE1 'RIAA Yearend Statistics (Millions Shipped) After Transposing';
PROC PRINT DATA=RIAA.YEAREND2;
RUN;
```

```
RIAA Yearend Statistics (Millions Shipped) Before Transposing

Obs    YEAR       CD     CASSETTE    LP_EP    SINGLE

 1     1995     722.9      272.6      2.2      102.4
 2     1996     778.9      225.3      2.9      113.2
 3     1997     753.1      172.6      2.7      106.4
 4     1998     847.0      158.5      3.4       87.8
 5     1999     938.9      123.6      2.9       75.4
```

```
RIAA Yearend Statistics (Millions Shipped) After Transposing

Obs   CONFIGURATION   UNIT1995   UNIT1996   UNIT1997   UNIT1998   UNIT1999

 1      CD              722.9      778.9      753.1      847.0      938.9
 2      CASSETTE        272.6      225.3      172.6      158.5      123.6
 3      LP_EP             2.2        2.9        2.7        3.4        2.9
 4      SINGLE          102.4      113.2      106.4       87.8       75.4
```

The four analysis variables, CD, CASSETTE, LP_EP, and SINGLE, become observations in the transposed data. The ID variable, YEAR, is used to form the variable names for the transposed data. The input variable names become the values of the output variable CONFIGURATION.

23

SQL

The SQL proc lets you execute SQL statements that operate on SAS data. SQL, Structured Query Language, is the closest thing there is to a standard language for retrieving data from databases. SQL is not exactly a programming language, because it does not describe a sequence of actions. Instead, an SQL statement describes an outcome.

The simplest and most common action in SQL is to extract a subset of data from a table. The SQL statement to accomplish this is a description of the data to extract. SQL statements can often accomplish the same results as SAS statements, but because SQL's approach is so different, the SQL statements may look nothing like the SAS statements that do the same thing. For some tasks that involve combining data from several tables, the SQL approach can be more direct and concise than the SAS approach. There are other reasons why you might use SQL code in a SAS program. If you use SAS/ACCESS to connect to a database management system, you can write SAS SQL statements to access a database or to combine database data with SAS data. Or, if you move an existing project into the SAS environment, you can incorporate the project's existing SQL code into the new SAS programs.

IBM developed SQL in the 1970s as a database interface based on the relational database model, which IBM also developed around that same time. SQL has been used in nearly every relational database management system beginning with the original release of Oracle in 1979. It has been the subject of a series of ANSI and ISO standards since 1986. However, no implementation of SQL has ever followed any standard exactly. SAS SQL is mostly compliant with the 1992 SQL standard.

Lexicon

The prerelease version of SQL was called *SEQUEL* (for Structured English Query Language), and some people still pronounce SQL as "sequel."

SAS was originally developed around the same time as SQL, and both use many of the same concepts of data organization. However, some of the most important data objects are called different names in the SAS and SQL environments. The following table translates between SAS and SQL terminology.

SAS Term	SQL Term	Description
SAS data file	table	a file containing organized data
variable	column	a distinct data element
observation	row	an instance of data, including one value for each variable or column
missing	null	a special value of a data element that indicates that a value is not available
integrity constraint	constraint	a specific restriction on the value of a data element

Lexicon

The SAS and SQL terms for data objects differ because of the different purposes for which they were originally envisioned. The SAS terms *variable, observation,* and *missing* are the terms used in the field of statistics. The SQL terms *table, column,* and *row* are based on relational database theory.

The `PROC SQL` step starts with the `PROC SQL` statement, which indicates that the statements that follow are SQL statements. The `PROC SQL` statement can set options that affect the execution of the SQL statements. These options can be changed between SQL statements using the RESET statement. SQL uses an interpretive style, executing each statement as soon as it reaches it. You can write global statements to execute between the SQL statements; the settings of a TITLE statement or other global statement take effect for the following SQL statement. Use the QUIT statement, if necessary, to mark the end of the `PROC SQL` step. Because the proc executes each statement immediately, it ignores the RUN statement.

The syntax of the `PROC SQL` step is summarized as:

```
PROC SQL options;
statement
. . .
QUIT;
```

Each statement after the `PROC SQL` statement can be an SQL statement, a RESET statement, or a global statement. The QUIT statement that marks the end of the step is optional.

Among the various differences in style between SAS and SQL syntax, there is one difference that is especially important to notice. In SQL, a reference to a table, column, or any other data object is an expression. Therefore, in any list of objects in an SQL statement, you must write commas as separators between list items. This contrasts with the SAS style, in which a reference to an object is merely a name, and in lists, the names are separated only by spaces.

Query Expressions

SQL syntax is built around specific kinds of expressions, the most important of which is the query expression. A query expression defines a set of data in the shape of a table. The data of a query expression could be the entire contents of a table, but more often, it is a smaller amount of data extracted from one or more tables.

Lexicon

Strictly speaking, the simple form of the query expression that is described here is a *table expression*. More complicated query expressions use set operators to combine two or more table expressions.

The SELECT and WHERE Clauses

A minimal query expression contains the word SELECT, a list of columns, the word FROM, and a table. These terms form a SELECT clause, to indicate what table to draw data from, what columns to use, and the order of the columns. This is an example of a SELECT clause:

```
SELECT ADDRESS, TITLE, SIZE FROM MAIN.PAGES
```

The table is the SAS dataset MAIN.PAGES; the columns that are selected are the variables ADDRESS, TITLE, and SIZE in that SAS dataset.

This kind of query expression returns all the rows from the indicated table. To return only selected rows, add a WHERE clause with the condition for selecting rows. The syntax for the WHERE condition is the same as for any WHERE condition in SAS software. This is an example of a query expression with a WHERE clause:

```
SELECT ADDRESS, TITLE, SIZE FROM MAIN.PAGES
WHERE SIZE <= 81920
```

With the WHERE condition, the query expression returns only those rows where the value of the column SIZE is no more than 81,920.

A SELECT clause and a WHERE clause are sufficient to form most query expressions. The following table summarizes the syntax of these clauses:

Terms	Meaning
SELECT *column*, *column*, . . .	Selects columns and indicates their order
FROM *table*, *table*, . . .	The table that contains the columns
WHERE *condition*	Condition for selecting rows (optional)

The order of the clauses in a query expression is important. It is a syntax error if the clauses are in the wrong order.

SELECT Statement

A query expression can be used by itself as a statement. This kind of SQL statement is called a SELECT statement. It produces a table report of the data that the query expression selects. This is an example of a `PROC SQL` step that executes a SELECT statement:

```
PROC SQL;
SELECT ADDRESS, TITLE, SIZE FROM MAIN.PAGES
  WHERE SIZE <= 81920;
```

As with any SAS statement, an SQL statement ends in a semicolon.

A SELECT statement can also contain an ORDER BY clause to indicate the order of the rows. Write the sort order with commas between the columns. Use the modifier DESC after a column to sort in descending order of that column. This is an example of a SELECT statement with an ORDER BY clause:

```
SELECT ADDRESS, TITLE, SIZE FROM MAIN.PAGES
  WHERE SIZE <= 81920
  ORDER BY TITLE, SIZE DESC;
```

Print Output

Like the PRINT and REPORT procs, the SELECT statement produces print output in table form. This example demonstrates the print output of the SELECT statement:

```
PROC SQL;
SELECT * FROM MAIN.CITYS;
QUIT;
```

```
STATE                    POP     URBAN
--------------------------------------
Ohio                10847040      81.4
Michigan             9295895      82.8
```

The SQL table output is similar to the output that the PRINT proc produces with the NOOBS and WIDTH=FULL options. It is similar to the output of the REPORT proc with the HEADLINE and WRAP options. PROC PRINT and PROC REPORT steps and output are shown here for comparison:

```
PROC PRINT DATA=MAIN.CITYS NOOBS WIDTH=FULL;
RUN;
PROC REPORT DATA=MAIN.CITYS NOWD HEADLINE;
RUN;
```

```
STATE                          POP              URBAN

Ohio                      10847040               81.4
Michigan                   9295895               82.8
```

```
  STATE                    POP     URBAN
  --------------------------------------
  Ohio                10847040      81.4
  Michigan             9295895      82.8
```

List of Values

In a SELECT clause, use the keyword DISTINCT before the list of columns to eliminate duplicates from the selected data. The result is a list of the distinct combinations of values of the selected columns. If you list only one column with the DISTINCT option, it produces a list of the different values of that column.

The following example shows the effect of the DISTINCT keyword in a query. The first query shows the value of TYPE for every row in the table. The second query, with the word DISTINCT added, shows only the two different values of TYPE, and it arranges them in sorted order.

```
SELECT TYPE FROM MAIN.LETTER;
SELECT DISTINCT TYPE FROM MAIN.LETTER;
```

```
TYPE
---------
Vowel
Consonant
Consonant
Consonant
Vowel
Consonant
...
```

```
TYPE
---------
Consonant
Vowel
```

Combining Tables

In relational database theory, it is expected that related columns are often stored in separate tables. Queries often combine columns from two or more tables. To write a query that draws columns from multiple tables:

- List the tables, separated by commas, after the keyword FROM.
- After each table name, write an alias. An alias is a one-word name for the table that is used in the query. Programmers usually use single letters for table aliases — often the letters A, B, and so on.
- In the query, identify columns with two-level names that combine the table alias and the column name. For example, the column STATE in the table whose alias is A is referred to as A.STATE.
- In the WHERE expression, include the condition for combining rows between the two tables. Most often, the requirement for combining rows is that the key variables match. For example, if the key variables that connect two tables are DAY and STATE, the WHERE condition could be `A.DAY = B.DAY AND A.STATE = B.STATE`. Writing the WHERE condition correctly is critical for a multi-table query (assuming the query uses tables that have multiple rows). If the WHERE condition is incorrect, the result of the query could be an enormous number of rows or no rows at all.

The query below is an example. It draws the columns NAME, AGE, and SUBSPECIES from the table MAIN.TIGER and adds the column NATIVE from the table MAIN.TSUBS. It shows only rows for which the values of SUBSPECIES match between the two tables. The selected column names appear as the column headings in the output.

```
SELECT I.NAME, I.AGE, I.SUBSPECIES, T.NATIVE
FROM MAIN.TIGER I, MAIN.TSUBS T
WHERE I.SUBSPECIES = T.SUBSPECIES;
```

```
NAME                  AGE  SUBSPECIES        NATIVE
-------------------------------------------------------------
Leah                    7  Bengal            Bay of Bengal
Max                     5  Indochinese       SE Asia
Pierce                  2  Siberian          SE Siberia
Princess                3  Bengal            Bay of Bengal
Stick                   8  Sabertooth        N Eurasia
```

Creating Files From Query Expressions

In a SELECT statement, the results of a query are converted to an output object. Query results can also be stored as data. The CREATE TABLE statement creates a table with the results of a query. The CREATE VIEW statement stores the query itself as a view. Either way, the data identified in the query can be used in later SQL statements or in other SAS steps.

A SELECT statement is converted to a CREATE TABLE statement by adding terms to the beginning of the statement. The added words are CREATE TABLE, the name of the table, and AS. This is an example of a CREATE TABLE statement:

```
CREATE TABLE WORK.PAGE80 AS
  SELECT ADDRESS, TITLE, SIZE FROM MAIN.PAGES
  WHERE SIZE <= 81920
  ORDER BY TITLE, SIZE DESC;
```

```
NOTE: Table WORK.PAGE80 created, with 25 rows and 3 columns.
```

The new table WORK.PAGE80 is a SAS data file. The log note that describes it is the similar to the note that ordinarily describes a new SAS data file, but it uses the SQL words `table`, `columns`, and `rows` in place of the SAS words `SAS data set`, `variables`, and `observations`.

The CREATE VIEW statement creates an SQL view. It is the same as the CREATE TABLE statement except that the word VIEW replaces the word TABLE. A view can be used the same way a table is used. However, a view does not operate the same way as a table. When a table is created, the query is executed and the resulting data is stored in a file. When a view is created, the query itself is stored in the file. The data is not accessed at all in the process of creating a view. The log note says only that a new view has been stored:

```
NOTE: SQL view WORK.CABLES has been defined.
```

The query is executed, and the number of rows and columns determined, only when a later program reads from the view.

Query expressions depend on librefs to identify the tables from which they draw data. The libref definitions of an SQL view can be stored in the view itself, so that the view is self-contained. At the end of the CREATE VIEW statement, write a USING clause that contains LIBNAME statements. Write the LIBNAME statements as that statement is normally written, but with commas, rather than semicolons, separating multiple LIBNAME statements.

Column Expressions and Modifiers

Most columns in queries are simply columns of a table. However, this is only one of the possibilities for writing a column expression in a query expression. The table below lists various ways that a column expression can be formed.

Expression	Description
column	A column of a table.
table alias.column	A column in the indicated table.
*	All columns of all tables in the query.
*table alias.**	All columns of the indicated table.
constant	A constant value.
expression	An expression that computes a value using columns, constants, operators, and functions.

When SAS variables are used as columns in a query, the SQL proc uses the format and label attributes of the columns in the resulting output object, table, or view. However, the SQL proc does not use the FORMAT or LABEL statements that other procs use to change these attributes. Instead, to change the appearance of a column, use the FORMAT= and LABEL= column modifiers. List the modifiers after the column name or expression (not separated by commas). This is an example of a SELECT statement that uses column modifiers:

```
SELECT
  START FORMAT=DATE9. LABEL='Start Date',
  END FORMAT=DATE9. LABEL='End Date'
  FROM MAIN.EVENTS
  ORDER BY START, END;
```

The INFORMAT= and LENGTH= column modifiers can also be used to set those two attributes. These attributes are especially useful when creating a table.

When a query column is not a table column, or when the same column name is used more than once in the same query, it is useful to assign an alias to it. The alias can be used as a name for the query column anywhere else in the same query. For example, aliases can be used in column expressions and WHERE expressions. The alias is also the name of the column in the print output or in the table or view created from the query. To assign an alias, write the keyword AS and the alias at the end of the column definition (after the column modifiers, if there are any). This example includes two column definitions with aliases:

```
SELECT
  TEMPERATURE/1.8 + 32 AS TEMPERATURE_FAHRENHEIT,
  PRECIPITATION AS RAIN
  FROM MAIN.WEATHER
```

```
WHERE RAIN > 0 AND TEMPERATURE_FAHRENHEIT >= 32
;
```

If you do not provide an alias for a column that is computed as an expression, the column is displayed without column headings. If you create a table or view that includes unnamed columns, the SQL proc generates names for the columns based on the column numbers.

SQL Expressions

Column expressions and WHERE expressions are two examples of SQL expressions that result in single values. These SQL expressions are also used in several other places in SQL syntax. The rules of syntax for SQL expressions are mostly the same as that of SAS expressions, with these differences:

- The WHERE operators are used.
- Comparison operators do not use the : modifier.
- The NOT operator has the lowest priority of any SQL operator.
- SQL expressions cannot use the queue functions or those that refer to specific data step objects.
- The INPUT and PUT functions do not use error control terms.
- SQL also supports the COALESCE function. This function returns the first non-null (nonmissing) value among its arguments. The arguments can be any number of columns of the same data type.
- The CASE operator of SQL allows logic within an expression that resembles the SELECT block of the data step. This is an example of a column definition that uses a CASE expression and an alias:

```
CASE
  WHEN AMOUNT < 0 THEN 'Credit'
  WHEN AMOUNT > 0 THEN 'Debit'
  ELSE 'No Balance'
  END
AS BALANCE
```

Dataset Options and Reserved Words in SQL Statements

When SAS datasets are used as tables in SQL statements, you can use any of the usual dataset options. As always, write the dataset options in parentheses after the SAS dataset name.

The SQL standards restrict the use of all words that SQL uses as keywords. In standard SQL, these reserved words cannot be used as the names of SQL objects. SAS SQL reserves only a few words. The names CASE and USER cannot be used as column names. If these are names of SAS variables that you want to use in a query, use the RENAME= dataset option to change their names.

Do not use SQL keywords as table aliases. Most programmers use single letters as table aliases; no single letter is a reserved word. It is also safe to use a letter or word with a numeric suffix, such as T1, as a table alias; no reserved word has a numeric suffix.

Summary Statistics

SQL can calculate summary statistics for a column. Write the statistic, then the column name in parentheses. For example, SUM(TRAFFIC) calculates the sum of the column TRAFFIC. The syntax is the same as a function call, but it is computed as a column statistic as long as the function name is a statistic and the argument is a column or column alias.

The available statistics are the standard set of SAS statistics, other than SKEWNESS and KURTOSIS. In addition to the usual SAS names for the statistics, AVG can be used as an alias for MEAN and COUNT and FREQ as aliases for N. Use the expression COUNT(*) to count the rows in a table.

Use aliases to create column names for statistic columns. Without aliases, statistic columns are displayed without column headings. This is an example of a query that uses summary statistics as columns:

```
TITLE1 'RIAA Yearend Statistics (Units Shipped)';
SELECT
  MIN(CD) FORMAT=COMMA14. AS MinCD,
  MIN(CASSETTE) FORMAT=COMMA14. AS MinCassette,
  MIN(LP_EP) FORMAT=COMMA14. AS MinLP_EP,
  MIN(SINGLE) FORMAT=COMMA14. AS MinSingle,
  MEAN(CD) FORMAT=COMMA14. AS AvgCD,
  MEAN(CASSETTE) FORMAT=COMMA14. AS AvgCassette,
  MEAN(LP_EP) FORMAT=COMMA14. AS AvgLP_EP,
  MEAN(SINGLE) FORMAT=COMMA14. AS AvgSingle,
  MAX(CD) FORMAT=COMMA14. AS MaxCD,
  MAX(CASSETTE) FORMAT=COMMA14. AS MaxCassette,
  MAX(LP_EP) FORMAT=COMMA14. AS MaxLP_EP,
  MAX(SINGLE) FORMAT=COMMA14. AS MaxSingle,
  COUNT(*) AS Years
  FROM RIAA.UNITS;
```

```
RIAA Yearend Statistics (Units Shipped)

         MinCD      MinCassette         MinLP_EP       MinSingle
         AvgCD      AvgCassette         AvgLP_EP       AvgSingle
         MaxCD      MaxCassette         MaxLP_EP       MaxSingle      Years
---------------------------------------------------------------------------
   286,500,000      123,600,000        1,200,000      75,400,000
   622,560,000      280,620,000        3,600,000     117,130,000
   938,900,000      442,200,000       11,700,000     191,500,000         10
```

A statistic can use the keyword DISTINCT before the column name. With DISTINCT, the statistic is applied to the set of distinct values in the column, rather than the values of all the rows. For example, COUNT(PLACE) counts rows in which PLACE has a value, but COUNT(DISTINCT PLACE) counts the different values of PLACE. As another example, MEAN(X) is the mean of X for all rows in the table, but MEAN(DISTINCT X) is the mean of the set of distinct values of X.

If all columns in a query are summary statistics, the result is one row that contains a summary of the rows that the query reads. If a query contains a combination of summary statistics and other expressions, the statistics are repeated in each row of the result. Combining summary statistics and detail data in this way is called *remerging*. A log note indicates the process of remerging in case you wrote a remerging query without realizing it.

Summary statistics and detail data can be used together in column expressions. The most common use of this is to calculate percents or relative frequencies, as in this example:

```
SELECT BROWSER, HIT FORMAT=COMMA9. AS HITS,
  HIT/SUM(HIT)*100 FORMAT=7.3 AS SHARE
  FROM MAIN.BROWSERS ORDER BY BROWSER;
```

```
BROWSER                          HITS     SHARE
-------------------------------------------
AOL 3                             811     2.117
Internet Explorer 2             2,667     6.962
Internet Explorer 3             1,085     2.832
Internet Explorer 4             8,515    22.229
Internet Explorer 5             5,377    14.037
Lynx                               89     0.232
Netscape 1                        163     0.426
Netscape 2                        707     1.846
Netscape 3                      4,098    10.698
Netscape 4                      5,950    15.533
Netscape 5                      8,182    21.360
Other                             532     1.389
Prodigy                            95     0.248
Sega Saturn                        35     0.091
```

Grouping

The GROUP BY clause in a query expression lets you divide the rows of the query into groups and apply summary statistics within those groups. The GROUP BY clause follows the WHERE clause, if there is one. It usually lists one or several columns. This is an example of a GROUP BY clause:

```
GROUP BY STATE, YEAR
```

The GROUP BY columns are essentially the same as class variables. They organize the data into groups. Statistics are calculated within the groups instead of being calculated for the entire set of data. In this example, statistics are calculated separately for each state and year.

GROUP BY items are usually table columns, but they can also be column expressions. If you use an integer as a GROUP BY item, it is used as a column number, and that column of the query is used for grouping. This makes it possible to group by computed columns that do not have names. A better approach, however, is to use aliases for the computed columns.

If there is a GROUP BY clause and all of the columns are summary statistics or GROUP BY items, the query results in only one row for each group. If the query contains a combination of summary statistics and other expressions, the summary statistics are repeated in each row of the group. If summary statistics are used to calculate percents, they are percents of the total for the group, rather than percents of the total for the entire set of data.

The simplest use of grouping is to create a frequency table, as shown in this example:

```
SELECT SYMBOL, NAME, COUNT(*) AS FREQUENCY
  FROM MAIN.SPLIT GROUP BY SYMBOL, NAME;
```

```
SYMBOL NAME                              FREQUENCY
--------------------------------------------------
AAPL   Apple Computer Inc                        1
AOL    America Online                            5
BRCM   Broadcom Corp                             2
CPB    Campbell Soup                             1
CRA    PE Corp - Celera Genomics Grp             1
HGSI   Human Genome Sciences Inc                 1
IBM    Intl Business Machines                    2
K      Kellogg Co                                1
QCOM   Qualcomm Inc                              2
RHAT   Red Hat Inc                               1
SUNW   Sun Microsystems                          4
TWX    Time Warner Inc                           1
YHOO   Yahoo Inc                                 4
```

You cannot use the WHERE clause to select groups or rows based on summary statistics of a group. That is because the WHERE clause is always evaluated separately for each individual row. Instead, when a condition contains summary statistics, write it in a HAVING clause. Write the HAVING clause after the GROUP BY clause. In a query that has no GROUP BY clause, a HAVING clause is applied to all the rows as one group.

The HAVING condition is used instead of the WHERE condition in queries that combine summary rows of tables. The criterion for matching summary rows must be written in the HAVING clause, rather than the WHERE clause, because the WHERE clause applies only to individual rows, not to the summary rows of groups.

Other Queries

SQL has several more features that make a much wider range of queries possible.

- *Table join operators* represent alternate ways to combine two tables: LEFT JOIN, RIGHT JOIN, and FULL JOIN.
- *Set operators* combine the results of two query expressions. The set operators are UNION, OUTER UNION, EXCEPT, and INTERSECT. The INTERSECT operator can modified by the CORRESPONDING and ALL modifiers.
- A *database query* is an expression in the form CONNECTION TO *database* (*query expression*). The query is passed to a database, and the results that are returned from the database are used within the SAS query expression. This *SQL pass-through* feature requires SAS/ACCESS and other statements within the PROC SQL step to connect to the database.
- *Subqueries* are queries written in parentheses and used as values or as tables within a query expression. This is an example of the use of

a subquery to determine the number of different combinations of values for two columns:

```
SELECT COUNT(*) AS N
  FROM (SELECT DISTINCT DEPT, VENDOR FROM CORP.SOURCE);
```

- An *INTO clause* in a query expression assigns the results of the query to macro variables.

DICTIONARY Tables

The special libref DICTIONARY contains tables that can only be used in SQL queries. These tables list objects in the SAS environment:

DICTIONARY.OPTIONS	System options
DICTIONARY.TITLES	Title and footnote lines
DICTIONARY.EXTFILES	Filerefs
DICTIONARY.MEMBERS	SAS files
DICTIONARY.CATALOGS	Catalogs
DICTIONARY.MACROS	Macros
DICTIONARY.TABLES	SAS data files
DICTIONARY.VIEWS	Views
DICTIONARY.COLUMNS	Variables in SAS datasets
DICTIONARY.INDEXES	Indexes

You can query these tables in a SAS program to get information about objects and settings in the SAS session. For example, this query returns information on the PAGENO system option, including its current value:

```
SELECT * FROM DICTIONARY.OPTIONS WHERE OPTION='PAGENO';
```

```
optname          setting          optdesc                              level
----------------------------------------------------------------------------------
PAGENO           1                Beginning page number for            Portable
                                  the next page of output
                                  produced by the SAS System
```

To get a list of the columns in a DICTIONARY table, use a `DESCRIBE TABLE` statement, such as:

```
DESCRIBE TABLE DICTIONARY.COLUMNS;
```

Database Management Actions

Other SQL statements are designed to manage data. These statements can create, describe, update, modify, and delete various objects, including tables and views and the indexes and integrity constraints of a table. The following table summarizes the database management actions that are available in SAS SQL.

Action	Object	Statement
Create	Table	CREATE TABLE *table* (*definition*, . . .);
		CREATE TABLE *table* LIKE *table*;
		CREATE TABLE *table* AS *query expression*;
	View	CREATE VIEW *view* AS *query expression*;
	Index	CREATE INDEX *index* ON TABLE (*column*, . . .)
	Constraint	ALTER TABLE *table* ADD CONSTRAINT *constraint rule*;
Describe	Table	DESCRIBE TABLE *table*;
	View	DESCRIBE VIEW *view*;
	Constraint	DESCRIBE TABLE CONSTRAINTS *table*;
Update	Table	UPDATE *table* SET *column*=*value*, . . . WHERE *condition*;
		INSERT INTO *table* SET *column*=*value*, . . . *or* VALUES (*value*, . . .) *or query expression*;
		DELETE FROM *table* WHERE *condition*;
Modify	Table	ALTER TABLE *table action*;
Delete	Table	DROP TABLE *table*;
	View	DROP VIEW *view*;
	Index	DROP INDEX *index* FROM *table*;
	Constraint	ALTER TABLE *table* DROP CONSTRAINT *constraint*;

Indexes and integrity constraints and the SQL statements for creating them are described in chapter 11, "Options for SAS Datasets."

Describing

The DESCRIBE statement creates a log note that provides a description of a view or table or the constraints of a table. The DESCRIBE VIEW statement shows the query program that is stored in a view, with a log note similar to the one shown in this example:

```
DESCRIBE VIEW SASHELP.VTABLE;
```

```
NOTE: SQL view SASHELP.VTABLE is defined as:

        select *
          from DICTIONARY.TABLES;
```

The DESCRIBE TABLE generates a log note in the form of a CREATE TABLE statement that might have originally defined the table. This is an example:

```
DESCRIBE TABLE RIAA.YEAREND;
```

```
NOTE: SQL table RIAA.YEAREND was created like:

create table RIAA.YEAREND( bufsize=4096 )
  (
   YEAR num,
   CD num,
   CASSETTE num,
```

```
  LP_EP num,
  SINGLE num
 );
```

For a table that has constraints, the DESCRIBE TABLE and DESCRIBE TABLE CONSTRAINTS statements generate the same "Alphabetic List of Integrity Constraints" that the CONTENTS proc generates.

Creating a Table

The CREATE TABLE statement creates a new table. Define the columns for the new table in a list of column definitions in parentheses after the table name in the CREATE TABLE statement:

CREATE TABLE *table* (*column definition*, . . .);

This form of the CREATE TABLE statement creates a table with no rows. You can then add rows to the table using other SQL statements.

This example creates the table MAIN.STOCK with the columns SYMBOL, DATE, and CLOSE:

```
CREATE TABLE MAIN.STOCK (SYMBOL CHAR(8), DATE DATE, CLOSE NUM);
```

```
NOTE: Table MAIN.STOCK created, with 0 rows and 3 columns.
```

Use the DESCRIBE TABLE statement with existing tables, as shown above, to see more examples of this kind of CREATE TABLE statement.

A column definition consists of the column name, its data type, and any column modifiers. Column modifiers declare attributes such as the format and label, as described earlier in this chapter.

The SAS data types are character and numeric. SQL syntax also provides several other data types, but all the data types are actually stored in SAS files as the numeric and character data types. Some data types use arguments to set the width of the value. The following table lists the data types that are available in SAS SQL.

SAS SQL Data Types

SQL Data Type	Aliases	Arguments[1]	SAS Data Type
REAL	DOUBLE PRECISION		Numeric
DECIMAL	NUM NUMERIC DEC FLOAT	(*width*, *decimal*)	Numeric
INTEGER	INT SMALLINT		Numeric
DATE			Numeric
CHARACTER	CHAR VARCHAR	(*width*)	Character

[1] The arguments are optional. The default width of a character column is 8.

To create a new table that has the same columns as an existing table, name the existing table in a LIKE clause. For example, this statement creates the table MAIN.DEST that has the same columns as the table MAIN.ORIGIN:

```
CREATE TABLE MAIN.DEST LIKE MAIN.ORIGIN;
```

To create a table that contains rows when you create it, use the AS clause of the CREATE TABLE statement, as described earlier in this chapter. The CREATE TABLE statement with an AS clause creates a new table with the results of a query expression.

Modifying and Updating a Table

You can change the data and structure of an existing table by adding, modifying, and deleting rows and columns.

Use the ALTER TABLE statement for actions on columns. The ALTER TABLE statement indicates the table name followed by the details of an action on the table. An ADD clause contains column definitions to add columns to the table. For example, this statement adds the numeric columns FORECAST and ERROR to the table CORP.REVENUE:

```
ALTER TABLE CORP.REVENUE
  ADD
  FORECAST NUMERIC FORMAT=COMMA14.,
  ERROR NUMERIC FORMAT=COMMA14.2;
```

```
NOTE: Table CORP.REVENUE has been modified, with 11 columns.
```

Similarly, a MODIFY clause contains column definitions with column modifiers to apply new attributes to existing columns.

A DROP clause contains a list of columns to remove from the table. This ALTER TABLE statement deletes the columns CENTER and REGION from the table CORP.REVENUE:

```
ALTER TABLE CORP.REVENUE
  DROP CENTER, REGION;
```

```
NOTE: Table CORP.REVENUE has been modified, with 9 columns.
```

The INSERT statement adds rows to a table. It can also be used to add rows to some kinds of SQL views. To add rows with specific values, use the VALUES clause in the INSERT statement with a list of values in parentheses. The example below adds a row to the table MAIN.STOCK, which was defined in an earlier example with the columns SYMBOL, DATE, and CLOSE. In the new row, SYMBOL has a value of 'CPB', DATE has a value of '30DEC1994'D, and CLOSE has a value of 21.07.

```
INSERT INTO MAIN.STOCK VALUES ('CPB', '30DEC1994'D, 21.07);
```

```
NOTE: 1 row was inserted into MAIN.STOCK.
```

You can list selected columns of the table after the table name in the INSERT statement. List the values in the same order in the VALUES clause. This example adds another row to MAIN.STOCK:

```
INSERT INTO MAIN.STOCK (SYMBOL, DATE, CLOSE)
   VALUES ('CPB', '31DEC1999'D, 38.69);
```

```
NOTE: 1 row was inserted into MAIN.STOCK.
```

Any columns that are not listed get null values in the new rows of the table.

Use multiple VALUES clauses to add multiple rows. For example:

```
INSERT INTO MAIN.STOCK (SYMBOL, DATE, CLOSE)
   VALUES ('AOL', '30DEC1994'D, 0.88)
   VALUES ('AOL', '31DEC1999'D, 75.88)
   VALUES ('TWX', '30DEC1994'D, 17.56)
   VALUES ('TWX', '31DEC1999'D, 72.31)
   ;
```

```
NOTE: 4 rows were inserted into MAIN.STOCK.
```

To add existing data to a table, use a query expression in the INSERT statement. Write the query expression so that its columns are in the same order as the columns of the table or the columns listed in the INSERT statement. This is an example:

```
INSERT INTO MAIN.STOCK SELECT SYMBOL, DATE(), CLOSE FROM MAIN.DAILY;
```

Another way to add rows to a table is with the SET clause. In a SET clause, each column is listed with an equals sign and a value for the column, much like an assignment statement. Any columns that are not listed get null values in the new row. Use multiple SET clauses to add multiple rows. This example adds four rows to a table:

```
INSERT INTO MAIN.STOCK
   SET SYMBOL='HGSI', DATE='30DEC1994'D, CLOSE=7.38
   SET SYMBOL='HGSI', DATE='31DEC1999'D, CLOSE=76.31
   SET SYMBOL='CRA', DATE='28APR1999'D, CLOSE=12.50
   SET SYMBOL='CRA', DATE='31DEC1999'D, CLOSE=74.50
   ;
```

The UPDATE statement modifies existing values in a table. It uses a SET clause written the same way as the SET clause of the INSERT statement. Usually, the UPDATE statement includes a WHERE clause so that changes are made in one specific row or a selected set of rows. The new values of the SET clause are applied to all rows that meet the WHERE condition. Columns that are not listed in the SET clause are not changed. If there is no WHERE clause, the new values are applied to every row in the table.

This example changes one specific value to another in one column of a table:

```
UPDATE CORP.SOURCE SET VENDOR='Time Warner Inc.'
   WHERE VENDOR = 'Warner Communications Corp.';
```

```
NOTE: 4 rows were updated in CORP.SOURCE.
```

The DELETE statement removes rows from a table. A WHERE clause identifies the rows to delete. Without a WHERE clause, the DELETE statement removes all rows from a table. This statement removes from the table MAIN.ACTIVE any rows for which the value of EXPIR is earlier than the current date returned by the DATE function:

```
DELETE FROM MAIN.ACTIVE WHERE EXPIR < DATE();
```

```
NOTE: 5 rows were deleted from MAIN.ACTIVE.
```

Deleting

Use the DROP statement to delete a table, view, or index. To delete tables, list them in the `DROP TABLE` statement, for example:

```
DROP TABLE WORK.TEMP1, WORK.TEMP2, WORK.TEMP3;
```

```
NOTE: Table WORK.TEMP1 has been dropped.
NOTE: Table WORK.TEMP2 has been dropped.
NOTE: Table WORK.TEMP3 has been dropped.
```

To delete views, list them in the `DROP VIEW` statement. To delete indexes from a table, list the indexes in the `DROP INDEX` statement followed by a FROM clause to identify the table. For example, this statement removes three indexes from the table CORP.CENTURY:

```
DROP INDEX PRIORITY, CONT, START FROM CORP.CENTURY;
```

```
NOTE: Index PRIORITY has been dropped.
NOTE: Index CONT has been dropped.
NOTE: Index START has been dropped.
```

Database Connections

The statements described here act on SAS files. It is also possible to take actions on an external database in the `PROC SQL` step. This is part of the SQL pass-through feature of SAS/ACCESS. The CONNECT statement establishes a connection to a specific database; the EXECUTE statement passes an SQL statement to the database for execution; the `CONNECTION TO` clause in a query passes a query to the database; and the DISCONNECT statement ends the connection to the database.

SQL Execution

SQL performance cannot be easily predicted from the appearance of the SQL statements. You get a more accurate idea by considering the processing that a statement requires. Options in the SQL proc control details of its actions and can help you deal with the performance issues of SQL.

Engine Requirements and Performance Issues

When SAS data files are used as SQL tables, they are still subject to the usual limits and considerations that apply to SAS data files. The file's storage device and its library engine must support the specific kind of action that is requested in an SQL statement. For example, you must have write access to a library to execute the `CREATE TABLE` and `DROP TABLE` statements; the engine must support update access to execute the UPDATE and DELETE statements. Some view engines support update access; statements such as UPDATE and DELETE can be executed for those views, but not for other views, such as data step views.

The speed of SQL actions tends to be similar to that of other proc steps and data steps doing similar things. For example, when you sort a table with an `ORDER BY` clause, that uses approximately the same computer resources as sorting with the SORT proc; forming groups with a `GROUP BY` clause is comparable to forming them with a CLASS statement in another proc step. Adding and dropping columns in a large table can be a substantial task, because the SAS supervisor has to completely rewrite the table, the same way it would if you used a data step to add or remove a variable.

Table joins are a special area of concern in SQL programming, whether in the SAS environment or elsewhere. An SQL table join produces, at least in theory, every combination of rows of the tables (an effect often described as a *Cartesian product*). It then reduces the number of resulting rows by applying the relevant conditions from the WHERE and HAVING clauses. The number of rows produced in a join is the product of the number of rows of each of the tables. For two large tables or for four or more small tables, this can be a very large number of rows. For example, if two tables each have one million rows, joining the two tables generates one trillion rows. The same is true when you join four tables that each have 1,000 rows, or six tables that each have 100 rows. Table joins on this kind of scale can take a very long time to execute, or they might execute quickly, depending on other details of the processing that should be carefully considered. Similar issues may arise when a query contains subqueries or views.

The processing time for a query is not necessarily proportional to the number of rows read or generated, because the SAS supervisor optimizes queries to eliminate some unnecessary work. Joins can sometimes execute much faster when the WHERE conditions are written in a certain way and the appropriate indexes exist for the tables. Investigate these issues if a query takes too long to run. Several performance options of the SQL proc can reduce the risk that a poorly designed query may accidentally run for a very long time.

Options

Options for the `PROC SQL` step can be initialized in the `PROC SQL` statement. They can be changed in the RESET statement, which can appear between any two SQL statements in the step. The following table describes the options.

Option	Description
EXEC	Executes SQL statements.
NOEXEC	Checks the syntax of SQL statements, but does not execute them.
PRINT	Prints the results of SELECT statements.
NOPRINT	Executes SELECT statements, but does not print the results. This can be useful when you use a SELECT statement to assign the results of a query to macro variables.
NUMBER	Prints a column called Row that contains row numbers.
NONUMBER	Removes the Row column.
DOUBLE	Writes blank lines between rows in the print output.
NODOUBLE	Does not write blank lines between rows.
FLOW=*width*	Sets the width of character columns and flows longer character values on multiple lines.
FLOW=*min max*	Sets the minimum and maximum width of character columns. The SAS supervisor adjusts the column widths to make effective use of the width of the page.
FLOW	Equivalent to FLOW=12 200.
NOFLOW	Writes long character values consecutively, without flowing them.
FEEDBACK	Shows log messages with the query code that results when the SQL interpreter expands view references and wild-card references in a query.
NOFEEDBACK	Does not show expanded query code.
SORTMSG	Generates log messages about sort operations.
NOSORTMSG	Does not show messages about sort operations.
SORTSEQ=*collating sequence*	The collating sequence for sorting.
DQUOTE=ANSI	Treats text in double quotes as names, following the ANSI standards for SQL syntax.
DQUOTE=SAS	Treats text in double quotes as character values, following the rules of SAS syntax.
ERRORSTOP	Stops executing SQL statements after an error occurs.
NOERRORSTOP	Continues to execute SQL statements after an error occurs.
LOOPS=*n*	Limits the number of loop iterations in the execution of a query. Use this option especially for untested query expressions to limit the computer time and resources that an improperly constructed query might use.
INOBS=*n*	Limits the number of input rows from a table. This option is especially useful for debugging and testing queries.
OUTOBS=*n*	Limits the number of output rows from a query.

continued

Option	Description
PROMPT	Prompts the interactive user with the option to continue or stop when a query reaches the limit of the LOOPS=, INOBS=, or OUTOBS= option.
NOPROMPT	Stops executing when a query reaches the limit of the LOOPS=, INOBS=, or OUTOBS= option.
STIMER	Writes performance statistics in the log for each SQL statement. This requires the STIMER system option.
NOSTIMER	Writes performance statistics in the log for the SQL step as a whole. This requires the STIMER system option.

Checking for Errors

The VALIDATE statement lets you check the syntax of a query expression without executing it:

```
VALIDATE query expression;
```

SQL statements also generate several automatic macro variables that help you keep track of SQL performance and errors. The macro variable SQLOBS indicates the number of rows generated by a query or otherwise processed by an SQL statement. The macro variable SQLOOPS counts the number of row iterations required to execute a query.

The macro variable SQLRC is an error code. A value of 0 indicates that the SQL statement completed successfully. A positive value indicates a problem. The following table describes the various possible values of SQLRC.

Value	Meaning
0	The statement completed successfully.
4	There was something questionable about the statement. A warning message was issued.
8	Execution stopped because the statement contained an error.
12	There was a bug in the SQL interpreter.
16	The statement used data objects incorrectly.
24	Execution stopped because of an operating system failure.
28	There was a bug in SQL execution.

This example shows the value of SQLRC after an incorrect coded query:

```
SELECT COUNT(*);
%PUT &SQLRC;
```

```
8
```

Two more automatic macro variables, SQLXRC and SQLXMSG, contain error codes and messages from the SQL pass-through facility.

24

Creating Formats and Informats

Formats and informats are important routines in the SAS environment. They do the conversions between text values that are stored in text files and displayed on-screen and data values that can be used in SAS programs and stored in SAS data files. They can also be used in SAS programs for other kinds of conversions and translations. Base SAS includes hundreds of formats and informats, many of which are described in chapter 13. In addition, it includes the FORMAT proc, which lets you create three kinds of formats and informats.

The PROC FORMAT Step

In the FORMAT proc, each statement creates one format or informat. There are three statements, which create three different kinds of formats and informats.

- The VALUE statement creates a value format. A value format translates individual values or ranges of values to specific text values. Whenever a variable has a limited set of values, you can define a value format for it to get complete control over the way the variable is formatted. Another use for value formats is to group the values of a variable into ranges or categories.
- The INVALUE statement creates a value informat. A value informat can translate specific input text values to specific data values. Whenever an input field holds a limited set of values, you can define a value informat that converts the field values into the corresponding data values.
- The PICTURE statement creates a picture format. The "picture" of a picture format is a code string that defines the way each character position is used for formatting numeric values.

Example: Writing Yes and No

This is a simple example of a step that defines a value format:

```
PROC FORMAT;
VALUE ANSWER
   1 = 'Yes'
   0 = 'No'
```

```
  ;
RUN;
```

The VALUE statement defines the numeric format ANSWER. This format has a default width of 3, because that is longest text value in the format definition. The ANSWER format converts the value 1 to the text `Yes` and the value 0 to the text `No`.

Formats and informats that are defined in a program are stored as catalog entries. By default, they are stored in the WORK.FORMATS catalog and are available for use for the remainder of the SAS session. Use the stored formats and informats the same way that you use the SAS System's formats and informats. This data step demonstrates the use of the ANSWER format:

```
DATA _NULL_;
  DO I = 0 TO 1;
    PUT I F1. +4 I ANSWER.;
    END;
RUN;
```

```
0     No
1     Yes
```

The first column in each line of the output shows the value as it normally appears; the second column shows the value formatted with the ANSWER format. As a result of the way the format was defined in the VALUE statement, the value 0 is formatted as `No` and the value 1 is formatted as `Yes`.

Names

When you create formats and informats, you have to give them unique names. The names of stored formats and informats are subject to more restrictions than the names of most SAS objects. Observe these rules and cautions when you name a format or informat:

- Do not use a name that conflicts with a SAS System format or informat; this results in an error condition, and the FORMAT proc does not create the new format or informat.
- If you reuse the same name, the new format or informat supersedes the previously stored format or informat of the same name. The FORMAT proc generates a message saying that the informat or format "is already on the library." What this means is that the previously existing informat or format is being replaced by the new one.
- The name of a character format or informat must begin with a dollar sign ($).
- A format name can be no longer than 8 characters. This includes the dollar sign at the beginning of a character format name.
- An informat name can be no longer than 7 characters. This includes the dollar sign at the beginning of a character informat name.
- The name cannot end in a digit.

- If you create an informat and a format that work together, doing the opposite conversions, it is a good idea to give them the same name.

Statement Terms

Most of the terms of the VALUE, INVALUE, and PICTURE statements are the same. Each statement begins with a section that identifies the routine:

```
keyword name (options)
```

Any number of sections may follow to indicate specific conversions. These sections follow the general form:

```
range = label
```

For clarity, it is usually best to write each section of the statement on a separate line.

In the previous example, the VALUE statement contains the ranges 1 and 0 and their respective labels, 'Yes' and 'No'.

Ranges

A range is a set of values that may be used as input to the routine. For a format, a range represents values of a variable. For an informat, a range represents input text values. A range can be a specific value, a consecutive interval of values, a special range keyword, or any combination of these. The values in the ranges of numeric formats are numeric values; range values for informats and for character formats are character values.

The following example of a numeric value format shows several different ways that ranges can be written:

```
VALUE LEVEL
  0 = 'Zero'
  1 = 'Singular'
  2, 3, 4, 5, 6, 7, 8 = 'Multiple'
  .5 = 'Half'
  LOW -< 0 = 'Negative'
  0 <-< .5, .5 <-< 1 = 'Fractional'
  ._ - .Z = 'Missing'
  OTHER = 'Positive'
  ;
```

In each line, the terms to the left of the equals sign indicate a range. The values that can appear in ranges are constant values and the special values LOW, HIGH, and OTHER. LOW indicates the lowest possible value, not including missing values. HIGH indicates the highest possible value. OTHER includes all values that do not fall into any other range. You can write a list of values, with commas separating them.

A hyphen between two values indicates an interval. To write an interval, write the lower value first, then a hyphen, then the higher value. The interval includes all values from the lower value to the higher value. To exclude one of the endpoints from the interval, write a less than sign between the endpoint value and the hyphen. For example, the interval LOW -< 0 includes the value LOW, but it does not include the value 0. The interval 0 <-< .5 includes all numbers between 0 and .5, but it does not include 0 or .5. A range can be a single interval, a list of intervals, or any combination of values and intervals.

The interval ._ - .Z includes all numeric missing values.

Ordinarily, ranges should not overlap. Each value should appear in only one range. If a value is included in more than one range, the routine uses the first range that contains the value. Numeric ranges are extended slightly, an effect called *fuzz*, so that they include values that nearly match. The default value by which ranges are extended is 1E-12. You can use the FUZZ= format option to indicate a different value for this.

The ranges do not have to span all possible values. If an input value to a format or informat does not fall into any of the ranges, the SAS supervisor uses the default format or informat to process the value. If you want to cover any values that are left over, use the special value OTHER as a range.

Labels

In the statements that create formats and informats, the values to the right of the equals signs are labels. The word *label* is only a technical term that is used to describe the workings of the FORMAT proc. It does not describe the way these values are used. That varies, depending on the kind of routine.

- In a value format, the label is the text value that the format produces.
- In a numeric value informat, the label is the numeric value that the informat produces.
- In a character value informat, the label is the character value that the informat produces.
- In a value informat, the label can be one of the keywords _ERROR_ or _SAME_.
- In a value format or informat, the label can be a reference to another format or informat.
- In a picture format, the label is a picture, a code string that indicates the use of the character positions when the numeric value is formatted.

Format and Informat Options

Write format and informat options in the VALUE, INVALUE, and PICTURE statements in parentheses after the name of the routine. These options control details of the way the routine works. Some of the options are the same in all three statements.

DEFAULT= The DEFAULT= option indicates the default width of the format or informat. This is the width that the SAS supervisor uses when the format or informat is used without a width argument.

If you do not indicate a default width in the DEFAULT= option, the FORMAT proc uses the maximum text length that appears in the statement.

For a format, this is the length of the longest label. For an informat, it is the length of the longest constant value used as the start or end of a range. If a numeric informat is defined with numeric ranges, the FORMAT proc gives it a default width of 12.

MIN= The MIN= option indicates the minimum width. This is the smallest value that the SAS supervisor allows as a width argument for the format or informat.

If you do not use the MIN= option, the FORMAT proc sets a minimum width of 1. The default width is always a valid width, even if it is less than the minimum width you set.

MAX= The MAX= option indicates the maximum width. This is the largest value that the SAS supervisor allows as a width argument for the format or informat. It is the length of the longest text that the format can produce or the informat can interpret. The maximum width cannot be greater than 40. If you do not use the MAX= option, the FORMAT proc sets a maximum width of 40.

The maximum width cannot be less than the minimum or default width. If the MAX= value is less than the MIN= value, the FORMAT proc ignores both options and sets a minimum width of 1 and a maximum width of 40. If the MAX= value is less than the DEFAULT= value, the proc increases the maximum width to make it equal to the default width.

FUZZ= The FUZZ= option indicates a fuzz value, a small number by which numeric ranges are extended. This option is especially useful for numeric formats. When a value does not fall into any specific range, but it is within the fuzz value of a start or end value that is included in a range, the value is included in that range. The default fuzz value is 1E-12.

NOTSORTED The NOTSORTED option tells the proc not to sort the ranges of the format or informat before storing it. Most formats and informats execute faster when sorted, but in some special cases, depending on how they are used, they might execute faster if they are not sorted.

Value informats also use the UPCASE and JUST informat options. Picture formats also use the ROUND format option.

Protecting Existing Entries

One option in the PROC FORMAT statement affects the way the proc creates formats and informats. With the NOREPLACE option, the FORMAT proc does not replace existing catalog entries. If you attempt to reuse a name, the proc keeps the previous format or informat and does not create the new one. Use the NOREPLACE option when you need to ensure that a new format or informat does not replace an existing one.

Value Formats

A value format converts specific ranges to specific constant text values. Use the VALUE statement to create a value format.

The VALUE Statement

The VALUE statement indicates the format name, format options, and the ranges and labels that make up the format:

```
VALUE format name (format options)
range = label
. . .
;
```

The format options that can be specified for a value format are the DEFAULT=, MIN=, MAX=, FUZZ=, and NOTSORTED options that are described earlier in this chapter.

A value format can be either a character format or a numeric format. To define a character value format, use a dollar sign as the first character of the format name and use character values to define the ranges for the format. To define a numeric value format, use numeric values to define the ranges. For either type of value format, each label is a character constant that indicates the specific text value that the format produces for data values in the associated range.

Example: Substitute Text

A value format can be used simply to substitute other text for the specific values of a variable. This example creates the format QTT for use with a variable that uses the numbers 1, 2, 3, and 4 for the quarters of a year. The format converts the numeric values to the words commonly used to identify the quarters.

```
VALUE QTT
  1 = 'First Quarter'
  2 = 'Second Quarter'
  3 = 'Third Quarter'
  4 = 'Fourth Quarter';
```

This is an example of using the QTT format:

```
DATA _NULL_;
  DATE = '31JAN2001'D;
  YEAR = YEAR(DATE);
  QUARTER = QTR(DATE);
  PUT QUARTER : QTT. YEAR;
RUN;
```

```
First Quarter 2001
```

Example: Grouping

Value formats are often used to group values — especially to group numeric values into ranges. This example creates the format WR that groups numeric values into three ranges.

```
VALUE WR
  LOW -< 40 = 'Light'
  40 - 84 = 'Medium'
  84 <- HIGH = 'Heavy';
```

Formats such as WR can be useful with BY and class variables in proc steps. Class groups are formed based on the formatted values of the class variables. When BY groups are formed, consecutive BY groups that have the same formatted BY variable values are grouped together into a single BY group. Use a FORMAT statement in the proc step to associate a grouping format such as WR with a variable. The example below uses the WR format with the class variable WTKG.

```
PROC MEANS MIN MEAN MAX DATA=MAIN.DEMO;
  CLASS WTKG;
  FORMAT WTKG WR.;
RUN;
```

With the WR format, the proc groups the data into three class groups corresponding to the three ranges of the WR format, rather than forming a separate class group for each separate value of WTKG.

Using Existing Formats

Instead of supplying a specific text value for a range, a value format can call another format to handle that range. This is especially useful for numeric ranges and for the special range OTHER. Write the label as a format enclosed in the symbols (| and |), which are formed with parentheses and the vertical bar character. Alternatively, on computers that use the ASCII character set, you can use the bracket characters, [and], to enclose the format. These two forms are equivalent:

```
range = (|format|)
range = [format]
```

The example below defines the value format PC. This format calls the COMMA9.2 format to write values with commas and two decimal places. However, for values that are too large to write this way, and for negative values, it calls the BEST9. format to write the values without commas.

```
VALUE PC (MIN=9 DEFAULT=9)
  0-<100000 = (|COMMA9.2|)
  OTHER = (|BEST9.|);
```

The step below demonstrates the output of the PC format.

```
DATA _NULL_;
  DO VALUE = -1, 0, 1000., 1000000;
    PUT VALUE PC9.;
    END;
RUN;
```

```
       -1
     0.00
 1,000.00
  1000000
```

The special range OTHER includes all values that are not specifically included in a range. Often, as in the preceding example, it is a good idea to use a format as a label for the OTHER range. If you do not mention the OTHER range in the definition of a format, the value format uses a default format for any values that do not fall into one of its ranges. For those values, a character value format calls the standard character format, $F; a numeric value format calls the BEST format.

Example: Calling Attention to Missing Values

By default, ordinary numeric formats write standard missing values as a period. You can use the MISSING= system option to change this to any other single character. There may be times, however, when you want to use more than one character to write a missing value. In some reporting, for example, it might be necessary to write missing values as a word such as "Unknown."

This value format, NKA, writes standard missing values as `**Unknown**` and calls the standard numeric format to write nonmissing values:

```
VALUE NKA (MIN=11 DEFAULT=11)
  . = '**Unknown**'
  OTHER = (|F11.|);
```

Writing missing values as a word can help to call attention to missing values. Missing values might be overlooked in this table of numbers, formatted with the standard numeric format:

```
        833     1250444              .          18        1
          1   209946605              .           .        1
          .         833              .          15        0
          0         833           1071          18        1
```

The missing values are more visible when the same table is formatted with the NKA format:

```
        833     1250444    **Unknown**          18        1
          1   209946605    **Unknown**  **Unknown**       1
**Unknown**         833    **Unknown**          15        0
          0         833           1071          18        1
```

Value Informats

A value informat converts specific input text values to specific data values. A numeric value informat creates numeric values; a character value informat creates character values. Value informats can also create error conditions for input text values that are not valid.

The INVALUE Statement

The INVALUE statement that defines a value informat is nearly the same as the VALUE statement that defines a value format. It indicates the informat name, informat options, and the ranges and labels that make up the informat:

```
INVALUE informat name (informat options)
range = label
. . .
;
```

Informat options include the DEFAULT=, MIN=, MAX=, FUZZ=, and NOTSORTED options, and two others, UPCASE and JUST. The UPCASE and JUST informat options apply effects to the input values before the informat compares the input values to the character values that define the ranges of the informat. The UPCASE option converts lowercase letters to uppercase. The JUST option left-aligns the value, removing leading spaces.

Both numeric and character values can be used to define the ranges for a value informat. The informat compares the actual input text values against character range values. It applies the standard numeric informat to the input text value to generate the numeric value that it compares against numeric range values.

A value informat can generate either character or numeric values. To define a character value informat, use a dollar sign as the first character of the informat name and use character values as the labels. To define a numeric value informat, use numeric values as the labels.

Example: Reading Yes and No

An example at the beginning of the chapter defined a value format ANSWER that writes the numeric values 1 and 0 as the words `Yes` and `No`. This example defines a corresponding value informat, also called ANSWER. The informat ANSWER reads the words and produces the corresponding numeric values.

```
INVALUE ANSWER (UPCASE JUST MAX=32)
  'Y', 'YES' = 1
  'N', 'NO' = 0
  ' ', '.' = .
  ;
```

In defining this informat:

- The UPCASE and JUST informat options are used because spacing and case do not affect the meaning of words. These two options allow the informat to correctly interpret such input values as "`yes`" and "`  No  `".

- The abbreviations Y and N are also included in the ranges for the informat. This allows the same informat to read fields that contain these abbreviations.
- A standard missing value is assigned for fields that are blank or contain only a period. This allows the informat to read missing values in an input field.
- Any other input text would not be valid for this informat. If it encounters any other input text value, the informat generates an error condition.
- The MAX= option sets the maximum width that can be used with the informat. Setting this option allows the informat to read fields that are wider than necessary.

When you define an informat and a format that work with the same kinds of fields and values, such as the ANSWER informat defined here and the earlier ANSWER format, it is helpful to give them the same name.

Example: Roman Numerals

This statement defines the ROMAN informat to read small Roman numerals:

```
INVALUE ROMAN (UPCASE JUST MAX=32) 'I' = 1 'II' = 2 'III' = 3 'IV' = 4 'V' = 5
  'VI' = 6 'VII' = 7 'VIII' = 8 'IX' = 9 'X' = 10 'XI' = 11 'XII' = 12 ' ', '.' = .;
```

Example: Encoding and Decoding

It sometimes happens that a variable has only a few values, and those values are relatively long character values. A variable LOCATION that has the values Daytona Beach, Valley Forge, Vancouver, and Tijuana would be an example of this. For efficiency, you might want to store the values as one-character codes in SAS data files that you create. You can do this with a value informat that converts the text values to the codes and a value format that does the reverse. These statements define an informat and format, both called $LOC, for this purpose:

```
INVALUE $LOC (UPCASE JUST MAX=32)
  'DAYTONA BEACH' = 'D'
  'VALLEY FORGE' = 'E'
  'VANCOUVER' = 'P'
  'TIJUANA' = 'T'
  ' ', '.' = ' '
  ;
VALUE $LOC (MIN=13 MAX=32)
  'D' = 'Daytona Beach'
  'E' = 'Valley Forge'
  'P' = 'Vancouver'
  'T' = 'Tijuana'
  ;
```

In a data step that creates a SAS dataset with the variable LOCATION:

- Define LOCATION as a character variable with a length of 1.

- Use the $LOC informat to read the input field.
- Use a FORMAT statement to associate the $LOC format with the variable LOCATION. Use an INFORMAT statement to associate the $LOC informat with the variable.

The variable LOCATION uses 92 percent less storage space when stored this way, and there is a corresponding reduction in the time it takes to read and write the variable in a SAS data file. When you work with a coded variable this way, remember these points:

- The sort order of the variable is the sort order of the code values.
- Use the code values in comparisons, for example, in WHERE conditions.
- If the SAS datasets are stored permanently, the informat and format must also be stored permanently. The informat and format are necessary for the variable to be used correctly.

Error Conditions and Existing Informats

In addition to using specific values as labels, value informats also use other informats as labels. This works the same way that value formats use other formats as labels. In addition, value informats can use the special labels _SAME_ and _ERROR_. The label _SAME_ leaves a value unchanged. In the case of a numeric informat, it applies the standard numeric informat to the input text to produce a numeric value. The label _ERROR_ creates an error condition. Use the label _ERROR_ for specific values that are not valid as input values for the informat.

SAME and other informats are most often used as labels for the special range OTHER. Input text values that do not fall into any range of a value informat result in an error condition. If you do not want the informat to create error conditions, associate the range OTHER with the label _SAME_ or an appropriate informat.

Example: Flagging Errors

Sometimes certain specific values are not considered valid when they appear in an input field. You can use a value informat to reject those values while accepting other values that are entered. This statement defines the informat RSM, which treats the values –1, 98, and 99 as errors, but allows all other numeric values.

```
INVALUE RSM (MAX=32)
  -1, 98, 99 = _ERROR_
  OTHER = _SAME_;
```

For all values other than the ones it treats as invalid data, the informat RSM works the same way as the standard numeric informat.

Example: Interpreting Notes

A value informat can be used to interpret notes that appear in an input field. Suppose, for example, that an input file contains the note "n/a" in place of the value 0. The informat NA, defined in the example below, interprets n/a as 0 and calls the COMMA informat to interpret all other values.

```
INVALUE NA (MAX=32)
  'n/a' = 0
  OTHER = (|COMMA32.|);
```

Picture Formats

Picture formats are numeric formats that write the digits of a numeral along with other characters that are used as punctuation. A picture format can do anything a numeric value format can, but it can do more with the use of code strings that are called pictures. Picture formats depend on the picture processor, the routine that converts a picture to the output text of the format.

The PICTURE Statement

There are only a few differences between the VALUE statement that defines a value format and the PICTURE statement that defines a picture format. The labels in a picture format are pictures, rather than the actual formatted values that appear in the VALUE statement. Each picture shows how to format the numbers of the associated range. A picture can have picture options that control the way the picture is used.

```
PICTURE format name (format options)
range = picture (picture options)
. . .
;
```

Picture formats can use all of the format options that value formats use and one additional format option, ROUND. Normally, the picture processor truncates values to form the integer values that it works with. With the ROUND option, it rounds to the nearest integer value instead of truncating. This option affects the formatting of values that have a fractional part of .5 or greater.

Pictures

A picture is a character string that is used as a model for the text value that the format produces. Some of the characters in the picture are literal characters that are written unchanged in the output. Others are code characters that represent positions where the picture processor can substitute parts of the value that is being formatted.

When a picture is used to format numbers, the digit characters in the picture are placeholder characters called *digit selectors*. Each digit selector represents a digit position that the picture processor uses for one of the digits of the value. All other characters in the picture are literal characters. The picture must start with a digit selector.

This is an example of a picture:

'11,111'

If this picture is used to format the value 12345, the picture processor substitutes the digits of the value for the digit selectors, resulting in the formatted text value `12,345`.

There are two kinds of digit selectors. The digit selector 0 is treated differently from the nonzero digit selectors when determining whether to write a value with leading zeroes. The position of a nonzero digit selector always contains a digit from the value, which might be a leading zero. But the position of a zero digit selector at the beginning of the picture is not used for a leading 0.

This means that, for example, the picture '11,111' is different from the picture '00,000' when formatting numbers of fewer than five digits. The picture '00,000' formats the number 1234 without a leading zero, as `1,234`. The picture '11,111' formats the same number with a leading zero, as `01,234`.

A different kind of picture is used for time data. These pictures use code sequences called *picture directives*. The picture processor substitutes formatted calendar and clock elements for the picture directives. For example, in the picture '%0y-%0m', the picture directive %0y indicates the 2-digit year number, and the picture directive %0m indicates the month number. The picture produces formatted text values such as `00-09` and `04-12`. Use the DATATYPE= picture option to indicate what kind of value the picture is formatting — a SAS date, SAS time, or SAS datetime value.

Picture Options

The options associated with a picture can control various details of the way the picture processor puts the picture together with the value that is being formatted. The table on the next page describes the picture options.

Actions of the Picture Processor in Formatting a Number

The picture processor combines the value with the picture to produce the output text value for the format. Its actions when it formats a number can be divided into these four stages:

1. *Integer value.* The picture processor takes the absolute value of the number to convert negative numbers to positive values. It multiplies the positive value by the multiplier, then truncates the result to get a positive integer value. If the ROUND option is used, it rounds to the nearest integer instead of truncating.

Picture Options

Option	Value	Description
MULTIPLIER= MULT=	*integer*	Multiplier: the factor multiplied by the value to get the integer value that the picture processor uses. The default multiplier is the power of 10 that is required to create the number of decimal places indicated in the picture.
PREFIX=	*'string'*	Prefix characters: placed in the first digit positions in the picture, if the positions are otherwise unused.
FILL=	*'character'*	Fill character: placed in unused digit positions at the beginning of the picture, after the prefix character. The default fill character is a space.
DECSEP=	*'character'*	The decimal point character. The default is '.'.
DIG3SEP=	*'character'*	The three-digit separator character. The default is ','.
NOEDIT		Treats the picture as a constant value, rather than as a picture. This is the default for a picture that does not contain any digit selectors or picture directives.
DATATYPE=		The kind of data that the value represents:
	DATE	A SAS date value.
	TIME	A SAS time value.
	DATETIME	A SAS datetime value.
		The default is that the value represents a number.
DATATYPE=	*language*	The language used for localization of calendar and clock elements. The default is to localize according to the value of the DFLANG= system option when the format executes.

2. *Digit string.* The picture processor converts the integer value to a numeral. The number of digits that the picture processor can use from the numeral is the number of digit selectors in the picture. It forms a digit string of this length from the numeral by adding leading zeroes or by removing digits from the beginning of the numeral, if necessary.
3. *Digits.* The picture processor writes the digits of the digit string in the digit positions of the picture. However, it does not write a leading zero in the position of a zero digit selector at the beginning of the picture. If all the digit positions are used, the picture processor stops at this point.
4. *Fill.* If there is a prefix, and there are enough character positions for the prefix to the left of the first digit, the picture processor writes the prefix in the digit positions immediately before the first digit. Finally, it writes the fill character in any remaining character positions to the left of the first digit and the prefix, if there is one.

Example: Dollars and Cents

This example creates a picture format, PRC, that writes small prices as cents, with four decimal places, and larger prices as dollars and cents. If a price is 0, the format writes it as the word "Free."

```
PICTURE PRC
  0 = 'Free'
  0 <-< .1 = '1.1111c' (MULT=1000000)
  .1 - 9999.99 = '001.11' (PREFIX='$');
```

The following statements demonstrate the use of the PRC format.

```
DO PRICE = 0, .0005, .05, .10, .75, 5, 100;
  PUT PRICE PRC.;
  END;
```

```
   Free
0.0500c
5.0000c
  $0.10
  $0.75
  $5.00
$100.00
```

Picture Directives

The DATATYPE= picture option tells the picture processor to treat the value as a SAS date, SAS time, or SAS datetime value. The picture processor derives the calendar and clock elements of the value and inserts them in the picture at the positions indicated by picture directives. The table on the following page lists the available picture directives.

Enclose the pictures in single quotes so that the SAS supervisor does not misinterpret the percent signs as macro references. If necessary, write trailing spaces in the picture so that the length of the picture is long enough to hold the value that will result.

Picture Directives

Calendar or Clock Element	Directive[1]	Examples
The four-digit year	%Y	2001, 2016
The two-digit year[1]	%y %0y	1, 16 01, 16
The month number[1]	%m %0m	3, 7, 11 03, 07, 11
The month name[2]	%B	September
The month name, abbreviated[2]	%b	Apr Aug Oct
The week number of the year (starting week 1 on the first Sunday of the year)[1]	%U %0U	1, 14, 27, 40 01, 14, 27, 40
The day of the year[1]	%j %0j	1, 92, 184 001, 092, 184
The day of the month[1]	%d %0d	5, 12, 19, 26 05, 12, 19, 26
The number of the day of the week (starting from Sunday=1)	%w	2, 3, 4, 5, 6
The name of the day of the week[2]	%A	Sunday
The name of the day of the week, abbreviated[2]	%a	Mon, Tue, Wed
The day half[2]	%p	AM, PM
The hour of the 24-hour clock[1]	%H %0H	9, 17 09, 17
The hour of the 12-hour clock[1]	%I %0I	9, 5 09, 05
The minute[1]	%M %0M	0, 15, 30, 45 00, 15, 30, 45
The second[1]	%S %0S	0, 15, 30, 45 00, 15, 30, 45
The character % (percent sign)	%%	%

[1] Picture directives that contain a 0 write numbers with leading zeroes. Other picture directives write numbers without leading zeroes.

[2] These calendar and clock elements are localized according to the LANGUAGE= picture option or the DFLANG= system option.

Examples of Date and Time Formats

This picture format writes dates with the day of the month and the name of the month, for example, `1 August`:

```
PICTURE DMON (MIN=5 DEFAULT=12 MAX=32) OTHER = '%d %B      ' (DATATYPE=DATE);
```

This format writes a length of time as minutes and seconds, with a period in between (a European style of notation):

```
PICTURE MPS (MIN=5 DEFAULT=5 MAX=32) OTHER = '%M.%0S' (DATATYPE=TIME);
```

The picture directive `%0S` is used in order to write the seconds with leading zeroes.

Managing Stored Formats and Informats

By default, the formats and informats you create are stored in the WORK library and are available for use only in the same SAS session. It is possible to store formats and informats permanently, so that you do not have to create them again for each SAS session in which they are needed.

Format Catalogs

Each format and informat you create in the FORMAT proc is stored as a separate entry. The entry name is taken from the name of the routine. The entry type indicates the type of routine. The following table shows the entry types and the way the entry names correspond to the routine names.

Routine Type	Entry Type	Routine Name
Character informat	INFMTC	$*entry name*
Numeric informat	INFMT	*entry name*
Character format	FORMATC	$*entry name*
Numeric format	FORMAT	*entry name*

The libref LIBRARY is set aside for use as a format library. Formats and informats that you store in the catalog LIBRARY.FORMATS are automatically available for use in the SAS session. Use a LIBNAME statement near the beginning of the SAS session — perhaps in the autoexec file — to define the LIBRARY libref. To add new formats and informats to the format library, use the `LIBRARY=LIBRARY` option in the `PROC FORMAT` statement.

The LIBRARY= option can indicate any available libref. All formats and informats created in the step are stored in the FORMATS catalog in the indicated library. To store formats and informats in a catalog with a name other than FORMATS, use the two-level catalog name in the LIBRARY= option. It is possible to use the CATALOG proc to copy stored formats and informats from one catalog to another. If you need to rename or delete stored formats and informats, use the CATALOG proc to rename or delete the entries.

Searching for Formats and Informats

When you use a stored format or informat in a program, the SAS supervisor looks for it in the available format catalogs. By default, it looks first in WORK.FORMATS and next in LIBRARY.FORMATS. You can add other catalogs to the search sequence with the FMTSEARCH= system option. The value of this option is a list of librefs and/or catalogs that indicates the order in which the SAS supervisor should search for formats and informats. If you list only a libref, the SAS supervisor searches the FORMATS catalog of that library. The default catalogs WORK.FORMATS and LIBRARY.FORMATS are always searched first and second unless you list them in a different sequence in the option.

This is an example of the use of the FMTSEARCH= system option:

```
OPTIONS FMTSEARCH=(PROJECT SASUSER.MYFORMAT);
```

With this option, the search proceeds in the following sequence:

1. WORK.FORMATS
2. LIBRARY.FORMATS
3. PROJECT.FORMATS
4. SASUSER.MYFORMAT

The response of the SAS supervisor when it does not find a format or informat depends on the FMTERR system option. If this option is on, the SAS supervisor generates an error condition when it cannot find a format or informat. If the option is off, it uses the default format or informat in place of the missing format or informat.

Search errors occur most often when a stored format is used as the format attribute of a variable in a permanent SAS dataset. If SAS datasets that you move to another location contain variables that depend on stored formats or informats, be sure to also move the required formats and informats.

Information About Formats and Informats

The FORMAT proc can print detailed information about stored formats and informats. Use the FMTLIB option in the PROC FORMAT statement along with the LIBRARY= option to identify the format catalog:

```
PROC FORMAT FMTLIB LIBRARY=libref or catalog;
RUN;
```

Use the PAGE option instead of or in addition to the FMTLIB option to have the proc print each format and informat on a separate page.

To print only selected formats and informats from the catalog, use the SELECT or EXCLUDE statement. List specific entries to use in the SELECT statement, or list specific entries to exclude in the EXCLUDE statement. The SELECT and EXCLUDE statements in the FORMAT proc use a special kind of one-level name. Include the dollar sign at the beginning of a character format or informat name as part of the name. Write an at-sign before the name of an informat. In a SELECT or EXCLUDE statement, these would be names of formats:

```
ANSWER WR $LOC LEVEL
```

These would be names of informats:

```
@ANSWER @$LOC @NA @RSM
```

Control Datasets

The FORMAT proc does not let you edit a stored format or informat directly. However, it can convert stored formats and informats to a SAS dataset, called a *control dataset*. You can change the control dataset and have the FORMAT proc convert it back into the formats and informats. It is also possible to create a new format or informat by using data step logic to create a control dataset.

To create a control dataset from a format catalog, use the CNTLOUT= option in the PROC FORMAT statement with the name of the new control dataset. Use the LIBRARY= option to identify the format catalog. Use the SELECT or EXCLUDE statement, if necessary, to include only selected entries in the control dataset. This step creates a control dataset WORK.CONTROL of the informat RSM and the format DMON, then prints the control dataset:

```
PROC FORMAT LIBRARY=LIBRARY  CNTLOUT=WORK.CONTROL;
  SELECT @RSM DMON;
RUN;
PROC PRINT DATA=WORK.CONTROL HEADING=HORIZONTAL;
RUN;
```

Obs	FMTNAME	START	END	LABEL	MIN
1	DMON	**OTHER**	**OTHER**	%d %B	5
2	RSM	-1	-1	_ERROR_	1
3	RSM	98	98	_ERROR_	1
4	RSM	99	99	_ERROR_	1
5	RSM	**OTHER**	**OTHER**	_SAME_	1

Obs	MAX	DEFAULT	LENGTH	FUZZ	PREFIX	MULT	FILL	NOEDIT	TYPE
1	32	12	12	0		0		1	P
2	32	12	12	0		0		0	I
3	32	12	12	0		0		0	I
4	32	12	12	0		0		0	I
5	32	12	12	0		0		0	I

Obs	SEXCL	EEXCL	HLO	DECSEP	DIG3SEP	DATATYPE	LANGUAGE
1	N	N	O	.	,	DATE	
2	N	N	I				
3	N	N	I				
4	N	N	I				
5	N	N	O				

The control dataset created by the FORMAT proc contains most or all of the variables that are described in the following table. Some variables might not be included if they are not used in the selected formats and informats.

Variables in a Control Dataset

Variable Name	Type	Description
TYPE	Character	A code indicating the type of routine
FMTNAME	Character	The name of the routine
DEFAULT	Numeric	The default width
MIN	Numeric	The minimum width
MAX	Numeric	The maximum width
LENGTH	Numeric	The text length
FUZZ	Numeric	The fuzz value
START	Character	The start value for the range interval
END	Character	The end value for the range interval
SEXCL	Character	A code indicating whether the start value is excluded from the interval: Y (yes) or N (no)
EEXCL	Character	A code indicating whether the end value is excluded from the interval: Y (yes) or N (no)
HLO	Character	Codes indicating special ranges and options
LABEL	Character	The label or picture
MULT	Numeric	The multiplier
NOEDIT	Character	A code indicating whether the NOEDIT option applies to the picture: 1 (yes) or 0 (no)
PREFIX	Character	The prefix
FILL	Character	The fill character
DECSEP	Character	The decimal point character
DIG3SEP	Character	The 3-digit separator
DATATYPE	Character	A code indicating the data type for the picture: blank, DATE, TIME, or DATETIME
LANGUAGE	Character	The language of calendar and clock elements

Control Dataset TYPE Codes

Code	Type of Routine
C	Character value format
N	Numeric value format
J	Character value informat
I	Numeric value informat
P	Picture format

Control Dataset HLO Codes

Code	Description
F	Label is format or informat
H	End value of range interval is HIGH
L	Start value of range interval is LOW
O	Special range OTHER
I	Range interval for informat defined with numeric values
S	NOTSORTED option
R	ROUND option
U	UPCASE option
J	JUST option
N	Format or informat has no ranges

The MAXSELEN= and MAXLABLEN= options of the PROC FORMAT statement affect the lengths of range and label variables in the output control dataset. The MAXSELEN= option sets the length of the START and END variables. The MAXLABLEN= option sets the length of the LABEL variable. These options also affect the print output produced by the FMTLIB option.

To edit a stored format or informat, create a control dataset and edit the control dataset. Then use the CNTLIN= option in the PROC FORMAT statement to convert the control dataset back into a format or informat. The CNTLIN= option can also be used with a SAS dataset that you create with a data step.

To be used as an input control dataset, a SAS dataset does not have to have all the variables of the control dataset that the FORMAT proc creates. At a minimum, a control dataset must contain the FMTNAME, START, and LABEL variables. The TYPE variable is necessary if any of the routines is not a value format. If a SAS dataset does not have the correct variable names of a control dataset, use the RENAME= dataset option in the PROC FORMAT statement to change the variable names.

If the control dataset represents more than one routine, the observations for each routine must be grouped together. If necessary, sort the control dataset by the variables TYPE and FMTNAME.

In the PROC FORMAT statement, indicate the input control dataset in the CNTLIN= option and the format catalog in the LIBRARY= option:

```
PROC FORMAT CNTLIN=SAS dataset LIBRARY=libref or catalog;
RUN;
```

Formats and informats created from the input control dataset replace formats and informats of the same names in the format catalog unless you use the NOREPLACE option. If you need to process only some of the formats and informats that are in the control dataset, use the WHERE= dataset option on the control dataset with a condition that selects values of the FMTNAME and TYPE variables.

The example below creates a numeric value informat MNAME in the catalog LIBRARY.FORMATS. In the control dataset, WORK.MON, the variables TYPE and FMTNAME indicate a numeric informat with the name MNAME. The value of HLO indicates the UPCASE and JUST options. The values of START and LABEL in each observation supply the ranges and the corresponding labels. The MNAME informat reads the name of a month and produces the month number as the resulting value. For example, it interprets `January` as the number 1.

```
DATA WORK.MON;
  RETAIN TYPE 'I' FMTNAME 'MNAME' HLO 'UJ';
  * LABEL is month number. ;
  DO LABEL = 1 TO 12;
    * START is left-aligned, uppercase month name. ;
    START = LEFT(UPCASE(PUT(MDY(LABEL, 1, 0), MONNAME9.)));
    OUTPUT;
    END;
RUN;
PROC FORMAT CNTLIN=WORK.MON LIBRARY=LIBRARY;
RUN;
```

Glossary

across variable, *n.* in the REPORT proc, a variable whose values form groups of observations for analysis, with each value appearing in a separate column in the report.

address, *n.* a number, name, or code that indicates the location of an object.

algorithm, *n.* a sequence of actions for completing a specifically defined task.

alias, *n.* an alternate name for an object.

ampersand, *n.* the character "&", used as the prefix of a macro variable reference.

analysis variable, *n.* in a proc step, one of the primary set of input variables. Analysis variables are listed in the VAR statement of a proc step and are often used to calculate statistics.

application, *n.* 1. a specific area of use for a technology. 2. a program that provides its own primary sequence of execution, as opposed to a program that derives its primary sequence of execution from another program, such as the operating system or the SAS application. 3. applications program.

applications program, *n.* a program that a user executes to accomplish a task. *Also,* **application program**. *Compare* systems program.

argument, *n.* 1. a value provided to a routine, such as function, format, etc., by the program that calls the routine. 2. a value provided as a component of the syntax of any part of a program.

array, *n.* a named list of variables in a data step. An array is defined in the ARRAY statement. Any of the variables, called **elements** of the array, can be referred to by a reference to the array and a **subscript** value, a numeric value that distinguishes the separate elements of the array.

ASCII, *n.* 1. the standard character encoding used on most computers and in most communication between computers. The ASCII character set associates specific character forms with the numbers 0–127. *adj.* 2. using ASCII encoding.

ASCII file, *n.* a text file that contains ASCII characters.

assignment statement, *n.* a statement that assigns the value of an expression to a variable. The statement can be identified by the equals sign that is used as its second term. The value of the expression on the right of the equals sign is assigned to the variable on the left of the equals sign.

asterisk, *n.* the character "*", used to indicate multiplication and crossing.

at-sign, *n.* the character "@", used in column pointer controls.

attribute, *n.* 1. a property of an object that indicates a quality of the data that the object contains. 2. a fixed quality of a variable, as its name, type, length, informat, format, or label.

audit file, *n.* a special SAS data file that records changes to observations of a SAS data file.

audit trail, *n.* the association between a SAS data file and an audit file by which changes are recorded in the audit file.

automatic variable, *n.* a variable that is created automatically in the execution of a program, that is, without the program specifically requesting it.

background, *adj.* executing with no direct connection to any display or user actions. *Compare* foreground.

backslash, *n.* the character "\", used to separate directory levels in file systems derived from MS-DOS.

base, *n.* 1. the number to which an exponent is applied; the left operand of the exponentiation operator. 2. the number used in defining a system of numerals, written as the numeral 10, as the number ten in the decimal system. 3. in a floating point form, the number, usually 2, to which the exponent is a applied to indicate the approximate magnitude of the value.

base SAS, *n.* the principal SAS product, encompassing the most generally useful features of SAS software. It includes the SAS supervisor, engines, functions, informats, formats, programming and Explorer windows, etc.

batch, *adj.* 1. executing without user interaction. 2. of a SAS program, running as a separate SAS session, separate from the SAS interactive environment.

binary, *adj.* 1. having to do with the quantity two. 2. using a set of two digits, 0 and 1, to represent numbers or other information. 3. (of an operator) having two operands. One operand appears on each side of the operator. 4. (of a field or variable) having a value that can be understood by considering the separate bits. 5. (of a file) considered as a sequence of bytes.

bit, *n.* 1. a binary digit. 2. the smallest possible unit of information, having only two possible values.

bitfield, *n.* a data value treated as a sequence of bits.

bitwise, *adj.* operating on the separate bits of a value.

blank, *adj.* 1. containing only space characters. *n.* 2. a space character, especially in the EBCDIC character set.

block, *n.* consecutive statements that are treated as one statement for control flow purposes.

Boolean, *adj.* the mathematical treatment of problems of logic, using 1 to represent true and 0 to represent false.

brace, *n.* either of the characters "{" (**left brace**) and "}" (**right brace**), used to enclose array subscripts.

bracket, *n.* either of the ASCII characters "[" (**left bracket**) and "]" (**right bracket**), used to enclose format and informat references in the FORMAT proc.

branch, *v.* 1. to cause execution to continue at another place, or conditionally at one of several places in the program. *n.* 2. an instance of branching.

bug, *n.* a behavior of a program that differs from the program's intended or documented behavior. **buggy**, *adj.*

BY group, *n.* a group of consecutive observations read from a SAS dataset which have the same values of one or more key variables, as identified by a BY statement.

BY variable, *n.* a key variable identified in a BY statement. BY variables indicate the sort order or groups of observations in an input SAS dataset.

byte, *n.* 1. eight bits, considered as a unit. 2. an amount of information equal to 8 bits.

byte stream, *n.* data organized as a sequence of bytes.

C, *n.* a programming language, characterized by concise syntax, portability, efficiency, and the capacity for both low-level and high-level programming. C has been the predominant general-purpose programming language since the 1980s and has influenced the syntax of most programming languages developed since that time.

call, *v.* to cause (a function, routine, macro, or other program unit) to execute. *n.* 2. an instance of calling.

CALL routine, *n.* a routine used in a CALL statement. CALL routines have arguments and often alter the values of variables used as arguments.

catalog, *n.* a SAS file that can contain various kinds of data objects (called **entries**).

cell, *n.* the intersection of a row and a column in a table or grid.

centiles, *n.* approximate percentiles of the key of an index, stored in the index and used to optimize the processing of the index.

character, *n.* 1. a symbol that can be displayed or that has some special meaning in the structure of a file or data stream, usually represented as one byte. 2. a byte, considered as a character. *adj.* 3. organized as a sequence of characters: one of the two data types of SAS software. 4. in or of the character data type. 5. resulting in a character value. 6. used with character variables.

character set, *n.* a scheme for translating character data to the specific characters that are displayed for a character value. SAS data uses the ASCII or EBCDIC character set, depending on the operating system.

code, *n.* 1, a system for representing information in digits or other symbols. 2. a value that represents information in a particular code. 3. source code. *v.* 4. to write source code. 5. to represent in a code; encode.

collating sequence, *n.* the sequence in which character values are sorted or compared.

colon, *n.* the character ":", which follows a statement label.

column, *n.* 1. a vertical arrangement of data values, characters, text, etc. 2. a character position in a record. 3. a data element in SQL data.

command, *n.* text that tells a computer program to take a specific, immediate action, especially when typed by a user.

command line, *n.* 1. the complete command, including options, that is necessary to request a specific action. 2. a place where a command can be entered.

comment, *n.* text written in a program file, usually containing a description of the program, that is marked to not be executed as part of the program. A **delimited comment** is enclosed in the symbols "/*" and "*/". A **comment statement** begins with "*".

compile, *v.* to translate a program into machine language, or a similar code, so that it can be executed. **compiler**, *n.* **compilation**, *n.*

compile-time, *adj.* occurring during the process of compiling.

compute, *v.* to determine a value by a series of actions. **computation**, *n.*

computed variable, *n.* a variable whose values are computed by statements in a program.

concatenate, *v.* to combine two or more objects of the same type end to end to form a new object of the same type. **concatenation**, *n.*

condition, *n.* an expression representing a logical true or false value, especially when used to control whether statements execute or not. **conditional**, *adj.*

constant, *n.* a value written in a program, used by the program and, not changing in the course of executing the program. *Also,* **constant value**.

control character, *n.* a character in text data that is not displayed, but indicates something about the structure of the data.

constraint, *n.* integrity constraint.

control flow, *n.* the order of execution of the statements in a program, or the program logic that determines the order of execution of statements.

counter, *n.* a variable that counts the number of occurrences of something.

counting number, *n.* a number used in counting; positive integer.

cross, *v.* to combine two objects in a way that combines each element of one object with each element of the other.

crosstabulation, *n.* a table that shows the results of crossing. **crosstabulate**, *v.*

database, *n.* an organized collection of data.

database management system, *n.* an application that organizes and manages databases. *Abbr.* **DBMS**

data error, *n.* an error condition in a program that occurs when an input data value is not valid, that is, when it is a value that the program is designed not to accept.

dataset data vector, *n.* the block of memory that holds the values of active variables in the current observation of an open SAS dataset. *Abbr.* **DDV**

dataset option, *n.* an option that affects the way a SAS dataset is read or written. Dataset options are written in parentheses after the SAS dataset name.

data step, *n.* a step that can create SAS datasets and in which programming can be done. A data step begins with a DATA statement. *Compare* proc step.

data type, *n.* a particular way of organizing binary data to represent a kind of value. SAS software uses two data types: character and numeric.

DBCS, double byte character set.

DBMS, database management system.

DDV, dataset data vector.

debug, *v.* to remove bugs.

debugger, *n.* an interactive program that allows a programmer to examine the values of variables in a program while the program is executing.

decimal, *adj.* using a system of 10 digits (0–9) in numerals to represent numbers.

decimal point, *n.* 1. the position in a numeral between the digits that represent the integer part (on the left) and the digits that represent the fractional part (on the right). 2. a period or other symbol used to indicate this position in a numeral.

declarative, *adj.* (of a data step statement) serving to define or modify objects, but not corresponding to any specific action in the execution of the program. *Also,* **declaration**

default, *n.* 1. a setting, action, or decision that holds in the place of an explicit decision, such as the value that an option has when the option is not written in a program statement. 2. an instance of using a default. *adj.* 3. used as a default. *v.* 4. to use a default.

delimiter, *n.* a character, such as a space, comma, or tab, that separates text objects. Delimiters may separate tokens in a program and fields in a record of a text data file. **delimit**, *v.* **delimited**, *adj.*

dialog, *n.* a user interface object that uses a dynamic that resembles a conversational exchange between the program and the user.

digit, *n.* a character that represents magnitude in a numeral, especially the ten digits (0–9) used in decimal numerals.

digit string, *n.* a character string containing digits, usually used as a code.

dimension, *n.* 1. an aspect of variability in an object. 2. a position for a subscript in the definition of an array.

direct access, *n.* 1. access of an object, such as an observation in a SAS data file, by the object's address. *adj.* 2. using, providing, or allowing direct access.

directory, *n.* a list of related files, which includes information about the files, such as their size and location, especially such a list as part of a file system.

display variable, *n.* in the REPORT proc, an input variable whose values for each observation are displayed in the report.

DO block, *n.* a control flow device, beginning with a DO statement and ending with an END statement, that allows several statements to be treated as a unit for control flow purposes.

document, *n.* 1. a collection of information to be viewed or otherwise used by people, especially when stored as a computer file. Also, **document file**. *v.* 2. to write documentation.

documentation, *n.* information about the structure and use of a computer program, data file, etc. Documentation of a computer program may be contained in comments in the program file, separate documents, or both. *n.*

DO loop, *n.* a DO block that is modified by terms in the DO statement to cause the statements in the block to execute repeatedly.

dollar sign, *n.* the character "**$**", used in SAS syntax to indicate the character data type.

domain, *n.* the set of values that can be used as a specific argument to a function.

dot, *n.* a period used to separate parts of a name or similar object, as in a file name or variable name.

double byte character set, *n.* 1. a character set in which some or all characters are encoded in two bytes. *adj.* 2. using or allowing the use of such a character set. *Abbr.* **DBCS**

double precision, *n.* 1. a floating point form that uses 8 bytes of data to represent a numeric value. SAS programs and SAS datasets use double precision for all numeric values. *adj.* 2. represented in double precision.

EBCDIC, *n.* 1. the character encoding used on IBM mainframe computers. *adj.* 2. using EBCDIC encoding.

element, *n.* 1. an object that is a part of a composite object. 2. one of the variables in an array.

end, *v.* (of a program) to stop executing, returning control to the calling program. This is something that all programs, except for operating systems, are expected to do.

engine, *v.* a routine that manages access to SAS files.

entry type, *n.* a code that indicates the kind of data that is stored in an entry. The entry type is the last level of the four-level name of an entry.

error, *n.* 1. an instance in which a program fails to run or runs incorrectly because the program or its input data is incorrect. *See* data error, error condition, syntax error, semantic error, logical error. 2. the difference between the digital representation of a value and its exact value, or between a predicted value and the actual value. *See* rounding error.

error code, *n.* a return code in which a nonzero value indicates an error condition.

error condition, *n.* an unintended state of a program that results from an error in the execution of the program.

error message, *n.* a message, especially one that the SAS supervisor writes in the log, to indicate the presence of an error.

executable, *adj.* (of a data step statement) indicating a specific action or a part of the flow of control in the execution of the program. Executable statements can have statement labels and can be controlled by control flow statements.

exponent, *n.* 1. in exponentiation, the number of times to multiply the base; the right operand of the exponentiation operator. 2. in scientific notation, the number of times to multiply by 10. 3. in a floating point format, the number of times to multiply the base to arrive at the approximate magnitude of the value.

exponentiation, *n.* the process of multiplying a number (the **base**) a fixed number of times (indicated by the **exponent**); indicated by the operator "**".

extension, *n.* a suffix of a file name, often three letters, indicating the file type. A period separates the extension from the rest of the file name.

FDB, file data buffer.

field, *n.* a division of a record that contains one data value.

file, *n.* a named group of related stored data. I/O routines treat each file as a separate location in storage.

file data buffer, *n.* the block of memory that holds an image of the current record of an open text file. *Abbr.* **FDB**

fileref, *n.* a short name temporarily associated with a text file or directory, as by a FILENAME statement.

file type, *n.* a particular kind of data that is stored in a file, or a particular way of organizing data to form a file.

fixed-field, *adj.* (of a record) divided into fields located at fixed character positions.

floating point, *n.* 1. a binary method for representing numeric data in which the order of magnitude is indicated by one number (the **exponent**) and the detail of the magnitude is indicated by another number (the **mantissa**). *adj.* 2. represented in floating point.

footnote line, *n.* a line of text defined in a FOOTNOTE statement and printed at the bottom of the page in a print file.

foreground, *adj.* executing with a simultaneous display or other user interaction.

format, *n.* 1, a routine that converts data values to text values to display or write to a text file. 2. a reference to a format, which may include width and decimal arguments. 3. such a reference, used as an attribute of a variable to indicate the default format for use with the variable. 4. the appearance, layout, or structure of a field, file, page, etc. 5. a specific way to structure or arrange data in a field, file, etc. *v.* 6. to use a format to convert (a value) to text. 7. to apply a particular appearance, layout, structure, etc., when creating a field, file, page, etc.

format option, *n.* an option set for a value or picture format when it is created in the FORMAT proc.

formatted, *adj.* (of a value) having been converted to text with the use of a format.

function, *n.* a routine that returns a value, usually computed from the values of one or more arguments.

generation, *n.* one of the versions of a generation dataset.

generation dataset, *n.* consecutive versions of a SAS dataset stored under the same name.

gigabyte, *n.* an amount of data equal to 1,000 or 1,024 megabytes, *Abbr.* **GB**

group variable, *n.* in the REPORT proc, a variable that groups observations for analysis and determines the order of rows.

hard-code, *v.* to write data values in a program's source code. **hard-coded**, *adj.*

header, *n.* 1. descriptive text at the top of a table column. 2. dissimilar data, such as identifying information, located at the beginning of a data file. 3. the block of data in a SAS data file that contains data set attributes and variable attributes.

heading, *n.* text at the top of a page, table, etc.

hexadecimal, *adj.* using a set of 16 digits (0–9, A–F) to represent numbers.

hierarchical, *adj.* structured in consecutive levels with each object in each level associated with an object in the next higher level. **hierarchy**, *n.*

high-level, *adj.* written in a way that conforms more closely to the programmer's understanding of the events of a program. *Compare* low-level.

host, *n.* the class of computer and operating system on which the SAS application executes.

HTML, *n.* (Hypertext Markup Language.) a standard language used to format text and place pictures in web pages.

implied decimal point, *n.* a decimal point whose location is determined by the structure of a field rather than by a period or other symbol.

index, *n.* 1. a number assigned sequentially to one of a set of objects. 2. a structure stored with a SAS data file that lists the locations of observations having each different value of one or more key variables. *v.* 3. to create an index on or for.

index variable, *n.* 1. a variable that takes on different values for repeated executions of a loop. 2. a variable used as the subscript of an array.

infinite loop, *n.* a loop whose structure has no provision for ending.

informat, *n.* 1. a routine that interprets input text values to create data values. 2. a reference to an informat, which may include width and decimal arguments. 3. such a reference, used as an attribute of a variable to indicate the default informat for use with the variable.

informat option, *n.* an option set for a value informat when it is created in the FORMAT proc.

initialization option, *n.* a system option that can be set only at the beginning of a SAS session.

input, *n.* the process of transferring data into the computer's memory from storage, the keyboard, or another device.

instruction, *n.* a unit of machine language, executed by a processor as a single action, usually in a single clock cycle.

integer, *n.* 1. a number with no fractional part; a whole number or the negative of a whole number. *adj.* 2. using only integers. 3. representing numbers in binary form, using each bit as a binary digit, except for one bit sometimes used as a sign bit.

integrity constraint, *n.* a rule that restricts the data values that can be entered in a SAS data file.

interactive, *n.* (of a program) responding to user actions or input.

interleave, *v.* to combine sorted SAS datasets in a way that uses all the observations of each SAS dataset in the same sorted order. Interleaving is done with the SET and BY statements of the data step.

interpret, *v.* 1. to determine the meaning of; to convert data to a more usable form. 2. to read a program and carry out the actions it indicates.

interpreter, *n.* a program that reads a program written in a particular programming language and carries out the actions it indicates.

invoke, *v.* to call by name. **invocation**, *n.*

I/O, *n.* input and output, considered together.

item store, *n.* a SAS file that contains output objects or other objects used by ODS.

iteration, *n.* a repetition of an action or a sequence of actions, as in a loop.

job, *n.* the execution of a program or a sequence of programs, or another sequence of actions, treated as a separate process by the operating system.

key, *n.* 1. one or several variables whose values identify observations or indicate a sort order. *adj.* 2. used in the key.

keyword, *n.* a word that has a specific syntactical meaning. *Compare* name.

kilobyte, *n.* an amount of data equal to 1,000 or 1,024 bytes. *Abbr.* **kB**, **KB**, **K**

label, *n.* 1. a character string, usually descriptive, that is associated with an object such as a variable or a SAS dataset. 2. statement label.

leading, *adj.* at the beginning of a text value.

leading space, *n.* a space or one of several spaces at the beginning of a text value.

leading zero, *n.* a zero digit or one of several zero digits written at the beginning of a numeral.

length, *n.* the number of bytes or characters occupied by or used in a variable, value, record, or other object or the return value of a function.

level, *n.* one of several parts of a name separated by periods. SAS files and some variables have two-level names, such as WORK.OPEN. Entries have four-level names.

library, *n.* 1. a group of related files, especially when stored under a common name. 2. SAS data library.

library engine, *n.* an engine that provides access to SAS data libraries, SAS data files, and catalogs.

libref, *n.* a short name used to identify a SAS data library in a SAS session. A libref is defined in the LIBNAME statement and is the first level of the name of a SAS file.

line, *n.* 1. a sequence of characters arranged horizontally. 2. a record in a program file or other text file.

linear, *adj.* (of a storage device) allowing only sequential access.

linear search, *n.* the process of finding an object in a file or array by checking each object in sequence until the correct object is found.

Linux, *n.* an operating system, designed to be mostly compatible with Unix, developed as an open source project in the late 1990s.

list input, *n.* input of a variable from a text file, reading the value from a field that is delimited by spaces or another delimiter character. List input is indicated in an INPUT statement by a variable name with no associated informat or modifiers.

list output, *n.* output of a variable to a text file, writing the value with leading and trailing spaces removed and skipping a character position after the value. List output is indicated in a PUT statement by a variable name with no associated format.

log, *n.* a print file that the SAS supervisor writes with program lines and messages about the execution of the program. The fileref LOG identifies the log.

logic, *n.* 1. the systematic use of reasoning to arrive at true conclusions. 2. the aspects of a program by which it takes different actions in different situations or for different conditions of data.

logical, *adj.* 1. applying logic or used in the application of logic. 2. according to the meaning of an object as it is used or applied. 3. containing a value that indicates true or false.

logical error, *n.* a shortcoming in the design of a program that causes the program, when it runs, to produce results different from the results that were intended.

lookup, *n.* table lookup.

loop, *n.* 1. a control flow structure that executes a set of actions repeatedly. *v.* 2. to execute actions repeatedly in a loop.

low-level, adj. written or structured in a way that corresponds more closely to the sequence of instructions a computer follows in carrying out a program. *Compare* high-level.

macro, *n.* 1. a stored object that contains macro language code. Macros are defined by the %MACRO and %MEND statements and stored as entries of type MACRO. *adj.* 2. of or having to do with macro language.

macro language, *n.* a language that can be used in a SAS program to apply conditional logic in generating SAS statements; the preprocessor language of SAS software.

macro processor, *n.* the part of SAS software that interprets macro language.

macro variable, *n.* a macro language object that contains a text value and is kept in memory. A macro variable reference is written with an ampersand followed by the macro variable name.

mainframe, *n.* 1. a mainframe computer. *adj.* 2. of or having to do with a mainframe computer.

mainframe computer, *n.* a large, central multiuser computer characterized by expandability and high I/O capacity.

mantissa, *n.* a number that indicates the detail of the magnitude of a value, but not its order of magnitude. In scientific notation, the mantissa is multiplied by a power of 10 to determine the magnitude. In most floating point formats, the mantissa is multiplied by a power of 2 to determine the magnitude.

match, *v.* 1. to be equal. 2. to have equal key values. 3. to associate two or more matching objects. *n.* 3. the condition of matching. 4. values, observations, or other objects that match.

match merge, *v.* 1. to merge in a way that matches objects according to their key values. *n.* 2. an instance of match merging.

megabyte, *n.* an amount of data equal to 1,000 or 1,024 kilobytes. *Abbr.* **MB**

member, *n.* 1. a SAS file belonging to a SAS data library. 2. a file belonging to a library or directory.

member type, *n.* a code indicating the file type of a SAS file, as DATA for a SAS data file, VIEW for a view, and CATALOG for a catalog.

memory, *n.* the use of electrical devices to maintain digital values, serving as the data space for the objects of a program while the program is executing. Memory is distinguished from storage in that memory is electrical, while storage is electromechanical. Memory is about one million times as fast as storage.

merge, *v.* to combine SAS datasets in a way that combines their observations. *See* match merge.

Microsoft Windows, *n.* any of several operating systems published by Microsoft, widely used on desktop computers, characterized by its graphical user interface and compatibility with MS-DOS. *Abbr.* **Win**

missing, *adj.* indicating the absence of a value. In numeric data, nans are used for missing values. In character data, a blank or null value is considered a missing value. *See* standard missing value, special missing value.

MS-DOS, *n.* an operating system published by Microsoft, widely used on desktop computers in the late 1980s and incorporated into some versions of Microsoft Windows. The file system of MS-DOS has influenced the style of file systems for most subsequent operating systems.

multiuser, *adj.* intended for use by several users at the same time.

MVS, *n.* an IBM mainframe operating system, an early version of OS/390.

name, *n.* a word used to identify an object.

named input, *n.* a process of assigning reading input data in which a variable name and equals sign is written before a value.

named output, *n.* output to a text file that writes variables by writing the name of the variable, an equals sign, and the value of the variable.

nan, *n.* (not a number.) a floating point bit pattern that represents an error condition or a missing value rather than a number.

native, *adj.* 1. (of a SAS dataset) in a form designated by SAS software. 2. in a form that conforms to the standards of the hardware or operating system of a computer.

natural language, *n.* language as used for communication among people, as distinguished from computer languages.

negative sign, *n.* a minus sign or sign bit that indicates a negative number.

nonblank, *adj.* (of a character or character value) visible; not a space, or containing at least one character that is not a space.

nonmissing, *adj.* containing a value other than a missing value.

nonprint file, *n.* a text file that does not contain print control characters.

note, *n.* a log message that provides information about the routine execution of a program.

null, *n.* 1. a control character represented as a zero byte value. Null characters are used in some data to indicate the end of a character value. 2. a missing value in SQL. 3. a text value that contains no characters (that is, having a length of 0). *adj.* 4. equal to or containing a null.

numeral, *n.* a representation of a number in characters.

numeric, *n.* 1. containing, representing, or using numbers. 2. using a double precision format to represent numbers: one of the two SAS data types.

object, *n.* a data value or collection of data created or used by a program; something that occupies or is associated with space in memory.

observation, *n.* 1. an instance of values of the variables of a SAS dataset. 2. one iteration of the observation loop. 3. the input data of one iteration of the observation loop. 4. one item of a sample.

observation loop, *n.* the automatic loop of the data step or another loop in a program that repeats certain actions as many times as necessary to process the data read from an input file.

ODS, (Output Delivery System.) the part of the SAS System that converts output objects to text or to other document formats, such as HTML and PDF.

operand, *n.* a value used with an operator.

operating system, *n.* a program that controls the hardware of a computer and allows other programs to execute.

operator, *n.* a token that represents a mathematical or other action to be taken on one or two values. The values used by an operator are called operands and are indicated by expressions written next to the operator.

option, *n.* a value, set in a statement or command, that controls details of the resulting actions.

order variable, *n.* in the REPORT proc, a variable that determines the order of rows.

OS, *n.* 1. operating system. 2. an IBM mainframe operating system, an early version of OS/390.

OS/2, *n.* an operating system, developed by IBM for its PS/2 line of computers and used on various desktop computers.

OS/390, *n.* the operating system used on most IBM mainframe computers, noted for its security features and scalability.

output, *n.* the process of transferring data from the computer's memory to storage, the display, or another device.

output object, *n.* a two-dimensional, partially formatted representation of the results of a step, which is converted to print output or another document format by ODS.

Output Delivery System, *n.* ODS.

pad, *v.* to extend text to a certain length by adding spaces, zeroes, etc., to the beginning or end. Character values are padded with trailing spaces when they are assigned to variables whose length is longer than the length of the value. With the PAD option, input records that are shorter than the logical record length are padded with trailing spaces.

page, *n.* 1. an area of a printed sheet of paper or a segment of a print file intended to be seen at one time. 2. web page.

panel, *n.* one of several divisions of a page, window, etc., used to display information.

parameter, *n.* 1. a value that controls some of the actions of a program. 2. a numeric attribute of a probability distribution or mathematical model. 3. an argument of a macro.

parenthesis, *n.*, *pl.* **parentheses**. either of the characters "(" (left parenthesis) and ")" (right parenthesis), used for grouping and to enclose arguments of functions.

parse, *v.* to divide the text of a computer program, data file, natural language, etc. into the meaningful units that make it up according to a set of rules. **parser**, *n.*

PDF, *n.* (Portable Document Format.) a document file format, developed by Adobe Systems to maintain the integrity of typography and page layout when exchanging documents between computers.

PDV, *n.* program data vector.

percent sign, *n.* the character "%", used as the prefix of a macro reference, macro function call, or macro keyword.

Perl, *n.* a programming language invented in 1987, widely used for Internet text processing applications.

picture, *n.* a character string that contains a combination of literal characters and formatting codes, used by the picture processor to create a formatted value in a picture format.

picture format, *n.* a numeric format that formats a number by combining its digits with other characters according to a particular pattern. Picture formats are defined in the PICTURE statement of the FORMAT proc.

picture option, *n.* in a picture format, an option that indicates actions associated with the processing of a particular picture.

picture processor, *n.* the routine that applies a picture to a numeric value to create a formatted value for a picture format.

pointer, *n.* an indication of the active location in a file, window, etc.

pointer control, *n.* a term that moves the pointer to a different column or line. Pointer controls are used in the INPUT and PUT statements.

population, *n.* the complete set that a statistical study seeks to understand, describe, or measure.

port, *v.* to move programs, data, etc., from one environment to another.

portable, *adj.* 1. able to be ported easily. 2. allowing the use of programs, data files, etc., that can be ported easily. **portability**, *n.*

positive sign, *n.* a plus sign or sign bit that indicates a positive number.

power, *n.* the result of multiplying a number by itself; the mathematical effect of an exponent.

print control character, *n.* a control character in a print file that separates pages or positions a line on a page.

print file, *n.* a text file that contains control characters that format the text as pages.

print output, *n.* text that a program writes to a print file.

priority, *n.* 1. the state or condition of occurring prior to something else. 2. an attribute of an operator that determines whether it is executed before or after other operators in the same expression.

proc, *n.* one of the primary routines of SAS software designed for use in SAS programs. A proc is executed in a proc step.

procedure, *n.* 1. a sequence of actions, especially one indicated by a routine in structured programming. 2. proc.

proc step, *n.* A step in a SAS program, indicated by a PROC statement, that executes a proc. The proc step identifies the proc and supplies details of its execution.

program, *n.* 1. a document that sets forth a sequence of actions to be taken, especially by a computer. *v.* 2. to write a program.

program data vector, *n.* a block of memory that holds the values of all variables in a step. *Abbr.* **PDV**

pseudo-variable, *n.* the use of a part of character variable, buffer, etc. as a variable.

quantile, *n.* 1. one of a set of values that mark the division of a distribution into equal parts, as a median, quartile, percentile, etc. 2. Also, **quantile rank**. one of the equal parts of a distribution between successive quantiles.

quote, *n.* 1. also, **quotation mark**, **quote mark**. either of the characters """ (**double quote**) or "'" (**single quote**), used in a matched pair to enclose a character constant or any of various other kinds of constant values. *v.* 2. to enclose in quotes. 3. to cause to be treated as a constant value.

random number, *n.* 1. a number selected randomly. 2. also, **pseudo-random number**. a number generated by a mathematical process that simulates the essential qualities of a random number.

range, *n.* 1. a set of values considered together, as in the definition of a format. 2. the set of values that may correctly be used for an array subscript. 3. the difference between the highest and lowest value in a sample.

recode, *v.* to replace one code with another.

record, *n.* a sequence of characters or bytes in a file, treated as a unit in I/O processing. When a text file is displayed, a record is displayed as a line. In a SAS data file, a record is an observation. In a catalog, a record is an entry.

record format, *n.* 1. the rule or standard that defines the way a file is divided into records. 2. record layout.

record layout, *n.* a document that defines the location and contents of the fields in a record. *Also,* **record format**.

release, *n.* a minor change in SAS software, identified by a fractional number, such as 8.1.

remerge, *v.* in an SQL query expression, to combine summary statistics with the detail data from which they were computed.

return, *v.* (of a program) at the end of processing, to provide (a resulting value) to the calling program.

return code, *n.* a numeric code returned by a function that indicates a quality of the success of a function's processing.

return value, *n.* the value that a function provides as the outcome of its processing.

round, *v.* to change a numeric value to a nearby value that belong to a specific set, such as integers or double precision values.

rounding error, *n.* the difference between an exact quantity and the approximation of it in a digital representation of the quantity.

routine, *n.* a program that is intended for use as part of another program; subprogram.

RTF, *n.* (rich text.) a document file format used for exchange of data with Microsoft Word.

run time, *n.* 1. the time at which a program runs. 2. the length of time it takes a program to run. *adj.* 3. occurring at run time. Also, **runtime**.

sample, *n.* a subset of a population, especially one used to study the population.

SAS, *n.* 1. SAS Institute Inc., a Cary, NC-based software developer. 2. a system of integrated software published by SAS Institute Inc., used for working with data. 3. the programming language of this software. *adj.* 4. of or having to do with SAS.

SAS application, *n.* 1. the application that embodies the SAS System. 2. an application executed within the SAS environment.

SAS data file, *n.* a SAS dataset that physically stores data values in a file. *Compare* view.

SAS data library, *n.* a group of related SAS files that are accessed using a common name (a **libref**).

SAS dataset, *n.* a data file organized in a way that makes it able to be used in the special I/O syntax of the SAS System, as in the DATA and SET statements and as input to procs. A SAS dataset organizes data logically in table form, with variables and observations. However, the physical structure in which the data is stored could be in any form.

SAS date, *n.* a SAS date or point in time expressed as the number of days since the beginning of 1960.

SAS datetime, *n.* a point in time expressed as the number of seconds since the beginning of 1960.

SAS file, *n.* a file that uses a file format defined by SAS software, such as a SAS dataset or catalog.

SAS process, *n.* an independent thread of execution of a SAS program or command in a SAS session, started by a STARTSAS statement or command.

SAS product, *n.* one of the separately licensed software products that make up the SAS System.

SAS session, *n.* an instance of execution of the SAS application. Librefs, macro variables, and various other objects in the SAS environment are maintained for the duration of a SAS session.

SAS supervisor, *n.* The central program of the SAS System, encompassing language processing, messages, memory and file management, and other core functionality.

SAS time, *n.* a time of day expressed as the number of seconds since midnight.

scanning control, *n.* a symbol, as ":" or "&", in an INPUT statement that indicates a delimited field.

scientific notation, *n.* any of various ways of writing numbers as a number multiplied by a stated power of 10.

SCL, *n.* (SAS Component Language.) a programming language of the SAS System used principally for interactive applications.

scope, *n.* the area in which a name is recognized.

semantic error, *n.* an incorrect reference to an object that prevents a syntactically correct statement from executing. Using the wrong number of arguments for a function and referring to an array that has not been declared are examples of semantic errors.

semicolon, *n.* the character ";"; used to mark the end of statements.

sequential, *adj.* accessing the data of a file in the order in which it is stored.

server, *n.* 1. a program that supplies resources requested by other programs or computers. 2. the computer on which a server program runs. 3. a computer designed to be used as a server.

session. *See* SAS session.

session option, *n.* a system option that can be changed during the course of a SAS session.

sign, *n.* 1. an indication that a number is positive or negative, as the symbol "–" or "+" in a numeral, or a bit used as a sign in a binary field. 2. the state of being positive or negative.

signed, *adj.* having a sign.

simulation, *n.* a mathematical representation of a real-world, hypothesized, or imaginary phenomenon as a system of discrete objects.

single-user, *n.* intended for use by one user at a time.

slash, *n.* the character "/", used to indicate division and to separate a list of objects from a list of options.

sort, *v.* 1. to put the elements of an object in order; especially, to put the observations of a SAS dataset in order according to the values of one or more key variables. *n.* 2. an instance of sorting. 3. an algorithm for sorting.

source, *n.* 1. source code. 2. an input file or other means of supplying data to a program.

source code, *n.* text written to represent a computer program in a programming language.

space, *n.* a character displayed as a horizontal displacement between other characters, used as a separator between words.

special character, *n.* a character that is displayed as a visible character form other than a letter, digit, or underscore. Special characters used in SAS statements include these:

" # $ % & ' ( ) * + , -
. / : ; < = > @ { | }

special missing value, *n.* any of 27 distinct missing values in SAS numeric data. A special missing value is written in a SAS program as a period followed by a letter or underscore. Formats write a special missing value as the letter, in uppercase, or underscore.

SQL, *n.* (Structured Query Language.) a standardized language for retrieving data from databases.

standard missing value, *n.* the numeric missing value represented in SAS programs and output as a period. A standard missing value is the missing value generated by SAS processing when a numeric value is not available for any reason.

startup option, *n.* a system option that can be set only at the beginning of a SAS session or process.

standard print file, *n.* a print file that contains the print output of most proc steps. The fileref PRINT identifies the standard print file.

statement, *n.* a meaningful unit of a program, usually representing one action or defining one object. A statement in a SAS program ends in a semicolon.

statement label, *n.* a name that identifies a point in a program. A statement label is written before an executable statement in a data step and is followed by a colon.

statement option, *n.* an option that is part of the syntax of a statement, affecting details of the way the statement executes.

statistic, *n.* a value that is computed from the values of a sample.

statistics, *n.* the mathematical study of the significance of measurements of differences and similarities between groups. Statistics attempts to distinguish measured differences that indicate a real, underlying phenomenon from those that occur as the result of random chance.

step, *n.* a separately executed section of a SAS program, starting with a DATA or PROC statement and usually including other statements.

step boundary, *n.* the beginning or end of a step in a SAS program.

stochastic, *adj.* 1. random. 2. using random numbers.

storage, *n.* 1. electromechanical devices, usually using disks, that contain digital information. 2. the space for data in these devices.

string, *n.* consecutive characters or other objects, considered as a unit.

subscript, *n.* a numeric value that selects an element of an array. In a one-dimensional array, one subscript identifies a specific array element. In a multidimensional array, a separate subscript is required for each dimension of the array. Subscript expressions are written inside braces after the array name to form an array reference.

subset, *n.* 1. a set that is contained within another set. *v.* 2. to select some of the observations of a SAS dataset according to a condition of the data.

sum statement, *n.* a statement that adds a quantity to a numeric variable. A sum statement is identified by the plus sign used as its second term. It adds the value of the expression on the right of the plus sign to the variable on the left of the plus sign.

supervisor. See SAS supervisor.

symbol, *n.* 1. a special character or a sequence of special characters that is a meaningful unit of a program. 2. token. 3. macro variable.

symbol table, *n.* the block of memory that contains the values of macro variables.

syntax, *n.* the rules that determine the correct formation of statements in a programming language. **syntactical**, *adj.* **syntactically**, *adv.*

syntax error, *n.* an error in the syntax of a statement that prevents the statement from executing.

system, *n.* 1. a group of interrelated objects. 2, operating system.

system option, *n.* an option that affects the way the SAS System operates. System options can be set at SAS startup and in the OPTIONS statement of a SAS program.

systems program, *n.* a program created to manage a computer.

ab, *n.* a control character that indicates horizontal displacement.

table, *n.* a two-dimensional arrangement of data, in rows and columns, especially as a way of storing or displaying data.

table lookup, *n.* determining a value that is associated with a key value by finding the key value and the associated value in a table.

tabulate, *v.* to create a table for the display of data. **tabulation**, *n.*

term, *n.* a meaningful part of a statement, consisting of one or more tokens.

tab-delimited, *adj.* using the tab character as a delimiter between fields in a record.

terminal, *n.* a device at the end of a communications link, especially such a device used by a person to interact with a computer.

ternary, *adj.* (of an operator) having three operands. The BETWEEN-AND operator is a ternary operator.

test, *v.* 1. to carry out actions to determine a particular quality of something. 2. to run a computer program in a way intended to reveal any behaviors of the program that are inconsistent with its design or documentation. *n.* 3. an instance of testing. 4. an expression that carries out a test of a variable.

text, *n.* the use of characters to represent words, sentences, program statements, and other information for people to read.

text editor, *n.* a program that allows a user to edit text files.

text file, *n.* 1. a file containing characters that are used as text, such as a text data file or a program file. 2. any file accessed with INPUT and PUT statements and other SAS software features designed for files that contain text data.

title line, *n.* a line of text defined in a TITLE statement and printed at the top of the page in a print file.

token, *n.* the smallest meaningful division of a program, consisting of one or more characters that form a word, constant, symbol, etc.

trailing, *adj.* at the end of a text value.

trailing space, *n.* a space or one of several spaces at the end of a text value. The SAS supervisor disregards trailing spaces when it makes character comparisons.

truncate, *v.* 1. to shorten an object by omitting part of it. 2. to omit characters from the end of a character string. 3. to convert a real number to an integer by rounding toward 0, removing the fractional part of the number. 4. to reduce the precision of a floating point number by erasing one ore more of the least significant bytes of the mantissa.

type, *n.* a particular standard for representing digital data. *See* data type, file type, member type, entry type, SAS dataset type.

unary, *adj.* (of an operator) having one operand.

underscore, *n.* the character "_", used to form words in the SAS language.

Unix, *n.* any of various operating systems based on standards of software compatibility published by The Open Group. The original Unix was an operating system developed by AT&T in 1969. Unix has substantially influenced the development of operating systems since that time.

unsigned, *adj.* not having a sign; assumed to be positive.

user, *n.* a person who interacts with a computer, taking actions that lead to immediate actions of the computer.

user interface, *n.* 1. the manner in which a computer program interacts with a user. 2. the aspects or components of a computer program that interact with a user.

utility program, *n.* a program run by a user in order to manage a computer.

value, *n.* any of the various states that are possible for a variable, option, etc., or within a particular data type.

value format, *n.* a format that converts specific values or ranges of values to specific formatted values. Value formats are created in the VALUE statement of the FORMAT proc.

value informat, *n.* an informat that interprets specific input text values as specific data values. Value informats are created in the INVALUE statement of the FORMAT proc.

variable, *n.* 1. a named object in a computer program that uses a location in memory as a particular data type, able to take on different values in the course of running the program. 2. one of the distinct values in each observation of a SAS dataset, identified by a name and having other attributes that are stored in the SAS dataset.

version, *n.* a major change in SAS software, identified by a whole number, usually encompassing several releases.

vertical bar, *n.* the character "|", two of which form the concatenation operator.

view, *n.* a SAS dataset that stores a program for accessing its data, rather than storing the data itself.

view engine, *n.* an engine that assembles the data of a view according to the program stored in the view.

warning, *n.* a log message that indicates that the interpretation or execution of a program may differ from what the programmer intended. *Also,* **warning message**.

web page, *n.* a display of formatted text, usually including graphics, from a standardized document format designed for the delivery of such information across a network to dissimilar computers, allowing for links to other such documents anywhere on the network. Web pages are typically stored as HTML files.

whitespace, *adj.* (of a character) invisible; not representing any character form, as the space, tab character, etc.

whole number, *n.* a number that indicates the result of counting, as 0, 1, 2, 3, 4, etc.

width, *n.* 1. the number of bytes or characters in a field. 2. an argument to an informat or format indicating the number of bytes to process.

window, *n.* a rectangular area of a computer screen in which a program displays related information.

word, *n.* 1. a combination of letters and/or other characters that forms a meaningful unit of a program, serving as the name of an object or having a specifically defined meaning as a part of a statement. SAS words are formed of letters, underscores, and digits. The first character of a SAS word must be a letter or underscore. 2. token.

word wrap, *n.* the process of dividing text into lines of approximately a certain length, breaking the text between words.

wrap, *v.* to do word wrap to.

XHTML, *n.* a simplified version of HTML implemented as an XML application.

XML, *n.* (Extensible Markup Language.) a standard language used to mark text objects as data elements, especially in documents used for exchange of data.

Index

The Author

Rick Aster is a SAS programmer whose dedication and hard work have changed the way the world works with data. In his own work, Aster has applied his SAS programming skills to public finance, drug safety, corporate mergers, insurance, inventory control, biotechnology, and human resources, among many other subject areas. Projects he has worked on include customer data warehouses for some of the world's largest banks and an econometric model to pinpoint the local impacts of global economic changes. In his consulting, teaching, and writing, Aster has always stood for bringing the information power of SAS programming to everyone. His disciplined, creative approach, numerous innovations, and outspoken advocacy have helped to bring SAS programming the recognition it deserves — and have brought Aster recognition as the leading expert in the SAS programming field. *Professional SAS Programming Logic* is his fifth book on the subject. Aster lives in a farmhouse in Valley Forge, Pennsylvania. He is a musician, distance runner, philosopher, and sports photographer.